Example of the three solutions $\dfrac{dx}{dt} = f(x, t) = 1 - t - 4x \qquad x(0) = 1$

$$i := 0 \ldots 10 \qquad \Delta t := 0.1$$

A relative large time increment has been selected

$$t_i := i \cdot \Delta t$$

Define differential equation and initial value

$$f(x, t) := 1 - t + 4 \cdot (x) \qquad x_0 := 1$$

Euler formula

$$x_{i+1} := x_i + f(x_i, t_i) \cdot \Delta t$$

Heun formula

$$f(x1, t) := 1 - t + 4 \cdot (x1) \qquad x1_0 := 1$$

$$x1_{i+1} := x1_i + \frac{1}{2} \cdot (f(x1_i, t_i) + f(x1_i + f(x1_i, t_i) \cdot \Delta t, t_i + \Delta t)) \cdot \Delta t$$

Runge-Kutta method

$$f(x2, t) := 1 - t + 4 \cdot (x2)$$

$$f2(x2, t) := f\left(x2 + \frac{\Delta t}{2} \cdot f(x2, t), t + \frac{\Delta t}{2}\right)$$

$$f3(x2, t) := f\left(x2 + \frac{\Delta t}{2} \cdot f2(x2, t), t + \frac{\Delta t}{2}\right)$$

$$f4(x2, t) := f(x2 + \Delta t \cdot f3(x2, t), t + \Delta t) \qquad x2_0 := 1 \qquad t_0 := 0$$

$$x2_{i+1} := x2_i + \frac{\Delta t}{6} \cdot (f(x2_i, t_i) + 2 \cdot f2(x2_i, t_i) + 2 \cdot f3(x2_i, t_i) + f4(x2_i, t_i))$$

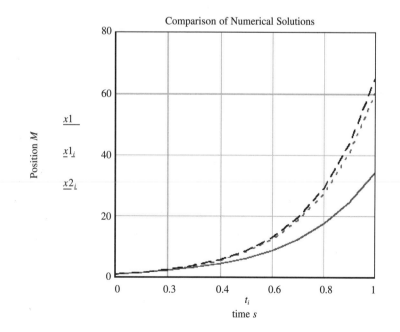

For a time increment of 0.001, the Euler method would be the same as the Runge-Kutta but would require 1000 steps to obtain motion for one second.

ENGINEERING MECHANICS

DYNAMICS

COMPUTATIONAL EDITION

Robert W. Soutas-Little
Michigan State University

Daniel J. Inman
Virginia Polytechnic Institute and State University

Daniel S. Balint
Imperial College London

THOMSON

Australia · Canada · Mexico · Singapore · Spain · United Kingdom · United States

THOMSON

Engineering Mechanics: Dynamics, **Computational Edition**
by Robert W. Soutas-Little, Daniel J. Inman, and Daniel S. Balint

Publisher:
Chris Carson

Developmental Editor:
Hilda Gowans

Permissions Coordinator:
Rose MacLachlan

Production Services:
RPK Editorial Services

Copy Editor:
Shelly Gerger-Knechtl

Proofreader:
Erin Wagner

Indexer:
Rose Kernan

Production Manager:
Renate McCloy

Creative Director:
Angela Cluer

Interior Design:
Carmela Pereira

Cover Design:
Andrew Adams

Compositor:
International Typesetting and
Composition

Printer:
Courier-Westford

Cover Image Credit:
Zac Macauley/The Image Bank/
Getty Images

Library Congress Control Number:
2007901590

ISBN-10: 0-534-54885-7

ISBN-13: 978-0-534-54885-8

North America
Thomson Learning
1120 Birchmount Road
Toronto, Ontario MIK 5G4
Canada

Asia
Thomson Learning
5 Shenton Way #01-01
UIC Building
Singapore 068808

Australia/New Zealand
Thomson Learning
102 Dodds Street
Southbank, Victoria
Australia 3006

Europe/Middle East/Africa
Thomson Learning
High Holborn House
50/51 Bedford Row
London WCIR 4LR
United Kingdom

Latin America
Thomson Learning
Seneca, 53
Colonia Polanco
11560 Mexico D.F.
Mexico

Spain
Paraninfo
Calle/Magallanes, 25
28015 Madrid, Spain

CONTENTS

PREFACE

When first examining this text, one might think that it is a mathematical text or an advanced dynamics text, but it is an introductory text taking full advantage of the computer tools that are currently available and familiar to many students. The study of dynamics of particles and rigid bodies is the foundation of Newtonian mechanics. In the past, students have been expected to use only introductory differential equations to analyze dynamics problems and, in many cases, treat the problems in a quasi-static solution. We define a *quasi-static solution* as one that determines the acceleration only at an instant of time or a specific position. If one considers what one obtains from this type of solution, one realizes that little is known about the motion of the object and no understanding is obtained of the dynamic parameters. In the study of planar kinematics of rigid bodies, examination of the four-bar linkages in both sample problems and homework problems in many of the current texts show that the systems could not complete a full cycle. Is the solution given in these texts wrong? No, because it only asks for the velocities and accelerations of the links at the position shown. It does not, however, prepare the student for the design or analysis of real linkages. We have tried in this text, wherever possible, to present a full understanding of the motion. Does this come at a cost? Yes, it demands that the students use all the mathematical and conceptual tools available to them.

This text has been developed over the past decade to present a comprehensive introduction of dynamics, with emphasis on modeling, development of the differential equations of motion, and complete solution of these equations. Many, if not most, of the differential equations of dynamics are nonlinear and, therefore, analytical solutions are difficult to obtain. This text presents all the traditional topics in dynamics guiding the student through the basic methods of modeling, constructing free-body diagrams and writing the equations of constraint and the equations of motion. These differential equations are then solved analytically or numerically. Most of the dynamic problems one encounters in engineering involve the solution of **vector differential equations**, many of which will be nonlinear. We are not proposing that the dynamics course be turned into an advanced differential equations course, but that the emphasis be placed on the understanding of kinematics, modeling, expressing the constraints to the motion and, most important, formulation of the differential equations of motion. Numerical methods of solution of these equations are shown. Individual instructors can limit the problems to formulation of these equations or include the numerical solutions. The numerical solutions are obtained by use of computational software programs. The text is supported by four computational manuals; Mathcad®, MATLAB®, Maple® and Mathematica®. Although the use of computers has been tried previously in dynamics courses, many of these attempts suggested use of basic languages such as C++ or other programming languages. Unless the students had taken a computer science course, few attempted to solve these problems. Many of the new texts are suggesting the use of computational software to numerically integrate the nonlinear initial value differential equations and graph the resulting motion. Examination of the answers supplied at the end of this text show such graphs. One popular text also encourages the use of computational software and provides a web site to obtain instruction on the use of Mathcad and MATLAB and another includes some instruction on use of MATLAB in an appendix. We feel that the separate manuals provide the best manner to integrate the use of the software by solving the sample problems in the text, where appropriate, in each software program. This allows the instructor or student the choice of which software to use. Other texts do not support this approach to teaching dynamics and to make our approach very obvious, we have titled the text "Computational Edition."

However, if a dynamics text were to concentrate only on the formulation of the differential equations of motion and not their solution, students might question the value of the course. But students can be introduced to numerical methods of solution to both linear and nonlinear differential equations. All of the commonly available computational software packages (MATLAB, Mathcad, Maple and Mathematica) include methods of solution of linear and nonlinear differential equations. We have detailed in the text the ***Euler method*** and make reference to the more accurate Runge-Kutta method. Most software packages include the Runge-Kutta method and individual instructors may wish to use this method. We included only development of the Euler method to avoid an appearance of a "black box" approach. The Euler method is easy to present and is taught in many advanced high school math courses.

Including numerical solutions allows the investigation of many design problems and allows the student to calculate such things as golf shots, ski jumps with wind resistance, and motion in viscous media. Car crashes are analyzed in the study of rigid bodies and we have found students become very excited about real-life examples. We have tested this approach for the past 10 years and have found that the students gain a greater understanding of dynamics when full solutions are obtained. Current engineering students are very computer literate and many have been introduced to computational software in high school. Students also report that listing competence in these programs on their resume impresses employers. We have learned that using this approach requires little additional class time and the manuals are self-standing. Homework problems are marked such that student and instructor know which problems can be solved "by hand," which can be aided by use of the software, and which require numerical solution. Students also have told us that they wish that they had taken the dynamics course either concurrently or before taking the basic course in ordinary differential equations. They said that dynamics gave meaning to the solution of the differential equations.

The biggest change in courses taught in this manner was in the type of examinations that were given. Since a greater emphasis has been placed on modeling and writing the equations of motion while deemphasizing the numerical work, examinations needed to reflect this change. Therefore, many exam questions ask that problems be solved to a point where only numerical calculations remain. Typically, one take-home problem that required full numerical solution of the equations of motion was included in the exams, allowing the student to use computational software. Our approach, of using computational software is simply the next step in the evolution from slide-rule and trig tables to calculators and now to computers. It also has allowed the examination and solution of a much wider range of practical problems.

To prepare students for this type of approach to the study of dynamics, the first chapter is very mathematical in nature and does depend upon the student having a solid understanding of vector algebra. The following chapter takes advantage of this beginning and focuses on kinematics and modeling the problem. When kinetics of a particle is considered, the students are in the position to solve the differential equations of motion, greatly enriching their understanding of dynamics. The organization of the text follows the standard organization of all texts in dynamics. This is a tried presentation and we do not wish to change it. Special topics are introduced in each chapter that use the increased capabilities of the mathematical analysis. The Heaviside step function is introduced early in the text in order to allow forces to be started and stopped during the motion (for example, fuel burns on a rocket or landing in water or a mat or net).

We realize that this is a new way to present dynamics, but have shown through testing in many classes, that it adds greatly to the study of dynamics. Many individuals have contributed to the development of this text. We cannot individually thank the many students who have given us "feedback" and made many useful suggestions, but we can thank the many faculty who have helped. We apologize in advance for any persons we may have missed, but here in alphabetical order is a partial list.

Dean Nicholas Altiero, *College of Engineering, Tulane University*
Professor Ryan Elliott, *University of Minnesota*
Dr. Scott Hendricks, *Virginia Tech*
Professor Daniel C. Kammer, *University of Wisconsin*
Professor Tom Mase, *Cal Poly, San Luis Obispo*
Professor Darren Mason, *Alma College*
Professor Tom Pence, *Michigan State University*
Professor Wendy Reffeor, *Grand Valley State University*
Dr. Joseph Sater, *Write State University*
Professor Richard A. Scott, *University of Michigan*
Professor Henry Scanton, *RPI*
Bill Stenquist, Editor, *McGraw Hill*

and from Thomson Engineering

Chris Carson
Hilda Gowans
Rose Kernan, *RPK Editorial Services*

Please contact us with any comments or corrections:

Robert Soutas-Little, soutas@egr.msu.edu
Dan Inman, dinman@vt.edu
Dan Balint, d.balint@imperial.ac.uk

Chapter 1

KINEMATICS OF A PARTICLE

The ski jumper can be modeled as a particle accelerating due to gravitational attraction and deceleration due to air resistance proportional to the velocity of the jumper.

1.1 INTRODUCTION

We will start our study of dynamics by considering the motion of a particle. The study of motion without consideration of the causes of that motion is called *kinematics* and is divided into two parts: the analysis of bodies that can be modeled as particles and the analysis of rigid bodies with a definite geometry. If a body is modeled as a particle, its position in space is given by three coordinates. These coordinates are relative to a fixed coordinate system, called a *primary inertial reference system,* and are the coordinates of the point that the particle occupies in space at any instant of time. This position is expressed by a *position vector* **r** from the origin to the particle's spatial position. The motion of the particle can be described by expressing the coordinates as a function of time, and the curve that the particle traces in space is called the path or the *trajectory of the motion.* The coordinate system in which the motion of the particle is described may be, for example, rectilinear (x, y, z), cylindrical (r, θ, z), or spherical (θ, R, ϕ), as shown in Figure 1.1. Other special coordinates will be developed in this chapter to express a particle's motion.

Being a point, a particle can only translate; it makes no sense to specify the rotation or orientation of a point. Therefore, the particle has at most *three degrees of freedom,* one for each coordinate direction. The motion of the particle may be constrained or unconstrained. The particle is said to be *constrained* if it is not free to move in one of the coordinate directions; that is, a constraint reduces the number of degrees of freedom of the particle. The study of kinematics of a particle is based on the theory of differential equations. Major portions of this chapter will be devoted to methods of solving the differential equations of motion. In Chapter 2, these differential equations will be obtained from Newton's laws. Therefore, we will first master the solution of differential equations of motion in general and then concentrate on obtaining Newton's differential equations and solving them by the methods we have learned. We will initially try to find analytical solutions to the differential equations, but will solve many using numerical techniques. Before the development of computational software packages, most of the differential equations of dynamics could not be solved, but this has changed, and today engineers are expected to be conversant with computational software.

The basic concepts of kinematics, or the geometry of motion, will be introduced by constraining the particle to move in a single direction—that is, with only one degree of

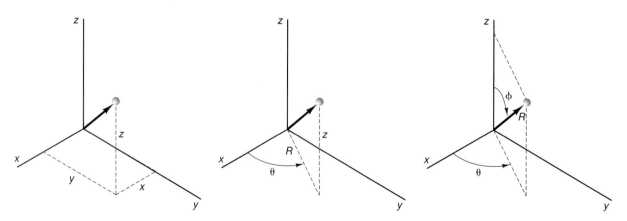

Figure 1.1

freedom. This type of motion, when expressed in rectilinear coordinates, is called *rectilinear motion*. Such motion is one dimensional, and the position vector has only one component. The motion can be described in scalar notation, and we will delay our development of vector kinematics until it is needed.

1.2 RECTILINEAR MOTION OF A PARTICLE: SINGLE DEGREE OF FREEDOM

Figure 1.2

Rectilinear motion of a particle—that is, motion in a straight line—is a one dimensional motion and may be described using only scalar notation. Specifying the position of the particle in space requires establishing a single rectilinear coordinate with proper origin and sense of direction. As always, the coordinate system is established by the individual analyzing the problem, and thus there is no right or wrong coordinate system. However, the choice can make the mathematical details of the solution easier in many cases and, in some cases, can aid in physical interpretation of the results. A detailed explanation of the choice of coordinates will be given in cases that simplify either the mathematical analysis or the physical interpertation. Figure 1.2 shows a coordinate system used to designate the position of a particle in rectilinear motion.

The position of the particle A is given by the coordinate of A's displacement from the origin, x_A. If this displacement is written as a position vector to the particle, it has the form

$$\mathbf{r} = x\hat{\mathbf{i}} \tag{1.1}$$

where $\hat{\mathbf{i}}$ is a unit vector in the x-direction. The vector notation is not needed in this case, but will be shown to facilitate transition to two- and three-dimensional motion. By establishing the coordinate system as shown, the position of the particle A relative to the origin of the coordinate system is specified. In the position shown in Figure 1.2, the particle is $1\frac{1}{2}$ units from the origin in the positive direction. Note that the *sense* of the positive direction in the coordinate system has been chosen with positive to the right and negative to the left. The designation of units depends on the choice of measurement system. In the English or U.S. customary system, the units of length are inches, feet, yards, and miles. This system is more difficult to use and is, in fact, not used by engineers and scientists elsewhere in the world. Its continued use in the United States is due to some individuals' conceptual familiarity with these lengths and to the resistance of some parts of industry to change over to the international standard. The S.I. (or international standard) unit measures distance in meters, centimeters (not used in most scientific literature) millimeters, or kilometers. The exact choice of units does not matter, but it is important that units be used in a consistent and well-defined manner. It is also important to develop the ability to change from one system of units to a different system.

Now suppose that there is a particular change in position during a specific change in time. For example, suppose that at time t_1 the particle is at position x_1, and at time t_2, the particle is at position x_2, as shown in Figure 1.3.

The change in position is denoted as Δx and is seen to be

$$\Delta x = x_2 - x_1 \tag{1.2}$$

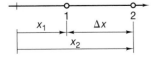

It is important to note that Δx does not depend upon the location of the origin of the coordinate system, as the same value of Δx could be obtained in a different coordinate system. The change in time between the two observations, denoted Δt, is

Figure 1.3

$$\Delta t = t_2 - t_1 \tag{1.3}$$

Now, the **average velocity**, or rate of change of position, during this period of time is defined as

$$v = \frac{\Delta x}{\Delta t} \tag{1.4}$$

The units of velocity are length per unit of time and are given in feet per second (ft/sec), or meters per second (m/s) for scientific calculations. More common measures of velocity are miles per hour (mph) or kilometers per hour (km/h) for land travel (cars, trucks, or trains) and knots (1 knot = 1.152 mph) for air or sea travel (planes or boats). Jet planes often express velocity in terms of Mach number; that is, Mach 1 is the speed of sound at the air density of a specific altitude.

The actual movement Δx could have occurred very quickly, and the two position observations may have been separated over a long length of time. In that case, the average velocity is a poor measurement of the actual velocity at any instant of time, called the **instantaneous velocity**. For example, suppose that the sampling rate, or time between observations, was once every hour. During that period, the observed motion was 2 m, so the average velocity is 2 m/h. If, however, the position was measured every half hour, and no movement occurred between the first two observations, and the 2-m movement occurred between the second and third observations, the average velocity during the first half hour is 0 m/h and during the second half hour is 4 m/h. It is obvious that a higher sampling rate, or a smaller time interval between observations, yields a more accurate estimate of the instantaneous velocity.

The previous discussion shows how little information is conveyed by giving the distance covered in a specific time. Consider if you are told that an individual walks a 1000 meters in an hour. However, if you were given position–time plots, you could accurately describe the motion. There are three plots shown next, designated by $x_1(t)$, $x_2(t)$ and $x_3(t)$.

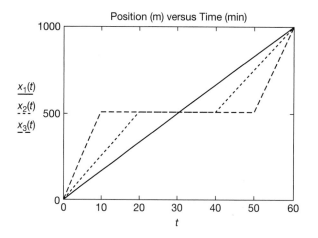

In the plot $x_1(t)$, it is obvious that the individual moved at a constant rate of 1 km/hr. In the movement designated by $x_2(t)$, the individual moved the first 500 m in 20 minutes (at a rate of 1.5 km/hr), rested for 20 minutes, and then moved the last 500 m in the last 20 minutes. The final movement plot shows that the individual moved the first 500 m in the first 10 minutes (at a rate of 3 km/hr), rested for 40 minutes, and then covered the final 500 m in 10 minutes.

This very simple example shows how important it is in any dynamic analysis to obtain the position–time information in either a graphical or equation form. We will stress in every case that to understand any motion, we must have complete position–time information. Many introductory dynamics texts seek solution at an instant of time and solve the problem in a quasi-static sense.

From this point of view, the most accurate estimate of the instantaneous velocity would be obtained if one could average over the smallest interval of time. Newton (1642–1727) and Liebniz (1646–1716) independently developed the field of calculus, in which the derivative is the ultimate reduction of the time interval. The two mathematicians defined the instantaneous velocity as the *limit* of the average velocity as the time interval approaches zero. Thus the instantaneous velocity is

$$v = \lim_{\Delta t \to 0} \frac{\Delta x}{\Delta t} = \frac{dx}{dt} \tag{1.5}$$

or the derivative of x with respect to t. The derivative is the measure of rate of movement, the instantaneous velocity or the intent to move. Often introduced, for convenience, is the Newtonian notation of placing a dot over a quantity to indicate differentiation with respect to time. Hence,

$$\dot{x} = \frac{dx}{dt} \tag{1.6}$$

is used to denote the velocity, v.

Again the velocity in rectilinear motion can be written as a vector in the form

$$\mathbf{v} = v\hat{\mathbf{i}}$$

The velocity is formally defined as the rate of change of the position x with respect to time. The position is considered a function of time $x(t)$, and the velocity, the derivative of that function, is also a function of time $v(t)$. The velocity may be constant, may change over time either increasing or decreasing in magnitude or become negative indicating movement in the negative direction. The distance that the particle moves is the sum of all movement, both positive and negative and is equal to the integral of the absolute value of the velocity over time. To measure this change in velocity, the concept of acceleration is introduced. In a formal sense, the acceleration is defined as the time rate of change of velocity. Average acceleration is defined as the change in the velocity divided by the corresponding change in time or

$$a \approx \frac{\Delta v}{\Delta t}$$

The instantaneous acceleration is thus

$$a = \lim_{\Delta \to 0} \frac{\Delta v}{\Delta t} = \frac{dv}{dt} \tag{1.7}$$

The acceleration is the derivative of the velocity with respect to time,

$$a = \frac{dv}{dt} = \dot{v} \tag{1.8}$$

or the second derivative of position with respect to time,

$$a = \frac{d^2 x}{d^2 t} = \ddot{x} \tag{1.9}$$

If velocity is viewed as the rate of change of position x, or the slope of the position versus time curve at that instant and the acceleration as the rate of change of velocity v,

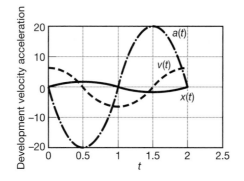

Figure 1.4   Position, velocity, and acceleration for $x = 2 \sin(\pi t)$

visual determination of the derivatives may be obtained from a plot of position versus time, or an *x–t* plot, and from a plot of velocity versus time, or a *v–t* plot, as shown in Figure 1.4.

Examination of the position and velocity curves shows that the velocity is the slope of the position curve. Therefore, when the position curve is at a maximum or minimum, the change in position, and the velocity is zero. Obviously, the velocity will be at a maximum or minimum when the slope of the position curve is a maximum. In a similar manner, the acceleration is zero when the velocity is at a maximum or minimum, and the acceleration is at a maximum or minimum when the slope of the velocity curve is greatest. These are simple observations and consistent with the derivative of a function but can be very useful in visually analyzing very complex motions.

1.3   CLASSIFICATION OF KINEMATICS PROBLEMS

There are essentially two basic problems in kinematics: the *inverse problem* and the *forward problem,* or *direct dynamics problem.* From Newton's law, the forces acting on an object determine the object's acceleration. Hence, the acceleration is often given, and one seeks to obtain the resulting velocity and displacement. This type of problem is called the direct (or forward) problem and involves integrating the given functional form of the acceleration—that is, a differential equation—to find the remaining two quantities of motion (velocity and position). The other type of kinematics problem occurs when an object's position or velocity can be measured as a function of time. In this case, it is necessary to differentiate the given position or velocity to determine the acceleration of the particle. This type of problem is called the inverse dynamics problem.

In most problems of dynamics, the forces applied to the body are specified, and therefore, the acceleration (by Newton's second law) is known as a function of time, position, and/or velocity. The problem is then formulated in terms of a differential equation, which may be linear or nonlinear. The solution involves integrating this equation to determine the velocity and the position at any time. Because of the nonlinearity of the equations, many problems have historically been solved only for the acceleration at a given position or time. In that case, the solution gives only the state of the dynamic variables at that time and does not predict the further motion of the object. In the past, many dynamic problems were only partially solved, but with the availability of modern computational software and computers, these nonlinear problems may now be solved completely.

1.4 THE INVERSE DYNAMICS PROBLEM

If the position of an object is given or measured as a function of time, the velocity and acceleration can be determined by differentiating the position function using Eqs. (1.5) and (1.8). This type of approach is called the inverse dynamics problem and is mathematically simpler than the forward, or direct, dynamics problem.

As an example of the inverse dynamics problem, consider a displacement given as a function of time, say,

$$x(t) = \sin(\pi t)e^t$$

The velocity and acceleration are obtained by successive differentiation of this function:

$$v(t) = e^t[\sin(\pi t) + \pi \cos(\pi t)]$$

$$a(t) = e^t[2\pi \cos(\pi t) + (1 - \pi^2)\sin(\pi t)]$$

In this case, the position is given analytically as a continuous function of time. The motion of the particle can then be graphed and examined in detail using a graphical calculator or computational software. (See Computational Supplement.)

Modern motion analysis equipment allows measurement of position data to occur 50 to 200 times a second (frame rates are 50 to 200 Hz) using high-speed video cameras. This rate of measurement means that the position data are not known as a continuous analytical function, but are determined at discrete times. Higher frame rates can be obtained for use in measurements of high-velocity movements. These systems are routinely used in clinical studies of human motion and are finding increased applications in governmental, industrial, and medical facilities. Other experimental apparatuses, such as proximity probes, strain gauges, and lasers, are used to collect position data. An early experiment performed on the NASA space shuttle used both a video-based and a laser-based motion analysis system to track the position of a solar panel. As previously noted, most experimental systems that collect position data do so not as a continuous function of time, but at specific intervals of time. The velocity and acceleration are obtained by numerical differentiation of the data and are thus subject to increased noise in their calculation. As an example, suppose that the correct position function of an object is

$$g(t) = 8\left[\sin\frac{\pi t}{40}\right]$$

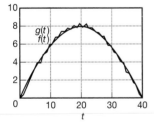

Figure 1.5

and the position data $f(t)$ have noise of a random nature and a maximum magnitude of 0.3 units, as shown in Figure 1.5. (This example does not contain actual data, but instead was generated by a file using a random number function.) Now, we numerically differentiate both $g(t)$ and $f(t)$ to see the effect of the noise on the differential. The differentiated functions $dg(t)$ and $df(t)$ are shown in Figure 1.6. It is clearly seen in this data file that the differentiated data are too noisy to be reliable. If the file contained the actual position data of a particle with time, the velocity data shown in Figure 1.6 would not be reliable. To completely describe the motion, the data would have to be differentiated a second time to obtain acceleration. In practice, the position data are *digitally filtered* to smooth out their values and to obtain more reliable differentiation in the presence of noise. Students interested in these techniques should consult texts on numerical analysis.

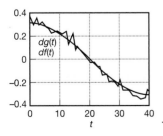

Figure 1.6

MATHEMATICS WINDOW 1.1

Numerical Differentiation   The easiest method to differentiate numerical data is a two-point differentiation. This computes the slope of the function (the derivative at the midpoint between two points) using the formula

$$f'_{i+\frac{1}{2}} = \frac{1}{\Delta t}\left(f_{i+1} - f_i\right)$$

Δt is the time difference between data points. For example, if the frame rate were 100 Hz, Δt would be 0.01. This differentiation is accurate to the order of Δt. The accuracy can be increased to the order of Δt^2 by using three-point formulas. There are three formulas for the three-point differentiation; one to be used on the first point in the file, one for all the central points, and finally a formula for the last point in the file. The three formulas are

$$f'_{i-1} = \frac{1}{2\Delta t}\left(-3f_{i-1} + 4f_i - f_{i+1}\right)$$

$$f'_i = \frac{1}{2\Delta t}\left(-f_{i-1} + f_{i+1}\right)$$

$$f'_{i+1} = \frac{1}{2\Delta t}\left(f_{i-1} - 4f_i + 3f_{i+1}\right)$$

The first formula is the forward-difference formula usually used only at the beginning of the file, the second is the central-difference formula used for most of the file, and the last is the backward-difference formula to be used at the end of the file.

The accuracy may be further increased by using the five-point formulas that are accurate to Δt^4.

$$f'_{i-2} \approx \frac{1}{12\Delta t}\left(-25f_{i-2} + 48f_{i-1} - 36f_i + 16f_{i+1} - 3f_{i+2}\right)$$

$$f'_{i-1} \approx \frac{1}{12\Delta t}\left(-3f_{i-2} - 10f_{i-1} + 18f_i - 6f_{i+1} + f_{i+2}\right)$$

$$f'_i \approx \frac{1}{12\Delta t}\left(f_{i-2} - 8f_{i-1} + 8f_{i+1} - f_{i+2}\right)$$

$$f'_{i+1} \approx \frac{1}{12\Delta t}\left(-f_{i-2} + 6f_{i-1} - 18f_i + 10f_{i+1} + 3f_{i+2}\right)$$

$$f'_{i+2} \approx \frac{1}{12\Delta t}\left(3f_{i-2} - 16f_{i-1} + 36f_i - 48f_{i+1} + 25f_{i+2}\right)$$

As before, the first two are forward-difference formulas to be used at the beginning of the file, the third formula is the central-difference formula to be used on most of the file, and the last two are backward-difference formulas to be used at the end of the file.

Derivation of these formulas may be found in texts on numerical analysis.

Sample Problem 1.1

An automobile moves along a straight, level section of road such that the displacement is $x(t) = 0.4t^3 + 8t + 10$, where t is in seconds and x is given in feet. Calculate the time it takes the car to reach a speed of 60 mph from its initial state at $t = 0$. How far does the car travel during this time, and what is the value of the acceleration when the car reaches 60 mph?

Solution

The velocity and acceleration at any time may be obtained by differentiating the displacement:

$$x(t) = 0.4t^3 + 8t + 10 \text{ ft}$$
$$v(t) = 1.2t^2 + 8 \text{ ft/s}$$
$$a(t) = 2.4t \text{ ft/s}^2$$

The initial velocity (the velocity when the time t is zero) is 8 ft/s, or 5.45 mph. The speed of 60 mph is equal to 88 ft/s, so the time taken to reach this speed may be found by solving the velocity equation:

$$88 = 1.2t^2 + 8$$

or $t = 8.16$ s when the automobile reaches a speed of 60 mph. The position of the automobile at this time is

$$x(8.16) = 0.4(8.16)^3 + 8(8.16) + 10 = 292.6 \text{ ft}$$

Since the automobile started at $x(0) = 10$ ft and moved only in the positive direction, the total distance traveled is

$$d = \int_0^{8.16} |v(t)| \, dt = 282.6 \text{ ft}$$

The acceleration at this time is

$$a(8.16) = 2.4(8.16) = 19.6 \text{ ft/s}^2$$

The solution and the total time history for this motion can be obtained using computational software and plots of displacement, velocity, and acceleration.

Sample Problem 1.2

During walking, the center of mass of an individual rises and falls, following a sinusoidal motion $y(t) = C \cos(2\pi t - \pi) + y_0$ where y_0 is the height of the center of mass when the individual is standing and C is the amplitude of the displacement of the center of mass. Determine the vertical velocity and acceleration of the center of mass, and graph the change of vertical displacement, velocity, and acceleration for a time of 1 s.

Solution

The displacement is given by

$$y(t) = C \cos(2\pi t - \pi) + y_0$$

Differentiating with respect to time yields the velocity:

$$v(t) = -2\pi C \sin(2\pi t - \pi)$$

The acceleration is obtained by differentiating the velocity:

$$a(t) = -4\pi^2 C \cos(2\pi t - \pi)$$

The plots of displacement, velocity, and acceleration versus time with $C = 10$ cm and $y_0 = 100$ cm are as follows:

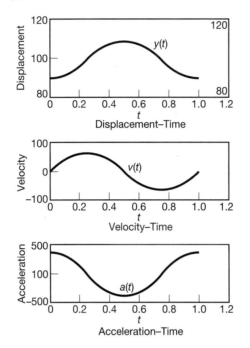

Note that the acceleration is out of phase with the displacement. During walking, the center of mass is at its lowest point when the individual has both feet in contact with the ground (double stance) and is highest when only one foot is in contact with the ground (single stance). The only forces acting on the individual are gravitational attraction and the ground reaction force under the feet. By Newton's second law, the ground reaction force must be greater than body weight when the body is in double stance and less than body weight when the body is in single stance.

Problems

1.1 An automobile is known to reach 60 mph from rest in an interval of time of 9.2 s. Assume that the acceleration of the car is constant over the interval, and calculate the value of the car's acceleration in ft/s².

1.2 The motion of a car sliding along a straight line is given by the function $x(t) = -5t^2 + 88t + 500$ ft, where time is measured in seconds. Determine the acceleration and velocity of the car as functions of time, and determine the time at which the velocity is zero.

1.3 A particle moves in a straight line according to the rule $x(t) = t^3 - 2t + 5$, where $x(t)$ is given in meters and where t is given in seconds. Determine the position, velocity, and

acceleration of the particle at $t = 0$ and $t = 3$ s. How far has the particle moved during this 3-s period?

1.4 The motion of a particle is defined by the equation $x(t) = t^2 - 2t + 3$ ft, where t is in seconds. Calculate the value of the time when the velocity is zero. Does this particle move with constant or variable acceleration?

1.5 The motion of a particle is described by $x(t) = 3 \sin 2t$ m, where t is measured in seconds. Calculate the time t at which the acceleration of the particle is zero.

1.6 Two trucks start moving in a straight line from rest at the same time, as shown in Figure P1.6. Truck A moves according to the equation $x_A(t) = 3t^2 + 6t$ ft, and truck B moves according to the equation $x_B(t) = 3t^3 + 2t$ ft.

a) After $t = 1$ s, which truck is ahead, which truck has the largest velocity, and which truck has the largest acceleration?

b) Make the same comparison at $t = 2$ s.

c) Calculate the time at which both trucks have moved the same distance—that is, when one truck passes the other truck.

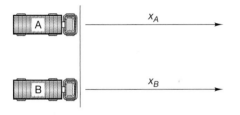

Figure P1.6

1.7 The oscillation of a mass hanging on a spring is described by the displacement $x(t) = 3e^{-t}\sin(10t)$. Calculate the velocity and acceleration of the mass. See Figure P1.7.

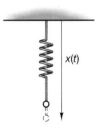

Figure P1.7

1.8 The position of a particle moving along a straight line is given by $x(t) = t^3 - 6t^2 - 15t + 40$ ft.

a) Calculate the value of time for which the velocity is zero.

b) Calculate the new position at this time and the distance traveled from rest.

c) Calculate the value of the acceleration of the particle at this time.

1.9 In Figure P1.9, a mass moves against a spring according to the equation $x(t) = 0.3 \sin(2t)$ m.

a) Calculate the velocity and acceleration of the mass as functions of time.

b) Calculate the total distance traveled by the mass during the time interval $t = 0$ to $t = \pi/2$ s.

c) Determine the position of the mass at $t = \pi/2$ s, and explain why this value is different from the total distance traveled.

d) Plot $x(t)$, $v(t)$ and $a(t)$ for $t = 0$ to $t = \pi/2$ s.

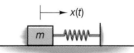

Figure P1.9

1.10 A popular way to compare the performance of various sports cars is to measure the time it takes a car to accelerate from rest to 60 mph and then slam on the brakes and stop. (The process is called 0–60–0 times.) This method provides a measure of both acceleration and braking capability. A crude description of such a motion is given by $x(t) = 342[1 - \cos(0.26t)]$ ft for $0 < t < 12.2$ s. Plot the displacement of the car, and compute and plot the velocity and acceleration of the car during this test.

1.11 A motion analysis system is used to collect the following position data for a conveyor belt, using a 100-Hz camera (100 measurements each second):

$t(s)$	$x(mm)$
0	8
0.01	9
0.02	11
0.03	13
0.04	14
0.05	15
0.06	17
0.07	18
0.08	22
0.09	27
0.10	32
0.11	37
0.12	41
0.13	44
0.14	46
0.15	48
0.16	49
0.17	49
0.18	48
0.19	47
0.20	46

These data are plotted in Figure P1.11. Determine the approximate velocity of the belt using the approximation $v_{i+1} = (x_{i+1} - x_i)/\Delta t$, and the acceleration using the approximation $a_{i+1} = (v_{i+1} - v_i)/\Delta t$. Note that the first and last data points of the velocity and the first two and last two data points of the acceleration are lost. Sketch the functions of velocity and acceleration versus time.

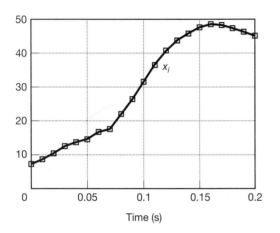

Figure P1.11

🖥 **1.12** Using the data in Problem 1.11, approximate the velocity by $v_i = (x_{i+1} - x_{i-1})/2\Delta t$. This function is called the *central difference approximation*. Sketch the velocity curve.

1.13 The position function of a ball thrown up into the air from a roof is given by

$$y(t) = -4.905t^2 + 20t \text{ m}$$

Determine the position, velocity, and acceleration of the ball when 5 seconds have elapsed. How far has the ball traveled?

For Problems 1.14–1.19, the trajectory of a particle in rectilinear motion is given by a function of time. (All distances are in meters.) Determine the position, velocity, and acceleration of the particle. What are the initial position, velocity, and acceleration, that is, when $t = 0$.

1.14 $x(t) = e^{-ct}\sin \omega t$

1.15 $x(t) = 3t^3 - 2t^2 + 5$

1.16 $x(t) = 2t^2\cos \pi t$

1.17 $y(t) = 3t^2 - 20$

1.18 $x(t) = e^{-0.1t}[3\cos(2t) + \sin(2t)]$

1.19 $y(t) = 5t - e^{-t}(3t)$

1.20 Position data as a function of time are shown in Figure P1.20. Sketch the velocity and acceleration curves.

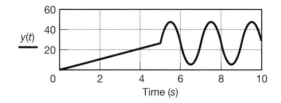

Figure P1.20

1.21 If a well-tuned sports car can accelerate from 0 to 60 mph in 5 seconds, determine the average acceleration. Express the acceleration in m/s^2, ft/s^2, $\%g$. (The quantity g is the acceleration of a particle in free fall near the earth's surface.)

1.22 Approximately what is the maximum average running speed that a human can achieve? State the basis for your estimation.

🖥 **1.23** A laboratory device called a laser vibrometer is used to measure the velocity of a cutting tool, resulting in the velocity-versus-time record shown in Figure P1.23. Using this graph, estimate the acceleration versus time and the displacement versus time for the cutting tool. Assume that the initial position of the tool is zero at a time t equal to zero, and estimate the position at $t = 1$ s.

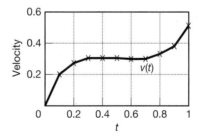

Figure P1.23

🖥 **1.24** A proximity probe is used to record the position versus time of a moving part. For the position record shown in Figure P1.24, determine the corresponding velocity and acceleration records.

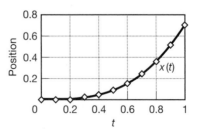

Figure P1.24

1.5 THE DIRECT DYNAMICS PROBLEM: RECTILINEAR MOTION WHEN THE ACCELERATION IS GIVEN

When the direct dynamics problem is solved, the acceleration may be a known function of time obtained either from experimental data or by the use of Newton's second law when known forces given as functions of time are applied to the particle. Alternatively, the acceleration can be specified as a function of position—for example, when a spring acts on a particle. In other applications, involving viscous forces, the acceleration is specified as a function of the velocity. The most general case occurs when the acceleration is given as a function of two or all three of the variables v, x, and t. The problem is mathematically formed as a differential equation—that is, an equation with an unknown function (the displacement) and containing one or more of its derivatives. This equation is solved (integrated) by the use of the techniques of ordinary differential equations. If the acceleration is expressed in terms of a linear relationship of the velocity, displacement, and time, then the resulting differential equation is linear, and the solution is straightforward. However, if the acceleration is related to the velocity or displacement in a nonlinear manner, then the resulting differential equation will also be nonlinear, and the solution is obtained by numerical integration.

First, consider the general case when the acceleration is given as a function of velocity, displacement, and time: $a = f(v,x,t)$. In this case, a second-order linear or nonlinear differential equation will be formed. [Problems of this type will be considered in detail in Chapter 9, when the vibration of systems of particles is presented. A brief discussion of differential equations is presented here, and a complete presentation of methods of solution can be found in texts such as Boyce and DiPrima's *Elementary Differential Equations and Boundary Value Problems* (New York: John Wiley & Sons, Inc.).]

Many methods for obtaining a solution of a particular differential equation are not intuitively obvious, and mathematicians have had to develop them ever since Newton (1642–1727) and Leibniz (1646–1716) independently first discovered calculus. Newton classified first-order differential equations according to their form—that is, $dx/dt = f(t)$, $dx/dt = f(x)$, or $dx/dt = f(t, x)$. He developed a method of solution in the form of infinite series for cases in which $f(t, x)$ is a polynomial in t and x. The Bernoulli brothers, Jakob (1654–1705) and Johann (1667–1748), developed methods to solve differential equations and extended the applications of differential equations to physics. Jakob Bernoulli first used the term "integral" in 1690 in a paper on the solution of a particular differential equation. One problem that was solved by both Bernoulli brothers, Leibniz, and Newton was the *brachistochrone* problem (shortest time)—that is, the determination of the path down which a particle must slide under the influence of gravity, but neglecting friction, for the fastest decent between two points A and B lying in a vertical plane where B is not directly beneath A. This path is defined by the nonlinear differential equation

$$y\left[1 + \left(\frac{dy}{dx}\right)^2\right] = c$$

Newton did not hear of the problem until it was presented to him after he had come home from a long day at the mint. He solved the problem after dinner and communicated his solution anonymously to the Royal Society. When Bernoulli saw the solution, he exclaimed "Ah! I recognize the lion by his paw."

Daniel Bernoulli (1700–1782), the son of Johann, continued the study of differential equations, and his name is associated with the famous Bernoulli equation in fluid mechanics.

Differential equations were also of special interest to the greatest mathematician of the 18th century, Leonhard Euler (1707–1783). Euler's interests ranged over all the fields of mathematics and its areas of application, and he was the most prolific mathematician of all time, even though he was blind for the last 17 years of his life. He made numerous contributions to the field of dynamics and was the first to develop a numerical method for solving differential equations. Another who studied differential equations was Joseph Louis Lagrange (1736–1813), who became a professor of mathematics at the age of 19. He is most famous for his work *Mécanique analytique,* published in 1788, a comprehensive treatise on Newtonian mechanics. The Lagrangian approach to dynamics is based upon the calculus of variations and is presented in most advanced courses in dynamics.

By the end of the 18th century, many methods for solving ordinary differential equations existed, and mathematicians turned their interest to other mathematical topics. The classical methods presented here are based on the work of the 18th century mathematicians, but computers have opened up almost all differential equations to numerical solution. Differential equation theory had its birth in mechanics, and a student must have a basic understanding of differential equations if dynamics is to be understood.

1.5.1 CLASSIFICATION OF DIFFERENTIAL EQUATIONS

There are a number of ways to classify differential equations. These classifications aid in the selection of methods of solution of the equation. Two types of differential equations are encountered in engineering: *ordinary differential equations* and *partial differential equations*. If a differential equation has only one independent variable, it is called an ordinary differential equation. For example, the acceleration of a particle moving in rectilinear motion takes the form

$$\frac{d^2x}{dt^2} = F\left[t, x(t), \frac{dx(t)}{dt}\right] \tag{1.10}$$

The only independent variable in this equation is time (t), and the equation is an ordinary differential equation. You will encounter many physical phenomena in which the equation that describes them will have more than one independent variable—for example, when the unknown function is dependent upon position in the xy-plane. Laplace's equation,

$$\frac{\partial^2 u(x,y)}{\partial x^2} + \frac{\partial^2 u(x,y)}{\partial y^2} = 0 \tag{1.11}$$

is an example of such an equation. This equation is called a *partial differential equation*. Other examples of partial differential equations are the wave equation and the diffusion equation.

The *order* of a differential equation is the order of the highest derivative that appears in the equation. Therefore, Eq. (1.10) is a second-order ordinary differential equation, and Eq. (1.11) is a second-order partial differential equation. Differential equations are also classified as linear or nonlinear. The differential equation of rectilinear motion, Eq. (1.10), is linear if F is a linear function of $x(t)$ and $dx(t)/dt$. For example, a particle moving through a fluid may be subject to forces and, therefore, accelerations that are proportional to the velocity or the velocity squared. In the first case the differential equation would be linear, and in the second case the differential equation would be nonlinear.

1.5.2 SEPARABLE FIRST-ORDER SCALAR DIFFERENTIAL EQUATIONS

When the acceleration of a particle with a single degree of freedom is given as a function of only one other dynamic variable—time, position, or velocity—the solution can be obtained by direct integration after the resulting scalar differential equation is separated. This type of differential equation is called a *separable differential equation*. The solution of a linear differential equation can be obtained by other methods, as is shown in Sample Problem 1.5. If the acceleration is given as a nonlinear function of only one other dynamic variable, separation is a useful approach. Even in these cases, however, the problem can be difficult to solve, and the solution may have to be obtained using numerical techniques. Three special cases will be considered, in which the acceleration is a function of only time, displacement, or velocity.

1. **The acceleration is given as a function of time, $a = f(t)$**   In this case, the acceleration is given as a function of time only, and the time appears as the independent variable. To complete the description of the dynamic state, the other two dynamic variables, velocity and position, are determined as functions of time. This case, discussed in more detail in Chapter 3, will arise when the forces acting on the particle are known functions of time, and by Newton's second law, $\mathbf{F} = m\mathbf{a}$, the acceleration will be a function of time. The first integral expressing the velocity in terms of time is called the impulse–linear momentum relationship. This approach is the most straightforward, as there will be a unique velocity and position for each time, the velocity is obtained by integration of the acceleration, and the position is obtained by integration of the velocity. The constants of integration are determined by specification of the particle's initial conditions (i.e., the initial position and velocity). Using the definition of the acceleration yields

$$a = \frac{dv}{dt}$$

or

$$dv = a\,dt \tag{1.12}$$

The differential equation can be solved by taking the definite integral of both sides with explicit limits (the lower limits being the initial conditions) or using indefinite integrals and determining the constants of integration by substitution of the initial conditions—that is, the conditions at time zero—into the integrated equations. Using definite integrals, we find that the solution is

$$\int_{v_0}^{v} dv = \int_{0}^{t} a(t)\,dt \tag{1.13}$$

In the form of an indefinite integral, the velocity can be written as

$$v(t) = \int a(t)\,dt + C_1 \tag{1.14}$$

The position is determined next, by a similar integration of the definition of velocity

$$v = \frac{dx}{dt}$$

as

$$\int_{x_0}^{x} dx = \int_{0}^{t} v(t)\,dt \tag{1.15}$$

or as an indefinite integral,

$$x(t) = \int v(t)dt + C_2 \tag{1.16}$$

The constants of integration can be determined from the initial conditions—that is, when $t = 0$. The constants C_1 and C_2 frequently are equal to the initial velocity and the initial position, respectively. The definite integrals in Eqs. (1.13) and (1.15) give the same results for the velocity and the displacement.

2. **The acceleration is given as a function of position, $a = f(x)$**   In this case, position is the independent variable, acceleration is specified as a function of position, and the velocity and time are also to be determined as functions of position. This situation arises when the force is a known function of the position, and therefore, by Newton's second law, the acceleration is a known function of the position. Since the position is the independent variable and the particle can occupy the same position at different times in the motion, the acceleration, velocity, and time may not have unique dependencies upon position. That is, for a given position, the particle may have passed that position during different times and may have different accelerations or velocities with each passing. Acceleration is defined as the time derivative of velocity, but by the use of the chain rule of differentiation, the acceleration can be expressed as

$$a = \frac{dv}{dt} = \frac{dv}{dx}\frac{dx}{dt} = v\frac{dv}{dx}$$

or

$$a\,dx = v\,dv \tag{1.17}$$

Forming the definite integral of both sides of Eq. (1.17), where $a(x)$ is a known function of position [that is, $a = f(x)$], yields

$$a = f(x)$$

$$\int_{v_0}^{v} v\,dv = \int_{x_0}^{x} f(x)dx \tag{1.18}$$

where v_0 and x_0 are the initial velocity and initial position, respectively. The integral on the left side can now be evaluated to determine the velocity as a function of position:

$$\frac{v^2 - v_0^2}{2} = \int_{x_0}^{x} f(x)dx$$

$$v(x) = \sqrt{v_0^2 + 2\int_{x_0}^{x} f(x)dx} \tag{1.19}$$

(This case will be investigated in more detail in Chapter 3, and this first integral will form the basis of the principle of the equivalence of work and kinetic energy.)

Once the velocity has been determined as a function of position, the time can also be expressed as a function of position. The definition of the velocity is

$$v = \frac{dx}{dt}$$

Rewriting this definition gives

$$dt = \frac{dx}{v(x)} \tag{1.20}$$

and integrating both sides yields

$$\int_0^t dt = t(x) = \int_{x_0}^x \frac{dx}{v(x)}$$

The expression for time as a function of displacement is usually inverted such that the displacement is given as a function of time. The displacement–time relation can then be differentiated to yield the velocity as a function of time. The velocity–time function can also be obtained by substituting the displacement–time function into the velocity–displacement relationship, Eq. (1.19).

3. **The acceleration is given as a function of velocity,** $a = f(v)$ In this case, the velocity will be the independent variable, and the position and time will be determined as functions of velocity. Eq. (1.17), obtained using the chain rule of differentiation, yields

$$a = v\frac{dv}{dx}$$

Assuming that $a = f(v)$, this equation may be rewritten as

$$dx = \frac{vdv}{f(v)} \tag{1.21}$$

Definite integration of both sides yields the desired expression for the displacement as a function of the velocity:

$$\int_{x_0}^x dx = \int_{v_0}^v \frac{vdv}{f(v)}$$

$$x(v) = x_0 + \int_{v_0}^v \frac{vdv}{f(v)} \tag{1.22}$$

Time may now be obtained from the definition of acceleration as follows:

$$a = \frac{dv}{dt}$$

Substituting $f(v)$ for a and solving for t, we obtain

$$dt = \frac{dv}{f(v)} \tag{1.23}$$

Integration of both sides gives time as a function of velocity:

$$\int_0^t dt = \int_{v_0}^v \frac{dv}{f(v)} \quad \text{or} \quad t(v) = \int_{v_0}^v \frac{dv}{f(v)}$$

In this case, the time and displacement are expressed as functions of the velocity and are not necessarily single-valued functions. That is, for a given velocity, there may be more than one time when the particle has that velocity. It is usually required to invert the time–velocity relationship to obtain the velocity as a function of time. This relationship can then be used to obtain the displacement–time function.

For the simple case of rectilinear translation—one degree of freedom—the dynamic variables can be expressed as functions of time, position, or, when needed, velocity. In all three of these cases, the acceleration, velocity, and position can be expressed as functions of time by algebraic manipulations of the preceding equations.

Sample Problem 1.3

The acceleration of a particle, given as a function of time, is $a(t) = A \sin(pt)$ m/s², where A and p are unspecified, but are assumed to be known constants. If the initial velocity is denoted as v_0 and the initial displacement as x_0, determine the velocity and displacement as functions of time in terms of A, p, v_0 and x_0.

Solution   Since the acceleration is a function of time, this situation is an example of Case 1. The given acceleration is

$$a(t) = A \sin(pt) \text{ m/s}^2$$

Integrating using an indefinite integral yields

$$v(t) = -\frac{A}{p} \cos(pt) + C_1 \text{ m/s}$$

Where C_1 is the constant of integration that needs to be obtained using the initial velocity when $t = 0$, that is,

$$v(0) = v_0$$

Substituting $t = 0$ into the velocity equation and solving for C_1 yields

$$C_1 = v_0 + \frac{A}{p} \text{ m/s}$$

The general expression for the velocity is

$$v(t) = \frac{A}{p}(1 - \cos(pt)) + v_0 \text{ m/s}$$

Indefinitely integrating the expression for the velocity yields

$$x(t) = \left(\frac{A}{p} + v_0\right)t - \frac{A}{p^2}\sin(pt) + C_2 \text{ m}$$

The constant of integration C_2 is evaluated using the initial condition

$$x(0) = x_0$$

Substituting $t = 0$ into the expression for displacement yields

$$C_2 = x_0$$

Thus, the general expression for the displacement is

$$x(t) = \left(\frac{A}{p} + v_0\right)t - \frac{A}{p^2}\sin(pt) + x_0 \text{ m}$$

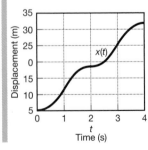

If $A = 20$ m/s², $p = \pi$ (1/s), $v_0 = 0.5$ m/s, and $x_0 = 5$ m, the displacement—time graph for the first 4 seconds, as in the accompanying figure.

Sample Problem 1.4

A boy standing on a 10-m-high platform throws a ball straight up with an initial velocity of 5 m/s. (See accompanying figure.) If the acceleration of the ball due to gravitational attraction is constant and equal to 9.81 m/s² downward (i.e., the only force acting on the ball is the gravitational attraction of the earth), determine (a) the ball's velocity and height above the ground at any instant of time, (b) the maximum height the ball will reach and the time when it reaches this height, and (c) the time the ball takes to reach the ground and the velocity when it strikes the ground.

Solution   The acceleration is a constant and, therefore, can be considered as a function of time, position, or velocity. The direction x is taken to be positive upward, and the origin is taken to be ground level. The acceleration is constant and downward and is written as

$$a(t) = \frac{d^2x}{dt^2} = -9.81 \text{ m/s}^2$$

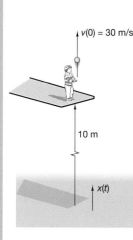

Writing the acceleration as $a(t) = dv(t)/dt$ and integrating with respect to time gives the velocity as

$$\int_{v_0 = 30}^{v} dv = -\int_0^t 9.81 dt$$

$$v(t) - 5 = -9.81t$$

$$v(t) = 5 - 9.81t \text{ m/s}$$

(a) The displacement measured from the ground can be found by integrating the velocity with respect to time:

$$\int_{x_0 = 10}^{x} dx = \int_0^t (5 - 9.81t) dt$$

$$x - 10 = 5t - 4.905t^2 \text{ m}$$

$$x(t) = 10 + 5t - 4.905t^2$$

(b) The maximum height will occur when the velocity is zero. Setting $v(t) = 0$ yields the time at the maximum height:

$$t = 5/9.81 = 0.51 \text{ s}$$

Substituting this value for time into the displacement expression yields

$$x(0.51) = 10 + 5(0.51) - 4.905(0.051)^2 = 11.27 \text{ meters}$$

(c) The displacement of the ball when it hits the ground is zero; therefore, we may find the time that the ball hits the ground by setting $x(t) = 0$:

$$0 = 10 + 5t - 4.905t^2$$

Solving the quadratic equation yields the solution $t = +2.03$ s or $t = -1.01$ s.

The positive time value is when the ball hits the ground after it is thrown, while the negative value is when it would have left the ground if the equations of motion were valid for negative time. The velocity when it hits the ground is

$$v(2.03) = 5 - 9.81(2.03) = -14.91 \text{ m/s}$$

The total distance the ball traveled was 11.27 m up and 21.27 m down, or a total distance of 32.54 m.

The following graphs of the velocity and displacement with time are useful to view the motion.

$$v(t) := 5 - 9.81 \cdot t$$

$$x(t) := 10 + 5 \cdot t - 4.905 \cdot t^2 \qquad\qquad v(0.51) = -3.1 \cdot 10^{-3}$$

$$t := 0, 0.01 \ldots 2.54$$

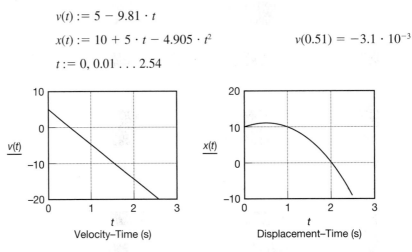

Velocity–Time (s)   Displacement–Time (s)

Consider the following three separate cases, where the acceleration is given as a function of only one of the other dynamic variables and with the initial conditions at time $= 0$ given for each case:

1. $a(t) = 3te^{-t}$ m/s^2 $v_0 = 0$; $x_0 = 0$
2. $a(x) = -2x + 1$ m/s^2 $v_0 = 4$ m/s; $x_0 = 5$ m
3. $a(v) = a_0 - cv$ m/s^2 $v_0 = 0$; $x_0 = 0$

Determine the other dynamic variables (v and x) in terms of the appropriate independent variable.

Solution

1. Treat t as an independent variable, and write $dv = a(t)dt = 3te^{-t}dt$. Using indefinite integrals, we obtain

$$a(t) = 3te^{-t}$$

$$v(t) = \int 3te^{-t}dt + C_1$$

$$v(t) = -3e^{-t}(t + 1) + C_1$$

where C_1 is the constant of integration. The initial condition is $v(0) = 0$; therefore:

$$0 = -3 + C_1$$

$$C_1 = 3 \text{ m/s}$$

and the velocity as a function of time is

$$v(t) = -3e^{-t}(t + 1) + 3 \text{ m/s}$$

The displacement may be found by integrating the velocity:

$$x(t) = \int [-3e^{-t}(t + 1) + 3]dt + C_2$$

$$x(t) = 3e^{-t}(t + 1) + 3e^{-t} + 3t + C_2$$

C_2 is, or course, the constant of integration. The initial condition is $x(0) = 0$; therefore,

$$C_2 = -6 \text{ m}$$

and

$$x(t) = 3e^{-t}(t + 2) + 3t - 6 \text{ m}$$

Plots of the dynamic postion and velocity are as follows:

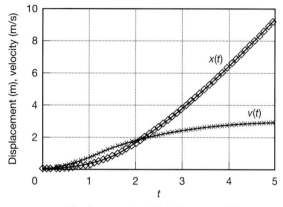

Displacement and Velocity versus Time

2. The acceleration is given as a function of displacement; therefore, Eq. (1.17) will be used:

$$a(x)dx = v\,dv$$

$$\int_5^x [-2x + 1]dx = \int_4^v v\,dv$$

$$-x^2 + x + 20 = \frac{v^2 - 16}{2}$$

$$v(x) = \sqrt{2x(1 - x) + 56} \text{ m/s}$$

The time is expressed as a function of displacement. By use of Eq. (1.20), we obtain

$$\int_0^t dt = \int_5^x \frac{dx}{v(x)}$$

$$t = \int_5^x \frac{dx}{\sqrt{2(x - x^2) + 56}}$$

$$= -\frac{1}{\sqrt{2}} \sin^{-1}\frac{-4x + 2}{\sqrt{4(56)(2) + 2^2}} + \frac{1}{\sqrt{2}} \sin^{-1}\frac{(-18)}{21.26}$$

$$t = -\frac{0.321\pi}{\sqrt{2}} - \frac{1}{\sqrt{2}} \sin^{-1}\frac{2 - 4x}{21.26}$$

This expression can be solved for x to yield the displacement as a function of time as follows:

$$x = \frac{1}{2} + 5.315 \sin(\sqrt{2}t + 0.321\pi) \text{ m}$$

The velocity can be written in terms of time by substitution of x as a function of time into the expression of v as a function of displacement, or by differentiation of the displacement as a function of time. The latter gives

$$v(t) = 5.315(\sqrt{2}) \cos{(\sqrt{2}t + 0.321\pi)} \text{ m/s}$$

Note, as a check, that at time $t = 0$, the displacement is 5 and the velocity is 4. The acceleration can be expressed in terms of time by differentiation of the velocity:

$$a(t) = -2(5.315) \sin{(\sqrt{2}t + 0.321\pi)} \text{ m/s}^2$$

The displacement–time and velocity–time graphs are as follows:

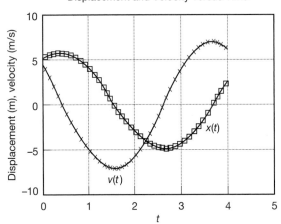

Displacement and Velocity versus Time

Examining the motion, one can note that the movement is sinusoidal in nature, although the acceleration is a linear function of the displacement. The displacement ranges from -4.635 to $+5.635$ m. If the solution were left in terms of displacement as an independent variable, the range of this variable would have to be determined by examining the position where the velocity was zero.

3. The acceleration is given as a function of velocity, and from Eq. (1.22), the displacement as a function of velocity is

$$a dx = v dv$$

$$\int_0^x dx = \int_0^v \frac{v dv}{(a_0 - cv)} = \frac{1}{c^2}[a_0 - cv - a_0 \ln(a_0 - cv)]_0^v$$

$$x = \frac{1}{c^2}\left[a_0 \ln\left(\frac{a_0}{a_0 - cv}\right) - cv\right] \text{ m}$$

Time can be expressed as a function of the velocity by using Eq. (1.23):

$$\int_0^t dt = \int_0^v \frac{dv}{a_0 - cv}$$

$$t = -\frac{1}{c}\ln\left(\frac{a_0 - cv}{a_0}\right)$$

Inverting this equation yields an expression for the velocity as a function of time:

$$v(t) = \frac{a_0}{c}(1 - e^{-ct}) \text{ m/s}$$

As time continues, the velocity approaches a constant, called the terminal velocity, given by

$$v(\infty) \approx \frac{a_0}{c} \text{ m/s}$$

Note that e^{-ct} goes to zero as t goes to infinity.

Consider an object dropped in the Earth's gravitational field such that

$$a_0 := 9.81 \text{ m/s}^2$$

$$c := 0.1 \text{ s}^{-1}$$

$$v(t) := \frac{a_0}{c} \cdot (1 - e^{-c \cdot t}) \qquad \frac{a_0}{c} \cdot 3.6 = 353.16 \frac{\text{km}}{\text{hr}}$$

$$353.16 \cdot 0.624 = 220.372 \text{ mph}$$

$$t := 0, 0.01 \ldots 120$$

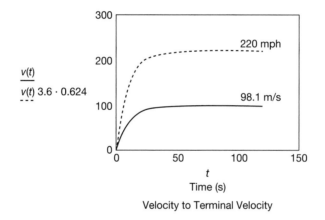

Velocity to Terminal Velocity

Determine the acceleration, velocity, and displacement as functions of time if the acceleration is given as

$$a(x) = 3x^2 - 8 \text{ ft/s}^2$$

and the initial position is $x_0 = 0$. The initial velocity is $v_0 = 2$ ft/s.

Solution   The equation of motion can be written in differential form as

$$\frac{d^2x}{dt^2} - 3x^2 = -8$$

This second-order differential equation is nonlinear. However, it is separable. Using Eq. (1.19) yields

$$v(x) = \pm\sqrt{4 + x^3 - 8x} \text{ ft/s}$$

The time, using Eq. (1.23), can be expressed as

$$t = \int_0^x \frac{dx}{\pm\sqrt{4 + x^3 - 8x}} \text{ s}$$

This integral is difficult to evaluate, and at this point in the analysis, the range of values of *x* is not known. The maximum and minimum values of *x* will occur when the velocity is zero. To determine these points, plot the cubic function under the square-root sign using computational software. The velocity will be zero at these two points, and the motion will oscillate between the minimum and maximum values of *x*. The velocity will be positive as the particle goes in the positive direction, and negative as the particle goes from its maximum position to its minimum position. The time parameter can be obtained by numerical integration, and time will be adjusted to begin when the particle is at its minimum position. Although a solution can be obtained by determining these endpoints, an alternative and easier approach is to obtain the solution by numerical integration of the original second-order nonlinear equation. The full development of the solution by numerical integration is shown in the Computational Supplement. The resulting velocity and displacement plots are as follows:

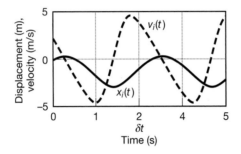

The displacement and velocity graphs were generated by numerical integration of the non-linear differential equation.

1.5.3 SPECIAL RECTILINEAR MOTIONS

Two unique types of motion will be examined as special cases of the rectilinear motion of a particle. These cases are quite common and lead to simple differential equations describing the motion.

1. **Uniform rectilinear motion**   This motion occurs when the acceleration of the particle is zero, i.e.,

$$a = \frac{dv}{dt} = 0 \tag{1.24}$$

If the initial velocity is v_0, integration of $a(t) = 0$ over time yields a constant velocity:

$$v = v_0 \tag{1.25}$$

This equation shows that if the acceleration is zero, the velocity is a constant; that is, uniform rectilinear motion is exhibited. If x_0 denotes the initial displacement (the displacement when $t = 0$), then integrating the constant velocity in Eq. (1.25) yields

$$x = v_0 t + x_0 \tag{1.26}$$

Hence, if the acceleration of the particle is zero, the particle moves with constant velocity in a straight line, and the displacement increases linearly with time. This is a form of Newton's first law, which states that in the absence of net external forces, the change in the linear momentum of a particle is zero. (Linear momentum is the product of mass and velocity and will be discussed in detail in Chapter 3.)

2. **Uniformly accelerated rectilinear motion**   In this case, the acceleration acting on the particle is constant. An example of this type of motion is a mass falling in a gravitational field with no air resistance—that is, free fall. This was the type of motion that Galileo termed "natural motion" and was the first detailed study of dynamics. The constant acceleration is denoted as

$$a = a_0 \tag{1.27}$$

This expression is a special case of a separable differential equation. The acceleration could be treated as a function of time—that is, $a(t) = a_0 t^0$—or as a function of position—that is, $a(x) = a_0 x^0$. Considering a as a function of time, integration of the constant acceleration yields

$$v = a_0 t + v_0 \tag{1.28}$$

where v_0 again denotes the initial velocity. Eq. (1.28) is then integrated to obtain

$$x = \frac{a_0 t^2}{2} + v_0 t + x_0 \tag{1.29}$$

where x_0 denotes the initial displacement at the initial time, $t = 0$.

In this case, the acceleration was treated as a function of time. Since the acceleration is a constant, it could equally well have been treated as a function of displacement, yielding

$$a = \frac{v\,dv}{dx}$$

$$\int_{x_0}^{x} a_0 \, dx = \int_{v_0}^{v} v \, dv$$

$$a_0(x - x_0) = \frac{1}{2}(v^2 - v_0^2)$$

$$v^2 = 2a_0(x - x_0) + v_0^2 \tag{1.30}$$

This expression for velocity as a function of position is valid only for the case of constant acceleration.

Sample Problem 1.7

During a "locked-up" skid, a car has a constant negative acceleration (braking deceleration). If there is a 1.75-s reaction time before braking (the average time observed for drivers), and the braking deceleration is 22.5 ft/s², determine the distance to stop if the initial speed is (a) 60 mph; (b) 45 mph; and (c) 30 mph.

Solution   The conversion between mph and ft/s is 1 mph = 88/60 ft/s, or 1 mph = 1.467 ft/s. The distance traveled during the reaction time, or the initial position of the car when braking starts, is

$$x_0 = 1.467 \, (1.75) \, v_0 = 2.567 \, v_0 \text{ ft/s}$$

The final velocity after braking is zero, so Eq. (1.51) can be written as

$$x_f = \frac{1}{2a_0}(v_f^2 - v_0^2) + x_0$$

With the given values,

$$a_0 = -22.5 \text{ ft/s}^2$$
$$v_0 = 1.467 \ (v_0 \text{ mph}) \text{ ft/s}$$
$$v_f = 0$$

The last expression gives the stopping distance as a function of the initial velocity by

$$x_f = x_0 + \frac{1}{2a_0}(-v_0^2)$$

$$x_f = 1.75 \times 1.467 \times v_0 + \frac{1.467^2}{2 \times 22.5} v_0^2$$

For each of the given initial velocities, the corresponding stopping distance is:

a) $x_f = 326$ ft
b) $x_f = 212$ ft
c) $x_f = 120$ ft

A plot of stopping distance is shown for various initial velocities. This type of information is included in most driving instruction manuals. Note that the stopping distance does not depend on the car's velocity in a linear way, as it involves a term that is proportional to the square of the velocity. Therefore, the distance to stop at 60 mph is not twice the distance to stop at 30 mph, but much more.

$$x_f(v_0) := 1.75 \cdot 1.467 \cdot v_0 + \frac{1}{2 \cdot 22.5} \cdot (1.467 \cdot v_0)^2$$

$$v_0 := 0, 5 \ldots 60$$

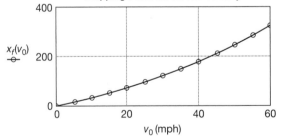

Problems

Problems 1.25–1.32: In each of these problems, determine the order of the given differential equation, and indicate whether it is linear or nonlinear. If it is nonlinear, indicate the nonlinear terms.

1.25 $m\dfrac{d^2x}{dt^2} = -c\dfrac{dx}{dt} - kx + \sin(\pi t)$, where m, c, and k are constants.

1.26 $\dfrac{dv}{dt} = 8t^2$

1.27 $\dfrac{dv}{dt} = 9.81 - 0.01v$

1.28 $\dfrac{d^2\theta(t)}{dt^2} = \sin[\theta(t)]$

1.29 $\dfrac{d^2\theta}{dt^2} = -c\dfrac{d\theta}{dt}\left|\dfrac{d\theta}{dt}\right| + g\sin\theta$, where c and g are constants.

1.30 $\dfrac{dv}{dt} = g - cv^2$, where g and c are constants.

1.31 $\dfrac{d^2x}{dt^2} = g \sin \beta$, where g and β and constants.

1.32 The angle β in Problem 1.31 is a function of time $[\beta(t) = 3t^2]$.

1.33 The acceleration of a particle moving in rectilinear motion is given as $a(t) = 3t^2 - 4t$ ft/s². If the initial position and velocity of the particle are zero, determine the velocity and position as functions of time. Sketch the position–time relationship for the first 3 s.

1.34 The acceleration of a particle is given as $a(t) = 40t \cos(\pi t)$ m/s²; determine the velocity and position. The initial position of the particle undergoing this acceleration is 5 m, and the initial velocity is 3 m/s. Sketch the position as a function of time.

1.35 An Olympic sprinter can reach 10 m/s from rest over a distance of 15 m. Assuming constant acceleration, determine the value of the sprinter's acceleration. Compare this acceleration to that of a car that can go from 0 to 60 mph in 6 s.

1.36 A ball is thrown straight up with an initial velocity of 10 m/s and is subjected to gravitational attraction of 9.81 m/s². Determine (a) the maximum height reached by the ball and (b) the time required for the ball to return to its initial position.

1.37 A particle accelerator consists of a 2-m long tube. Determine the constant acceleration required to accelerate the particle from 10^4 m/s to 5×10^6 m/s as it passes through the accelerator. Determine the time that the particle is in the accelerator.

1.38 A particle attached to a linear spring has an acceleration given by $a(x) = -kx$. If the initial position is x_0 and the initial velocity is zero, determine the position–time function.

1.39 An elevator in a 200-m high building runs with a constant acceleration (and deceleration) of 2.5 m/s² until it reaches its constant operating speed of 3 m/s. Determine the time for the elevator to go from the bottom floor to the top floor of the building.

1.40 The acceleration of the motor mount of an automobile engine is given by $a(x) = -c^2x$, where x is the displacement of the mount. Determine the value of c such that the velocity decreases from 30 mm/s at $x = 0$ to zero at $x = 10$ mm.

1.41 A body modeled as a particle is attached to a nonlinear spring, and the acceleration for a positive displacement is given by $a(x) = -cx^2$. (This is the type of spring that models tendons in the human body.) If the initial position of the particle is zero and the initial velocity of the particle is v_0, determine an expression for the velocity–position relationship.

1.42 The acceleration of a machine part is measured to be $a(v) = -v$ mm/s². Given the initial velocity of 750 mm/s at $t = 0$, determine the distance the part travels before coming to rest.

1.43 After the rocket engines are stopped, the acceleration of a spacecraft is given by $a(y) = -g_0 \dfrac{R^2}{(R + y)^2}$, where g_0 is the acceleration of gravity on the surface of the earth (9.81 m/s²), and R is the radius of the Earth (6370 km), and y is the distance above the earth. If the engines are stopped when $y = 40$ km and the velocity at this time is 6000 m/s, determine the maximum height attained by the spacecraft.

1.44 In Problem 1.43, determine the escape velocity (the velocity at engine shutdown for the spacecraft to reach an altitude approaching infinity) if the engines are stopped at $y = 40$ km.

1.45 A car hydroplaning on wet pavement undergoes a deceleration of $a(v) = -cv$, where $c = 0.4$ s⁻¹. If the initial velocity of the car is 100 km/hr, determine the distance–time relationship for the sliding car.

1.46 The acceleration of a mass is determined to be $a(t) = 5 \sin(20t)$ m/s². At time $t = 0$, the position of the mass is 1 m, and it has a velocity of 3 m/s. Determine expressions for the velocity and displacement as functions of time.

1.47 A particle moves with an initial velocity of 0.6 ft/s and experiences an acceleration of $a = -v^3$ ft/s². Calculate the velocity after 4 s.

1.48 A diver eases into the water from a boat and sinks into the water under gravitational attaction, resisted by damping, with acceleration $a(v) = g - cv^2$. Determine the diver's position and velocity as functions of time.

1.49 A ball is dropped from a window 30 ft above the ground. Determine the time required for the ball to hit the ground and the velocity when it hits the ground.

1.50 The ball in Problem 1.49 is thrown upward with an initial velocity of v_0. If the ball hits the ground 1.71 s later, determine the initial velocity of the ball.

1.51 An automobile in a locked-up skid on dry pavement can decelerate at a constant rate of $0.7g$. Determine the distance for the car to come to a stop if its initial velocity is (a) 60 mph, (b) 45 mph, and (c) 30 mph.

1.52 In Problem 1.51, the deceleration is reduced to $0.4g$ on wet pavement. Determine the time required for the car to stop if the initial velocity is (a) 60 mph, (b) 45 mph, and (c) 30 mph.

1.53 If a race car driver does not spin his wheels, he can accelerate the car at $0.6g$. Determine the time required for the car to reach the speed of 200 mph, starting from rest. What distance does the driver cover in that time?

1.54 Two drag cars start at the same place and time. Car A has an acceleration of $0.9g$, and car B has an acceleration of $0.85g$. How far ahead of B will A be when A crosses the finish line 1 km from the start?

1.55 A rocket produces the acceleration–time curve shown in Figure P1.55 as it completes the burn of the first and second stages if the initial velocity and position are zero.

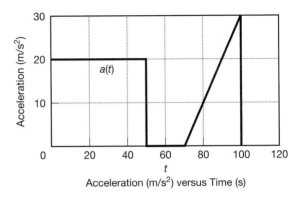

Figure P1.55

Determine the position and velocity of the rocket after 120 s.

1.5.4 SOLUTION OF A LINEAR FIRST-ORDER DIFFERENTIAL EQUATION BY USE OF AN INTEGRATING FACTOR

We can increase the number of analytical solutions of a first-order differential equation by considering the acceleration to be an explicit function of the velocity and time only. Therefore, in Eq. (1.10), $F[t,v]$ is not an explicit function of $x(t)$. The acceleration can be written as a first-order ordinary differential equation for the velocity. The differential equation has the form

$$\frac{dv}{dt} = F[t, v] \tag{1.31}$$

If, in addition, Eq. (1.31) is a linear differential equation, it can be written as

$$\frac{dv}{dt} + p(t)v = f(t) \tag{1.32}$$

This form of linear first-order differential equation can be solved by the use of an *integrating factor*, defined as

$$\lambda(t) = e^{\int p(t)dt} \tag{1.33}$$

The general solution of Eq. (1.32) is

$$v(t) = \frac{\int \lambda(t)f(t)dt + C}{\lambda(t)} \tag{1.34}$$

where C is a constant generally evaluated from the known value of v when $t = 0$. Equation 1.34 is the so-called *initial-value problem*. If $f(t)$ is identically zero, Eq. (1.34) reduces to

$$v(t) = Ce^{-\int p(t)dt} \tag{1.35}$$

If the function $p(t)$ is identically zero, Eq. (1.34) reduces to

$$v(t) = \int f(t)dt + C \tag{1.36}$$

We previously examined Eq. (1.36) when we considered the case of the acceleration as a function of time only.

<table>
<tr><td>**Sample Problem 1.8**</td></tr>
</table>

The acceleration of a particle moving in rectilinear motion is given as $a(v,t) = a_0 - cv$. Determine the velocity as a function of time if the initial velocity is zero.

Solution   The equation of motion can be written as

$$\frac{dv}{dt} + cv = a_0$$

The differential equation is in the form of Eq. (1.32) and can be solved using an integrating factor, namely,

$$\lambda(t) = e^{ct}$$

From Eq. (1.34), the velocity is

$$v(t) = \frac{\int a_0 e^{ct}dt + C}{e^{ct}}$$

$$v(t) = e^{-ct}\left[\frac{a_0}{c}e^{ct} + C\right]$$

$$v(t) = \frac{a_0}{c} + Ce^{-ct}$$

Evaluating the constant of integration under the condition that $v(0) = 0$ yields

$$v(t) = \frac{a_0}{c}[1 - e^{-ct}]$$

This solution agrees with the one obtained by separation in Sample Problem 1.5(c).

1.5.5 SECOND-ORDER LINEAR DIFFERENTIAL EQUATIONS

The scalar differential equation for rectilinear motion can always be written as a second-order ordinary differential equation, given as Eq. (1.10) and repeated here:

$$\frac{d^2x}{dt^2} = F\left[t, x, \frac{dx}{dt}\right] \tag{1.37}$$

This equation is a *linear* second-order differential equation if it can be written in the following form:

$$\frac{d^2x}{dt^2} + p(t)\frac{dx}{dt} + q(t)x = f(t) \tag{1.38}$$

Equation (1.38) is called *homogeneous* if the function $f(t)$ is zero for all values of t. If this is not the case, the equation is called *nonhomogeneous*. The nonhomogeneous differential solution is solved as the sum of the two general solutions of the homogeneous equation and a specific solution of the nonhomogeneous equation, called the *particular solution*. There are many special forms of Eq. (1.38) that can be solved for specific analytical solutions. It is beyond the scope of this text to discuss the general theory of differential equations; the interested reader should consult the many texts on the subject. In general, we will seek a numerical solution to Eq. (1.37), whether the equation is linear or nonlinear. This solution will form a general numerical approach to the initial-value problem of dynamics—that is, the solution of Eq. (1.37) when the values of x and v are known at $t = 0$. However, in one form of Eq. (1.38) that arises frequently in the dynamics problem of vibrations, $p(t)$ and $q(t)$ are constants, and Eq. (1.38) can be written as

$$m\frac{d^2x}{dt^2} + c\frac{dx}{dt} + kx = f(t) \tag{1.39}$$

Consider first the homogeneous part of Eq. (1.39):

$$m\frac{d^2x}{dt^2} + c\frac{dx}{dt} + kx = 0 \tag{1.40}$$

Based upon our experience with the linear first-order differential equation, we will assume a solution in the form of an exponential integrating factor:

$$x(t) = Ae^{\lambda t} \tag{1.41}$$

Substitution of this solution form into Eq. (1.40) yields the *characteristic equation* for the value of λ:

$$(m\lambda^2 + c\lambda + k)Ae^{\lambda t} = 0 \tag{1.42}$$

This equation holds only if the quadratic expression for λ inside the brackets is zero. Using the general solution to the quadratic equation, we find that the two solutions for λ are

$$\lambda = \frac{-c \pm \sqrt{c^2 - 4mk}}{2m} \tag{1.43}$$

If m, c, and k are real numbers, the discriminant $c^2 - 4mk$ will determine the form of the solution. If the discriminant is positive, the roots are real and unequal. If the discriminant is zero, the roots are real and equal, and if the discriminant is negative, the roots are imaginary and unequal. When the discriminant is negative, the solution can be written in terms of trigonometric functions. (See Sample Problem 1.9.)

This solution will be examined in detail in Chapter 9. Note that if the acceleration had been proportional to the square of the velocity instead of being linearly dependent upon the velocity, the resulting differential equation would have been the nonlinear equation

$$\frac{d^2x}{dt^2} + c\left(\frac{dx}{dt}\right)^2 + kx = \sin \lambda t \tag{1.44}$$

In that case, the integrating factor does not work, and the equation must be solved numerically.

Sample Problem 1.9

In Sample Problem 1.5 (part 2), $a(x) = -2x + 1$, with initial conditions $v(0) = 4$ m/s and $x(0) = 5$ m. The acceleration is a linear function of the displacement, and therefore, the problem can also be solved as a second-order linear differential equation using time as the independent variable.

Solution

The acceleration is $a(x) = -2x + 1$, which may be written

$$\frac{d^2x}{dt^2} + 2x = 1$$

The homogeneous part of this equation has constant coefficients and may be solved by assuming a solution in the form of an exponential. The characteristic equation is

$$\lambda^2 + 2 = 0$$

The roots of this equation are both imaginary:

$$\lambda_1 = i\sqrt{2}$$

$$\lambda_2 = -i\sqrt{2}$$

Here, $i = \sqrt{-1}$ and the homogeneous part of the solution is

$$x_h = Ae^{i\sqrt{2}t} + Be^{-i\sqrt{2}t}$$

The following exponential relation, called Euler's equation, can be used to write the exponentials in the form of trigonometric functions:

$$e^{i\theta} = \cos\alpha + i\sin\alpha \qquad \text{or} \qquad e^{-i\theta} = \cos\theta - i\sin\theta$$

If the constant B is written as the complex conjugate of A, the homogeneous solution takes the form

$$x_h = C_1\cos\sqrt{2}t + C_2\sin\sqrt{2}t$$

This form of the solution may be easily verified by substution into the differential equation. The particular solution of the differential equation is

$$x_p = 1/2$$

The initial conditions are used to determine the constants of integration:

$$C_1 = 4.5 \text{ and } C_2 = 2.829$$

Using the trigometric identity

$$\sin(A + B) = \sin(A)\cos(B) + \cos(A)\sin(B)$$

we can rewrite the solution as follows:

$$x(t) = \frac{1}{2} + 5.315(0.532\sin\sqrt{2}t + 0.847\cos\sqrt{2}t)$$

$$x(t) = \frac{1}{2} + 5.315\sin(\sqrt{2}t + 0.321\pi)$$

This solution agrees with the solution obtained in Sample Problem 1.5(part 2). However, that solution was obtained by first finding the time as a function of displacement and then inverting that relation. The solution is most easily understood when the dynamic parameters are expressed as functions of time. If the acceleration can be expressed as a linear function of velocity and position, it is always better to solve the second-order linear differential equation. If the acceleration is a nonlinear function of velocity or position, the technique of separation of the first-order nonlinear differential equation is preferable.

1.5.6 NUMERICAL SOLUTION OF DIFFERENTIAL EQUATIONS

When the acceleration is a nonlinear function of the velocity or position, the resulting differential equation is nonlinear. In this case, an exact expression for the solution may not be obtainable by using standard methods of differential equations. Problems of this type are very common in dynamics, and the equations of motion for most systems will be nonlinear. However, numerical techniques are available for solving these equations and may be programmed on computers or calculators. Indeed, such methods are tools in all computational software programs. We will present one method here that is easily used and that will give accurate results in most cases. This numerical solution uses the Euler, or tangent-line, method. Consider the first-order differential equation

$$\frac{dy}{dt} = f(y, t) \tag{1.45}$$

with the initial condition

$$y(0) = y_0 \tag{1.46}$$

The derivative of the function y is the slope of the y–t curve and is given by the differential equation (1.45). The first numerical solution of a differential equation was performed by Euler in 1768, and the Euler, or tangent-line, method is based on the observation that the derivative is the slope of the curve and is tangent to the curve. The Euler method is a first-order Runge-Kutta method, and most computational software packages have higher order routines with increased accuracy. However, we will present most of our numerical solutions of the differential equations using the Euler method to enhance the conceptual understanding of the process. (Consult the Computational Supplement for more details.)

Suppose the exact solution of the differential equation is $y = g(t)$, as shown in Figure 1.7. Now, the differential dy/dt is tangent to the curve $y(t)$ at any point on the curve and, therefore, is the slope of the curve. Hence, the value y_1 is related to the value y_0 by

$$y_1 = y_0 + \left(\frac{dy}{dt}\right)_0 (\Delta t) \tag{1.47}$$
$$= y_0 + f(y_0, t_0)\Delta t$$

where Eq. (1.45) has been used to replace the slope dy/dt evaluated at the point (y_0, t_0). A numerical solution is obtained by using the slope at one point to obtain the value of the function for the next point. Eq. (1.47) can be written for the $(i + 1)$th value of y as

$$y_{i+1} = y_i + f(y_i, t_i)\Delta t \tag{1.48}$$

The solution can be developed over a range of time from 0 to $n\Delta t$ by taking the integer i to range from 0 to n. Note that Eqs. (1.47) and (1.48) are the definition of the derivative before the limit is taken.

Now consider the dynamic differential equations written as two first-order coupled differential equations;

$$a = \frac{dv}{dt} = f(v, x, t)$$
$$v = \frac{dx}{dt} \tag{1.49}$$

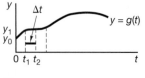

Figure 1.7

with initial conditions

$$v(t = 0) = v_0 \qquad (1.50)$$
$$x(t = 0) = x_0$$

The Euler method for the numerical solution of Eq. (1.49) yields the matrix equation

$$\begin{bmatrix} v_{i+1} \\ x_{i+1} \end{bmatrix} = \begin{bmatrix} v_i + f(v_i, x_i, t_i)\Delta t \\ x_i + v_i\Delta t \end{bmatrix} \qquad (1.51)$$

In this manner, the velocity and position are obtained for $n + 1$ values corresponding to $n + 1$ times $i\Delta t$.

As the step size Δt approaches zero, the numerical solution approaches the exact solution. The question remains as to what is the appropriate step size to use in a given case. There is no general answer to this question, and it may depend on the speed of the computer that is employed to generate the numerical solution. If the motion is to be analyzed over a period from 0 to 4 seconds, and the step size is $\Delta t = 0.001$, or one millisecond, then the value of n would be 4000, and 8000 calculations would have to be performed to obtain the values of the velocity and position. Current personal computers can accomplish this task in a matter of seconds. The easiest way to check if you have selected an appropriate step size is to reduce the step size and see whether graphs of the velocity and position change. If there is no significant change, the solution has converged.

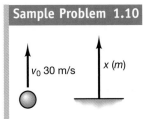

Sample Problem 1.10

A ball is thrown upward against gravitational attraction and air resistance proportional to the square of the velocity. Air resistance always opposes the motion; that is, it has the opposite sign of the velocity. The acceleration can be specified as $a(v) = -g - cv^2 \operatorname{sign}(v)$, where $c = 0.001$ (1/m) and $g = 9.81$ m/s². If the ball is thrown straight up with an initial velocity of 30 m/s, determine the velocity–displacement relationship.

Find the maximum height the ball reaches with and without air resistance.

Solution

The positive direction of the rectilinear motion is upward. (Note that the gravitational attraction term is negative, and the air resistance always opposes motion.) The sign (positive or negative) of the velocity is

$$\operatorname{sign}(v) = \frac{v}{|v|}$$

the velocity squared can be written as $v^2 = |v||v|$. Therefore, the acceleration is

$$a(v) = -g - cv|v|$$

This expression is valid for both the upward trajectory and the downward trajectory of the ball, as the air resistance is downward when the velocity is positive and upward when the velocity is negative. The differential equation for the ball during the upward motion is

$$v\frac{dv}{dx} = -(g + cv^2)$$

This differential equation is separable, and integration yields

$$\int_{v_0}^{v} -\frac{vdv}{(g + cv^2)} = \int_{0}^{x} dx$$

$$-\frac{1}{2c}\ln\left(\frac{g + cv^2}{g + cv_0^2}\right) = x$$

The height is a maximum when $v = 0$. Substituting to determine x yields

$$x_{\text{max}} = 43.89 \text{ m}$$

If there is no air resistance, the differential equation reduces to

$$v \frac{dv}{dx} = -g$$

The equation is again separable, and the solution for the position as a function of velocity is

$$\frac{1}{2}(v^2 - v_0^2) = -gx$$

The maximum height without air resistance is

$$x_{\text{max}} = 45.87 \text{ m}$$

or approximately 2 meters higher than the maximum height with air restistance.

The acceleration for the downward motion of the ball after it attains maximum height (with air resistance) is

$$a(v) = -g + cv^2$$

The initial values for this differential equation are $v_0 = 0$ and $x_0 = 43.89$ m, or the maximum height. Thus, for the downward motion, the particle's position is given by

$$\int_0^v \frac{u\,du}{(-g + cu^2)} = \int_{43.89}^x ds$$

$$x = 43.89 + \frac{1}{2c} \ln\left(\frac{-g + cv^2}{-g}\right)$$

The velocity at $x = 0$ is $v = 28.71$ m/s. This value is obtained by finding the value of v that is a root of the equation $x = 0$. (See the Computational Supplement for a numerical solution of the given equation.)

Problems

For Problems 1.56–1.58, determine the velocity and displacement as a function of time for the given acceleration of a particle.

1.56 $a(v,t) = -v + e^t$ m/s^2, where $v(0) = 0$ and $x(0) = 1$ m

1.57 $a(v, t) = -v + t$ m/s^2, where $v(0) = 0$ and $x(0) = 1$ m

1.58 a) $a(v,t) = -tv + e^t$ ft/s^2, where $v(0) = 10$ ft/s and $x(0) = 0$

b) $a(v,t) = -t^2v + 1 + 3t$ m/s^2, where $v(0) = 5$ m/s and $x(0) = 1$ m

1.59 Determine the velocity and displacement of a particle subjected to an acceleration given by $d^2x/dt^2 = -5dx/dt + 4x + 3t$ m/s^2 with initial conditions $v_0 = 0$ and $x_0 = 0.5$ m.

1.60 Detemine the equation for the velocity and displacement subjected to an acceleration given by $d^2x/dt^2 = -(c/m)(dx/dt) - (k/m)x$ m/s^2 with initial conditions $v(0) = 0$, $x(0) = 5$ m, $c = 5$ kg/s, $m = 1$ kg, and $k = 4$ kg/s^2.

1.61 Assume that the acceleration given in Problem 1.60 describes the suspension system of a 1000-kg automobile with a spring constant of 400,000 N/m. Select a damping coefficient c for the shock absorbers such that the oscillations damp out after two cycles for initial conditions $v(0) = 0$ and $x(0) = 0.01$ m.

1.62 When a tendon in the human body moves, it is characterized by an acceleration as: $a(v, x) = -cv - kx^2$ mm/s^2, where $c = 900/s$ and $k = 4000$ N/mm.

Determine the response of the tendon if it is stretched 20 mm and released. (*Note:* The tendon cannot go into compression, that is the displacement cannot be negative.)

1.63 A fast-closing release catch is made from a damped cubic spring such that the acceleration of the catch when released from an initial position of 10 mm is: $a(v, x) = -90v - 100x^3$ mm/s^2. Determine numerically the time required to close the catch.

1.64 Adjust the damping coefficient $c = 90/s$ in Problem 1.63 to a new value such that the system will oscillate two times before stopping.

1.65 Design a nonlinear damped spring so that when released from rest, it will damp out after one oscillation by choosing the numerical values for c, k, and x_0. The acceleration of the system is given by $a(v, x) = -cv|v| - kx|x|$ m/s^2 when released from a position x_0.

1.6 CURVILINEAR MOTION OF A PARTICLE

The general movement of a particle in space is called curvilinear motion. The position of a particle in space is specified by a *position vector* $\mathbf{r}(t)$ whose tail begins at the origin of an inertial reference coordinate system and whose head terminates at the particle. We will choose an orthogonal rectangular right-handed coordinate system to describe the motion, as shown in Figure 1.8, but cylindrical, spherical, or other coordinate systems will be used when appropriate. We will first develop the kinematic equations for a particle in rectangular coordinates and then repeat the development in other coordinate systems in later sections. The position vector to the particle at time t is designated by $\mathbf{r}$ in Figure 1.8 and, at a later time t', by $\mathbf{r}'$. During the change in time, Δt, the particle moves along a curve in space an amount Δs, and the corresponding change in the position vector is $\Delta \mathbf{r}$. As before, the average speed of the particle is given by $\Delta \mathbf{r}/\Delta t$, where

Figure 1.8

$$\Delta t = t' - t$$

and the change in the position vector is

$$\Delta \mathbf{r} = \mathbf{r}' - \mathbf{r}$$

However, unlike the situation in the rectilinear motion case, the change in position Δx must now be described by a vector $\Delta \mathbf{r}$. The velocity is the limit of the average change in the position vector as Δt approaches 0 and is formally defined as the vector

Figure 1.9

$$\mathbf{v} = \lim_{\Delta t \to 0} \frac{\Delta \mathbf{r}}{\Delta \mathbf{t}} = \frac{d\mathbf{r}}{dt} \tag{1.52}$$

On the other hand, if Δs is the magnitude of $\Delta \mathbf{r}$, the speed is the magnitude of the velocity vector, given by

$$\frac{ds}{dt} = \lim_{t \to 0} \frac{\Delta s}{\Delta t} = \left| \lim_{t \to 0} \frac{\Delta \mathbf{r}}{\Delta t} \right|$$

Speed is a scalar equal to the rate of change of position measured along the path of motion, while velocity is a vector equal to the rate of change of the position vector. Note that the magnitude of the velocity vector is equal to the speed ds/dt, and the velocity vector is always tangent to the curvilinear path of motion, as shown in Figure 1.9.

Although the velocity vector is always tangent to the path of motion, it will change in magnitude and direction as the particle moves along the path. This path of motion is called the *trajectory* of the particle in space. If the velocity vectors are plotted from a common

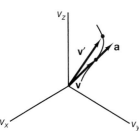

Figure 1.10

point in space, the termini, or heads, of the velocity vectors will trace out a path in "velocity space," called a *hodograph*. This path is illustrated in Figure 1.10.

The acceleration vector is defined as the time rate of change of the velocity vector, or the limit of

$$\mathbf{a} = \lim_{\Delta t \to 0} \frac{\Delta \mathbf{v}}{\Delta t} = \frac{d\mathbf{v}}{dt} \tag{1.53}$$

In rectangular Cartesian coordinates, the unit vectors $\hat{\mathbf{i}}, \hat{\mathbf{j}},$ and $\hat{\mathbf{k}}$ form the base vectors of the coordinate system in the x, y, and z directions, respectively. The position vector can be written in terms of the rectangular components as

$$\mathbf{r} = x(t)\hat{\mathbf{i}} + y(t)\hat{\mathbf{j}} + z(t)\hat{\mathbf{k}} \tag{1.54}$$

where $x(t)$, $y(t)$, and $z(t)$ are the coordinates, expressed as a function of time. The velocity vector is defined as the time derivative of the position vector. The rules of vector differentiation are shown in Mathematics Window 1.2.

MATHEMATICS WINDOW 1.2

$$\frac{d}{dt}(\mathbf{A} + \mathbf{B}) = \frac{d\mathbf{A}}{dt} + \frac{d\mathbf{B}}{dt}$$

$$\frac{d}{dt}(\alpha \mathbf{A}) = \frac{d\alpha}{dt}\mathbf{A} + \alpha\frac{d\mathbf{A}}{dt}$$

$$\frac{d}{dt}(\mathbf{A} \cdot \mathbf{B}) = \frac{d\mathbf{A}}{dt} \cdot \mathbf{B} + \mathbf{A} \cdot \frac{d\mathbf{B}}{dt}$$

$$\frac{d}{dt}(\mathbf{A} \times \mathbf{B}) = \frac{d\mathbf{A}}{dt} \times \mathbf{B} + \mathbf{A} \times \frac{d\mathbf{B}}{dt}$$

The velocity vector is expressed as the sum of products of the x, y, and z components with their respective unit vectors. This vector represents the sum of the products of scalar functions of time with the unit vectors and is written

$$\mathbf{v}(t) = \frac{dx}{dt}\hat{\mathbf{i}} + x\frac{d\hat{\mathbf{i}}}{dt} + \frac{dy}{dt}\hat{\mathbf{j}} + y\frac{d\hat{\mathbf{j}}}{dt} + \frac{dz}{dt}\hat{\mathbf{k}} + z\frac{d\hat{\mathbf{k}}}{dt} \tag{1.55}$$

If the coordinate system is not rotating, the unit vectors are not functions of time, and the second term in the differentiation of the product will be zero. This will not be true if the unit base vectors rotate or change direction during the motion. The magnitude of the unit base vectors is unity and therefore is not a function of time, but the direction can change in some coordinate descriptions. The velocity and acceleration in nonrotating rectangular coordinates can now be obtained by successive differentiation of the position vector $\mathbf{r}$:

$$\mathbf{r} = x\hat{\mathbf{i}} + y\hat{\mathbf{j}} + z\hat{\mathbf{k}}$$

$$\mathbf{v} = \dot{x}\hat{\mathbf{i}} + \dot{y}\hat{\mathbf{j}} + \dot{z}\hat{\mathbf{k}} \tag{1.56}$$

$$\mathbf{a} = \ddot{x}\hat{\mathbf{i}} + \ddot{y}\hat{\mathbf{j}} + \ddot{z}\hat{\mathbf{k}}$$

Eq. (1.56) can also be written in matrix notation as

$$\begin{bmatrix} v_x \\ v_y \\ v_z \end{bmatrix} = \frac{d}{dt} \begin{bmatrix} x(t) \\ y(t) \\ z(t) \end{bmatrix} = \begin{bmatrix} \dot{x} \\ \dot{y} \\ \dot{z} \end{bmatrix}$$
$$\begin{bmatrix} a_x \\ a_y \\ a_z \end{bmatrix} = \frac{d^2}{dt^2} \begin{bmatrix} x(t) \\ y(t) \\ z(t) \end{bmatrix} = \begin{bmatrix} \ddot{x} \\ \ddot{y} \\ \ddot{z} \end{bmatrix} \tag{1.57}$$

where it is understood that the first element corresponds to the *x*-component, the second element corresponds to the *y*-component, and the third element corresponds to the *z*-component. This form of the velocity and acceleration clearly shows the scalar equations as components of the vector equation.

The kinematics of a particle in curvilinear motion can be described by the *inverse dynamics problem* if the position vector is a known function of time. This vector can be differentiated to determine the velocity and acceleration. As discussed in Section 1.2, problems of this nature arise when motion analysis systems are used to collect three-dimensional position data at specific time increments. When the position vector is known at discrete time intervals, differentiation can cause noise in the velocity and acceleration data.

The dynamics problem is in the form of a vector differential equation if the acceleration vector is a known function of the velocity and position vectors and of time. As in the case of rectilinear motion, if the acceleration is only a function of time, the velocity and position can be determined by integration. If the acceleration vector is a function of position or velocity only, the vector differential equation may be separable. If the acceleration is a constant and the initial position and velocity are known, the solution is straight forward, as is shown for the special case of projectile motion.

1.6.1 VECTOR DIFFERENTIAL EQUATION

Differential equations are also classified by the number of unknown functions that are involved. In Eq. (1.10), there is only one unknown function $x(t)$ and the single differential equation is sufficient to determine the unknown function. For motion in three dimensions, the acceleration is a function of the velocity and position vectors and of time, and is denoted

$$\mathbf{a} = F(t, \mathbf{r}, \mathbf{v}) \tag{1.58}$$

Eq. (1.58) is a vector differential equation, and there are three unknown functions—$x(t)$, $y(t)$, and $z(t)$—and their derivatives. However, this vector differential equation can be separated into three scalar differential equations to form a *system of differential equations*. This system may take a form such that each scalar equation involves only one unknown function and its derivatives. In that case, the three equations are independent of each other and may be solved separately. However, in many dynamic problems, the equations are *coupled*; that is, two or more unknown functions appear in at least one of the equations. The solution of this system of differential equations is more difficult, and we will, in general, seek numerical solutions instead of analytical solutions. A detailed discussion of systems of differential equations is beyond the scope of this text; however, we will discuss the numerical solution of such systems of equations.

We will first write Eq. (1.58) as a matrix equation, with each element of the matrix corresponding to the scalar component in the x-, y-, and z-coordinate. Assuming that the system of differential equations is coupled, Eq. (1.58) will take the form

$$\begin{bmatrix} \dfrac{dv_x}{dt} \\[2mm] \dfrac{dv_y}{dt} \\[2mm] \dfrac{dv_z}{dt} \end{bmatrix} = \begin{bmatrix} a_x(t, x, y, z, v_x, v_y, v_z) \\ a_y(t, x, y, z, v_x, v_y, v_z) \\ a_z(t, x, y, z, v_x, v_y, v_z) \end{bmatrix} \tag{1.59}$$

Note that each of the acceleration functions on the right side of Eq. (1.59) is a general function of the components of the position vector and the velocity vector, and also a general function of time. These functions may be linear or nonlinear. If an acceleration component contains only the corresponding position and velocity components of a single coordinate, that differential equation is not coupled to the system and may be solved independently by the methods shown in previous sections. The Euler method will be used to solve the general coupled initial-value problem, Eq. (1.59), when the values of the position vector and the velocity vector are specified for time t equal to zero. To use this numerical method, you must first specify the number of steps and the size of the time increment. The more nonlinear and coupled the vector differential equation, the smaller is the time increment needed for convergence. If the integer n is used to count steps, n will start at zero, and the last step is designated as N:

$$n = 0, 1, 2, \ldots, N \tag{1.60}$$

We will designate the time increment by Δt, and for this case, we will use an increment of 1 ms, or $1/1000$ s:

$$\Delta t = 0.001 \tag{1.61}$$

The time at any step is

$$t_n = n \, \Delta t \tag{1.62}$$

To simplify the notation, we will use vx to denote v_x, and similar notation for the other velocity components. In matrix notation, we express the initial value of each component of the position vector and the velocity vector—that is, the value at position $n = 0$:

$$\begin{bmatrix} vx_0 \\ x_0 \\ vy_0 \\ y_0 \\ vz_0 \\ z_0 \end{bmatrix} = \begin{bmatrix} vx(0) \\ x(0) \\ vy(0) \\ y(0) \\ vz(0) \\ z(0) \end{bmatrix} \tag{1.63}$$

Eq. (1.63) gives the starting values for the Euler method, and the value for each successive time is given by

$$\begin{bmatrix} vx_{n+1} \\ x_{n+1} \\ vy_{n+1} \\ y_{n+1} \\ vz_{n+1} \\ z_{n+1} \end{bmatrix} = \begin{bmatrix} vx_n + a_x(t_n, x_n, y_n, z_n, vx_n, vy_n, vz_n) \times \Delta t \\ x_n + vx_n \times \Delta t \\ vy_n + a_y(t_n, x_n, y_n, z_n, vx_n, vy_n, vz_n) \times \Delta t \\ y_n + vy_n \times \Delta t \\ vz_n + a_z(t_n, x_n, y_n, z_n, vx_n, vy_n, vz_n) \times \Delta t \\ z_n + vz_n \times \Delta t \end{bmatrix} \tag{1.64}$$

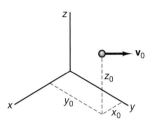

Figure 1.11

The acceleration is computed at each step and is used to determine the velocity at the next step. The velocity is used to determine the position at the next step.

1.6.2 PROJECTILE MOTION

Consider the motion of a particle under only the influence of the gravitational attraction of the earth. If the motion is close to the surface of the earth, the acceleration of the particle will be constant during the movement and directed toward the center of the earth. This type of motion is called *projectile motion*, and examples can be found when a ball is thrown or hit, when water particles are sprayed from a hose, and in many other common particle movements. We will neglect the resistance of the air in the first model, and no other forces will be assumed to be acting on the particle. Aristotle called this type of movement "violent motion," as he could not understand why the object would continue to move in that direction when there were no apparent forces acting in that direction. Newton was able to explain that if, indeed, there were no horizontal forces acting on the particle, the velocity of the particle in the horizontal direction would be constant, and the particle would accelerate only in the vertical direction, due to the gravitational attraction. If the particle is projected from an initial position with an initial velocity, as shown in Figure 1.11, then we can show that the horizontal component of the initial velocity will remain constant at the initial value.

The particle will move in a parabolic manner in a plane formed by the initial velocity and the vertical axis. This conceptualization allows projectile motion to be modeled as motion in a two-dimensional plane (planar motion). The absence of acceleration in the horizontal direction will be discussed in detail in Chapter 2, when we relate force to acceleration. A mathematical treatment of the gravitational attraction also will be given in Chapter 2, where we apply Newton's second law to particle acceleration. Here, it can be noted that Newton's second law is

$$\mathbf{F} = m\mathbf{a}$$

If only gravitational attraction is acting on the particle, and the z-axis is chosen to be vertical to the earth's surface, as shown in Figure 1.11, the force acting on the particle is

$$\mathbf{F} = -mg\hat{\mathbf{k}} = m\mathbf{a}$$

The acceleration is

$$\mathbf{a} = -g\hat{\mathbf{k}}$$

The acceleration in the vertical direction of the particle in free fall is designated by g. The constant g is actually the gravitational attraction acting on a unit mass close to the earth, as will be shown in Chapter 2. When an object is in free fall, the acceleration will be equal to g, a constant value of 9.81 m/s^2, or 32.2 ft/s^2. The constant g is frequently termed the *gravitational acceleration*. However, the object will accelerate at that value only if it is in free fall and air resistance and other factors are neglected. The acceleration can be written in vector or matrix form (with the vertical direction aligned with the z-axis) as

$$\begin{bmatrix} a_x \\ a_y \\ a_z \end{bmatrix} = \begin{bmatrix} 0 \\ 0 \\ -g \end{bmatrix} \tag{1.65}$$

The three scalar differential equations are linear and not coupled. Therefore, the velocity is obtained by integration, yielding

$$\begin{bmatrix} v_x(t) \\ v_y(t) \\ v_z(t) \end{bmatrix} = \begin{bmatrix} A_1 \\ B_1 \\ -gt + C_1 \end{bmatrix} \tag{1.66}$$

where A_1, B_1, and C_1 are constants of integration. These constants are evaluated in terms of the initial velocity, which is

$$\begin{bmatrix} v_x(0) \\ v_y(0) \\ v_z(0) \end{bmatrix} = \begin{bmatrix} \dot{x}_0 \\ \dot{y}_0 \\ \dot{z}_0 \end{bmatrix} \tag{1.67}$$

Substituting Eq. (1.67) into Eq. (1.66) for $t = 0$ yields

$$\begin{bmatrix} v_x(t) \\ v_y(t) \\ v_z(t) \end{bmatrix} = \begin{bmatrix} \dot{x}_0 \\ \dot{y}_0 \\ -gt + \dot{z}_0 \end{bmatrix} \tag{1.68}$$

Eq. (1.68) is integrated to obtain the displacements, yielding

$$\begin{bmatrix} x(t) \\ y(t) \\ z(t) \end{bmatrix} = \begin{bmatrix} \dot{x}_0 t + A_2 \\ \dot{y}_0 t + B_2 \\ -\dfrac{gt^2}{2} + \dot{z}_0 t + C_2 \end{bmatrix} \tag{1.69}$$

The three constants of integration are evaluated from the initial-position vector

$$\begin{bmatrix} x(0) \\ y(0) \\ z(0) \end{bmatrix} = \begin{bmatrix} x_0 \\ y_0 \\ z_0 \end{bmatrix} \tag{1.70}$$

Substituting Eq. (1.70) into Eq. (1.69) yields the position vector

$$\begin{bmatrix} x(t) \\ y(t) \\ z(t) \end{bmatrix} = \begin{bmatrix} \dot{x}_0 t + x_0 \\ \dot{y}_0 t + y_0 \\ -\dfrac{gt^2}{2} + \dot{z}_0 t + z_0 \end{bmatrix} \tag{1.71}$$

The acceleration, velocity, and displacement vectors can be written in vector notation as

$$\begin{aligned} \mathbf{a} &= -g\hat{\mathbf{k}} \\ \mathbf{v} &= \dot{x}_0\hat{\mathbf{i}} + \dot{y}_0\hat{\mathbf{j}} + (\dot{z}_0 - gt)\hat{\mathbf{k}} \\ \mathbf{r} &= (\dot{x}_0 t + x_0)\hat{\mathbf{i}} + (\dot{y}_0 t + y_0)\hat{\mathbf{j}} + \left(\dot{z}_0 t + z_0 - \dfrac{gt^2}{2}\right)\hat{\mathbf{k}} \end{aligned} \tag{1.72}$$

where

$$\mathbf{v}(0) = \dot{x}_0\hat{\mathbf{i}} + \dot{y}_0\hat{\mathbf{j}} + \dot{z}_0\hat{\mathbf{k}}$$
$$\mathbf{x}(0) = x_0\hat{\mathbf{i}} + y_0\hat{\mathbf{j}} + z_0\hat{\mathbf{k}}$$

As previously discussed, if the coordinate axes are chosen such that the initial velocity vector lies in a coordinate plane—that is, $\dot{y}_0 = 0$—the problem may be treated as a two-dimensional case. If the object is thrown with an upward trajectory, the particle will continue to rise until the vertical component of its velocity is zero, or, as shown in Eq. (1.72) until the time when $t = \dot{z}_0/g$, after which the velocity in the vertical direction will continue to increase in a downward direction. In many cases, it is desirable to determine when and where the particle will hit a particular plane. It is important to realize that the velocity

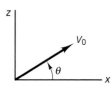

Figure 1.12

is not zero when the particle hits the ground, even though the z-component of the position vector is zero at that time. The changes in velocity due to the impact with the ground will be treated in Chapter 3.

If the only force acting on a projectile is gravitational attraction, projectile problems can always be formulated as problems of planar motion in which the plane of motion is formed by the vertical and the initial-velocity vector. Choosing the xz-plane as this plane, we find that the scalar components of the displacement vector in Eq. (1.72) are

$$x = (v_0 \cos \theta)t + x_0 \qquad z = -gt^2/2 + (v_0 \sin \theta)t + z_0 \qquad (1.73)$$

where θ is the angle between the initial-velocity vector and the horizontal, as shown in Figure 1.12. Projectile problems relate the position of the particle at any time to the initial conditions at the initial time, zero. Seven possible variables are related by the two scalar equations in Eq. (1.73). Let us designate the final position of interest as (x_f, z_f) at time t_f. The seven possible variables are x_f, z_f, t_f, x_0, z_0, v_0 and θ. The problems can always be formulated as two nonlinear algebraic equations in two unknowns. This means that five of the seven variables must be known. For example, the initial and final position and the initial velocity may be specified, and Eq. (1.73) consists of two nonlinear algebraic equations for the angle θ and the final time t_f.

Sample Problem 1.11

A football player wishes to punt a football 65 yards and have a hang time of 5 s. What initial velocity V_0 must the ball have as it leaves his foot?

Solution   Assume that the ball leaves the punter's foot at the same height at which the receiver catches the ball. The coordinate system will be chosen with an origin at the punter's foot. The initial position is (0,0), and the position when the ball hits the receiver's hands is $x_f = 3(65) = 195$ ft and $y_f = 0$. The time elapsed when the ball reaches the receiver is 5 s. First, we formulate the problem in scalar notation:

$$\ddot{x} = 0 \qquad\qquad \ddot{y} = -g$$
$$\dot{x} = V_0 \cos \theta \qquad\qquad \dot{y} = -gt + V_0 \sin \theta$$

$$x = V_0 \cos \theta t + x_0 \qquad y = -\frac{gt^2}{2} + V_0 \sin \theta t + y_0$$

Substituting the initial position, the final position, and the time at the final position yields two nonlinear algebraic equations for V_0 and θ:

$$195 = V_0 \cos \theta (5)$$
$$0 = -16.1(5)^2 + V_0 \sin \theta (5)$$

The first equation can be written solving for V_0 in terms of θ, and then substituting into the second equation yields

$$\tan \theta = \frac{16.1(5)^2}{195}$$

$$\theta = 64.2° \qquad\qquad V_0 = 89.6 \text{ ft/s, or 61 mph}$$

This punt is realistic, and the ball leaves the punter's foot traveling at over 60 mph. Think of the force on the opposing player who blocks the punt.

Consider a particle moving in a helical motion at a constant pitch. The position vector can be given as

$$\mathbf{r}(t) = (R \cos \omega t)\hat{\mathbf{i}} + (R \sin \omega t)\hat{\mathbf{j}} + (pt)\hat{\mathbf{k}}$$

where R is the radius of the helix, p is the pitch of the helical curve in space, and ω determines the time to complete one cycle around the helix. Determine the velocity and acceleration of the particle at any time in terms of the constants R, p, and ω.

Solution   Differentiating the position vector with respect to time yields the velocity and acceleration:

$$\mathbf{v}(t) = (-R\omega \sin \omega t)\hat{\mathbf{i}} + (R\omega \cos \omega t)\hat{\mathbf{j}} + p\hat{\mathbf{k}}$$

$$\mathbf{a}(t) = (-R\omega^2 \cos \omega t)\hat{\mathbf{i}} + (-R\omega^2 \sin \omega t)\hat{\mathbf{j}}$$

The magnitudes of both the velocity and acceleration vectors are constant for this motion and are equal to

$$|\mathbf{v}(t)| = \sqrt{(R\omega)^2 + p^2}$$

$$|a(t)| = R\omega^2$$

If the punter on a football team can kick the ball with an initial velocity of 80 ft/s, what angle of the initial-velocity vector with the ground will result in the longest punt, and how far will the ball go before hitting the ground?

Solution   Assume that the origin of a coordinate system is at the kicker's foot. The actual height of the point of contact between the ball and the punter's foot is above ground level, but this will make only a minor difference in the calculation. Integration of the acceleration yields

$$\mathbf{a}(t) = -32.2\hat{\mathbf{j}}$$

$$\mathbf{v}(t) = 80 \cos \theta \hat{\mathbf{i}} + (-32.2t + 80 \sin \theta)\hat{\mathbf{j}}$$

$$\mathbf{r}(t) = 80t \cos \theta \hat{\mathbf{i}} + (-16.1t^2 + 80t \sin \theta)\hat{\mathbf{j}}$$

The ball will hit the ground when $y \approx 0$, or when $t = (80/16.1) \sin \theta$. The value of x at this point is

$$x = (80^2/16.1) \sin \theta \cos \theta = (80^2/32.2) \sin 2\theta$$

The maximum value of x would be when $\theta = 45°(\sin 2(45°) = 1)$, and for this angle, $x = 198.8$ ft, or 66.25 yds.

The punter kicks the football into a heavy rain such that the air resists the motion in a manner proportional to the velocity squared. If the initial velocity of the ball is 80 ft/s and the ball leaves the punter's foot at a 45° angle, determine the distance the ball travels. The acceleration is $\mathbf{a} = -g\hat{\mathbf{j}} - cv|v|$, where $c = 0.001$ (ft^{-1}).

Solution   The acceleration can be written in scalar form as

$$a_x = -cv_x\sqrt{v_x^2 + v_y^2}$$

$$a_y = -g - cv_y\sqrt{v_x^2 + v_y^2}$$

The two scalar differential equations are coupled and may be solved numerically using Euler's method. The numerical details using Mathcad software are as follows:

$$i = 0 \ldots 3400$$

$$c := 0.001$$

$$\Delta t = 0.001$$

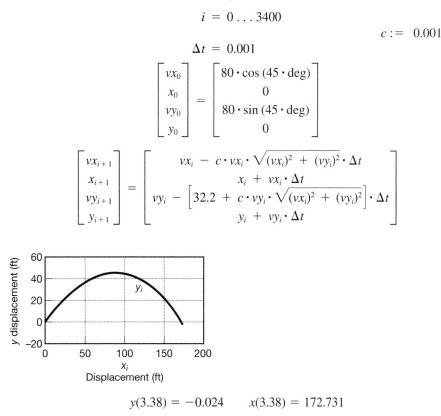

$$\begin{bmatrix} vx_0 \\ x_0 \\ vy_0 \\ y_0 \end{bmatrix} = \begin{bmatrix} 80 \cdot \cos{(45 \cdot \deg)} \\ 0 \\ 80 \cdot \sin{(45 \cdot \deg)} \\ 0 \end{bmatrix}$$

$$\begin{bmatrix} vx_{i+1} \\ x_{i+1} \\ vy_{i+1} \\ y_{i+1} \end{bmatrix} = \begin{bmatrix} vx_i - c \cdot vx_i \cdot \sqrt{(vx_i)^2 + (vy_i)^2} \cdot \Delta t \\ x_i + vx_i \cdot \Delta t \\ vy_i - \left[32.2 + c \cdot vy_i \cdot \sqrt{(vx_i)^2 + (vy_i)^2} \right] \cdot \Delta t \\ y_i + vy_i \cdot \Delta t \end{bmatrix}$$

$$y(3.38) = -0.024 \qquad x(3.38) = 172.731$$

The football travels 173 feet, or 55 yards.

Problems

For Problems 1.66–1.69, write the equivalent system of scalar equations from the vector equation of motion. Classify the system of differential scalar equations that come from the vector differential equation as linear or nonlinear, and coupled or uncoupled.

1.66 $\dfrac{d^2\mathbf{r}}{dt^2} = -g\hat{\mathbf{j}}$, where g is a constant.

1.67 $\dfrac{d^2\mathbf{r}}{dt^2} = -c\mathbf{v} - g\hat{\mathbf{j}}$, where c and g are constants.

1.68 $\dfrac{d^2\mathbf{r}}{dt^2} = -c\mathbf{v}^2 \text{sign}(\mathbf{v})$, where $\text{sign}(\mathbf{v}) = \dfrac{\mathbf{v}}{|\mathbf{v}|}$. Also,

note that $\mathbf{v}^2 = |\mathbf{v}||\mathbf{v}|$, where $|\mathbf{v}| = \sqrt{v_x^2 + v_y^2 + v_z^2}$

1.69 $\dfrac{d^2\mathbf{r}}{dt^2} = 3t^2\hat{\mathbf{i}} - \sin{(\pi t)}\hat{\mathbf{j}} + xz\hat{\mathbf{k}}$

1.70 A golfer wishes to hit an iron to the green as shown in Figure P1.70. If the golfer can give the ball an initial velocity of 130 ft/s, what angle of loft, θ, should the iron have, and how long will the ball be in the air?

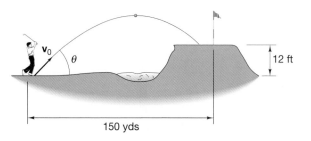

Figure P1.70

1.71 Gravel is unloaded using a conveyor belt moving at 2 m/s, as depicted in Figure P1.71. Determine the distance d from the end of the conveyor belt to where the gravel will land.

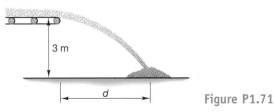

Figure P1.71

1.72 A water sprinkler, shown in Figure P1.72, oscillates for 30° to either side of its center while it discharges water at a velocity of 20 ft/s. If the sprinkler is set on a 10° slope, determine the distances d_1 and d_2 covered by the water spray.

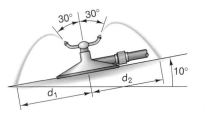

Figure P1.72

1.73 At what angle with the horizontal should a gardener hold a hose for the water to reach a maximum distance?

1.74 A batter makes contact with a ball 4 ft above the ground, and the ball leaves the bat with a velocity of 140 ft/s. If the ball leaves the bat at a 20° angle with the horizontal, determine the distance the ball will fly before it hits the ground.

1.75 In Problem 1.74, determine the maximum height above the ground reached by the ball.

1.76 A boy wants to throw a ball through a 1-m opening in a wall, as represented in Figure P1.76. If the ball is thrown at an angle of 30° to the horizontal, determine the minimum and maximum velocity that the ball must be thrown in order for it to pass through the opening.

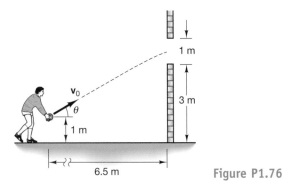

Figure P1.76

1.77 In Problem 1.76, if the ball is thrown at a velocity of 10 m/s, determine the minimum and maximum initial angle θ at which the ball must be thrown in order for it to pass through the opening.

1.78 A ski jumper comes off a jump in a horizontal direction with a velocity of 25 m/s, as shown in Figure P1.78. Neglecting air resistance, determine the distance d the jumper travels before landing on the 45° slope.

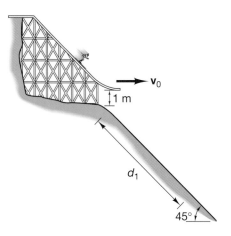

Figure P1.78

1.79 In Problem 1.78, determine the distance d that the ski jumper travels if the wind resistance during flight is proportional to his velocity such that $\mathbf{a} = -g\hat{\mathbf{j}} - c\mathbf{v}$, $c = 0.04$.

1.80 In Problem 1.78, determine the distance d that the ski jumper travels if the wind resistance during flight is proportional to the velocity squared $\mathbf{a} = -g\hat{\mathbf{j}} - c\mathbf{v}|\mathbf{v}|$, $c = 0.002$.

1.81 A professional golfer can hit a drive 300 yards with a driver that has a 9° loft. Determine the velocity of the ball as it leaves the driver.

1.82 If the golf ball leaves the 9°-loft driver with a velocity of 200 mph, determine the distance the ball will travel on a level fairway.

1.83 Determine the distance the golf ball in Problem 1.82 will travel if wind resistance is proportional to the velocity $\mathbf{a} = -g\hat{\mathbf{j}} - c\mathbf{v}$, $c = 0.05$.

1.84 A basketball player shoots a ball at a position 20 ft away from the basket, as shown in Figure P1.84. If the ball leaves his hands 7 ft above the floor and he shoots

45° from the horizontal, determine the required initial velocity of the ball if it hits the basket.

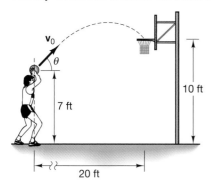

Figure P1.84

💻**1.85** In Problem 1.84, if the initial velocity of the ball is 30 ft/s, determine the angle from the horizontal at which the ball must travel in order for the player to make the shot.

💻**1.86** A professional tennis player can serve the ball at a velocity of 120 mph. If the ball leaves the racket 10 ft above the ground and in a horizontal direction, how far in front of the server will the ball hit the ground?

💻**1.87** A young girl wants to throw a ball onto a level roof 15 ft above the ground. If the ball leaves her hand 4 ft above the ground at an angle of 80°, and the roof is 10 ft wide, determine (a) the minimum initial velocity required for the ball to land on the roof and (b) the maximum velocity necesary so that the ball does not go over the roof.

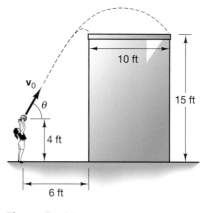

Figure P1.87

1.7 NORMAL AND TANGENTIAL COORDINATES

When the path of motion of a particle is known, it is useful to separate the acceleration vector into two orthogonal parts: the change in magnitude of the velocity vector and the change in direction of the velocity vector. Problems of this nature arise in the design of highways, railroads, airplane wings, or cams whenever the path of motion of a particle is specified. The maximum safe speed of a car going around a curve on a highway is determined using analytical methods of this type. We will first consider the case of planar motion—that is, when the particle remains in the *xy*-plane. We will develop a coordinate system such that one coordinate lies tangent to the path of motion, or in the direction of the velocity vector, and the other coordinate is normal or perpendicular to the path. Since the velocity vector is tangent to the path of motion, an acceleration component tangent to that path changes only the magnitude of the velocity vector. An acceleration component perpendicular to the velocity vector changes only the direction of the velocity vector. This separation of the acceleration into a component expressing the change in magnitude of the velocity and a component expressing the change in direction of the velocity requires the examination of some new mathematical concepts. We will develop a coordinate system that rides with the particle and is defined by two base unit vectors—one tangent to the path of motion at any instant and one normal to the path and directed toward the center of curvature of the path at that instant. This coordinate system not only translates along the path of motion, but also rotates to maintain the tangential and normal orientations. As we did in the rectangular coordinate system, we will use base unit vectors to define the coordinate direction. However, these vectors will change as the particle moves along the path of motion; that is, the tangential and normal base unit vectors will be functions of time.

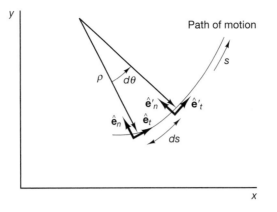

Figure 1.13

Consider a section of the path of motion, shown in Figure 1.13. Let ρ be the radius of curvature of the path of motion at the chosen point. Note that for a straight line, the radius of curvature becomes infinite. We define the coordinate unit vectors $\hat{\mathbf{e}}_n$ and $\hat{\mathbf{e}}_t$ as normal and tangential to the path of motion, respectively, as shown in Figure 1.13. The tangential unit vector $\hat{\mathbf{e}}_t$ is taken to be positive along the path of motion—that is, in the direction of the velocity vector, which is tangent to the path. The normal unit vector $\hat{\mathbf{e}}_n$ is always taken to be positive toward the center of curvature. The normal unit vector will change direction by $180°$ at a point of inflection on the path of motion; that is, the radius is infinite, because the line is straight at the point of inflection and the center of curvature "flips" sides. The parameter s measures the distance in meters or feet along the path of motion. These unit vectors ride with the particle and are changing with respect to the spatial coordinates x and y as time progresses. Note that the arc length ds is related to ρ and $d\theta$ by

$$ds = \rho d\theta \qquad (1.74)$$

Differentiation of Eq. (1.74) with respect to time relates the speed, or magnitude of the velocity vector, to the radius of curvature and the angular velocity. The magnitude of the angular velocity is defined as

$$\omega = \frac{d\theta}{dt} \qquad (1.75)$$

The angular velocity measures the time rate of change of the angle θ, in rad/s, and is expressed as a vector acting perpendicular to the plane of motion when one works in three dimensions. The sense of the angular velocity vector is given by the "right-hand rule," as with moments of forces and cross products of vectors. The direction of the $\boldsymbol{\omega}$ vector in Figure 1.13 is in the positive $\hat{\mathbf{k}}$ direction: that is, $\boldsymbol{\omega} = \omega\hat{\mathbf{k}} = (d\theta/dt)\hat{\mathbf{k}}$. The three unit vectors $\hat{\mathbf{e}}_t$, $\hat{\mathbf{e}}_n$, and $\hat{\mathbf{k}}$ form an orthogonal set of base vectors for the right-handed coordinate system tangential, normal, and z, (t, n, z), respectively. Differentiating Eq. (1.74) yields

$$\frac{ds}{dt} = \rho\frac{d\theta}{dt} = \rho\omega \qquad (1.76)$$

Thus, the velocity is expressed either in terms of the speed along the path of motion or as a function of the radius of curvature and the angular velocity of the radius of curvature as it sweeps an arc segment. The normal and tangential vectors are shown at two points on

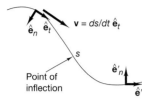

Point of
inflection

Figure 1.14

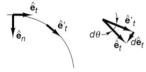

Figure 1.15

the path of motion in Figure 1.14. Since the velocity is in the tangential direction, the velocity vector is

$$\mathbf{v} = \frac{ds}{dt}\hat{\mathbf{e}}_t = \rho\frac{d\theta}{dt}\hat{\mathbf{e}}_t \tag{1.77}$$

The acceleration can now be determined by differentiation of the velocity vector with respect to time. Note that the derivative of the tangential unit vector is not zero, since that vector changes with time. Because $\hat{\mathbf{e}}_t$ changes direction as the particle moves along the path of motion, it is a function of time, even though its magnitude remains unity. The acceleration is thus

$$\mathbf{a} = \frac{d^2s}{dt^2}\hat{\mathbf{e}}_t + \frac{ds}{dt}\frac{d\hat{\mathbf{e}}_t}{dt} \tag{1.78}$$

The derivative of the unit vector $\hat{\mathbf{e}}_t$ can be determined by examining the change in $\hat{\mathbf{e}}_t$ as it moves along an infinitesimal arc. This change is shown enlarged in Figure 1.15 for easier examination. Note that $d\hat{\mathbf{e}}_t$ is the change in direction of $\hat{\mathbf{e}}_t$ as there can be no change in length of the unit vector. Therefore, the change must be perpendicular to the original unit vector and is in the normal direction. The magnitude of the change may be obtained by the arc length:

$$d\hat{\mathbf{e}}_t = |\hat{\mathbf{e}}_t|d\theta\hat{\mathbf{e}}_n$$

$$|\hat{\mathbf{e}}_t| = 1$$

Therefore:

$$\frac{d\hat{\mathbf{e}}_t}{dt} = \frac{d\theta}{dt}\hat{\mathbf{e}}_n \tag{1.79}$$

Using Eq. (1.75) in the form

$$\frac{d\theta}{dt} = \frac{1}{\rho}\frac{ds}{dt} = \frac{|\mathbf{v}|}{\rho}$$

We can also write the change in the unit vector as

$$\frac{d\hat{\mathbf{e}}_t}{dt} = \frac{1}{\rho}\frac{ds}{dt}\hat{\mathbf{e}}_n = \frac{|\mathbf{v}|}{\rho}\hat{\mathbf{e}}_n$$

The acceleration may now be written as

$$\mathbf{a} = \frac{d\mathbf{v}}{dt} = \frac{d^2s}{dt^2}\hat{\mathbf{e}}_t + \frac{v^2}{\rho}\hat{\mathbf{e}}_n \tag{1.81}$$

The first term in the preceding equation is the *tangential acceleration*, which is a measure of the change in magnitude, or speed, of the velocity vector. The second term is the *normal acceleration*, which is the change in direction of the velocity vector. The velocity can have a constant magnitude, but because of the change in direction, it will still have an acceleration, v^2/ρ. This normal acceleration easily is observable when one drives around a curve at a constant speed, or when one drives over the top of a hill or reaches the bottom of a hill.

In Chapter 2, we will see that forces are required to develop normal accelerations. When a car goes around a curve, friction between the tires and the road surface provides the force needed for the car to execute the curve and not slide sideways. When a vehicle goes over a hill, its weight provides the force for the required acceleration. If the radius of curvature of the hill is too small or the velocity of the vehicle is too great, the weight of the car will be insufficient to provide this acceleration, and the car will become airborne, or "fly the hill." When high-speed trains were introduced in Europe, the rail beds had to be redesigned in order to reduce the normal accelerations necesary for proper execution of hills and curves. This topic is still actively researched by manufacturers of high-speed locomotives.

1.7.1 CIRCULAR MOTION

Circular motion is a very important application of normal and tangential coordinates. In circular motion, the radius of curvature is constant, and only the angle $\theta(t)$ is needed to specify the position of the particle. The unit normal vector is directed inward along this radius, and the unit tangent vector is always tangent to the circle, as shown in Figure 1.16. From Eq. (1.75), the angular acceleration α is related to the linear acceleration in the tangential direction by

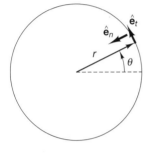

Figure 1.16

$$\frac{d^2s}{dt^2} = r\frac{d^2\theta}{dt^2} = r\alpha \tag{1.82}$$

The angular acceleration is expressed as a vector α, with its sense defined by the right-hand rule, as was done with the angular velocity vector $\boldsymbol{\omega}$. Using the angular velocity and angular acceleration defined by Eqs. (1.76) and (1.82), respectively, we can write Eq. (1.81) as

$$\mathbf{a} = r\alpha\hat{\mathbf{e}}_t + r\omega^2\hat{\mathbf{e}}_n \tag{1.83}$$

The acceleration can be written in matrix form as

$$\begin{bmatrix} a_t \\ a_n \\ a_z \end{bmatrix} = \begin{bmatrix} r\dfrac{d\omega}{dt} \\ r(\omega)^2 \\ 0 \end{bmatrix} \tag{1.84}$$

The motion is described by a single degree of freedom—that is, the angle θ. The definitions of angular velocity and angular acceleration are analogous to the definitions of rectilinear velocity and acceleration:

$$\alpha = \frac{d\omega}{dt} = \frac{d^2\theta}{dt^2} \qquad a = \frac{dv}{dt} = \frac{d^2x}{dt^2}$$

$$\omega = \frac{d\theta}{dt} \qquad v = \frac{dx}{dt} \tag{1.85}$$

Although this problem is not one dimensional, it is a problem involving only one degree of freedom and for this special case, the angular acceleration, velocity, and position form a system of equations that is analogous to the system of equations for rectilinear motion discussed in Section 1.5. The analogy is more transparent if the differential equations are placed side by side:

$$a = \frac{d^2x(t)}{dt^2} = F\left[t, x,(t),\frac{dx(t)}{dt}\right]$$

$$\alpha = \frac{d^2\theta(t)}{dt^2} = F\left[t, \theta,(t),\frac{d\theta(t)}{dt}\right] \tag{1.86}$$

The angular differential equation is either a linear or a nonlinear second-order differential equation, depending on whether F is a linear or nonlinear function of θ and ω. If the angular acceleration is a function of only one other dynamic variable, the differential equation can be written as a separable first-order differential equation, as in Section 1.5. Using the chain rule of differentiation, we obtain

$$\alpha = \frac{d\omega}{dt} = \frac{d\omega}{d\theta}\frac{d\theta}{dt} = \omega\frac{d\omega}{d\theta} \qquad (1.87)$$

As in rectilinear motion, we may consider three special cases in which the angular acceleration is either a function of only time, only the angle θ, or only the angular velocity ω. This development parallels the development of linear acceleration.

If the angular acceleration is a function of only time, the following relations result from repeated integration:

$$\alpha = f(t)$$

$$\omega(t) = \int \alpha(t)dt + C_1 \qquad (1.88)$$

$$\theta(t) = \int \omega(t) + C_2$$

If the angular acceleration is a function of only the angle θ, relations similar to Eqs. (1.17) to (1.20) are applicable using the first-order differential Eq (1.87):

$$\alpha = f(\theta)$$

$$\int \omega d\omega = \int \alpha(\theta)d\theta + C_1 \qquad (1.89)$$

$$\int dt = \int \frac{d\theta}{\omega(\theta)} + C_2$$

If the angular acceleration is a function of only the angular velocity, then Eq. (1.87) yields

$$\alpha = f(\omega)$$

$$\int d\theta = \int \frac{\omega d\omega}{\alpha(\omega)} + C_1 \qquad (1.90)$$

$$\int dt = \int \frac{d\omega}{\alpha(\omega)} + C_2$$

If the angular acceleration is a linear function of the angular velocity, then the integrating factor defined in Eq. (1.33) can be used to solve the resulting first-order linear differential equation,

$$\frac{d\omega}{dt} + p(t)\omega = f(t) \qquad (1.91)$$

The general form of the solution of this equation is

$$\omega(t) = \frac{\int \lambda(t)f(t)dt + C}{\lambda(t)} \qquad (1.92)$$

where $\lambda(t) = e^{\int p(t)dt}$ and C is the constant of integration.

Eq. (1.86) is solved using Euler numerical integration for most linear or nonlinear second-order differential equations, as was shown in Eqs. (1.50) and (1.51). For the angular acceleration $\alpha = f(t, \theta, \omega)$, with initial conditions $\theta(0) = \theta_0$ and $\omega(0) = \omega_0$, the solution for each increment n of Δt is

$$\begin{bmatrix} \omega_{n+1} \\ \theta_{n+1} \end{bmatrix} = \begin{bmatrix} \omega_n + f(t_n, \theta_n, \omega_n)\Delta t \\ \theta_n + \omega_n \Delta t \end{bmatrix} \qquad (1.93)$$

Most computational software packages are able to use higher-order Runge-Kutta methods to solve nonlinear second-order differential equations. (These methods are shown in the Computational Supplement.)

Sample Problem 1.15

A car, starting from rest, drives around a curve with a radius of 1000 feet. If the car undergoes a constant tangential acceleration of 8 ft/s², determine the total acceleration after 10 seconds.

Solution   The tangential acceleration is

$$\frac{d^2s}{dt^2} = 8 \text{ ft/s}^2$$

Integrating with respect to time gives the magnitude of the velocity at any time:

$$\frac{ds}{dt} = 8t + C \text{ ft/s}$$

C is, of course, a constant of integration. The car starts from rest; thus, the constant of integration is zero:

$$\frac{ds}{dt}(0) = C = 0$$

The velocity at 10 s is

$$\frac{ds}{dt}(10) = 8(10) = 80 \text{ ft/s}$$

The acceleration is

$$\mathbf{a} = \frac{d^2s}{dt^2}\hat{\mathbf{e}}_t + \frac{1}{\rho}\left(\frac{ds}{dt}\right)^2 \hat{\mathbf{e}}_n$$

Substituting the values of the radius, velocity, and tangential acceleration into this equation yields

$$\mathbf{a} = 8\hat{\mathbf{e}}_t + \frac{80^2}{1000}\hat{\mathbf{e}}_n$$

$$|\mathbf{a}| = \sqrt{8^2 + 6.4^2} = 10.245 \text{ ft/s}^2$$

Sample Problem 1.16	A particle moving along a circular path undergoes an angular acceleration given by $\alpha = -2\omega + 3t^2$ rad/s². If the particle starts at rest, determine the angular velocity and the angular postion as functions of time.

Solution The angular acceleration can be written as the linear first-order differential equation in Eq. (1.91):

$$\frac{d\omega}{dt} + 2\omega = 3t^2$$

subject to the initial condition $\omega(0) = \omega_0 = 0$. The integrating factor is $\lambda(t) = e^{\int 2dt} = e^{2t}$. Therefore, the general solution is

$$\omega(t) = \frac{\int 3t^2 e^{2t}dt + C}{e^{2t}} = \frac{3}{2}\left(t^2 - t + \frac{1}{2}\right) + Ce^{-2t}$$

Since the angular velocity is zero at $t = 0$, the constant is $C = -3/4$. Hence, the angular velocity is

$$\omega(t) = \frac{3}{2}\left(t^2 - t + \frac{1}{2}\right) - \frac{3}{4}e^{-2t}$$

The angular postion (in radians) can be obtained by integrating $\omega(t)$. Doing so, we obtain

$$\theta(t) = \frac{t^3}{2} - \frac{3t^2}{4} + \frac{t}{2} + \frac{3}{8}e^{-2t} + C_2$$

where C_2 is a constant of integration. If $\theta(0) = 0$, the value of this constant is $C_2 = -3/8$.

The angular velocity and angular position can be graphed as functions of time, thus providing a complete description of the particle's motion as it moves around in a circular path.

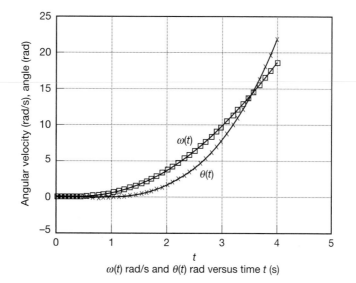

$\omega(t)$ rad/s and $\theta(t)$ rad versus time t (s)

Sample Problem 1.17   If the radius of the circular path in Sample Problem 1.16 is 1.2 meters, determine general functions of time for the tangential and normal acceleration and the magnitude of the total acceleration. Plot the results.

Solution   The angular velocity and acceleration in Sample Problem 1.16 are

$$\omega(t) = \frac{3}{2}\left(t^2 - t + \frac{1}{2}\right) - \frac{3}{4}e^{-2t} \text{ rad/s}$$
$$\alpha = -2\omega + 3t^2 \text{ rad/s}^2$$

Substituting the angular velocity into the equation for angular acceleration yields

$$\alpha(t) = -3t^2 + 3t - \frac{3}{2} + \frac{3}{2}e^{-2t} + 3t^2 = 3\left(t - \frac{1}{2}\right) + \frac{3}{2}e^{-2t} \text{ rad/s}^2$$

The tangential acceleration is

$$a_t(t) = r\alpha(t) \text{ m/s}^2$$

The normal acceleration is

$$a_n = r\omega^2(t) \text{ m/s}^2$$

The total acceleration is

$$\mathbf{a} = a_t\hat{\mathbf{e}}_t + a_n\hat{\mathbf{e}}_n \text{ m/s}^2$$

The magnitude of the acceleration vector is

$$|\mathbf{a}| = \sqrt{a_t^2 + a_n^2} \text{ m/s}^2$$

Note that for the first second the tangential acceleration is the dominant component, but after that time the normal acceleration becomes the largest component.

1.7.2 NORMAL AND TANGENTIAL COORDINATES IN THREE DIMENSIONS

Although mathematically more difficult, a development similar to the two-dimensional development can be made for a particle moving in three-dimensional space along a curved path, as shown in Figure 1.17.

The tangential direction is the direction of the velocity vector. The principal normal vector $\hat{\mathbf{e}}_n$ is defined as lying in the direction of the change in the tangential vector. The two unit vectors $\hat{\mathbf{e}}_t$ and $\hat{\mathbf{e}}_n$ form a plane called an *osculating plane* (from the Latin word "osculari,"

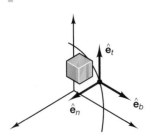

Figure 1.17

meaning "to kiss"). The third vector $\hat{\mathbf{e}}_b$ is called the *binormal vector* and is perpendicular to the osculating plane, thereby forming an orthogonal set of base vectors. The binormal vector is formed by the cross product of the other two vectors:

$$\hat{\mathbf{e}}_b = \hat{\mathbf{e}}_t \times \hat{\mathbf{e}}_n \tag{1.94}$$

The acceleration vector has only two components: the change in magnitude of the velocity vector in the tangential direction and the change in direction of the velocity vector in the principal normal direction. There is no acceleration in the binormal direction, and at any instant the particle is considered to be moving in the osculating plane. The binormal vector changes direction as the particle moves along the curve of the path of motion in space, and the time derivative can be written as

$$\frac{d\hat{\mathbf{e}}_b}{dt} = \frac{d\hat{\mathbf{e}}_b}{ds}\frac{ds}{dt} \tag{1.95}$$

The term $d\hat{\mathbf{e}}_b/ds$ is called the *torsion* of the curve in space.

Although not as directly applicable as the plane motion case, the osculating plane is useful for describing the path of motion in space when the position vector is given as a function of time. The plane of curvature and the radius of curvature can then be determined for any time by using the equations

$$\mathbf{r} = \mathbf{r}(t)$$

$$\mathbf{v} = \frac{d\mathbf{r}}{dt} \tag{1.96}$$

$$\mathbf{a} = \frac{d\mathbf{v}}{dt}$$

The tangential direction can be determined at any time by observing that

$$\hat{\mathbf{e}}_t = \frac{\mathbf{v}}{|\mathbf{v}|} \tag{1.97}$$

The tangential acceleration is

$$\mathbf{a}_t = (\mathbf{a}\cdot\hat{\mathbf{e}}_t)\hat{\mathbf{e}}_t \tag{1.98}$$

and the normal acceleration is

$$\mathbf{a}_n = \mathbf{a} - \mathbf{a}_t$$
$$\hat{\mathbf{e}}_n = \frac{\mathbf{a}_n}{|\mathbf{a}_n|} \tag{1.99}$$

The radius of curvature at any instant and the binormal unit vector are, respectively,

$$\rho = \frac{v^2}{|\mathbf{a}_n|} \tag{1.100}$$
$$\hat{\mathbf{e}}_b = \hat{\mathbf{e}}_t \times \hat{\mathbf{e}}_n$$

When the position vector is given as a function of time in rectangular coordinates, it is frequently useful to describe the motion in tangential, normal, and binormal coordinates and to identify the components of the acceleration in these coordinates. The acceleration component that increases the speed can be separated from the normal component that changes the direction of the velocity vector.

| Sample Problem 1.18 | Consider particle's position vector, given in meters: |

$$\mathbf{r}(t) = t^2\hat{\mathbf{i}} + 3t\hat{\mathbf{j}} - t^3\hat{\mathbf{k}}$$

Determine the base tangential, principal normal, and binormal vectors, and the radius of curvature when time equals 2 seconds.

Solution   Determination of the base vectors and the radius of curvature requires the systematic use of Eqs. (1.95) through (1.99). The vector development is as follows:

$$\mathbf{r}(t) = t^2\hat{\mathbf{i}} + 3t\hat{\mathbf{j}} - t^3\hat{\mathbf{k}} \text{ m}$$

$$\mathbf{v}(t) = 2t\hat{\mathbf{i}} + 3\hat{\mathbf{j}} - 3t^3\hat{\mathbf{k}} \text{ m/s}$$

$$\mathbf{a}(t) = 2\hat{\mathbf{i}} - 6t\hat{\mathbf{k}} \text{ m/s}^2$$

$$\mathbf{v}(2) = 4\hat{\mathbf{i}} + 3\hat{\mathbf{j}} - 12\hat{\mathbf{k}} \text{ m/s}$$

$$|\mathbf{v}(2)| = 13 \text{ m/s}$$

$$\hat{\mathbf{e}}_t = 0.308\hat{\mathbf{i}} + 0.231\hat{\mathbf{j}} - 0.923\hat{\mathbf{k}}$$

$$\mathbf{a}(2) = 2\hat{\mathbf{i}} - 12\hat{\mathbf{k}} \text{ m/s}^2$$

$$\mathbf{a}_t = (\mathbf{a} \cdot \hat{\mathbf{e}}_t)\hat{\mathbf{e}}_t = 3.601\hat{\mathbf{i}} - 2.701\hat{\mathbf{j}} - 10.792\hat{\mathbf{k}} \text{ m/s}^2$$

$$\mathbf{a}_n = \mathbf{a} - \mathbf{a}_t = -1.601\hat{\mathbf{i}} - 2.701\hat{\mathbf{j}} - 1.208\hat{\mathbf{k}} \text{ m/s}^2$$

$$|\mathbf{a}_n| = 3.364 \text{ m/s}^2$$

$$\hat{\mathbf{e}}_n = \frac{\mathbf{a}_n}{|\mathbf{a}_n|} = -0.476\hat{\mathbf{i}} - 0.803\hat{\mathbf{j}} - 0.359\hat{\mathbf{k}}$$

$$\rho = \frac{v^2}{|\mathbf{a}_n|} = \frac{(13)^2}{3.364} = 50.237 \text{ m}$$

$$\hat{\mathbf{e}}_b = \hat{\mathbf{e}}_t \times \hat{\mathbf{e}}_n = -0.824\hat{\mathbf{i}} + 0.550\hat{\mathbf{j}} - 0.137\hat{\mathbf{k}}$$

The radius of curvature may be determined as a function of time using computational software. (See the Computational Supplement.)

Problems

1.88 A car starts from rest drives on a circular path with a radius of 150 ft. The car has a constant tangential acceleration $a_t(t) = 12$ ft/s^2. If the car slips when the normal acceleration is 25 ft/s^2, determine the time before the car slips, and determine the total acceleration at this time.

1.89 Starting from rest, a particle moves on a 2-m radius circular path with a tangential acceleration $a_t(t) = 6\sin(\pi t)$ m/s^2. Develop an expression for the total acceleration as a function of time. Graph the velocity and total acceleration curves for the first 2 s.

1.90 In Problem 1.89, determine the time when the total acceleration is equal to 5 m/s^2.

1.91 An automobile goes over a hill of radius 200 m at a speed of 30 km/hr. Find the normal acceleration of the automobile.

1.92 A particle, starting from rest, moves in a circular path of radius 4 m. If the angular acceleration of the particle is $\alpha(t) = d^2\theta/dt^2 = 2t^2$ rad/s^2, determine (a) the total acceleration after 2 s and) (b) the total distance the particle traveled.

1.93 The angular acceleration of a particle moving in a circular path is given by $\alpha(t) = 3t^2 - 2t$ rad/s². If the initial angular velocity and angular position are zero, determine the angular velocity and angular position as functions of time.

1.94 The angular acceleration of a particle moving in a circular path is given by $\alpha(t) = t\cos(\pi t)$ rad/s². If the initial angular velocity is 2 rad/s and the initial position is 30°, determine an expression for the angular velocity and position.

1.95 If a particle moving in a circular path has an initial angular velocity of 3 rad/s and an angular acceleration of $\alpha(\omega) = -2\omega^2$ rad/s², determine the angular position of the particle after 10 s. Assume $\theta(0) = 0$.

1.96 The angular acceleration of a particle is $\alpha(\omega, t) = -0.01\omega + 4t$ rad/s². If the particle has an initial angular velocity of 2 rad/s, obtain an expression for the angular velocity and angular position. Assume $\theta(0) = 0$.

1.97 The angular acceleration of a particle is given by $\alpha(\omega, \theta) = 2\sin(\theta) - 0.4\omega$ rad/s². If the initial conditions are $\theta_0 = \pi/6$ rad and $\omega_0 = 0$, determine the angular position of the particle for the first 5 s.

1.98 The angular acceleration of a particle is given by $\alpha(\omega, \theta) = 3\cos(\theta) - 0.02\omega|\omega|$ rad/s². If the initial conditions are $\theta_0 = \pi/6$ rad and $\omega_0 = 0$, determine the angular position of the particle for the first 5 s.

1.99 A particle's position vector, given in meters,

$$r(t) = 3(\cos t)\hat{\mathbf{i}} + 3(\sin t)\hat{\mathbf{j}} + 4t\hat{\mathbf{k}} \text{ m}$$

Determine the base tangential, principal normal, and binormal vectors, and the radius of curvature at $t = 3$ s.

1.100 In Problem 1.99, determine the velocity and acceleration of the particle at any time t.

1.101 In Problem 1.99, determine the radius of curvature as a function of time. Describe the motion of the particle.

1.102 The position vector for a particle is given as the following function:

$$r(t) = t^2\hat{\mathbf{i}} + 3t\hat{\mathbf{j}} + 10(\sin t)\hat{\mathbf{k}} \text{ m}$$

Calculate the radius of curvature at (a) $t = 1$ s, (b) $t = 3$ s, and (c) $t = 5$ s.

1.8 RADIAL AND TRANSVERSE COORDINATES (POLAR COORDINATES)

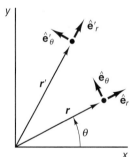

Figure 1.18

Frequently, it is useful to define position, velocity, and acceleration in terms of polar coordinates. The polar coordinate unit vectors change direction as a particle moves in a plane, as shown in Figure 1.18. The unit base vectors for polar coordinates are similar to the normal and tangential coordinates used in Section 1.6. The unit vector $\hat{\mathbf{e}}_r$ is directed in the positive r-direction, and the unit vector $\hat{\mathbf{e}}_\theta$ is directed perpendicular to the $\hat{\mathbf{e}}_r$ base vector in the positive θ-direction. The position vector is written as

$$\mathbf{r} = r\hat{\mathbf{e}}_r \tag{1.101}$$

Note that the base vectors are not functions of r; that is, an increase or decrease in r does not change the direction of these base vectors. The base vectors are functions of the angle θ, which is the angle that the position vector $\mathbf{r}$ makes with the x-coordinate axis. This functional dependency of the base vectors on the angle can be seen in Figure 1.19. Note that $d\hat{\mathbf{e}}_r$ must be in the $\hat{\mathbf{e}}_\theta$ direction—that is, perpendicular to $\hat{\mathbf{e}}_r$. As in the case of the tangential vector in Section 1.7, $\hat{\mathbf{e}}_r$ is a unit vector, and, of course, the length of the unit vector cannot change. The derivatives of the base vectors with respect to time are obtained by a method similar to that used for calculating the derivatives of the normal and tangential coordinate base vectors. That is,

$$d\hat{\mathbf{e}}_r = 1d\theta\hat{\mathbf{e}}_\theta$$

$$\frac{d\hat{\mathbf{e}}_r}{dt} = \frac{d\theta}{dt}\hat{\mathbf{e}}_\theta \tag{1.102a}$$

$$\dot{\hat{\mathbf{e}}}_r = \dot{\theta}\hat{\mathbf{e}}_\theta$$

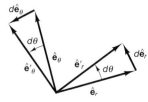

Figure 1.19

and similarly

$$d\hat{\mathbf{e}}_\theta = -1d\theta\hat{\mathbf{e}}_r$$

$$\frac{d\hat{\mathbf{e}}_\theta}{dt} = -\frac{d\theta}{dt}\hat{\mathbf{e}}_r \qquad (1.102b)$$

$$\dot{\hat{\mathbf{e}}}_\theta = -\dot{\theta}\hat{\mathbf{e}}_r$$

Now, the position vector can be differentiated to determine the velocity and acceleration in polar coordinates:

$$\mathbf{r} = r\hat{\mathbf{e}}_r$$

$$\mathbf{v} = \dot{r}\hat{\mathbf{e}}_r + r\dot{\theta}\hat{\mathbf{e}}_\theta$$

$$\mathbf{a} = \ddot{r}\hat{\mathbf{e}}_r + \dot{r}\dot{\theta}\hat{\mathbf{e}}_\theta + \dot{r}\dot{\theta}\hat{\mathbf{e}}_\theta + r\ddot{\theta}\hat{\mathbf{e}}_\theta - r\dot{\theta}^2\hat{\mathbf{e}}_r \qquad (1.103)$$

$$\mathbf{a} = (\ddot{r} - r\dot{\theta}^2)\hat{\mathbf{e}}_r + (r\ddot{\theta} + 2\dot{r}\dot{\theta})\hat{\mathbf{e}}_\theta$$

Differentiation of the products in Eq. (1.101) yields two terms in the velocity equation and five terms in the acceleration equation. The scalar components of the velocity and the acceleration are written as

$$v_r = \dot{r}$$

$$v_\theta = r\dot{\theta}$$

$$a_r = \ddot{r} - r\dot{\theta}^2 \qquad (1.104)$$

$$a_\theta = r\ddot{\theta} + 2\dot{r}\dot{\theta}$$

The inverse dynamics problem easily is solved when the magnitude of the position vector and angular velocity vector are given as functions of time. The velocity and acceleration of the particle can then be obtained by using Eq. (1.104) in straightforward differentiation. If, on the other hand, the acceleration components are given as functions of time, the problem is in the form of coupled nonlinear differential equations, and this direct dynamics problem is difficult to integrate and solve. Special numerical integration solutions are shown in Chapter 2. The coupled nature of the equations can be seen if Eq. (1.103) is written in matrix notation:

$$\begin{bmatrix} a_r \\ a_\theta \end{bmatrix} = \begin{bmatrix} \ddot{r} - r\dot{\theta}^2 \\ r\ddot{\theta} + 2\dot{r}\dot{\theta} \end{bmatrix} \qquad (1.105)$$

The acceleration is said to be coupled because both of its components depend upon r and θ.

For the special case of circular motion, where r is a constant, the velocity and acceleration components are simpler. Setting derivatives of r equal to zero in Eq. (1.104) yields

$$v_r = 0$$

$$v_\theta = r\dot{\theta}$$

$$a_r = -r\dot{\theta}^2 = -\frac{v_\theta^2}{r}$$

$$a_\theta = r\ddot{\theta} = \frac{dv_\theta}{dt} \qquad (1.106)$$

$$\mathbf{v} = v_\theta\hat{e}_\theta$$

$$\mathbf{a} = \frac{dv_\theta}{dt}\hat{e}_\theta - \frac{v_\theta^2}{r}\hat{e}_r$$

These equations can be compared to their counterparts in the normal and tangential coordinate system for the special case of circular motion. For this case, only the angle changes with time, and the direct dynamics problem easily is solved when the angular acceleration is specified as a function of time, angular velocity, and angular acceleration. This problem is a one-degree-of-freedom problem, and the resulting differential equation can be solved in the manner described in Section 1.7.

Sample Problem 1.19

Consider a particle's path of motion, given by

$$\theta(t) = \pi t \text{ rad}$$
$$r(t) = 2 \sin 3\theta(t) \text{ m}$$

Plot the path of motion, and determine the velocity and acceleration of the particle.

Solution

The path of motion is easily seen if it is plotted in x- and y-coordinates. The relationship between polar coordinates and the rectangular coordinates is given by

$$x(t) = r(t) \cos \theta(t) \text{ m} \quad \text{and} \quad y(t) = r(t) \sin \theta(t) \text{ m}$$

Substituting the specified functions for the polar coordinates into these equations gives

$$x(t) = 2 \sin (3\pi t) \cos (\pi t) \text{ m} \quad \text{and} \quad y(t) = 2 \sin (3\pi t) \sin (\pi t) \text{ m}$$

The plot of this motion shows that the particle follows the path of a "three-leaved rose."

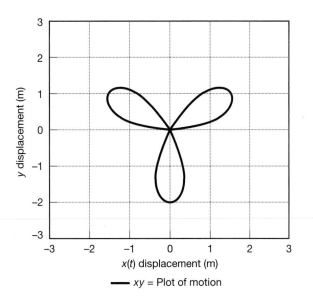

— xy = Plot of motion

The velocity and acceleration of the particle can be expressed in rectilinear coordinates by differentiation of the functions $x(t)$ and $y(t)$, or by expressing them in polar coordinates using Eq. (1.102) after $r(t)$ and $\theta(t)$ are differentiated. The values in polar coordinates are

$$\theta(t) = \pi t \quad r(t) = 2 \sin (3\pi t)$$
$$\dot{\theta}(t) = \pi \quad \dot{r}(t) = 6\pi \cos (3\pi t)$$
$$\ddot{\theta}(t) = 0 \quad \ddot{r}(t) = -18\pi^2 \sin (3\pi t)$$

Substituting these values into Eq. (1.102) yields

$$\mathbf{v} = 6\pi \cos(3\pi t)\hat{\mathbf{e}}_r + 2\pi \sin(3\pi t)\hat{\mathbf{e}}_\theta \text{ m/s}^2$$

$$\mathbf{a} = -20\pi^2 \sin(3\pi t)\hat{\mathbf{e}}_r + 12\pi^2 \cos(\pi t)\hat{\mathbf{e}}_\theta \text{ m/s}$$

Sample Problem 1.20

A spiral trajectory of a particle is given by the equations

$$\theta(t) = \pi t \text{ rad}$$
$$r(t) = e^{0.2\theta} \text{ m}$$

Plot the trajectory in space, and determine the r- and θ-components of the acceleration.

Solution   Again, the path of motion is best seen in xy-space. Using the polar-to-rectangular coordinate transformation gives

$$x(t) = e^{0.2\pi t} \cos(\pi t) \text{ m}$$

$$y(t) = e^{0.2\pi t} \sin(\pi t) \text{ m}$$

The trajectory for the first 5 seconds of motion is plotted in the following diagram:

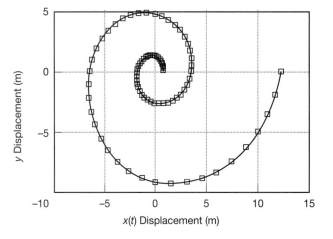

The path begins at $(1,0)$ and spirals out, as shown. The points are plotted at intervals of $\Delta t = 0.05$, and one can observe that as the velocity increases, the distance between points increases. Differentiating the coordinate functions yields

$$\dot{\theta}(t) = \pi \qquad \dot{r}(t) = 0.2\pi e^{0.2\pi t}$$

$$\ddot{\theta}(t) = 0 \qquad \ddot{r}(t) = 0.04\pi^2 e^{0.2\pi t}$$

The radial and transverse accelerations, obtained from Eq. (1.105), are

$$a_r = 0.04\pi^2 e^{0.2\pi t} - e^{0.2\pi t}\pi^2 = -0.96\pi^2 e^{0.2\pi t}$$

$$a_\theta = 2(0.2\pi^2)e^{0.2\pi t}$$

| Sample Problem 1.21 | A particle moves in a circular motion with a radius of 2 m and an acceleration given by the initial angular velocity and angle are |

$$\alpha(t) = \frac{d^2\theta}{dt^2} = 4\theta(t) \text{ rad/s}^2$$

$$\omega(0) = \frac{d\theta}{dt}(0) = 2 \text{ rad/s}$$

$$\theta(0) = 0 \text{ rad}$$

Solution   The angular acceleration is specified as a function of position only, and therefore the differential equation can be solved using separation-of-variable techniques, as shown in Sections 1.3 and 1.6. We get

$$\alpha = \frac{d\omega}{dt} = \frac{d\omega}{d\theta}\frac{d\theta}{dt} = \omega\frac{d\omega}{d\theta}$$

$$\int_{\theta(0)}^{\theta} \alpha(\theta)d\theta = \int_{\omega(0)}^{\omega} \omega d\omega$$

Substituting the expression for the acceleration and evaluating the integral yields

$$4\theta^2(t) = \omega^2(t) - 4$$

$$\omega(t) = 2\sqrt{\theta^2(t) + 1}$$

The angular velocity can now be written as

$$\omega(t) = \frac{d\theta(t)}{dt} = 2\sqrt{\theta^2(t) + 1}$$

Again, this is a separable first-order differential equation, which may be written as

$$dt = \frac{d\theta}{2\sqrt{\theta^2 + 1}}$$

Integrating yields

$$t = \frac{1}{2}\left[\ln\left(\theta + \sqrt{\theta^2 + 1}\right)\right]_0^{\theta} = \frac{1}{2}\left[\ln\left(\theta + \sqrt{\theta^2 + 1}\right)\right]$$

Multiplying by 2 and taking the exponential of both sides, we get

$$\theta + \sqrt{\theta^2 + 1} = e^{2t}$$

$$\sqrt{\theta^2 + 1} = e^{2t} - \theta$$

Squaring both sides of the equation gives

$$\theta^2 + 1 = e^{4t} - 2\theta e^{2t} + \theta^2$$

$$\theta = \frac{1}{2}(e^{2t} - e^{-2t}) = \sinh(2t)$$

An alternative method for solving the differential equation can be used, as the equation is an ordinary linear differential equation that may be solved directly with the use of an integrating factor. To use this method, write the differential equation in the following form:

$$\frac{d^2\theta}{dt^2} - 4\theta = 0$$

This expression is a second-order, homogeneous linear differential equation with constant coefficients. Assume a solution in the form

$$\theta(t) = Ce^{at}$$

Substituting the integrating factor into the differential equation yields the characteristic equation for a:

$$C(a^2 - 4)e^{at} = 0$$

Since taking the constant C as equal to zero would yield a trivial solution, and the exponential cannot be zero for all time, the term within the parentheses must be equal to zero. The two values of a that satisfy the characteristic equation are

$$a_1 = +2 \qquad a_2 = -2$$

The solution of the differential equation can then be written as

$$\theta(t) = C_1 e^{2t} + C_2 e^{-2t}$$

The exponential functions can be written in terms of hyperbolic functions using the identities

$$e^u = \cosh u + \sinh u$$
$$e^{-u} = \cosh u - \sinh u$$

Thus, the solution of the differential equation can be written as

$$\theta(t) = A \cosh(2t) + B \sinh(2t)$$

The angular velocity is obtained by differentiating the angular position with respect to time and is

$$\omega(t) = -2A \sinh(2t) + 2B \cosh(2t)$$

At time $t = 0$, the angular position is 0 and the angular velocity is 2 rad/s. Substituting these initial values into the expressions for angular position and angular velocity yields the values for the constants of integration:

$$A = 0 \qquad B = 1$$

The final solution for the angular motion is

$$\theta(t) = \sinh(2t) \text{ rad}$$

This equation agrees with the previous solution.

Sample Problem 1.22 Determine the path of motion of a particle if its acceleration is $a_r = 0.5$ m/s^2 and $a_\theta = 2$ m/s^2, and its initial velocity is $v_r = 1$ m/s and $v_\theta = 0.5$ m/s. At time equal to zero, the particle has initial coordinates $r = 0.2$ m and $\theta = 0$.

Solution The initial conditions can be expressed in matrix notation to prepare a numerical solution of the coupled nonlinear differential equations.

We can set the time increment to $\Delta t = 0.001$ s for the numerical integration, and if we integrate for N steps, the total time for the motion is $N\Delta t$. The initial values of position and velocity are

$$\begin{bmatrix} vr_0 \\ r_0 \\ \omega_0 \\ \theta_0 \end{bmatrix} = \begin{bmatrix} 1 \\ 0.2 \\ 2.5 \\ 0 \end{bmatrix}$$

where

$$\omega = \frac{d\theta}{dt} = \frac{v_\theta}{r}$$

The acceleration can be written as

$$\ddot{r} = 0.5 + r\omega^2$$

$$\alpha = \ddot{\theta} = \frac{2}{r} - \frac{2(vr)\omega}{r}$$

An Euler numerical integration takes the following form:

$$\begin{bmatrix} vr_{n+1} \\ r_{n+1} \\ \omega_{n+1} \\ \theta_{n+1} \end{bmatrix} = \begin{bmatrix} vr_n + (0.5 + r_n\omega_n^2)\Delta t \\ r_n + (vr_n)\Delta t \\ \omega_n + \left(\dfrac{2}{r_n} - \dfrac{2(vr_n)\omega_n}{r_n}\right)\Delta t \\ \theta_n + \omega_n\Delta t \end{bmatrix}$$

This process is repeated for $n = 0$ to $n = N$.

The path of motion can be graphed in rectangular coordinates as

$$x = r\cos(\theta)$$
$$y = r\sin(\theta)$$

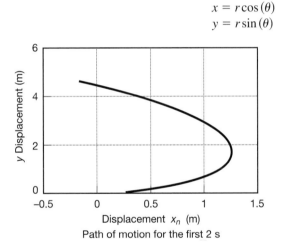

Path of motion for the first 2 s

Numerical details of the integration are shown in the Computational Supplement, in which, also, other methods of solving the differential equation are shown for each computational software package.

Problems

1.103 A particle rotates in a horizontal plane under a motion described by

$$r(t) = 0.5(1 + t^2)$$
$$\theta(t) = \pi t^2$$

where r is in meters and θ is in radians. Determine the velocity and acceleration of the particle, and plot the path of motion for the first 2 s.

1.104 In Problem 1.103, determine the total distance the particle has moved in the first 2 s. (*Hint: The magnitude of the velocity at any time is equal to* ds/dt. *This expression may be integrated to determine the distance traveled. Since the magnitude of the velocity is always positive, it follows that if the particle reverses direction, the integration in segments is not required.*)

1.105 The coordinates of a particle are given in meters and radians as

$$r(t) = 2$$
$$\theta(t) = \sin(\pi t)$$

Determine the velocity and acceleration vectors for the motion of the particle. Plot the path of the motion for the first 2 s.

1.106 In Problem 1.105, determine the total distance traveled by the particle in the first 2 s.

1.107 The second time derivative of the polar coordinates is given by

$$\ddot{r}(t) = 2e^{-t}$$

$$\ddot{\theta}(t) = \pi$$

where r is in meters and θ is in radians. If the particle starts from rest at a position $r = 1$ m and $\theta = 0$, determine the velocity and acceleration of the particle.

1.108 In Problem 1.107, plot the motion for the first 2 s, and determine the distance traveled by the particle during that time.

1.109 A machine part rotates at a constant rate of 2π rad/s while a small mass oscillates along the part. The radial displacement of the mass is given by $r(t) = r_0 + r_a \sin(2\pi t)$, where $r_a < r_0$. Determine the two components of the acceleration of the mass. (*Note:* Forces must be applied to the mass in a manner that is proportional to these components of acceleration in order to sustain this motion.)

1.110 In Problem 1.109, plot the motion in polar or xy-coordinates for one complete revolution, and determine the total distance the mass travels. (Assume $r_0 = 2$ m and $r_a = 1.5$ m.)

1.111 The acceleration of a particle is given by

$$a_r = 2t \text{ m/s}^2$$

$$a_\theta = \cos(\pi t)$$

Determine the polar coordinates of the particle for the first 2 s of the motion if the particle starts from rest at a position $r = 0.5$ m and $\theta = 0$ rad. (*Hint: Formulate the two coupled differential equations for $r(t)$ and $\theta(t)$, and solve them numerically.*)

1.112 In Problem 1.111, plot the trajectory of the particle for the first 2 s.

1.113 If the second derivatives of the polar coordinates of a particle's location are given as

$$\ddot{r}(t) = 3 - 0.01\dot{r}(t)$$

$$\ddot{\theta} = 0$$

where the initial conditions are

$$r(0) = 0.4 \text{ m} \qquad \dot{r}(0) = 0$$

$$\theta(0) = 0 \qquad \dot{\theta}(0) = 1.5 \text{ rad/s}$$

determine the components of the particle's acceleration in polar coordinates.

1.114 In Problem 1.113, plot the trajectory of the particle for the first 4 s.

1.115 The profile of a cam is given by $r(\theta) = 100 + 60 \cos(\theta)$ mm. If the cam follower is rotating at a constant angular velocity of $\dot{\theta} = \pi/4$ rad/s, determine the velocity and acceleration of the follower.

1.116 A collar slides along a horizontal bar, as shown in Figure P1.116. If the arm AB oscillates such that $\theta(t) = (\pi/4) \sin(\pi t)$ rad, determine the velocity and acceleration of the collar in polar coordinates.

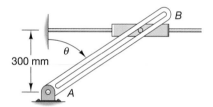

Figure P1.116

1.117 A collar slides around a 100-mm circular guide that is driven by the bar AB, as depicted in Figure P1.117. If the bar has a constant angular velocity of π rad/s, determine the acceleration of the collar in polar coordinates.

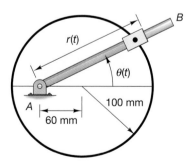

Figure P1.117

1.118 Determine the acceleration of the collar in Problem 1.117 if the bar has an angular acceleration given by $\alpha(t) = 0.5 \cos(t)$ rad/s^2 and if the particle starts from rest at $\theta = 0$.

*1.9 THREE-DIMENSIONAL COORDINATE SYSTEMS: CYLINDRICAL AND SPHERICAL COORDINATES

1.9.1 CYLINDRICAL COORDINATES

In the previous two sections, curvilinear coordinates were introduced to describe the two-dimensional motion of a particle. Here, this description will be expanded to include three-dimensional motion. In some kinematics problems, the three-dimensional motion of a particle is best described in cylindrical or spherical coordinates. Cylindrical coordinates are formed by adding the z-direction to the two-dimensional polar coordinates. Spherical coordinates make up a general three-dimensional coordinate system where the position vector $\mathbf{r}$ is described by a single coordinate in the R-direction. Both of these coordinate systems are orthogonal, but their base unit vectors are not constant, as is the case with three-dimensional rectangular coordinates. This dependency of the unit vectors on one or more of the coordinates has been discussed in Sections 1.7 and 1.8. The velocity and acceleration of a particle expressed in cylindrical coordinates are easily obtained from the relations presented in polar coordinates. To see how this is done, suppose the particle is located at the position shown in Figure 1.20, in cylindrical coordinates. In polar coordinates, the unit vectors $\hat{\mathbf{e}}_r$ and $\hat{\mathbf{e}}_\theta$ are functions of the coordinate θ. This dependency is given by

$$\frac{d\hat{\mathbf{e}}_r}{d\theta} = \hat{\mathbf{e}}_\theta$$
$$\frac{d\hat{\mathbf{e}}_\theta}{d\theta} = -\hat{\mathbf{e}}_r \qquad (1.107)$$
$$\frac{d\hat{\mathbf{e}}_z}{d\theta} = 0$$

The unit vector in the z-direction is a constant, as its magnitude is unity (one) and its direction does not change with changes in position of the particle. The position vector is written as

$$\mathbf{r} = r\hat{\mathbf{e}}_r + z\hat{\mathbf{e}}_z \qquad (1.108)$$

The velocity, defined as the derivative of the position vector with respect to time, is

$$v = \frac{d}{dt}\mathbf{r} = \frac{d}{dt}(r\hat{\mathbf{e}}_r + z\hat{\mathbf{e}}_z)$$
$$v = \frac{dr}{dt}\hat{\mathbf{e}}_r + r\frac{d\hat{\mathbf{e}}_r}{dt} + \frac{dz}{dt}\hat{\mathbf{e}}_z = \dot{r}\hat{\mathbf{e}}_r + r\dot{\theta}\hat{\mathbf{e}}_\theta + \dot{z}\hat{\mathbf{e}}_z \qquad (1.109)$$

Examination of the coefficients of the unit vectors shows that the scalar components of the velocity in this coordinate system are

$$v_r = \dot{r}$$
$$v_\theta = r\dot{\theta} \qquad (1.110)$$
$$v_z = \dot{z}$$

The acceleration is obtained by differentiation of Eq. (1.109). Note that the acceleration cannot be obtained by differentiation of the scalar relations given in Eq. (1.110), because the unit vectors are a function of time. Differentiating Eq. (1.109) yields

$$\mathbf{a} = \frac{d(\dot{r}\hat{\mathbf{e}}_r)}{dt} + \frac{d(r\dot{\theta}\hat{\mathbf{e}}_\theta)}{dt} + \frac{d(\dot{z}\hat{\mathbf{e}}_z)}{dt}$$
$$\mathbf{a} = [\ddot{r}\hat{\mathbf{e}}_r + \dot{r}\dot{\theta}\hat{\mathbf{e}}_\theta] + [\dot{r}\dot{\theta}\hat{\mathbf{e}}_\theta + r\ddot{\theta}\hat{\mathbf{e}}_\theta - r\dot{\theta}^2\hat{\mathbf{e}}_r] + \ddot{z}\hat{\mathbf{e}}_z \qquad (1.111)$$
$$\mathbf{a} = (\ddot{r} - r\dot{\theta}^2)\hat{\mathbf{e}}_r + (2\dot{r}\dot{\theta} + r\ddot{\theta})\hat{\mathbf{e}}_\theta + \ddot{z}\hat{\mathbf{e}}_z$$

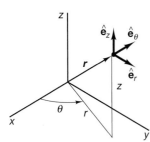

Figure 1.20

The scalar components of acceleration are

$$a_r = (\ddot{r} - r\dot{\theta}^2)$$
$$a_\theta = (2\dot{r}\dot{\theta} + r\ddot{\theta})$$
$$a_z = \ddot{z}$$

(1.112)

Note that the accelerations in r and θ are the same as those previously developed for polar coordinates.

Sample Problem 1.23

If an airplane descends for a landing in a helical pattern given by

$r = R_0$ m (a constant)
$\theta(t) = \omega_0 t$ rad, where ω_0 is the constant angular velocity
$z(t) = -pt$ m, where p is the tangent of the angle of descent

determine the velocity and acceleration of the plane.

Solution   The velocity of the airplane is obtained directly from Eq. (1.109):

$$\mathbf{v} = R_0\omega_0\hat{\mathbf{e}}_\theta - p\hat{\mathbf{e}}_z$$

From Eq. (1.111), the acceleration of the plane is

$$\mathbf{a} = -R_0\omega_0^2\hat{\mathbf{e}}_r$$

Note that a_θ and a_z are both zero, so the only acceleration of the airplane is the normal acceleration directed toward the helical axis of the path of descent.

1.9.2 SPHERICAL COORDINATES

The equations for the position, velocity, and acceleration of a particle in spherical coordinates are obtained in a manner similar to their derivation in cylindrical coordinates. However, the base unit vectors are functions of two coordinates, and therefore, their time derivatives are more complex, so the differentiation of a vector in spherical coordinates is tedious. The particle shown in Figure 1.21 is in a spherical coordinate system, where the coordinates are R, ϕ, and θ. The derivative of the unit vectors, with respect to the coordinates, are given by

$$\frac{\partial \hat{\mathbf{e}}_R}{\partial R} = 0 \qquad \frac{\partial \hat{\mathbf{e}}_R}{\partial \phi} = \hat{\mathbf{e}}_\phi \qquad \frac{\partial \hat{\mathbf{e}}_R}{\partial \theta} = \sin\phi\hat{\mathbf{e}}_\theta$$

$$\frac{\partial \hat{\mathbf{e}}_\phi}{\partial R} = 0 \qquad \frac{\partial \hat{\mathbf{e}}_\phi}{\partial \phi} = -\hat{\mathbf{e}}_R \qquad \frac{\partial \hat{\mathbf{e}}_\phi}{\partial \theta} = \cos\phi\hat{\mathbf{e}}_\theta$$

(1.113)

$$\frac{\partial \hat{\mathbf{e}}_\theta}{\partial R} = 0 \qquad \frac{\partial \hat{\mathbf{e}}_\theta}{\partial \phi} = 0 \qquad \frac{\partial \hat{\mathbf{e}}_\theta}{\partial \theta} = -\sin\phi\hat{\mathbf{e}}_R - \cos\phi\hat{\mathbf{e}}_\phi$$

Here, partial derivatives are used because each unit vector is a function of more than one variable. The time derivatives of the unit base vectors are

$$\frac{d\hat{\mathbf{e}}_R}{dt} = \dot{\phi}\hat{\mathbf{e}}_\phi + \sin\phi\dot{\theta}\hat{\mathbf{e}}_\theta$$

$$\frac{d\hat{\mathbf{e}}_\phi}{dt} = -\dot{\phi}\hat{\mathbf{e}}_R + \cos\phi\dot{\theta}\hat{\mathbf{e}}_\theta$$

(1.114)

$$\frac{d\hat{\mathbf{e}}_\theta}{dt} = -\sin\phi\dot{\theta}\hat{\mathbf{e}}_R - \cos\phi\dot{\theta}\hat{\mathbf{e}}_\phi$$

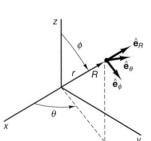

Figure 1.21

The position vector in spherical coordinates is simply

$$\mathbf{r} = R\hat{\mathbf{e}}_R \tag{1.115}$$

Using Eq. (1.114) to differentiate Eq. (1.115) yields the velocity

$$\mathbf{v} = \dot{R}\hat{\mathbf{e}}_R + R\dot{\phi}\hat{\mathbf{e}}_\phi + R\sin\phi\dot{\theta}\hat{\mathbf{e}}_\theta \tag{1.116}$$

The scalar components of the velocity, in spherical coordinates, are

$$
\begin{aligned}
v_R &= \dot{R} \\
v_\phi &= R\dot{\phi} \\
v_\theta &= R\sin\phi\dot{\theta}
\end{aligned}
\tag{1.117}
$$

The acceleration is obtained by differentiation of Eq. (1.116), yielding

$$
\begin{aligned}
\mathbf{a} = {}& \ddot{R}\hat{\mathbf{e}}_R + \dot{R}\dot{\phi}\hat{\mathbf{e}}_\phi + \dot{R}\sin\phi\dot{\theta}\hat{\mathbf{e}}_\theta \\
&+ \dot{R}\dot{\phi}\hat{\mathbf{e}}_\phi + R\ddot{\phi}\hat{\mathbf{e}}_\phi - R\dot{\phi}^2\hat{\mathbf{e}}_R + R\dot{\phi}\cos\phi\dot{\theta}\hat{\mathbf{e}}_\theta \\
&+ \dot{R}\sin\phi\dot{\theta}\hat{\mathbf{e}}_\theta + R\cos\phi\dot{\phi}\dot{\theta}\hat{\mathbf{e}}_\theta + R\sin\phi\ddot{\theta}\hat{\mathbf{e}}_\theta \\
&- R(\sin\phi\dot{\theta})^2\hat{\mathbf{e}}_R - R\sin\phi\cos\phi\dot{\theta}^2\hat{\mathbf{e}}_\phi
\end{aligned}
$$

Grouping terms in each coordinate direction, that is, the coefficients of the unit vectors, yields

$$
\begin{aligned}
\mathbf{a} = {}& [\ddot{R} - R\dot{\phi}^2 - R\sin^2\phi\dot{\theta}^2]\hat{\mathbf{e}}_R \\
&+ [R\ddot{\phi} + 2\dot{R}\dot{\phi} - R\dot{\theta}^2\sin\phi\cos\phi]\hat{\mathbf{e}}_\phi \\
&+ [R\ddot{\phi}\sin\phi + 2\dot{R}\dot{\theta}\sin\phi + 2R\dot{\phi}\dot{\theta}\cos\phi]\hat{\mathbf{e}}_\theta
\end{aligned}
\tag{1.118}
$$

The scalar components of the acceleration are

$$
\begin{aligned}
a_R &= \ddot{R} - R\dot{\phi}^2 - R\dot{\theta}^2\sin^2\phi \\
a_\phi &= R\ddot{\phi} + 2\dot{R}\dot{\phi} - R\dot{\theta}^2\sin\phi\cos\phi \\
a_\theta &= R\ddot{\theta}\sin\phi + 2\dot{R}\dot{\theta}\sin\phi + 2R\dot{\phi}\dot{\theta}\cos\phi
\end{aligned}
\tag{1.119}
$$

There are many more orthogonal curvilinear coordinates that can be used to describe the kinematics of a particle. The expressions for the velocity and acceleration can be derived directly from those coordinates, or they can be obtained from rotational transformation matrices. Special coordinate systems are selected if they are compatible with the constraints on the motion of the particle. For example, ellipsoidal coordinates could be used if the particle is constrained to move inside or on the surface of an ellipsoid. Here, we will use rectangular, cylindrical, or spherical coordinates, depending upon the constraints on the motion or in order to render the path of motion of the particle in the most conceptual manner. Constraints on the motion will be discussed in Section 1.12.

It is obvious that when the position vector is given or measured as a function of time in any of these coordinate systems, an indirect dynamics problem is easily solved by differentiation if the initial conditions are known. If, however, the accelerations are known or measured, the solution of a direct dynamics problem is very difficult, as it will involve coupled nonlinear differential equations. In general, there are no closed analytical solutions of these equations, and only numerical integration techniques are used to approach this type of problem.

Most studies of *celestial mechanics* are formulated in spherical coordinates. We can examine the earth's motion in this coordinate system. In 1543, Nicolaus Copernicus presented a theory that the earth was not the center of the universe, and that the sun and the stars **did not** rotate about the earth. This contradicted the teachings of Aristotle and Ptolemy that the earth was the immobile hub and conflicted with teachings that the Catholic Church felt were supported by the scriptures. Copernicus argued through mathematics (not observation) that the earth rotated on its axis once a day and that the earth and the moon rotated around the sun once a year. He published his work, *On the Revolutions of the Heavenly Spheres*, written in Latin and using mathematics. His book was meant for a scholarly audience and not the general public.

The general public felt that they knew that the earth did not rotate and argued that if it did, a ball tossed into the air would not fall right back into one's hands but land hundreds of feet away. Of course, they were not scholars, and relative motion and gravity were concepts that would have to wait for Galileo and Newton.

In 1626, Galileo was working on his book *Dialogue* in which he would argue the theory of Copernicus. He argued the theory of relative motion and showed that no experiment performed with ordinary objects on the surface of the earth could prove whether or not the earth was moving. He correctly reasoned that all objects on the surface would have the same global motion of the earth. He felt that only astronomical evidence and reasoning from observation could carry the argument. Six years later, in 1632, Galileo published *Dialogue on the Two Chief World Systems: Ptolemaic and Copernican*. Galileo stood trial for heresy before the Holy Office of the Inquisition in 1633, and the *Dialogue* was prohibited. It was not until 1728 that English astronomer James Bradley provided the first evidence for the earth's motion through space. In 1992, after 12 years of review, Pope John Paul II publicly endorsed Galileo's philosophy and termed his trial and house arrest a misunderstanding.

Let us examine the velocity of an object on the surface of the earth due to its daily rotation. The earth is not a sphere, as it is flattened at the poles, making it an oblate spheroid. For this approximation, we will model it as a sphere. The spherical radius of the earth is 3963 miles, and we will take the z-axis to have its origin at the center of the earth and passing through the north pole. The angular velocity due to the earth's rotation is $\dot{\theta} = \dfrac{2\pi}{24}$ rad/hr. The angle ϕ is equal to (90° minus the latitude of the point on the earth's surface). The degrees latitude in the northern hemisphere are treated as positive and those in the southern hemisphere are treated as negative. Therefore, the surface velocity of any point on the earth's surface is $v_\theta = R\sin\phi\dot{\theta}$. The velocity of a point on the equator is 1037 mph and at Cape Canaveral (28° latitude) is 916 mph. It is clear why rockets are launched to the east at the cape. We cannot calculate the absolute velocity of a point on the earth, as we have no universal inertial reference point. If we assume the sun to be at rest, then the additional velocity of the center of the earth is 66,700 mph. Earth is approximately 93,000,000 miles from the sun and makes a revolution about the sun in 365 days.

Sample Problem 1.24

A particle is released at the rim of a smooth hemispherical bowl of radius 200 mm with an initial velocity in the circumferential direction of $R\dot{\theta} = 1000$ mm/s, and it slides into the bowl under the influence of gravity ($g = 9810$ mm/s²). Write the kinematic equations of motion, and solve the resulting nonlinear differential equations using computational software.

Solution

The particle moves under gravitational attraction such that the acceleration in the ϕ and θ directions are:

1000 mm/s

200 mm

$$a_\phi = g\sin\phi$$
$$a_\theta = 0$$

The motion of the particle is constrained in the R direction such that $\ddot{R} = \dot{R} = 0$. We will take the origin of our spherical coordinates to be at the center of the circle formed by the rim of the bowl, and will start the particle at a position $R = 200$, $\phi = \pi/2$, $\theta = 0$, with an initial angular velocity $\dot{\theta} = 5$ rad/s. We expect the particle to execute a spiral motion as

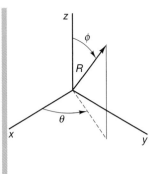

it slides in the bowl. From Eq. (1.119), the differential equations of motion in the ϕ and θ directions are

$$g \sin \phi = R\ddot{\phi} - R\dot{\theta}^2 \sin \phi \cos \phi$$
$$0 = R\ddot{\theta} \sin \phi + 2R \dot{\phi}\dot{\theta} \cos \phi$$

The first differential equation can be written in a form that is used for numerical differentiation:

$$\ddot{\phi} = \dot{\theta}^2 \sin \phi \cos \phi + \frac{g}{R} \sin \phi$$

The second differential equation can be writtten as an exact differential:

$$\frac{1}{\sin \phi} \frac{d}{dt} (\dot{\theta} \sin^2 \phi) = \ddot{\theta} \sin \phi + 2\dot{\theta}\dot{\phi} \cos \phi = 0$$

Therefore, we can determine a constant of the motion as

$$\dot{\theta} \sin^2 \phi = h$$

where h is a constant that is determined by the initial conditions. This constant can be related to a quantity called the *angular momentum*, which will be discussed in detail in Section 2.6. The angular velocity in the θ-direction can be written as

$$\dot{\theta} = \frac{h}{\sin^2 \phi}$$

Substitution into the first differential equation yields a single nonlinear differential equation for the motion of the particle:

$$\ddot{\phi} = \frac{h^2 \cos \phi}{\sin^3 \phi} + \frac{g}{R} \sin \phi$$

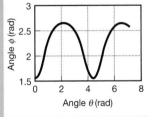

We can discover a lot about the motion of the particle by examining the expression for the angular velocity in the θ-direction. If the particle were to go to the bottom of the bowl, or $\phi = 180°$, the angular velocity would become infinite. Therefore, if h is not zero, the particle will move around the inside of the bowl in a spiraling motion. If h is zero, we have the differential equation for a pendulum, as we will show in Chapter 2.

  Therefore, the two coupled, nonlinear differential equations have been reduced to a single nonlinear differential equation that must be solved numerically. A graph of the motion giving the angle ϕ versus the angle θ may be drawn after a numerical solution is obtained. (Details of the numerical solution are shown in the Computational Supplement.) Notice that the particle oscillates by sliding up and down the bowl in a circular manner, without ever reaching the bottom. This motion appears to be perpetual, but in fact, some form of friction that damps out the motion is always present.

$$i = 0 \ldots 9000 \qquad \Delta t = 0.0001$$

Define the second derivatives of the coordinates:

$$dd\theta(d\theta, d\phi, \phi) = -2 \cdot d\theta \cdot d\phi \cdot \frac{\cos (\phi)}{\sin (\phi)}$$

$$dd\phi(d\theta, d\phi, \phi) = 49.05 \cdot \sin (\phi) + d\theta^2 \sin (\phi) \cdot \cos (\phi)$$

Initial conditions are

$$
\begin{bmatrix} d\theta_0 \\ \theta_0 \\ d\phi_0 \\ \phi_0 \end{bmatrix} = \begin{bmatrix} 5 \\ 0 \\ 0 \\ \dfrac{\pi}{2} \end{bmatrix}
$$

The Euler numerical solutions for the first 9 seconds is

$$
\begin{bmatrix} d\theta_{i+1} \\ \theta_{i+1} \\ d\phi_{i+1} \\ \phi_{i+1} \end{bmatrix} = \begin{bmatrix} d\theta_i + dd\theta(d\theta_i, d\phi_i, \phi_i) \cdot \Delta t \\ \theta_i + d\theta_i \cdot \Delta t \\ d\phi_i + dd\phi(d\theta_i, d\phi_i, \phi_i) \cdot \Delta t \\ \phi_i + d\phi_i \cdot \Delta t \end{bmatrix}
$$

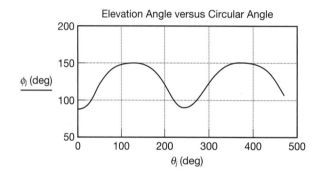

Elevation Angle versus Circular Angle

ϕ_i (deg) vs θ_i (deg)

Problems

1.119 The cylindrical coordinates of a particle are given by

$$r(t) = 1.5 \text{ m}$$
$$\dot{\theta}(t) = \pi \text{ rad/s}$$
$$z(t) = 0.5 \cos(2\pi t) \text{ m}$$

Determine the velocity and acceleration of the particle. Describe the trajectory of the particle in space.

1.120 In Problem 1.119, determine the total distance traveled in 2 s.

1.121 A particle spirals down a conical surface, as shown in Figure P1.121. If the cylindrical coordinates of the particle at any time are given by

$$r(t) = R - \frac{z}{h}R$$
$$\theta(t) = 2\pi t$$
$$z(t) = h - 0.1ht$$

then determine the velocity and acceleration of the particle for any time. How long does it take the particle to reach the bottom of the cone? When it reaches the bottom, how many times has it circled the cone?

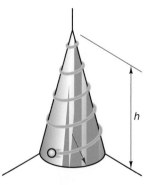

Figure P1.121

1.122 In Problem 1.121, if $R = 3$ m and $h = 5$ m, determine the total distance the particle travels from the top of the cone to the base.

1.123 The acceleration of a particle is given by

$$\mathbf{a}(t) = 3\hat{\mathbf{e}}_r + 0.1\hat{\mathbf{e}}_\theta - 2\hat{\mathbf{e}}_z \text{ m/s}^2$$

If the initial velocity of the particle is zero, and the initial position is $r = 1$, $\theta = 0$, and $z = 0$, use numerical methods to determine the trajectory of the particle during the first 3 s.

1.124 The spherical coordinates of a particle are given as

$$R(t) = 0.3 + 0.1t^2 \text{ m}$$
$$\theta(t) = \pi \sin(\pi t) \text{ rad}$$
$$\phi(t) = \frac{\pi}{2}(te^{-t}) \text{ rad}$$

Determine the velocity and acceleration of the particle.

1.125 In Sample Problem 1.24, determine the motion of the particle if the initial velocity in the circumferential direction is $R\dot{\theta} = 600$ mm/s and plot the trajectory.

1.126 In Sample Problem 1.24, show that the particle oscillates in the $R\phi$-plane if the initial velocity in the circumferential direction is zero. Use numerical methods to solve the nonlinear differential equation.

1.127 In Problem 1.126, assume small-angle oscillations; that is, $\phi = (\pi + \beta)$, where β is a small angle. The initial position of the particle is displaced from the bottom of the bowl by a small angle. Derive the linear differential equation of motion and analytically solve this equation.

1.128 The acceleration vector of a particle, in spherical coordinates, is

$$\mathbf{a} = 12\hat{\mathbf{e}}_R + 4\hat{\mathbf{e}}_\phi + 5\hat{\mathbf{e}}_\theta \text{ ft/s}^2$$

Determine the trajectory of the particle during the first second of motion if the particle starts from a point with coordinates $(2, \pi/2, 0)$ with an initial velocity of

$$\mathbf{v} = 2\hat{\mathbf{e}}_R + \hat{\mathbf{e}}_\phi - 4\hat{\mathbf{e}}_\theta \text{ ft/s}$$

1.10 RELATIVE RECTILINEAR MOTION OF SEVERAL PARTICLES

The motion of several particles along a straight line or parallel straight lines can be described in terms of their absolute movement in an inertial coordinate system or in terms of their movement relative to each other. In Figure 1.22, the relative position of B with respect to A is denoted as $x_{B/A}$ and defined as

$$x_{B/A} = x_B - x_A \tag{1.120}$$

where B/A is read as "B relative to A." The relative position of A with respect to B is the negative of the relative position of B with respect to A:

$$x_{A/B} = x_A - x_B = -x_{B/A} \tag{1.121}$$

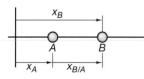

Figure 1.22

Examining Figure 1.22, we can see that this relationship is obvious. B is in front of A, and A is behind B, or in the negative direction from it, but in either case, the magnitude of the separation between the two points is the same. The velocity of B relative to A is obtained by differentiation of the relative position relationship, yielding

$$v_{B/A} = v_B - v_A \tag{1.122}$$

In a similar manner, the acceleration of B relative to A can be obtained by differentiation of the relative velocity relationship, resulting in

$$a_{B/A} = a_B - a_A \tag{1.123}$$

These definitions can also be viewed in a different manner. For example, if the absolute position, velocity, and acceleration of particle A are known, and the position, velocity, and acceleration of B relative to A are also known, then the absolute position, velocity, and acceleration of B can be obtained from Eqs. (1.121) through (1.123). In this situation, it is convenient to write the preceding relationships in the following form:

$$x_B = x_A + x_{B/A}$$
$$v_B = v_A + v_{B/A} \tag{1.124}$$
$$a_B = a_A + a_{B/A}$$

In many engineering applications, the relative motion between two or more particles is more important than the absolute motion of either particle. For example, an air controller is concerned about the relative motion of two airplanes under his control in order to maintain proper separation of the planes and to prevent collisions. Similarly, most spaceflights are based upon the motion of the spacecraft relative to the planets during flight. And although football normally is not thought of as relevant to engineering problems, the quarterback of a football team bases his passes on the motion of the ball relative to that of the receiver.

Sample Problem 1.25	Two cars begin moving from rest at the same time, but car A is 100 ft behind car B at the start. Car A maintains a constant acceleration of 8 ft/s², and car B has a constant acceleration of 6 ft/s². How much time is required for car A to overtake car B, and how much distance will car A have traveled when it overtakes car B?

Solution   The relative acceleration is

$$a_{B/A} = a_B - a_A = 6 - 8 = -2 \text{ ft/s}^2$$

Integrating this equation yields the relative velocity as

$$v_{B/A} = -2t + v_{B/A_0} = -2t \text{ ft/s}$$

where v_{B/A_0} is the initial relative velocity (zero in this case). Integrating again gives the relative position, that is,

$$x_{B/A} = -t^2 + x_{B/A_0} = (-t^2 + 100) \text{ ft}$$

where x_{B/A_0} is the initial relative position (100 ft). The relative position is zero when A overtakes B. Setting the relative position equal to zero yields

$$t = 10 \text{ s}$$

The distance that A has traveled in 10 s is given by

$$x_A = a_A t^2/2 = 8(100)/2 = 400 \text{ ft}$$

Note that the position of car A is behind that of car B until a time of 10 s, when the positions become equal. For times greater than 10 s, the position of A is in front of that of B.

Sample Problem 1.26	A surfer paddles her board out against the incoming waves; the constant velocity of her board relative to the water is $v_{B/W}$. The velocity of the waves is sinusoidal with time and decreases with distance from the beach according to the relationship

$$v_W = -Ve^{-cx}\sin(\omega t)$$

where V m/s is the maximum wave velocity, c m⁻¹ is the decay of the wave velocity, and ω s⁻¹ is the frequency of the wave. Determine the absolute velocity of the surfer and the time required for her to paddle out 150 m from the beach, in terms of the quantities v, c, and ω.

Solution   The absolute velocity of the surfer is equal to the velocity of the water plus the velocity of the board relative to the water. This relation can be written as

$$v_s = v_W + v_{B/W}$$

We will select x to be positive as one moves out from the beach, and therefore, the absolute velocity of the surfer is

$$\frac{dx}{dt} = -Ve^{-cx}\sin(\omega t) + v_{B/W}$$

This is a first-order nonlinear differential equation of the form

$$\frac{dx}{dt} = F(x,t)$$

Since the equation is nonlinear, we will solve it using numerical integration. (See Computational Supplement.) For the values

$$V = 6 \text{ m/s} \quad v_{B/W} = 1.5 \text{ m/s} \quad c = 1/100 \text{ m}^{-1} \quad \omega = \pi/4 \text{ s}^{-1} \quad \Delta t = 0.02 \text{ s}$$

The solution is found by

$$x_0 = 10 \text{ m}$$

Numerical integration for the subsequent values of x may be performed with

$$x_{n+1} = x_n + [-Ve^{-cx_n}\sin(\omega n\Delta t) + v_{B/W}]\Delta t$$

The displacement in meters is plotted against time in seconds from the numerical solution. From the numerical solution plotted, it requires 100 s for the surfer to paddle the 150 m from the beach when she starts paddling 10 m from the water's edge.

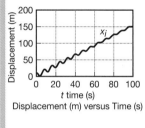

Displacement (m) versus Time (s)

Problems

1.129 A particle B moves relative to a particle A with a displacement of $x_{B/A} = 3t^2 + 2t + 3$ m. Both particles are moving along a straight line, and particle A is known to have a displacement of $x_A = 2t + 6$ m relative to a fixed origin. Calculate the position, velocity, and acceleration of B relative to the fixed origin, and determine the acceleration of B relative to A.

1.130 Two particles start from the same fixed reference at the same time and move according to the displacement $x_A = (3t + 2)$ m and $x_B = (6t^2 + 2)$ m, where t is measured in seconds. Calculate the velocity and acceleration of each particle and the time when they reach the same position. Also, calculate the time when they have the same velocity. Do they ever have the same acceleration?

1.131 Automobile B is moving at a constant speed of 15 m/s, as shown in Figure P1.131. When car B crosses the 100-m mark, car A starts out from the reference point, moving such that its position is given by $x_A = t^2$ m, until it reaches and remains at a constant velocity of

20 m/s. At what time does car A pass car B, and how far away from the origin do they pass each other?

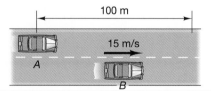

Figure P1.131

1.132 Particles A and B leave at the same time. Particle A moves with a displacement $x_A = 5 + 2t$ m, and particle B moves with a displacement $x_B = 12 + 18t - 4.9t^2$ m along the same line. Particle B moves out for a specific distance, reverses motion, and collides with particle A. Calculate the time when they collide and the relative velocity $v_{B/A}$ when they collide.

1.133 Two cars approach each other, each traveling at 60 mph on a head-on course. When they are 300 ft apart, the driver of each car slams on its brakes, producing a deceleration of 16 ft/s². Do the cars hit each other and, if so, what is their relative velocity at the time of impact, and what is the position of impact? Note that the relative velocity determines the resulting damage.

1.134 A police car A is moving at a constant speed of 45 mph when the policeman spots a car B 500 ft ahead moving at 58 mph. The police car accelerates at 4 ft/s², while at the same time, car B decelerates at 2 ft/s², with the hopes of avoiding a ticket. Determine the time and position at which the police car catches up with car B. What is the velocity of each vehicle at that time?

1.135 Two particles move relative to one another with an acceleration $a_{B/A} = 3$ m/s², and at $t = 0$, $v_{B/A} = 0$ and $x_{B/A} = 10$ m. If particle A moves with a displacement $x_A(t) = 3t^2 + 1$ m, determine the absolute acceleration of particle B and an expression for the position of particle B at any time.

1.136 A Boeing 747 leaves New York and flies towards London at a constant speed of 575 mph. A Concorde with a constant cruising speed of 1336 mph leaves New York 3 hours later and flies on the same straight flight path (but 20,000 ft higher). How long will it be before the Concorde overtakes the 747?

1.137 A pendulum B is released from rest and swings such that its angular position at any time is given by $\theta_B(t) = \theta_0 \cos\left(\sqrt{g/lt}\right)$ rad. Particle A is released from rest at a point 10 m from the vertical position of B and moves with a constant acceleration. Determine the

acceleration of A such that the particles will collide when B reaches the bottom of its swing. (See Figure P1.137.)

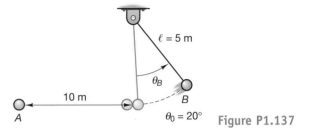

Figure P1.137

1.138 In Problem 1.137, determine the velocity of B relative to A at the time of collision.

1.139 A particle A is dropped from rest into a viscous medium at the same instant that a particle B 100 m below A is projected upward at an initial velocity of 30 m/s. Both particles are under the influence of gravitational attraction, and the fluid resistance is linearly proportional to the velocity, with a proportionality constant of 0.1N s/m. Determine the position of the particles when they collide and the relative velocity between the particles at the point of collision.

1.140 A swimmer swims against a river current that increases as the swimmer moves upstream. The velocity of the current is given by $v_c = 0.1 + 0.01x$ m/s, where x is measured from the point the swimmer starts her swim. If the swimmer can maintain a constant velocity of 1.5 m/s relative to the water, determine the distance the swimmer travels in the first minute of swimming.

1.141 In Problem 1.140, determine the time required for the swimmer to swim 100 m upstream.

1.11   GENERAL RELATIVE MOTION BETWEEN PARTICLES

Consider the movement of two particles in curvilinear motion. As in all previous cases, a coordinate system will be established and the particles located by position vectors. The position of one particle relative to the other is described by a relative position vector, as shown in Figure 1.23. The relative position for rectilinear motion was defined in Section 1.10 as

$$x_{B/A} = x_B - x_A$$

We will generalize that definition to

$$\mathbf{r}_{B/A} = \mathbf{r}_B - \mathbf{r}_A \qquad (1.125)$$

or

$$\mathbf{r}_B = \mathbf{r}_A + \mathbf{r}_{B/A}$$

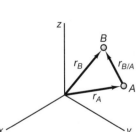

Figure 1.23

It can be seen that $\mathbf{r}_{A/B} = -\mathbf{r}_{B/A}$, which describes where A is relative to B. Differentiation of the relative position vector leads to the relative velocity and relative acceleration:

$$\mathbf{v}_{B/A} = \mathbf{v}_B - \mathbf{v}_A$$

or

$$\mathbf{v}_B = \mathbf{v}_A + \mathbf{v}_{B/A} \qquad (1.126)$$

and

$$\mathbf{a}_{B/A} = \mathbf{a}_B - \mathbf{a}_A$$

or

$$\mathbf{a}_B = \mathbf{a}_A + \mathbf{a}_{B/A}$$

When working with these relative position, velocity, or acceleration equations, we find that the use of components and their respective unit vectors often simplifies the work and reduces the chance of making errors.

Sample Problem 1.27

An airplane attempts to drop a paint marker onto a truck that is moving along the ground at a constant speed of 60 mph (88 ft/s) on a level road straight north. The plane is flying in a northwest direction, making its flight path 45° to the path of the truck. The plane is at an altitude of 2000 ft and flying level at a speed of 200 mph (293 ft/s). At what relative initial position should the plane be in order to ensure that the marker lands on the truck?

Solution Establish a coordinate system with x in the direction of north, y to the west, and z vertical, with the origin on the truck at the instant of release of the crate. The initial position of the truck will be taken to be the origin of the coordinate system. The acceleration of the marker after it is dropped is due to gravitational attraction, and the truck has zero acceleration. The absolute acceleration, velocity, and position of the marker and the truck are as follows:

Marker	Truck
$\mathbf{a}_p = -32.2\,\hat{\mathbf{k}}$	$\mathbf{a}_t = 0$
$\mathbf{v}_p = 293\cos 45\hat{\mathbf{i}} + 293\sin 45\hat{\mathbf{j}} - 32.2t\hat{\mathbf{k}}$	$\mathbf{v}_t = 88\hat{\mathbf{i}}$
$\mathbf{r}_p = (293\cos 45t + x_p)\hat{\mathbf{i}} + (293\sin 45t + y_p)\hat{\mathbf{j}} + (-16.1t^2 + 2000)\hat{\mathbf{k}}$	$\mathbf{r}_t = 88t\hat{\mathbf{i}}$

To determine the time taken for the marker to land on the truck, and to determine the initial position of the plane, set the two position vectors equal, or set the relative position vector equal to zero. The three scalar equations are

$$293\cos 45t + x_p = 88t \qquad \text{for the } x\text{-component}$$

$$293\sin 45t + y_p = 0 \qquad \text{for the } y\text{-component}$$

$$-16.1t^2 + 2000 = 0 \qquad \text{for the } z\text{-component}$$

The three nonlinear equations can be solved to yield

$$t = 11.15 \text{ s} \qquad x_p = -1328 \text{ ft} \qquad y_p = -2310 \text{ ft}$$

Sample Problem 1.28	A 30-ft-wide river flows south at a constant velocity v_r of 6 ft/s. If a boater wishes to take a route across the river and land at a point 40 ft upriver, what angle must he direct his boat if the boat's speed in calm water is 10 ft/s?

Solution   The absolute velocity of the boat can be expressed in terms of the absolute velocity of the water plus the velocity of the boat relative to the water:

$$\mathbf{v}_b = \mathbf{v}_w + \mathbf{v}_{b/w}$$

We do not know the magnitude of the absolute velocity of the boat, but from the figure, we do know that the direction of this velocity vector is

$$\hat{\mathbf{e}}_b = 0.6\hat{\mathbf{i}} + 0.8\hat{\mathbf{j}}$$

We also know that the velocity vector of the water is

$$\mathbf{v}_w = -6\hat{\mathbf{j}}$$

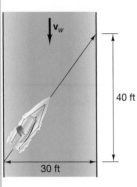

The magnitude of the velocity vector of the boat relative to the water is 10 ft/s, and we will assume that it is directed at some angle θ to the x-axis. We can then write the relative velocity equation as

$$v_b(0.6\hat{\mathbf{i}} + 0.8\hat{\mathbf{j}}) = -6\hat{\mathbf{j}} + 10(\cos\theta\hat{\mathbf{i}} + \sin\theta\hat{\mathbf{j}})$$

Equating the components yields

$$0.6v_b = 10\cos\theta$$
$$0.8v_b = -6 + 10\sin\theta$$

The solution is now in the form of two nonlinear simultaneous algebraic equations in the two unknowns v_b and θ. The first equation can be used to eliminate v_b from the second equation, resulting in a single transcendental equation for the angle θ. This equation may be solved by squaring both sides and eliminating the cosine term by using the trigonometric identity, or by the use of computational software. The resulting values of the two unknowns are

$$v_b = 4.53 \text{ ft/s} \quad \theta = 74.26°$$

The time required to travel 50 ft from one shore to the other is

$$t = \frac{d}{v_b} = \frac{50}{4.53} = 11.04 \text{ s}$$

Alternatively, these two equations could be solved by means of a symbolic code.

Sample Problem 1.29	In Sample Problem 1.28, we examined the effect of the velocity of a river on the velocity of a boat crossing the river. This type of problem occurs in many navigational situations involving aircraft and boats when the velocity of the air or water significantly affects the absolute velocity of the craft. However, the velocity of the air or water is usually not constant, increasing the complexity of the relative motion analysis.

Suppose a boat wishes to directly cross a 400-ft-wide river to the opposite bank, but the velocity of water flowing in the river channel has a parabolic profile in actual situations. The velocity of the boat in calm water is 10 ft/s. Determine the constant angle at which the boat should be pointed in order to directly cross the channel, and calculate the time that the boat takes to go across the river. Suppose the water flows in the channel as shown in the accompanying figure. If the origin of the coordinate system is as shown, then the velocity of the water is given by $\mathbf{v}_W = -v_W[1 - (x^2/d^2)]\hat{\mathbf{j}}$, where $v_W = 5$ ft/s and $d = 200$ ft.

Solution The relative velocity equation is

$$\mathbf{v}_B = \mathbf{v}_W + \mathbf{v}_{B/W}$$

The vectors are

$$\mathbf{v}_B = \frac{dx}{dt}\hat{\mathbf{i}} + \frac{dy}{dt}\hat{\mathbf{j}}$$

$$\mathbf{v}_W = -v_W\left[1 - \frac{x^2}{d^2}\right]\hat{\mathbf{j}}$$

$$\mathbf{v}_{B/W} = v_{B/W}(\cos\theta\,\hat{\mathbf{i}} + \sin\theta\,\hat{\mathbf{j}})$$

Substituting into the relative velocity equation yields

$$\mathbf{v}_B = \frac{dx}{dt}\hat{\mathbf{i}} + \frac{dy}{dt}\hat{\mathbf{j}} = -v_W\left[1 - \left(\frac{x}{d}\right)^2\right]\hat{\mathbf{j}} + v_{B/W}(\cos\theta\,\hat{\mathbf{i}} + \sin\theta\,\hat{\mathbf{j}})$$

Separating the vector equation into its scalar components gives

$$\frac{dx}{dt} = v_{B/W}\cos\theta$$

$$\frac{dy}{dt} = -v_W\left[1 - \left(\frac{x}{d}\right)^2\right] + v_{B/W}\sin\theta$$

We now have two coupled nonlinear first-order differential equations. The first equation can be solved by integration. Noting that $x(0) = -d$, we get

$$x(t) = v_{B/W}\cos\theta t - d$$

This result can be substituted into the second differential equation, decoupling the two equations. The second equation becomes

$$\frac{dy}{dt} = -v_W\left[1 - \left(\frac{v_{B/W}[\cos\theta]t}{d} - 1\right)^2\right] + v_{B/W}\sin\theta$$

$$\frac{dy}{dt} = v_W\left[\left(\frac{v_{B/W}\cos\theta}{d}\right)^2 t^2 - 2\left(\frac{v_{B/W}\cos\theta}{d}\right)t\right] + v_{B/W}\sin\theta$$

Since the right-hand side of the differential equation is a function only of t, it may be integrated with the initial condition $y(0) = 0$, resulting in

$$y(t) = v_W\left[\left(\frac{v_{B/W}\cos\theta}{d}\right)^2\frac{t^3}{3} - 2\left(\frac{v_{B/W}\cos\theta}{d}\right)\frac{t^2}{2}\right] + v_{B/W}[\sin\theta]t$$

where the constant of integration is zero, due to the initial condition. We still have not determined the value of the boat angle θ, but this angle may be determined such that when $x(t_f) = d$, the corresponding value for y is $y(t_f) = 0$. The time taken to reach the opposite bank is designated as $t = t_f$. Solving for this time from the $x(t)$ function yields

$$t_f = \frac{2d}{v_{B/W}\cos\theta}$$

Equating $y(t)$ to zero at this time gives

$$0 = v_{B/W}\sin\theta + \frac{4v_W}{3} - 2v_W$$

$$\sin\theta = \frac{2v_W}{3v_{B/W}}$$

For the specified values $d = 200$ ft, $v_W = 5$ ft/s, and $v_{B/W} = 10$ ft/s, the angle θ and the time to cross the river t_f are

$$\theta = 19.47° \qquad t_f = 42.4 \text{ s}$$

The actual motion of the boat is shown from an aerial view in the accompanying figure by plotting $y(t)$ versus $x(t)$ from the time $t = 0$ to $t = t_f$.

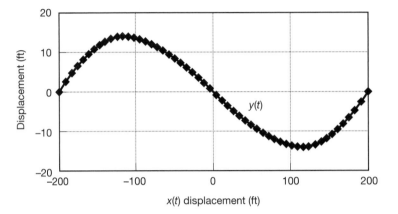

Note that the boat initially goes upriver when the current is slow, then is pulled to an online position at midstream, and then is pulled downstream until the river velocity decreases at the end. The direction of the boat's travel relative to the water is always equal to the angle θ. The other strategy that the boater could use would be to constantly correct the orientation of the boat with respect to the water, but this would be very difficult and, indeed, impossible in conditions of poor visibility.

1.11.1 NAVIGATION USING RELATIVE VELOCITY

Airline pilots have to plan relative velocity trajectories on long flight plans because of variations in the jet stream. For example, a jet-stream trajectory for a day in March is shown in Figure 1.24.

As the jet stream crosses the 3000-mile-wide North American continent, it dips to the south and then swings north. This pattern can be broken into three parts—that is, the wind

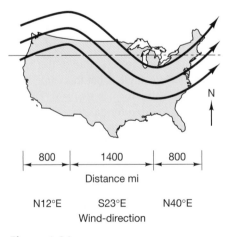

Figure 1.24

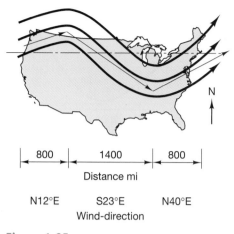

800 | 1400 | 800

Distance mi

N12°E   S23°E   N40°E

Wind-direction

Figure 1.25

is N12°E for the first 800 miles, S23°E for the next 1400 miles, and, finally, N40°E for the last 800 miles. This approximation is shown in Figure 1.25.

A pilot flying from San Francisco to Chicago (2000 miles due east) can now chart the bearing angle of the plane for the first 800 miles and the second 1200 miles using relative velocity. The absolute velocity of the plane is given by the relationship

$$\mathbf{v}_p = \mathbf{v}_W + \mathbf{v}_{P/W} \tag{1.127}$$

where $\mathbf{v}_p$ is the absolute velocity of the plane, $\mathbf{v}_w$ is the absolute velocity of the wind, and $\mathbf{v}_{P/W}$ is the velocity of the plane relative to the wind, or the airspeed of the plane. The bearing angle (the angle of the relative velocity vector) can be determined from this relationship. If the airspeed of the plane is 450 mph and the velocity of the jet stream is 80 mph, Eq. (1.127) can be written as follows:

$$v_p\hat{\mathbf{i}} = 80(\cos 12°\hat{\mathbf{i}} + \sin 12°\hat{\mathbf{j}}) + 450(\cos\theta\hat{\mathbf{i}} + \sin\theta\hat{\mathbf{j}}) \tag{1.128}$$

Separating Eq. (1.128) into two scalar equations, we may determine the velocity of the plane and the bearing angle from

$$v_P = 80\cos 12° + 450\cos\theta$$

$$0 = 80\sin 12° + 450\sin\theta$$

Solving the two equations for the two unknowns yields

$$\theta = -2° \qquad v_P = 528 \text{ mph}$$

The time taken to complete this leg of the journey is

$$t = x/v = 800/528 = 1.52 \text{ hr} = 1 \text{ hr, 30 min}$$

The bearing will have to change for the last 1200 miles to account for the change in the direction of the jet stream. Eq. (1.128) becomes

$$v_p\hat{\mathbf{i}} = 80(\cos 23°\hat{\mathbf{i}} - \sin 23°\hat{\mathbf{j}}) + 450(\cos\theta\hat{\mathbf{i}} + \sin\theta\hat{\mathbf{j}})$$

The new bearing angle and velocity of the plane are

$$\theta = 4° \qquad v_P = 523 \text{ mi/hr}$$

The flight time for the second leg of the trip is $t = 2.30 = 2$ hr, 18 min. Therefore, the total flight time is 3 hr, 48 min, not including the takeoff and landing times.

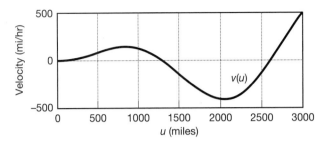

Figure 1.26   Jet stream pattern from west coast to east coast.

In fact, modern computers can compute flight patterns much more accurately than this. For example, the curve the jet stream makes as it crosses the United States can be approximated by the equation

$$y(x) = \Delta S\left[\frac{x}{X_1}\sin\left(\frac{\pi x}{X_2}\right)\right] \tag{1.129}$$

where ΔS is the southern deviation of the jet stream and the parameters X_1 and X_2 can be used to approximate the flow pattern in the eastern direction. For the wind pattern shown in Figure 1.24, ΔS is 400 miles, X_1 is 2000 miles and X_2 is 1300 miles. The jet-stream pattern from the west coast to the east coast (3000 miles in total distance) would then be approximated as shown in Figure 1.26. The slope of the curve is equal to the tangent of the angle that the jet stream makes with the horizontal west–east latitude. Differentiating the equation for the curve to determine the slope yields the angle that the jet stream makes, with NE being a positive angle $\theta(x)$. Thus,

$$\theta(x) = \tan^{-1}\left[\Delta S\left(\frac{1}{X_1}\sin\left(\frac{\pi x}{X_2}\right) + \frac{\pi x}{X_1 X_2}\cos\left(\frac{\pi x}{X_2}\right)\right)\right] \tag{1.130}$$

Now suppose a pilot wishes to fly his plane from San Franciso to Chicago, a 2000-mile flight due east. He wishes to set a constant northeastern bearing of $\beta = 2°$(north β degrees of east) for the entire flight and arrive on course in Chicago. This bearing would take the flight 70 miles north of Chicago, without consideration of the jet stream. The flight is to take place on the March day with the jet-stream pattern shown in Figure 1.24. If the constant airspeed of the aircraft is 450 mph and the velocity of the jet stream is 100 mph, we wish to determine the time of the flight and the absolute flight pattern relative to the surface of the earth. The relative velocity equation is

$$\mathbf{v}_p = \mathbf{v}_W + \mathbf{v}_{P/W} \tag{1.131}$$

where each of these velocity vectors may be written as

$$\mathbf{v}_p = \frac{dx}{dt}\hat{\mathbf{i}} + \frac{dy}{dt}\hat{\mathbf{j}}$$
$$\mathbf{v}_W = v_W(\cos\theta(x)\hat{\mathbf{i}} + \sin\theta(x)\hat{\mathbf{j}}) \tag{1.132}$$
$$\mathbf{v}_{P/W} = v_{P/W}(\cos\beta\hat{\mathbf{i}} + \sin\beta\hat{\mathbf{j}})$$

Using Eq. (1.130) and the values for the wind and airspeed of the aircraft, we get the two scalar differential equations for the actual flight path of the airplane:

$$\frac{dx}{dt} = v_W\cos\theta(x) + v_{P/W}\cos\beta$$
$$\frac{dy}{dt} = v_W\sin\theta(x) + v_{P/W}\sin\beta \tag{1.133}$$

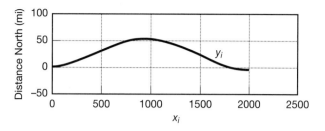

Figure 1.27 Flight pattern (y–north; x–east).

These two nonlinear differential equations are coupled and must be solved numerically. The time for the flight is 3 hrs and 45 min, not accounting for takeoff and landing. The flight pattern is shown in Figure 1.27. The maximum deviation from a flight pattern headed due east is 50 miles to the north.

Actual flights would receive midcourse corrections with the aid of ground control, and the flight would be flown in sections. For example, the airplane would fly the first 650 miles at a bearing southeast of $-2.5°$, the next 650 miles at a bearing northeast of $2.0°$, and the final 700 miles at a southeast bearing of $-4.5°$. The actual flight pattern in the three stages is shown in Figure 1.28.

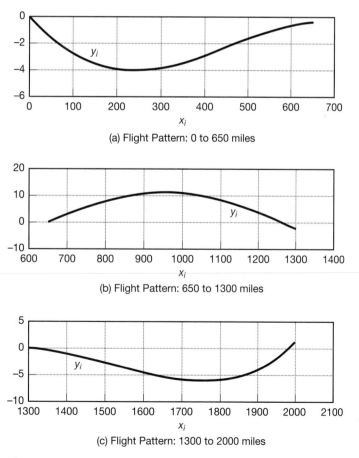

(a) Flight Pattern: 0 to 650 miles

(b) Flight Pattern: 650 to 1300 miles

(c) Flight Pattern: 1300 to 2000 miles

Figure 1.28

The maximum deviation from a flight pattern headed due east is 10 miles to the north, and the total air time is the same: 3 h and 45 min. Although the time of the flight did not change, the plane remained on a more directly east flight pattern. Airline pilots must also make flight corrections for weather conditions and other air traffic. (Details of the numerical solution of the equations for this problem are in the Computational Supplement.)

Sample Problem 1.30

A swimmer at the beach is caught in a riptide (a narrow, riverlike current fed by the long shore current and sets of waves). Although the riptide will not pull the swimmer under, she will become exhausted trying to swim directly to shore. If the riptide is 50 m wide and flows north at a rate of 0.5 m/s, and the velocity of the swimmer with respect to the water is 0.3 m/s, show that she can not swim directly west to the beach. Determine the minimum southwest angle at which she must swim in order to escape the riptide and reach the shore. How far north of her current position will she be when she reaches shore?

Solution Establish a coordinate system with the x-axis to the east and the y-axis to the north. Let θ be the angle to the southeast that the swimmer swims relative to the riptide. The velocity of the riptide and the relative velocity of the swimmer are

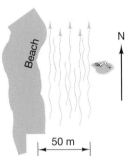

$$\mathbf{v}_{RT} = 0.5\hat{\mathbf{j}}$$

$$\mathbf{v}_{S/RT} = 0.3[-\cos\theta\hat{\mathbf{i}} - \sin\theta\hat{\mathbf{j}}]$$

The absolute velocity of the swimmer is

$$\mathbf{v}_S = \mathbf{v}_{RT} + \mathbf{v}_{S/RT} = -0.3\cos\theta\hat{\mathbf{i}} + [0.5 - 0.3\sin\theta]\hat{\mathbf{j}}$$

If the swimmer is to have an absolute velocity in direct line to the shore, her absolute velocity in the y-direction (north) would have to be zero. Therefore, the following relationships would have to hold

$$v_S = 0.3\cos\theta$$

$$0 = 0.5 - 0.3\sin\theta$$

The latter expression yields $\sin\theta = 1.66$, but the sine function is bounded between -1 and $+1$, so the expression cannot be true.

It is useful to make a vector diagram to understand why this situation arises and to conceptualize the minimum angle θ. Let β be the angle that the absolute velocity vector of the swimmer makes with a line due west.

The scalar components of the relative vector equation are

$$-v_S\cos\beta = -v_{S/RT}\cos\theta$$

$$v_S\sin\beta = v_{RT} - v_{S/RT}\sin\theta$$

We now have two nonlinear algebraic equations in three unknowns: v_S, θ, and β. Examining the vector diagram shows that the minimum value of β occurs when $\mathbf{v}_S$ is tangent to the circle of radius $v_{S/RT}$. Therefore, $\theta + \beta = 90°$. This expression is the third relationship to determine the three unknowns. Solving the three equations yields

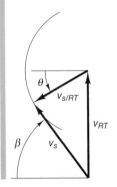

$$v_S = 0.4 \text{ m/s} \qquad \theta = 36.898° \qquad \beta = 53.112°$$

In this case, the right triangle in the vector diagram could be easily analyzed. The distance that the swimmer travels to the north is

$$y = 50 \tan \beta = 66.6 \text{ m}$$

The time required for her to swim to shore is

$$t = \frac{x}{v_x} = \frac{50}{0.4 \sin \beta} = 156 \text{ s}$$

Problems

1.142 Particle A moves at a constant velocity $\mathbf{v}_A = 3\hat{\mathbf{j}}$ m/s, and particle B starts from rest and accelerates at a constant rate $\mathbf{a}_B = 2[\cos (30°)\hat{\mathbf{i}} + \sin (30°)\hat{\mathbf{j}}]$ m/s². If, at the time B starts to move, the positions of A and B are $\mathbf{r}_A = 0$ and $\mathbf{r}_B = -4\hat{\mathbf{i}} + \hat{\mathbf{j}}$ m, determine whether the two particles collide. What is the minimum distance between the two particles, and at what time are the particles at that distance?

1.143 An airplane has an airspeed relative to the air of 200 mph. The pilot wants to fly a course straight east, and the wind is blowing at 50 mph in a N45°E direction. Determine the orientation of the plane so that it flies due east, and determine it absolute velocity.

1.144 A football is thrown downfield from a point 10 yds behind the line of scrimmage released at angle of 30 degrees with the horizontal and at a height of 6 ft. If a flanker leaves the line of scrimmage 2 s before the ball is thrown, 20 yds to the right of the quarterback and runs straight down field at a constant velocity of 20 ft/s, determine the velocity and the angle to the centerline of the field that the ball must be thrown for the receiver to make the catch with his hands 8 ft above the surface of the field. How much yardage is gained on the play?

1.145 A swimmer wants to swim directly across a 200-ft channel in which the water is flowing at 5 ft/s. If the swimmer's velocity relative to the water is 6 ft/s, determine the angle upstream that the swimmer should swim and the time required for the swimmer to complete the swim.

1.146 If a pilot is to fly a nonstop route from Seattle to Miami when the 50-mph jetstream flowing west to east has the profile shown in Figure P1.146, lay out a route

and the flight heading during each leg of the route. The airspeed of the plane is 300 mph.

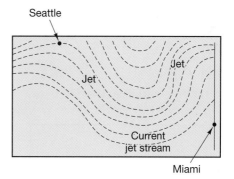

Figure P1.146

1.147 Determine the flight plan for the pilot in Problem 1.146 if the 100-mph jetstream pattern is as shown in Figure P1.147.

Figure P1.47

1.148 Determine the flight pattern in Problem 1.146 for a plane with a maximum airspeed of 200 mph to fly from New York City to San Franscisco.

1.149 Determine the flight pattern for the jetstream shown in Figure P1.147 for a flight straight south for 600 miles from Chicago to Memphis if the airspeed of the plane is 300 mph.

1.150 In Sample Problem 1.29, determine the angle at which the boat should be pointed so that it will reach a point 5 ft upstream, starting from a point directly across the river.

1.151 In Sample Problem 1.29, determine the angle at which the boat should be pointed in order for it to directly cross the river if the maximum velocity of the river is only 2 ft/s.

1.152 Solve Sample Problem 1.30 if the velocity of the riptide is 0.7 m/s.

1.153 A boy and girl start to run from the same spot. The boy runs in a zigzag manner such that his position at any time is given by $\mathbf{r}_B = (2 \sin \pi t)\hat{\mathbf{i}} + 5t\hat{\mathbf{j}}$, and the girl runs in a straight line such that her position is given by $\mathbf{r}_G = 4t(\cos 30°\hat{\mathbf{i}} + \sin 30°\hat{\mathbf{j}})$, as illustrated in Figure P1.153. Determine the velocity and acceleration of the girl relative to the boy. All units are meters.

$$t = 0, 0.1 .. 3$$

$$x_B(t) = 2 \cdot \sin(\pi \cdot t) \qquad y_B(t) = 5 \cdot t$$

$$x_G(t) = 4 \cdot t \cdot \cos(30°) \qquad y_G(t) = 4 \cdot t \cdot \sin(30°)$$

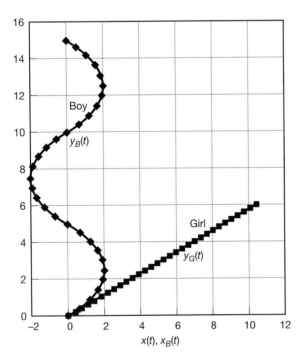

Figure P1.153

1.154 In Problem 1.153, determine the distance of the boy from the girl as a function of time, and plot the function.

1.155 In Problem 1.153, determine the velocity and acceleration of the boy relative to the girl if the position of the girl is given by $\mathbf{r}_G = 4t(\cos 30°\hat{\mathbf{i}} + \sin 30° \sin(\pi/2t)\hat{\mathbf{j}})$, as shown in Figure P1.155.

$$t = 0, 0.1 .. 3$$

$$x_B(t) = 2 \cdot \sin(\pi \cdot t) \qquad y_B(t) = 5 \cdot t$$

$$x_G(t) = 4 \cdot t \cdot \cos(30°) \qquad y_G(t) = 4 \cdot t \cdot \sin(30°) \sin\left(\frac{\pi}{2} \cdot t\right)$$

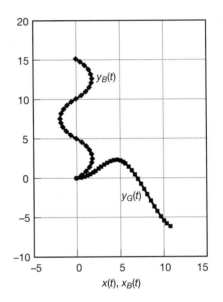

Figure P1.155

1.156 In Problem 1.155, determine the distance between the boy and girl as a function of time, and plot the function.

1.157 Particles A and B start from rest at the same point. If their accelerations are given by

$$\mathbf{a}_A = 2(0.5\hat{\mathbf{i}} + 0.612\hat{\mathbf{j}} - 0.612\hat{\mathbf{k}}) - 0.2\mathbf{v}_A \text{ m/s}^2$$

$$\mathbf{a}_B = 3t\hat{\mathbf{i}} + 2\cos(\pi t)\hat{\mathbf{k}} \text{ m/s}^2$$

determine the relative velocity and position between the two particles for the first 3 s of time.

1.158 In Problem 1.157, determine the distance between the particles at any time. Plot the distance as a function of time for the first 3 s.

1.159 Determine the relative position between the two particles in Problem 1.157 if particle A starts to move 1.5 s after particle B does.

1.12 DEPENDENT MOTIONS BETWEEN TWO OR MORE PARTICLES

Although not explicitly stated in the previous section, it is true that if a particle moves in pure rectilinear motion, the particle's motion must be constrained in some way. The most general motion that a particle can exhibit is translation in three perpendicular or orthogonal directions, called *curvilinear motion*. Constraining a particle to move in only one of these three directions allows the relationships between acceleration, velocity, displacement, and time to be considered as scalar functions. For example, a car, truck, or train modeled as a particle moving along a straight and level highway executes rectilinear motion due to the constraints imposed on it by the driver and the road. A box sliding down an inclined plane is constrained to move parallel to the surface of the inclined plane.

The weight pulled upwards by the rope-and-pulley system illustrated in Figure 1.29 can be *assumed* to move in one direction if no disturbance causes it to swing. This motion is constrained by the constant length of the rope. Therefore, if the end of the rope is pulled down a distance d, the weight must move up an equal distance, or, expressed mathematically, the change in position $\Delta x = x_2 - x_1$ is equal to the negative of d:

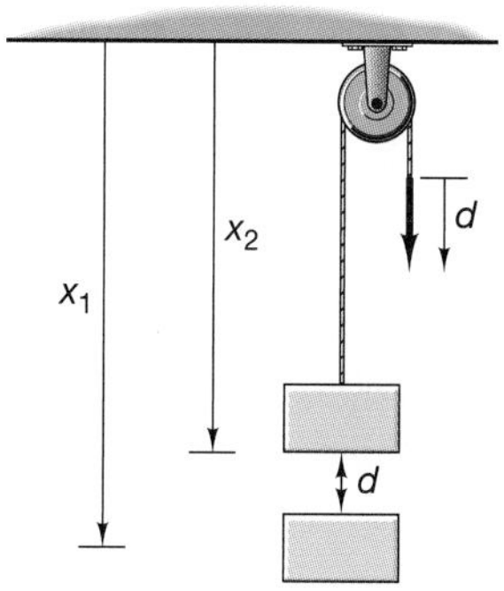

$$\Delta x = -d$$

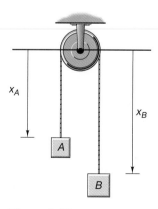

Figure 1.29

This constraint can be further understood if there are weights on both ends of the rope, as shown in Figure 1.30.

In the figure, a one-dimensional coordinate system has been established positive in the downward direction and with the origin at the center of the pulley. The length of the rope is assumed to be constant, and the *constraint on the system* can be written for *any* position of A and B as

$$x_A + x_B + \pi r = l \tag{1.134}$$

where r is the radius of the pulley and l is the length of the rope.

Since the length of rope wound over the pulley, πr, remains constant, the sum of the positions of A and B is a constant; that is,

$$x_A + x_B = l - \pi r = c \tag{1.135}$$

where c is a constant. This expression is the constraint equation in terms of the positions of the two weights. A change in position of A and the corresponding change in B are constrained as

$$\Delta x_A + \Delta x_B = 0$$

or $\tag{1.136}$

$$\Delta x_A = -\Delta x_B$$

It is obvious that if A goes down, B must go up an equal amount. Equation (1.136) expresses this obvious relation mathematically and allows the less obvious relationships that must exist between the corresponding velocities and accelerations to be determined. If Eq. (1.135) is differentiated, then the constraints on the velocities and accelerations are

$$v_A + v_B = 0$$
$$a_A + a_B = 0 \tag{1.137}$$

These values are the formal constraints on the motion, and in this case, if the motion of A is known, then the motion of B is fully described by the constraint equations.

Figure 1.30

Figure 1.31

As the mechanical system becomes more complex, the constraints on the motion are more difficult to formulate. Consider, for example, the pulley system used to move the safe shown in Figure 1.31. If the portion of the rope that wraps around the pulleys is ignored, since it remains constant (each portion equal to πr), the length of the rope can be written as

$$5x_A - x_T = l \tag{1.138}$$

Each length of the rope has been counted separately, and constant distances have been lumped as one constant. Therefore, a movement of the end of the rope, point T, is related to a movement of the safe by

$$5\Delta x_A = \Delta x_T \tag{1.139}$$

That is to say, the movement of the safe would be one-fifth of the movement of the end of the rope. The constraints on the velocities and accelerations are obtained by successively differentiating Eq. (1.138), yielding

$$5v_A - v_T = 0$$
$$5a_A - a_T = 0 \tag{1.140}$$

Since the tension in the rope is assumed constant, and since force is proportional to acceleration (Newton's second law), the force on the safe is five times the tension exerted on the rope, and the system has a "mechanical advantage" of 5. Even though the distance that point B moves is five times the distance that the safe moves, the advantage of such pulley systems is seen in the reduction of the force needed to move the safe. (A full discussion of the mechanical advantages of systems such as this is presented in Chapter 2.)

Another constrained system is shown in Figure 1.32. In the figure, two particles A and B are shown attached to a pulley system, and the two are constrained in their movements by the connecting cables—i.e., if particle A moves down, then particle B must move up. Note that the position coordinate x is positive in the downward direction. The constraint on movement in this example is the physical restriction that the lengths of the cables involved are constant. The length of the cable attaching the pulley to the ceiling is d, and the length of the cable between the pulley and B is c. Now we need to express the length of the cable that runs around the two pulleys and connects to A. The length of the main cable is l, and the constraint to the motion is obtained by adding the three sections of the cable and equating the sum to the total cable length l. This yields

$$(x_A - d) + (x_B - c - d) + (x_B - c) = l$$
$$x_A + 2x_B = l + d + 2c = \text{constant}$$

Differentiating, we then obtain

$$v_A + 2v_B = 0$$
$$a_A + 2a_B = 0 \tag{1.141}$$

The last two equations form the constraints on the velocity and acceleration of points A and B.

There is no general formula for constraint equations. A number of examples are given to help the reader understand the formulation of constraints. In the case of pulleys and masses connected by cables, the lengths of the cables are written in terms of the positions of the masses, and the constraints are that the lengths of the cables are constant.

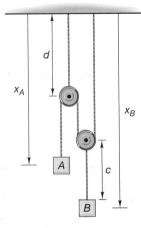

Figure 1.32

Sample Problem 1.31

Consider the block-and-pulley system illustrated in the accopanying figure. Treat the blocks as particles, and determine the velocity and acceleration of block *B* if block *A* is moving to the left with a velocity of 2 m/s and an acceleration (also to the left) of 0.3 m/s². Note that point *C* is fixed in space.

Solution

To avoid problems with signs in the equations, the origin of the coordinate system will be chosen to be outside the system, as shown (a good practice to follow). The length of the rope can be written as

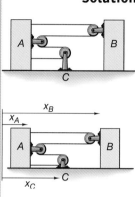

$$2(x_B - x_A) + 2(x_C - x_A) = l$$

where x_c is constant.

Differentiating this constraint equation gives the relationships between the velocities and accelerations of *A* and *B*:

$$2v_B - 2v_A - 2v_A = 0$$

$$v_B = 2v_A$$

$$a_B = 2a_A$$

Therefore, $v_B = 2v_A = 4$ m/s to the left, and $a_B = 2a_A = 0.6$ m/s² to the left.

Sample Problem 1.32

Consider the two-block-and-pulley system illustrated in the accompanying figure. In this case, the pulley is used to change the direction of motion as well as to provide mechanical advantage. However, the system is constrained such that both blocks move in rectilinear motion. Determine the velocity and acceleration of block *B* if block *A* has a velocity of 4 m/s to the left and an acceleration of 1 m/s² to the right.

Solution

The length of the rope, neglecting the constant amount around the pulley, is

$$x_A + 2y_B = l$$

Note the selection of positive signs for *x* and *y*. The constraints on the velocities and accelerations are found by taking successive derivatives of the constraint equation:

$$v_B = -\frac{v_A}{2}$$

$$a_B = -\frac{a_A}{2}$$

Therefore, substitution of the given velocity and acceleration yields

$$v_B = -2 \text{ m/s (up)}$$

$$a_B = \frac{1}{2} \text{ m/s}^2 \text{ (down)}$$

The selected coordinate system indicates that the velocity of *A* at the time shown is positive (to the left), while the acceleration of *A* is negative (to the right). Positive motion of *B* is down and is shown in the signs of the results. At the instant shown, block *B* is moving upward, but is slowing down, or decelerating.

Sample Problem 1.33   Determine the velocity and acceleration of block C in the mechanical system shown in the accompanying figure if the motor causes the cable velocity at point A to be

$$v_A(t) = t^2 + 2t$$

Solution   Note that in this case there are two ropes to consider, thus providing two equations of constraint related by the movement of the pulley B. The length of the rope that passes around pulley C is

$$x_C + (x_C - x_B) = l_1$$

The length of the second rope (going around pulley B and connecting to the motor) up to point A is

$$x_A + 2x_B = l_2$$

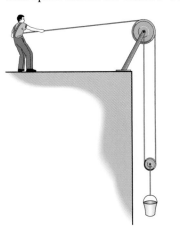

The two constraint equations can be combined to relate the position of A to that of C.

$$x_A + 4x_C = 2l_1 + l_2$$

This expression is differentiated twice to yield the velocity and acceleration of the block relative to the velocity and acceleration of the point on the cable A:

$$v_C = -\frac{v_A}{4} = -\frac{t^2}{4} - \frac{t}{2} \text{ mm/s}$$

$$a_C = -\frac{1}{2}(t + 1) \text{ mm/s}^2$$

The negative sign indicates that C is moving upwards and accelerating upwards as well. Since the rope to the motor is of finite length, the system will hit the top and stop. The top of the pulley is another constraint on the system, which occurs when x_B is zero. The system is then "at the end of its rope."

Problems

1.160 A window washer uses a pulley system to lower a bucket over the side of a building, as illustrated in Figure P1.160. If the bucket is to be lowered at 1 ft/s, at what speed should the window washer lower the rope?

1.161 Consider the block-and-pulley system illustrated in Figure P1.161, and determine a relationship between the magnitude of the acceleration of block A and the acceleration of block B.

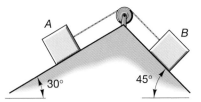

Figure P1.161

1.162 A sport utility vehicle is stuck in the mud and attaches its winch to a nearby tree in order to pull itself out, as depicted in Figure P1.162. The winch winds the

Figure P1.160

cable at a rate of 0.5 m/s. How fast does the truck move?

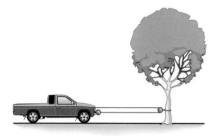

Figure P1.162

1.163 Block *B* moves upward with a speed of 10 ft/s, as shown in Figure P1.163. Calculate the corresponding speed of block *A*, and indicate whether it is moving up or down.

Figure P1.163

1.164 Block *A* moves to the right with an acceleration of 1 m/s² and a velocity of 3 m/s, as depicted in Figure P1.164. Determine a relationship between the velocity and acceleration of blocks *B* and *C* in terms of the motion of *A*.

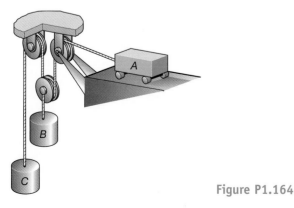

Figure P1.164

1.165 A common use of a pulley system is to lift a weight, as is shown in Figure P1.165. The tractor moves to the right with a constant velocity of 0.5 m/s. If the rope is 30 m long, determine the velocity of the weight when it is 10 m above the truck bed.

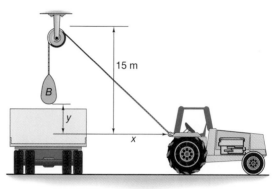

Figure P1.165

1.166 Develop the constraint relationship between the acceleration of *A* and the acceleration of *B* in Figure P1.166.

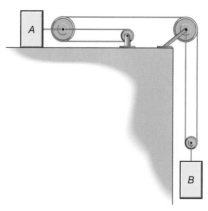

Figure P1.166

1.167 If the velocity of block *A* in Figure P1.167 is 2 m/s upward, determine the constraint on the velocities of blocks *B* and *C*.

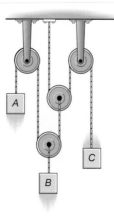

Figure P1.167

1.168 A motor is employed to raise a weight using the pulley system shown in Figure P1.168. If the motor turns a 200-mm-radius shaft at 3600 rpm, determine the velocity of the weight.

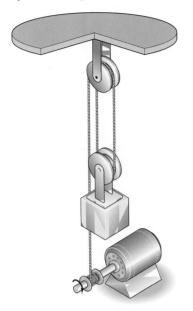

Figure P1.168

1.169 Develop the constraint equation relating the velocity of collar B to the velocity of block A in Figure P1.169.

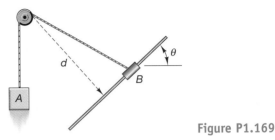

Figure P1.169

1.170 Develop the constraint equation relating the velocities of blocks A, B, and C in Figure P1.170.

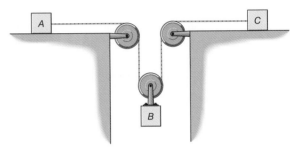

Figure P1.170

1.13 PARAMETRIC EQUATIONS IN KINEMATICS

In many engineering problems, it is useful to introduce a parameter that does not coincide with one of the coordinate directions. The equations of kinematics can then be formulated in terms of that parameter. Consider a general curve in a plane, as illustrated in Figure 1.33.

In the figure, the parameter s is a measure of the length of the path, and the unit tangent vector $\hat{\mathbf{t}}(s)$ is a function of this parameter; that is, it is a function of position along the path. By definition, the unit tangent vector is tangent to the path at every point. The unit tangent vector can be written parametrically in xy components as

$$\hat{\mathbf{t}}(s) = \cos\theta(s)\hat{\mathbf{i}} + \sin\theta(s)\hat{\mathbf{j}} \tag{1.142}$$

The angle that the tangent vector makes with the x-axis is a function of position along the path. The change in the position vector is

$$d\mathbf{r} = ds\,\hat{\mathbf{t}}(s) \tag{1.143}$$

Therefore, the velocity is

$$\mathbf{v} = \frac{d\mathbf{r}}{dt} = \frac{ds}{dt}[\cos\theta(s)\hat{\mathbf{i}} + \sin\theta(s)\hat{\mathbf{j}}] \tag{1.144}$$

and the acceleration is

$$\mathbf{a} = \frac{d\mathbf{v}}{dt} = \frac{d}{dt}\left\{\frac{ds}{dt}[\cos\theta(s)\hat{\mathbf{i}} + \sin\theta(s)\hat{\mathbf{j}}]\right\}$$
$$\mathbf{a} = \frac{d^2s}{dt^2}[\cos\theta(s)\hat{\mathbf{i}} + \sin\theta(s)\hat{\mathbf{j}}] + \left(\frac{ds}{dt}\right)^2\frac{d\theta(s)}{ds}[-\sin\theta(s)\hat{\mathbf{i}} + \cos\theta(s)\hat{\mathbf{j}}] \tag{1.145}$$

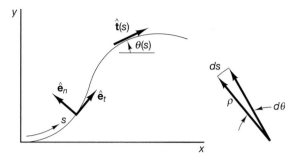

Figure 1.33

Examining Figure 1.33, we see that the term $d\theta/ds$ is equal to $1/\rho$, with a negative sign if the angle is decreasing and a positive sign if the angle is increasing. The unit vector in the second term is perpendicular to the unit tangent vector and is, therefore, a unit normal vector. This vector is always pointing in the increasing θ-direction. When normal and tangential coordinates were developed, the unit normal vector was chosen to point toward the center of curvature. The sign of the radius of curvature will cause the unit normal vector to point toward the center of curvature in this case. Therefore, Eq. (1.145) can be written as

$$\mathbf{a} = a_t\hat{\mathbf{e}}_t + a_n\hat{\mathbf{e}}_n \tag{1.146}$$

where

$$a_t = \frac{d^2s}{dt^2} \qquad \hat{\mathbf{e}}_t = \cos\theta(s)\hat{\mathbf{i}} + \sin\theta(s)\hat{\mathbf{j}}$$

$$a_n = \left(\frac{ds}{dt}\right)^2\frac{d\theta(s)}{ds} \qquad \hat{\mathbf{e}}_n = -\sin\theta(s)\hat{\mathbf{i}} + \cos\theta(s)\hat{\mathbf{j}}$$

It is apparent that any two-dimensional trajectory of a particle can be written as a function of a parameter of length, s, along the trajectory path. This parameter will be a function of time, its first derivative will be the magnitude of the velocity, and its second derivative will be the magnitude of the tangential acceleration component. Equations (1.146) and (1.81) are equivalent. The parameter s can be treated in a similar manner as was the position parameter in rectilinear motion. That is, the tangential acceleration can be expressed as a function of time, position, or speed. This expression will define completely the kinematics of the particle relative to the path of motion. The absolute velocity and acceleration vectors are given by Eqs. (1.144) and (1.145).

The origin of the path of motion has been chosen to coincide with the origin of the rectangular coordinate system, but could have been chosen to be anywhere in the plane. The rectangular coordinates of the particle at point s on the path of motion are

$$x(s) = x_0 + \int_0^s \cos\theta(\eta)d\eta$$

$$y(s) = y_0 + \int_0^s \sin\theta(\eta)d\eta \tag{1.147}$$

where x_0 and y_0 are the coordinates of the start of the path.

1.13.1 TRAJECTORIES EXPRESSED AS FUNCTION OF PARAMETER *S*

Certain curves for $\theta(s)$ are illustrated and the resulting trajectories shown using computational software as follows:

$\theta(s) = \dfrac{s}{R}$ (The angle is expressed in radians so that the function of s will be dimensionless.)

Circle:

$$\frac{d\theta}{ds} = \frac{1}{R} \text{ (A constant radius of curvature)}$$

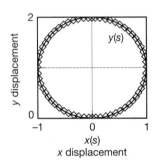

For $R = 1$:

$$\theta(s) = s$$

$$x(s) = \int_0^s \cos(\theta(\eta))\, d\eta \qquad y(s) = \int_0^s \sin(\theta(\eta))\, d\eta$$

Spiral:

$$\theta(s) = s - s^2$$

$$x(s) = \int_0^s \cos(\theta(\eta))\, d\eta \qquad y(s) = \int_0^s \sin(\theta(\eta))\, d\eta$$

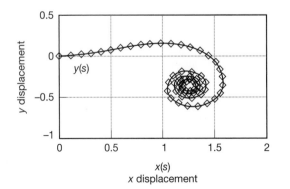

$$\theta(s) = 1 + s^2$$

$$x(s) = \int_0^s \cos(\theta(\eta))\,d\eta \qquad y(s) = \int_0^s \sin(\theta(\eta))\,d\eta$$

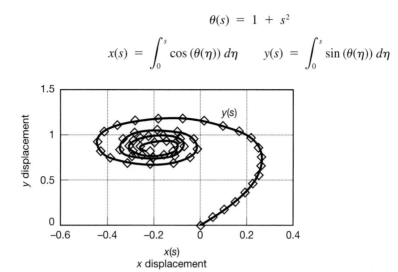

Incline:

$$\theta(s) = e^{-\frac{s}{2}}$$

$$x(s) = \int_0^s \cos(\theta(\eta))\,d\eta \qquad y(s) = \int_0^s \sin(\theta(\eta))\,d\eta$$

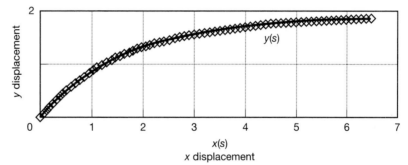

s-curve:

$$\theta(s) = 4 \cdot \sin\left(\frac{s}{2}\right)$$

$$x(s) = \int_0^s \cos(\theta(\eta))\,d\eta \qquad y(s) = \int_0^s \sin(\theta(\eta))\,d\eta$$

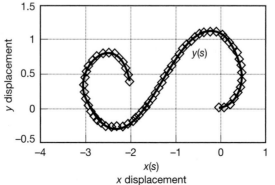

Other trajectories:

$$\theta(s) = \frac{1}{4} \cdot \sin\left(\frac{s}{2}\right)$$

$$x(s) = \int_0^s \cos(\theta(\eta))\, d\eta \qquad y(s) = \int_0^s \sin(\theta(\eta))\, d\eta$$

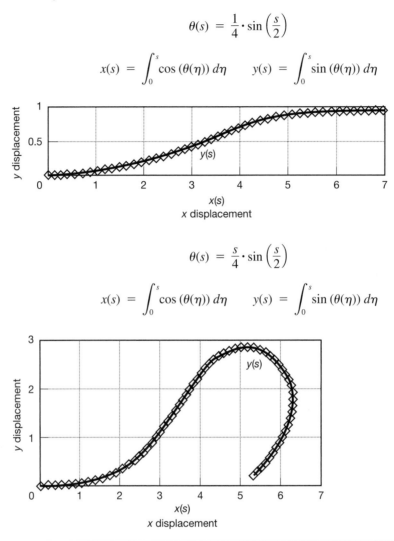

$$\theta(s) = \frac{s}{4} \cdot \sin\left(\frac{s}{2}\right)$$

$$x(s) = \int_0^s \cos(\theta(\eta))\, d\eta \qquad y(s) = \int_0^s \sin(\theta(\eta))\, d\eta$$

Sample Problem 1.34

A car starting from rest accelerates at a constant rate of 10 m/s², driving in a circle of radius 1000 m. Determine how far and how long the car will go before the normal acceleration component equals the tangential acceleration.

Solution   The parametric kinematic variables are

$$\frac{d^2 s}{dt^2} = 10 \qquad a_n = \frac{1}{\rho}\left(\frac{ds}{dt}\right)^2$$

$$\frac{ds}{dt} = 10t \qquad \left(\frac{ds}{dt}\right)^2 = 10(1000)$$

$$s = 5t^2 \qquad 100t^2 = 10{,}000$$

$$t = 10\ \text{s}$$
$$s = 500\ \text{m}$$

| Sample Problem 1.35 | Solve Sample Problem 1.34 for an automobile following a curved path given by |

$$\theta(s) = \frac{s}{400} \sin\left(\frac{s}{200}\right)$$

Solution The tangential acceleration is constant, and the velocity and displacement along the path of motion can be obtained by integration:

$$\frac{d^2s}{dt^2} = 10$$

$$\frac{ds}{dt} = 10t$$

$$s(t) = 5t^2$$

The reciprocal of the radius of curvature can be written as a function of position along the path, or as a function of time:

$$\frac{d\theta(s)}{ds} = \frac{1}{400} \sin\left(\frac{s}{200}\right) + \frac{s}{400} 200 \cos\left(\frac{s}{200}\right)$$

$$\frac{d\theta}{ds} = \frac{1}{400} \sin\left(\frac{t^2}{40}\right) + 2.5t^2 \cos\left(\frac{t^2}{40}\right)$$

The normal acceleration at any time t is

$$a_n = \left[\frac{1}{400} \sin\left(\frac{t^2}{40}\right) + 2.5t^2 \cos\left(\frac{t^2}{40}\right)\right](10t)^2$$

We wish to find the time when the normal acceleration is equal to the tangential acceleration of 10 m/s². This means that we have to determine the time that is the root of the transcendental equation

$$f(t) = a_n(t) - 10 = 0$$

This equation is difficult to solve, but the function may be plotted on a graphical calculator or by the use of computational software, as illustrated in the accompanying figure. The roots of $f(t) = 0$ may also be determined directly. (See the Computational Supplement.)

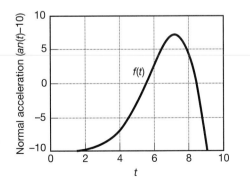

The first root is $t = 5.621$ s, which can be found by trial and error substitution for $f(t)$. The distance traveled is

$$s(5.621) = 5(5.621)^2 = 157.98 \text{ m}$$

1.13.2 PARAMETRIC EQUATIONS FOR THREE-DIMENSIONAL TRAJECTORIES

A general curve in space is shown in Figure 1.34. The unit tangent vector can be written as

$$\hat{\mathbf{t}}(s) = \cos\theta(s)\cos\beta(s)\hat{\mathbf{i}} + \sin\theta(s)\cos\beta(s)\hat{\mathbf{j}} + \sin\beta(s)\hat{\mathbf{k}} \tag{1.148}$$

The change in the position vector with time is the velocity and is equal to

$$\frac{d\mathbf{r}}{dt} = \frac{ds}{dt}\hat{\mathbf{t}} = \frac{ds}{dt}[\cos\theta(s)\cos\beta(s)\hat{\mathbf{i}} + \sin\theta(s)\cos\beta(s)\hat{\mathbf{j}} + \sin\beta(s)\hat{\mathbf{k}}] \tag{1.149}$$

The acceleration is obtained by differentiation of the velocity vector and is

$$\mathbf{a} = \frac{d^2\mathbf{r}}{dt^2} = \frac{d^2s}{dt^2}[\cos\theta(s)\cos\beta(s)\hat{\mathbf{i}} + \sin\theta(s)\cos\beta(s)\hat{\mathbf{j}} + \sin\beta(s)\hat{\mathbf{k}}]$$

$$+ \left(\frac{ds}{dt}\right)^2 \left\{ \begin{array}{l} \left[-\sin\theta(s)\cos\beta(s)\dfrac{d\theta}{ds} - \cos\theta(s)\sin\beta(s)\dfrac{d\beta}{ds}\right]\hat{\mathbf{i}} \\[2mm] + \left[\cos\theta(s)\cos\beta(s)\dfrac{d\theta}{ds} - \sin\theta(s)\sin\beta(s)\dfrac{d\beta}{ds}\right]\hat{\mathbf{j}} \\[2mm] + \left[\cos\beta(s)\dfrac{d\beta}{ds}\right]\hat{\mathbf{k}} \end{array} \right\} \tag{1.150}$$

In plane kinematics, the second term in Eq. (1.145) was shown to be the velocity squared, divided by the radius of curvature multiplied by the unit normal vector. A similar interpertation can be applied to Eq. (1.150). The bracketed vector in the second term multiplies the magnitude of the velocity squared and is not a unit vector. This vector, denoted by Γ,

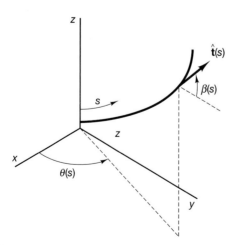

Figure 1.34

is perpendicular to the unit tangent vector, and its magnitude is the *curvature* of the path. The vector can be written as

$$
\Gamma = \begin{cases}
\left[-\sin\theta(s)\cos\beta(s)\dfrac{d\theta}{ds} - \cos\theta(s)\sin\beta(s)\dfrac{d\beta}{ds} \right]\hat{\mathbf{i}} \\[2mm]
+\left[\cos\theta(s)\cos\beta(s)\dfrac{d\theta}{ds} - \sin\theta(s)\sin\beta(s)\dfrac{d\beta}{ds} \right]\hat{\mathbf{j}} \\[2mm]
+\left[\cos\beta(s)\dfrac{d\beta}{ds} \right]\hat{\mathbf{k}}
\end{cases}
\tag{1.151}
$$

The square of the magnitude of Γ can be obtained from the scalar product of the vector with itself:

$$
\Gamma\cdot\Gamma = \left(\frac{d\theta}{ds}\right)^2 \cos^2\beta + \left(\frac{d\beta}{ds}\right)^2
\tag{1.152}
$$

The reciprocal of the magnitude of Γ is the radius of curvature in three dimensions, given by

$$
\frac{1}{\rho} = \sqrt{\left(\frac{d\theta}{ds}\right)^2 \cos^2\beta + \left(\frac{d\beta}{ds}\right)^2}
\tag{1.154}
$$

The physical interpretation of Eq. (1.150) is the same as that presented in Section 1.7 for the osculating plane.

Sample Problem 1.36

An airplane flying at a constant speed S is in a helical ascending pattern over New York. The parametric equations of the trajectory of the plane are

$$
\theta(s) = \frac{s}{R}
$$

$$
\beta(s) = p
$$

where R is the radius of the helix and p is the pitch of the helix. Determine the velocity and acceleration of the plane at any time.

Solution   The velocity vector, from Eq. (1.151), is

$$
\mathbf{v} = S\left\{ \cos\left(\frac{s}{R}\right)\cos p\,\hat{\mathbf{i}} + \sin\left(\frac{s}{R}\right)\cos p\,\hat{\mathbf{j}} + \sin p\,\hat{\mathbf{k}} \right\}
$$

The acceleration is

$$
\mathbf{a} = \frac{S^2}{R}\left\{ -\sin\frac{s}{R}\cos p\,\hat{\mathbf{i}} + \cos\frac{s}{R}\cos p\,\hat{\mathbf{j}} \right\}
$$

Problems

 1.171 A car drives at a constant speed of 20 m/s along a road defined by the unit tangent angle $\theta(s) = 4 \sin(s/2000)$ rad. Plot the layout of the roadway for the first 6 km, and determine the normal acceleration as a function of the car's position s along the roadway.

 1.172 A test driver drives along the spiral course shown in Figure P1.172, starting from rest and accelerating at a constant rate of 5 m/s². If the car will spinout when the total acceleration is 8 m/s², determine the distance along the path before the car spins out. The unit tangent vector angle is defined by $\theta(s) = 1 + (s/2000)^2$ rad. The first 3 km of the course are shown in Figure P1.172.

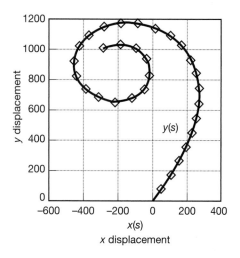

Figure P1.172

 1.173 A car drives at a constant speed of 60 mph around a 5000-ft curve described by a unit tangent vector angle $\theta(s) = (s/2000) + e^{-s/1000}$ rad, as illustrated in Figure P1.173. Determine the normal acceleration as a function of position s along the curve.

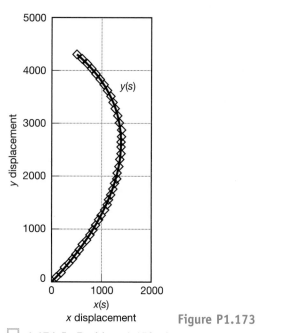

Figure P1.173

 1.174 In Problem 1.173, the curve occurs on a hill defined by $\beta(s) = \pi \sin[(\pi s)/5000]$ rad. Determine the normal acceleration of the car.

 1.175 In Problem 1.174, if the car starts from rest and accelerates at a constant rate of 1.5 ft/s² for the entire 5000 ft of the curve, determine the maximum normal acceleration and the velocity of the car at the end of the curve.

 1.176 Determine the radius of curvature as a function of s for the curve defined by Problem 1.174.

 1.177 A carnival ride moves in a circle while oscillating in the vertical direction. The unit tangent vector is defined by $\theta(s) = s/100$ and $\beta(s) = (\pi/2) \sin[(\pi s)/25]$ If the carnival car moves at a constant speed of 10 m/s, determine the normal acceleration of the car.

 1.178 In Problem 1.177, determine the total acceleration of the car if the speed of the car is given by $ds/dt = 10 - 5 \cos[(\pi s)/25]$ m/s.

Chapter 2 KINETICS OF PARTICLES

Race track curves must be banked to produce the normal acceleration so that the race cars traveling at 200 mph do not slide out.

2.1 INTRODUCTION

The study of the motion of particles is based upon axioms and laws of nature gained from experience, first formulated by Isaac Newton (1642–1727). Newton's laws were first published in his treatise, *Philosophiae naturalis principia mathematica,* in 1687. Although he was in his forties when this treatise was published, Newton had formulated most of the axioms earlier and, by the age of 23, had developed calculus and the binomial theorem. Newton's first two laws deal with the motion of a single particle, while the second two laws deal with the interaction between two particles. Euler (1707–1783) was able to show that it was possible to formulate the laws into a theory that was valid for the whole field of mechanics.

Newton's laws assume that one can define a frame of reference which is fixed in space, called an inertial frame of reference or a Newtonian frame of reference. Only in this frame of reference will Newton's second law be valid. For most purposes, a frame of reference attached to a star will serve as an inertial reference frame. However, any nonrotating reference frame moving at a constant velocity relative to an inertial frame also can be considered an inertial frame. Although the earth rotates, for most engineering problems, the errors incurred in assuming it to be a Newtonian frame of reference are small.

Newton's first three laws may be stated as follows:

1. *Every body or particle continues in a state of rest or in uniform motion (constant velocity) in a straight line, unless it is compelled to change that state by forces impressed upon it—that is, unless the external forces are not in equilibrium.*

2. *The change of motion of a body or particle is proportional to the net external force acting on the body or particle in the direction of the net external force.* In an inertial frame of reference, the time rate of change of the linear momentum of a body or particle is equal to the vector unbalance of forces. Linear momentum is defined as the vector product of the "inertial mass" and the velocity of the body or particle.

3. *If one body or particle exerts a force on a second body or particle, then the second exerts a force on the first that is equal in magnitude to, opposite in direction to, and collinear with the given force.*

The final law is the law of universal gravitational attraction:

4. *Any two particles are attracted toward each other with a force whose magnitude is proportional to the product of their gravitational masses and inversely proportional to the square of the distance between them.*

The concept of mass arises in two of the laws. In Newton's second law, inertial mass is considered to be a measure of a particle's resistance to acceleration. In Newton's fourth law, gravitational mass is defined as the property of the particle that influences its gravitational attraction. Newton further assumed that these two concepts of mass were equivalent.

If we write Newton's second law in mathematical form, the two definitions of mass can be compared. The mathematical form of this law is

$$\mathbf{F} = \frac{d}{dt}(m\mathbf{v}) \tag{2.1}$$

Since Newton assumed that the mass of the particle is constant, the second law may be rewritten as

$$\mathbf{F} = m\mathbf{a} \tag{2.2}$$

W

Figure 2.1

The linear momentum is defined by the vector $\mathbf{L} = m\mathbf{v}$, and so the second law may be written as

$$\mathbf{F} = \frac{d\mathbf{L}}{dt} \tag{2.3}$$

The force due to gravitational attraction (from Newton's fourth law) between a particle of mass m and the earth, with mass M and radius R, is

$$F = \frac{GMm}{R^2} \tag{2.4}$$

where

$$\begin{aligned} G &= 66.73 \pm 0.03 \times 10^{-12} \text{ m}^3/\text{kg}\cdot\text{s}^2 \\ &= 34.4 \times 10^{-9} \text{ ft}^4/\text{lb}\cdot\text{s}^4 \end{aligned} \tag{2.5}$$

is the universal gravitational constant.

The symbol g denotes a quantity relating the earth's properties and the gravitational constant and is given by

$$g = \frac{GM}{R^2} \tag{2.6}$$

Then, the gravitational force of attraction—that is, the weight of an object on the surface of the earth—is

$$F = W = mg \tag{2.7}$$

The value of g near the surface of the Earth is 9.81 m/s², or 32.2 ft/s².

Now, consider a mass in free fall in the earth's gravitational field, as shown in Figure 2.1. The only force acting on the object is W in the vertical direction, and Newton's second law with the downward direction taken as positive is

$$\begin{aligned} F = W &= ma \\ mg &= ma \\ a &= g \end{aligned} \tag{2.8}$$

Then, based upon the assumption that gravitational mass is equivalent to inertial mass, we can deduce that *an object in free fall will accelerate at a value exactly equal to* g. This discovery has led to the constant g being referred to as gravitational acceleration. At times, this reference may lead to confusion, as g is actually a measure of the earth's gravitational attraction on a mass m and can be interpreted as an acceleration only when an object is in free fall, neglecting the resistance of air.

2.1.1 EQUATIONS OF MOTION FOR A PARTICLE

Newton's second law may be written in vector form when the mass is constant:

$$\sum \mathbf{F} = m\mathbf{a}$$

If the vectors are expressed in rectangular coordinates, the law can be written as

$$\sum (F_x\hat{\mathbf{i}} + F_y\hat{\mathbf{j}} + F_z\hat{\mathbf{k}}) = m(a_x\hat{\mathbf{i}} + a_y\hat{\mathbf{j}} + a_z\hat{\mathbf{k}}) \tag{2.9}$$

The scalar forms of this vector equation are obtained by equating the coefficients of the unit vectors, yielding

$$\sum F_x = ma_x$$
$$\sum F_y = ma_y \qquad (2.10)$$
$$\sum F_z = ma_z$$

The equations of motion can be written in terms of the normal and tangential coordinates for planar motion as follows:

$$\sum \mathbf{F} = m\mathbf{a}$$
$$\sum F_t = ma_t = m\frac{dv}{dt} \qquad (2.11)$$
$$\sum F_n = ma_n = m\frac{v^2}{\rho}$$

The use of normal and tangential coordinates allows for the separation of the forces into those which cause an increase in speed and those which change the direction of the velocity. Equivalent forms of this law can be written in polar, cylindrical, and spherical coordinates.

2.2 SOLUTION STRATEGY FOR PARTICLE DYNAMICS

The analysis of the dynamics of a particle should be approached in a highly systematic manner in order for one to obtain the greatest understanding of the particle's motion. When the object to be studied is modeled as a particle, we assume that all of the forces acting on the object will be concurrent, and the mass is considered to be located at the point where all of the forces meet. When a car is modeled as a particle, forces are not considered to be acting on the front or rear wheels, but instead, all forces are modeled to be acting at the center of mass of the car. The orientation of the object in space is ignored, and it is assumed that the position of the object can be specified by a single position vector. In subsequent chapters, objects will be given size and shape, and then the position of the center of mass and the orientation of the object will be determined. A particle has only *three degrees of freedom,* representing the three possible translations.

In the previous chapter, the kinematics of a particle were examined, and the differential equations of motion were solved. These equations express the particle's acceleration in terms of velocity, position, and time, and may be either linear or nonlinear. Most of the nonlinear equations are solved by numerical integration, although it is always preferable to seek analytical solutions where possible. Newton's laws will now be used to study how forces cause these changes in motion. Dynamics is a subject that requires a conceptual, as well as an analytical, approach. Each problem should be examined carefully before one attempts to derive the differential equation of motion. The following steps will serve as a guide to the solution of these problems:

1. Try to form a conceptual appreciation of the problem. What will move? In which direction? Is the motion constrained in some manner so that the object moves only in a straight line or in a plane? Many of these questions may be answered by common sense. If a car is parked on a hill and the emergency brake is not set, will you look for the car at the top or the bottom of the hill? Before dismissing this example as silly, realize that many students are not bothered by a negative answer, which might indicate that the car slid up the hill. What you are trying to develop is the ability to "see" the mathematics or conceptualize the results. One of the difficulties in solving dynamics problems is that objects may move differently than one might first think.

2. Model the problem by creating a free-body diagram of each particle involved. This procedure is one of the most important skills obtained in the study of statics. Remember that without a correct model or free-body diagram, the problem cannot be solved correctly.

3. Establish a coordinate system to describe the motion of the object. The choice of coordinate system is yours. There is no wrong choice, but some choices will greatly simplify the equations of motion. For example, if the particle is moving along a circular path, normal and tangential coordinates or polar coordinates are better than rectangular coordinates to describe the motion.

4. Using the free-body diagram, express the force vectors in vector notation in the appropriate coordinate system. Write the equation of motion for each degree of freedom—that is, for each coordinate direction. For particle dynamics, a maximum of three equations of motion can be written for each particle.

5. Assess the problem by specifying what is known and what is not known. Count the number of unknowns and the number of equations. Do they balance? That is, are the number of equations equal to the number of unknowns? If there are more equations than unknowns, there is an error in the free-body diagram. If there are more unknowns than equations, there are constraints on the motion that can be expressed mathematically. For example, if a particle is constrained to move along a plane, then the acceleration perpendicular to that plane is zero. If there is friction acting between the particle and the surface it slides on, then the magnitude of the friction force is related to the normal force by the coefficient of kinetic friction.

6. Write the additional equations of constraint, or equations relating the forces to one another. If the object is sliding, the friction force is equal to the coefficient of kinetic friction times the normal force. At this point, a balance between the number of equations and unknowns should exist. If such a balance does not exist, there is no point in trying to solve the equations, as either you have made an error or a constraint or some other factor has been missed. Frequently, it may be that the problem has not been carefully read or an important piece of information has been overlooked. You may have missed a dimension on your model or failed to realize that a surface is specified to be smooth; that is, there is no friction.

7. Form the differential equations of motion by eliminating the unknown forces from the system of equations. The resulting differential equations should be of a form in which the acceleration is a function of the known forces, the velocity and position of the particle, and time.

8. Solve the differential equations of motion, and answer any questions concerning the resulting motions. If the differential equations are nonlinear, numerical integration will be required.

9. Review the results to be sure they make sense conceptually. In many cases, plotting the trajectories of the motion is of great use. These plots may be of a coordinate direction versus time, a velocity component versus time, or one coordinate versus another coordinate. In some cases, a plot of the velocity versus position is useful in interpreting the results.

2.2.1 REVIEW OF THE CONCEPTS OF STATIC AND KINETIC FRICTION

The concepts of dry friction, or Coulomb friction, were first introduced in our study of statics. However, the emphasis was on cases where friction prevented motion, and the limit of the friction force in preventing motion was given by the coefficient of static friction μ_s. At the point of impending slip, the *maximum friction force* before slip was given by the relationship

$$f_{\max} = \mu_s N$$

where N is the normal force between the two surfaces. In actuality, the friction force between two surfaces would never be exactly equal to this maximum value, as the system would be very unstable at that point. If two surfaces are pushed together with a normal force N, the friction force in the static situation will be equal and opposite to the applied force parallel to the contacting surfaces. As the applied force increases, the friction force increases until it reaches its maximum value, defined by the coefficient of static friction. At this point, slip occurs, and due to heat and the breakdown of the roughness, or asperities, of the contacting surface, the friction force drops to an almost-constant value, defined by the coefficient of kinetic friction μ_k. This situation is illustrated in the following diagram:

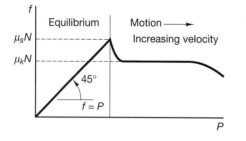

If motion occurs, the friction force is assumed to be constant during the motion and is given by

$$f = \mu_k N$$

where N is the normal force between the contacting surfaces. The value of f can never exceed the applied force parallel to the contacting surfaces; that is, friction always opposes motion. The value of the coefficient of kinetic friction is usually 20% lower than the value of the coefficient of static friction. When the applied force is resisted by a friction force having a value between that of kinetic friction and the maximum static friction, the system is unstable. For systems in this state, any perturbation or disturbance can cause motion. To analyze systems in which there is friction between the contacting surfaces (in reality, there is always friction between the surfaces), you must first find out whether motion will occur or whether there is sufficient static friction to resist motion. To find this out, you assume the acceleration to be zero and then examine the friction force for equilibrium. If this force is less than the maximum static friction defined by the coefficient of static friction, motion between the surfaces does not occur. If the equilibrium friction force is greater than the maximum static friction, motion does occur. However, if the latter is the case, then the friction force is assumed to be known and is equal to the kinetic friction defined by the coefficient of kinetic friction. The acceleration of the particle can then be determined. Friction always opposes motion, and when the motion changes such that the velocity changes sign, the friction force can be written

$$f = -\mu_k N \operatorname{sign}(v) = -\mu_k N \frac{v}{|v|}$$

where N is the normal force and v is the velocity of the particle. Note that the sign of the friction force is always the opposite of the sign of the velocity. The friction force enters into the differential equations of motion in this manner, and although analytical solutions of these differential equations may be difficult, numerical solutions can be obtained.

The 3000 lb car shown is driving down a 30° incline when the driver locks up all wheels in a panic stop. If the coefficient of kinetic friction between the tires and the pavement is 0.7, how far will the car skid before coming to a stop if its initial speed was 45 mph?

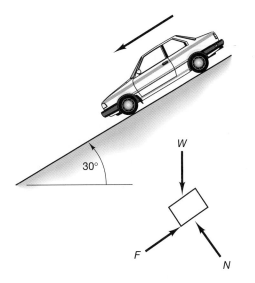

Solution

1. **Conceptualization**

   The car is modeled as a particle. The only forces acting on the car are its weight (due to gravitational attraction), the normal force between the tires and the ground, and the friction force. Since all tires on the car are skidding, the distribution of the car's weight on the front and rear tires is not needed, and so the car can be modeled as a single particle. The weight will cause the car to move down the hill, and the friction force will be the force that stops the car. Since the car is sliding, the friction force is related to the normal force by the coefficient of kinetic friction. The distance the car will slide is the required answer, and if the acceleration can be determined, the differential equation of motion can be solved for the velocity as a function of position. The initial velocity and the position of the car are known, and the car will stop when the velocity is zero.

2. **Modeling and the free-body diagram**

   The car is modeled as a particle and the free-body diagram is shown in the accompanying figure.

3. **Coordinate system**

   There are two logical choices of coordinate system. The x-axis could be chosen as horizontal and the y-axis as vertical, or the x-axis could be chosen as parallel to the incline and the y-axis as normal to it. The first option would make the problem more difficult, as there would be motion in two coordinate directions. The second choice would take advantage of the fact that the car is constrained to move down the incline, and thus, there is no movement in the direction normal to the incline. For this example, the problem will be solved in both coordinate systems to show the differences between the two methods.

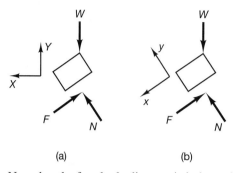

(a)   (b)

Note that the free-body diagram is independent of the choice of coordinate system, and the difference will appear only in the equations of motion and in the constraints on the motion. The coordinates for (a) have been designated with capital letters, and $\hat{\mathbf{I}}$ and $\hat{\mathbf{J}}$ are the unit vectors in the X- and Y-directions, respectively. The conventional notation has been used for (b).

4. **Expressing force vectors in the coordinate system**

   We then express the forces in both coordinate systems:

   a) $\mathbf{W} = -W\hat{\mathbf{J}}$, where $W = 3000$ lb
      $\mathbf{N} = N(\sin\theta\hat{\mathbf{I}} + \cos\theta\hat{\mathbf{J}})$, where $\theta = 30°$
      $\mathbf{F} = F(-\cos\theta\hat{\mathbf{I}} + \sin\theta\hat{\mathbf{J}})$

   b) $\mathbf{W} = W(\sin\theta\hat{\mathbf{i}} - \cos\theta\hat{\mathbf{j}})$
      $\mathbf{N} = N\hat{\mathbf{j}}$
      $\mathbf{F} = -F\hat{\mathbf{i}}$

   The two scalar equations of motion are

   a) $N\sin\theta - F\cos\theta = \dfrac{W}{g}a_x$

      $-W + N\cos\theta + F\sin\theta = \dfrac{W}{g}a_y$

   b) $W\sin\theta - F = \dfrac{W}{g}a_x$

      $-W\cos\theta + N = \dfrac{W}{g}a_y$

5. **Count of equations and unknowns**

   There are two scalar equations of motion, one for each degree of freedom. (Note, that the problem is solved twice in order to demonstrate the use of two different coordinate systems, and this does not double the number of equations, as it might appear to upon first glance.) The unknowns in either coordinate system are N, F, and the two scalar components of the acceleration. There is a total of four unknowns with only two equations. Hence, there must be more facts or constraints that we must use to balance the system of equations.

6. **Constraints on the motion and other facts**

   Since the car is in a locked-up slide, the friction force is equal to the coefficient of kinetic friction times the normal force:

   $$F = \mu_k N = 0.7\,N$$

The car is constrained to move on the surface of the incline; therefore, the direction of the acceleration vector is known. In our two coordinate systems, this constraint becomes

a) $\mathbf{a} = a(\cos 30°\hat{\mathbf{I}} - \sin 30°\hat{\mathbf{J}})$

b) $\mathbf{a} = a\hat{\mathbf{i}}$

A balance of equations and unknowns exists, and the differential equation of motion can be obtained. [Note that the constraint to the motion is a little more difficult to write in the coordinate system (a)].

Substituting for the friction force in the equations of motion, and using the constraint on the motion, we obtain two scalar equations in each coordinate system. In coordinate system (a), the solution becomes

$$N \sin 30° - 0.7N \cos 30° = \frac{W}{32.2} a \cos 30°$$

$$-W + N \cos 30° + 0.7N \sin 30° = -\frac{W}{32.2} a \sin 30°$$

Solving the two equations yields

$$N = 0.866W$$
$$a = -0.106g = -3.42 \text{ ft/s}^2$$

In coordinate system (b), the equations become

$$W \sin 30° - 0.7N = \frac{W}{32.2}a$$

$$-W \cos 30° + N = 0$$

Solving for N and a yields

$$N = 0.866W$$
$$a = -0.106g = -3.42 \text{ ft/s}^2$$

The answers are the same regardless of the choice of coordinate system, although the algebra is simpler in the second coordinate system.

Note that in this problem, the acceleration is independent of the mass of the car. The distance needed to stop a big car is no different than that required to stop a small car. Now that the acceleration of the car is known (deceleration due to braking is indicated as the negative of the acceleration), the length of the skid can be determined using the kinematics of a particle in linear translation.

The acceleration is a constant, and the differential equation can be written in a form that yields the stopping distance, without considering time as a variable. The differential equation will be separated as

$$a \, ds = v \, dv$$

$$\int_0^d a \, ds = \tfrac{1}{2}(v_f^2 - v_0^2)$$

Since the acceleration is constant during the skid, the distance to the point where the velocity is zero can be determined. The constant acceleration was determined to be

$$a = -3.42 \text{ ft/s}^2$$

and the final velocity when the car has stopped is

$$v_f = 0$$

The initial velocity was specified to be

$$v_0 = 45 \text{ mph} = 66 \text{ ft/s}$$

Substituting these values into the integrated equation of motion yields

$$-3.42 \, d = -1/2 \, (66)^2$$
$$d = 636.8 \text{ ft}$$

Thus, it would take a distance of over two football fields for this car to stop as it slides down the grade. If the initial speed was 30 mph, the car could stop in 283 ft. This big difference in stopping distance is due to the fact that the skid distance is related to the *square* of the velocity.

A 30° incline is a very steep slope, and most roads do not involve slopes of this magnitude. The coefficient of kinetic friction of 0.7 is a good estimate of the friction between tire rubber and dry pavement. A reasonable highway design problem requires examination of the stopping distance on dry pavement for different slopes or speeds. In this problem, the car has an initial velocity, and therefore, when the brakes are applied, the car will slide with the kinetic friction resisting movement. If the car had been parked on the incline and the coefficient of static friction were great enough, the car would not have started to slide down the hill. The angle of the incline necessary to initiate sliding is such that the tangent of this angle is equal to the coefficient of static friction. Since the coefficient of kinetic friction is less than that of static friction, once a parked car starts to slide, it will continue to accelerate down the incline. The approach to the parked car problem is to initially assume static equilibrium and see if there is sufficient friction available to prevent slipping. If the static friction is insufficient, the problem is modeled as a dynamics problem.

If you were an engineer working for the road commission, you would not want to solve this problem over and over again for different slopes, different initial speeds, and different coefficients of friction (different weather conditions). We have seen that the stopping distance is not a function of the mass or weight of the car but is a function of speed, slope, and coefficient of kinetic friction. Examining the solution in coordinate system *b* yields a general expression for the stopping distance:

$$s(\theta, v_0, \mu_k) = \frac{v_0^2}{2g(\mu_k \cos\theta - \sin\theta)}$$

Note that if $\mu_k \cos\theta \le \sin\theta$, then the vehicle cannot be stopped. This would occur if the slope is too great or if the slope was slippery.

This problem is investigated in detail in the Computation Supplement.

Sample Problem 2.2

A 200-N force is applied to a 5-kg box in an attempt to push it up a 30° incline. If the coefficient of static friction is 0.6 and the coefficient of kinetic friction is 0.5, will the box slide up the incline? If it slides, determine the length of time required to move the box 1 m.

Solution   A rectangular coordinate system has been chosen such that the *x*-axis is parallel to the incline and the *y*-axis is normal to the incline. The coordinate system could have been chosen in the horizontal and vertical directions, and that solution also will be shown. Note that the particle has been modeled in two dimensions only, as the motion occurs in one plane

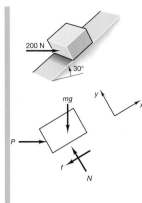

and there are no forces perpendicular to this plane. We will write each force in vector form before writing the equations of motion:

$$\mathbf{W} = mg(-\sin\theta\hat{\mathbf{i}} - \cos\theta\hat{\mathbf{j}}), \text{ where } mg = (5)(9.81) \text{ and } \theta = 30°$$

$$\mathbf{P} = P(\cos\theta\hat{\mathbf{i}} - \sin\theta\hat{\mathbf{j}}), \text{ where } P = 200 \text{ N}$$

$$\mathbf{f} = -f\hat{\mathbf{i}}$$

$$\mathbf{N} = N\hat{\mathbf{j}}$$

The two scalar equations for Newton's second law can now be written:

$$-mg\sin\theta + P\cos\theta - f = ma_x$$
$$-mg\cos\theta - P\sin\theta + N = ma_y$$

At this point in the solution, we have two equations to determine the four unknowns f, N, a_x and a_y. The motion is constrained such that the box cannot move in the y-direction. Therefore, the acceleration in that direction is

$$a_y = 0$$

The final equation involves the friction force. However, there are two cases to consider: The box moves and the box does not move. If the box does not move, the acceleration in the x-direction is zero, and the friction force must be less than or equal to the product coefficient of static friction and the normal force. We will examine this case first. Setting a_x equal to zero, we can write the first equation as

$$f = -mg\sin\theta + P\cos\theta = 148.68 \text{ N}$$

The normal force is obtained from the second equation:

$$N = mg\cos\theta + P\sin\theta = 142.48 \text{ N}$$

The maximum friction force available is $f_{max} = \mu_s N = 0.6 \,(142.48) = 85.49$ N. We have determined that, to prevent motion, the friction force would have to equal 148.68 N, and that the maximum static friction force is 85.49 N. Therefore, the box moves up the incline, and the friction force is determined by the coefficient of kinetic friction to be

$$f = \mu_k N = 0.5 \,(142.48) = 71.24 \text{ N}$$

The equation of motion in the x-direction is

$$a_x = 1/m(-mg\sin\theta + P\cos\theta - f) = 15.45 \text{ m/s}^2$$

The acceleration is a constant, and the differential equation of motion in the x-direction is

$$\frac{d^2x}{dt^2} = 15.45$$

Selecting the initial position and velocity to be zero, this equation can be solved by repeated integration to yield

$$a = 15.45$$
$$v = 15.45\,t$$
$$x = 7.73\,t^2$$

Therefore, the time required to move the box 1 m is

$$t = \sqrt{\frac{1}{7.73}} = 0.36 \text{ s, or } 360 \text{ ms}$$

The velocity at this time is

$$v = 15.45 \,(0.36) = 5.56 \text{ m/s}$$

If only the velocity was desired, the differential equation could be written in a separable form and the velocity determined directly as

$$a\,dx = v\,dv$$

$$\int_0^1 15.45dx = \int_0^v v\,dv$$

$$v = \sqrt{2(15.45)} = 5.56 \text{ m/s}$$

2.2.2 ALTERNATIVE SELECTION OF COORDINATE SYSTEM

If the *x*-coordinate had been selected in the horizontal direction and the *y*-coordinate in the vertical direction, the force vectors would have had the form

$$\mathbf{W} = -mg\hat{\mathbf{j}}$$
$$\mathbf{P} = P\hat{\mathbf{i}}$$
$$\mathbf{N} = N(-\sin\theta\hat{\mathbf{i}} + \cos\theta\hat{\mathbf{j}})$$
$$\mathbf{f} = f(-\cos\theta\hat{\mathbf{i}} - \sin\theta\hat{\mathbf{j}})$$

The constraint on the motion of the box moving up the incline is

$$\mathbf{a} = a(\cos\theta\hat{\mathbf{i}} + \sin\theta\hat{\mathbf{j}})$$

The equations of motion then become

$$P - N\sin\theta - f\cos\theta = ma\,\cos\theta$$
$$-mg + N\cos\theta - f\sin\theta = ma\,\sin\theta$$

Multiplying the first equation by $\cos\theta$ and the second by $\sin\theta$ and adding the two resulting equations yields

$$P\cos\theta - mg\sin\theta - f = ma$$

This expression is the same equation of motion as the one obtained in the other coordinate system.

Sample Problem 2.3

Consider a particle of mass *m* falling in the earth's atmosphere, and assume that air resistance is proportional to the velocity of the particle. Determine the velocity as a function of time.

Solution   This example is a problem of rectilinear motion, and *x* will be assumed to be positive in the downward direction. The free-body diagram of the particle is as follows:
The equation of motion is

$$ma(t) = m\frac{dv}{dt} = mg - cv$$

$$\frac{dv}{dt} + \frac{c}{m}v = g$$

$W = mg$

x

$F = cv$

Assume an initial condition of zero velocity at time equal to zero. The differential equation can be solved by separation of variables or as a nonhomogeneous first-order linear differential equation with constant coefficients. First consider a solution by separation of variables:

$$dt = \frac{dv}{g - \dfrac{cv}{m}}$$

$$\int_0^t dt = \int_0^v \frac{dv}{g - \dfrac{cv}{m}}$$

$$t = -\frac{m}{c}\ln\left(1 - \frac{cv}{mg}\right)$$

Solving for the velocity in terms of time yields

$$v(t) = \frac{mg}{c}\left(1 - e^{-\frac{c}{m}t}\right)$$

Since the exponential goes to zero as time continues, the velocity approaches a constant, an *asymptotic value* called the *terminal velocity* and equal to mg/c.

This problem was presented in Sample Problem 1.5 and solved both by separation of variables and as a linear first order differential equation with constant coefficients, using an integrating factor in the solution. The constant c is the air viscosity and is approximately 0.6 lb s/ft. (A velocity–time plot of the particle's fall in the atmosphere is shown in the Computational Supplement.)

Sample Problem 2.4

A better model of a particle falling in the earth's atmosphere includes the fact that the air resistance is proportional to the square of the velocity. The constant c, the drag coefficient, is proportional to the product of the density of the air and the cross-sectional area of the object, and is usually experimentally determined. Determine the velocity–time relationship and the terminal velocity for this case.

Solution The differential equation of motion is

$$m\frac{dv}{dt} = mg - cv^2$$

In this case, the equation is a nonlinear first-order differential equation that cannot be solved by an integrating factor. Since the acceleration is a function of the velocity only, it can be solved for separation of variables:

$$\int_0^t dt = \int_0^v \frac{dv}{g - \left(\frac{c}{m}\right)v^2}$$

Integrating both sides yields

$$t = \frac{1}{2\sqrt{\frac{cg}{m}}}\ln\left[\frac{g + \sqrt{\frac{cg}{m}}v}{g - \sqrt{\frac{cg}{m}}v}\right]$$

By introducing a constant k, we may invert this equation to express v as a function of time:

$$\frac{g - kv}{g + kv} = e^{-2kt}, \text{ where } k = \sqrt{\frac{cg}{m}}$$

$$kv(1 + e^{-2kt}) = g(1 - e^{-2kt})$$

$$v = \frac{g(1 - e^{-2kt})}{k(1 + e^{-2kt})}$$

as $t \to \infty$,

$$v \to \frac{g}{k} = \sqrt{\frac{mg}{c}}$$

For a mass of 5 slugs and a drag coefficient $c = 0.002$, the solution can be plotted as shown in the Computational Supplement. A velocity (ft/s)–time (s) plot for these values is shown in the following diagram:

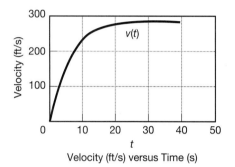

Velocity (ft/s) versus Time (s)

Note that the velocity approaches the terminal velocity after 30 to 40 seconds.

Sample Problem 2.5

a) Compare the distance that a ball can be thrown when air resistance is proportional to the velocity with the distance the ball can be thrown when air resistance is ignored. The ball has an initial velocity of v_0 at an angle θ with the horizontal.
b) Write the equations of motion if the air resistance is proportional to the square of the velocity, and discuss the difficulties of solving the equations.

Solution a) The drag force from air resistance is

$$\mathbf{F}_d = -c\mathbf{v}$$

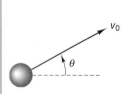

Note that the direction of the drag force is always opposite to the direction of the velocity vector; that is, the drag force always opposes motion. If the x-axis is horizontal and the y-axis is vertical, the vector equation of motion in rectangular coordinates is

$$m\frac{d\mathbf{v}}{dt} = -c\mathbf{v} - mg\hat{\mathbf{j}}$$

$$\mathbf{v}(0) = v_0 \cos\theta\hat{\mathbf{i}} + v_0 \sin\theta\hat{\mathbf{j}}$$

The two scalar differential equations are solved easily by the use of an integrating factor, yielding

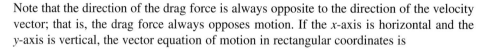

$$\frac{dv_x}{dt} + \frac{c}{m}v_x = 0 \qquad \frac{dv_y}{dt} + \frac{c}{m}v_y = -g$$

$$v_x = v_0 \cos\theta e^{-\frac{c}{m}t} \qquad v_y = \left(\frac{mg}{c} + v_0 \sin\theta\right)e^{-\frac{c}{m}t} - \frac{mg}{c}$$

Note that if the flight of the ball could continue indefinitely—that is, if the ball did not hit the ground—the terminal velocity would be

$$\mathbf{v} = -\frac{mg}{c}\hat{\mathbf{j}}$$

This result agrees with that of Sample Problem 2.3 when the change in sign of the y-coordinate is taken into account. The component of the velocity in the x-direction damps out, and the ball falls in the vertical direction at the terminal velocity. In this problem, the ball will hit the ground when $y = 0$, and any physical interpretation of the motion ends at that time. The velocity at that time is not zero, as anyone who has ever caught a ball knows.

The position of the ball at any time is obtained by integrating the two velocity components using the initial conditions $x(0) = 0$, and $y(0) = 0$, yielding

$$x(t) = -\frac{m}{c}v_0\cos\theta e^{-\frac{c}{m}t} + C_1 \quad y(t) = -\frac{m}{c}\left(\frac{mg}{c} + v_0\sin\theta\right)e^{-\frac{c}{m}t} - \frac{mg}{c}t + C_2$$

The constants of integration are obtained by substituting the initial conditions into the displacement equations when $t = 0$:

$$x(0) = 0 = -\frac{m}{c}v_0\cos\theta + C_1 \quad y(0) = -\frac{m}{c}\left(\frac{mg}{c} + v_0\sin\theta\right) + C_2$$

Solving for the constants, we find that the displacements become

$$x(t) = \frac{m}{c}v_0\cos\theta(1 - e^{-\frac{c}{m}t}) \quad y(t) = \left(\frac{m^2g}{c^2} + \frac{m}{c}v_0\sin\theta\right)(1 - e^{-\frac{c}{m}t}) - \frac{mg}{c}t$$

The upward flight of the ball ends when the velocity in the y-direction is zero. Therefore, if the maximum height that the ball attains is desired, the expression for v_y is set to zero, and the resulting equation for the time at maximum height is

$$\left(\frac{mg}{c} + v_0\sin\theta\right)e^{-\frac{c}{m}t_1} - \frac{mg}{c} = 0$$

$$t_1 = \ln\left(1 + \frac{cv_0}{mg}\sin\theta\right)$$

The maximum height is obtained by substituting t_1 into the equation for the y-coordinate.

The total distance in the x-direction that the ball travels before hitting the ground is obtained by determining the time t_2 when the y-coordinate is zero. This expression is obtained by setting the y-coordinate to zero, yielding

$$\left(\frac{m^2g}{c^2} + \frac{m}{c}v_0\sin\theta\right)\left(1 - e^{-\frac{c}{m}t_2}\right) - \frac{mg}{c}t_2 = 0$$

One root of this transcendental equation is zero, and the second root should be obtained numerically for specific values of mass, damping, and initial velocity. When t_2 is obtained, the $x(t)$ position can be evaluated at that time. An alternative numerical approach is to plot $y(t)$ versus $x(t)$ for the entire motion and determine the distance from this plot.

If air resistance is neglected, the two scalar differential equations with their solutions are

$$\frac{dv_x}{dt} = 0 \qquad \frac{dv_y}{dt} = -g$$

$$v_x = v_0\cos\theta \qquad v_y = (v_0\sin\theta) - gt$$

The displacement components are obtained by integrating the velocity components and evaluating the constants of integration, knowing that the initial displacement is zero. Upon so doing, we obtain

$$x(t) = (v_0 \cos \theta)\, t \qquad\qquad y(t) = (v_0 \sin \theta)t - \frac{gt^2}{2}$$

These two solutions are compared in the accompanying plot using the following values: $m = 0.5$ kg, $c = 0.01$ kg/s, $v_0 = 25$ m/s, and $\theta = 45°$.

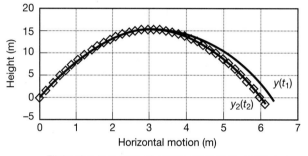

Flight plot, in meters, showing effect of air resistance

b) If the air resistance is taken to be proportional to the square of the velocity, the drag must always oppose the motion, so the drag force is $\mathbf{F}_d = -c\mathbf{v} \cdot \mathbf{v}\hat{\mathbf{e}}_v$. The unit vector $\hat{\mathbf{e}}_v$, may be written as $\hat{\mathbf{e}}_v = \dfrac{v}{|v|}$. The. vector equation of motion is

$$m\frac{d\mathbf{v}}{dt} = -c\mathbf{v} \cdot \mathbf{v}\hat{\mathbf{e}}_v - mg\hat{\mathbf{j}}$$

$$m\frac{d\mathbf{v}}{dt} = -c|\mathbf{v}|\mathbf{v} - mg\hat{\mathbf{j}}$$

The two scalar equations of motion for this condition are

$$m\frac{dv_x}{dt} + cv_x\sqrt{v_x^2 + v_y^2} = 0$$

$$m\frac{dv_y}{dt} + cv_y\sqrt{v_x^2 + v_y^2} = -mg$$

The solution is expressed as a system of two coupled nonlinear first order differential equations. The physics of the problem has been correctly presented, but great mathematical difficulties to obtaining a solution remain. Numerical solutions of this problem can be obtained by the use of computational software or a programmable calculator. The nonlinear equations are integrated numerically, step by step, using Euler's method, as follows:

$$x(t + \Delta t) = x(t) + v_x(t)\Delta t$$

$$y(t + \Delta t) = y(t) + v_y(t)\Delta t$$

$$v_x(t + \Delta t) = v_x(t) - \frac{cv_x(t)}{m}\sqrt{v_x^2(t) + v_y^2(t)}\,\Delta t$$

$$v_y(t + \Delta t) = v_y(t) - \left[\frac{cv_y(t)}{m}\sqrt{v_x^2(t) + v_y^2(t)} - g\right]\Delta t$$

Note that in the last equation the drag always opposes the motion: When the velocity is positive, the drag is negative, and when the velocity is negative, the drag is positive. If initial values are selected, this problem can be solved by step by step numerical integration. (Details are given in the Computational Supplement.)

Sample Problem 2.6

A spring slider mechanism is used to produce oscillatory motion. If the unstretched length of the spring which has a spring constant k, is l, and the slider of mass m is released from rest at the position shown in the accompanying diagram, write the equation of motion. Neglect the mass of the spring.

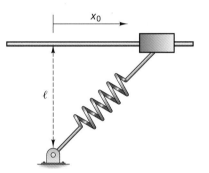

Solution   The free-body diagram of the slider at any position is first constructed as follows:

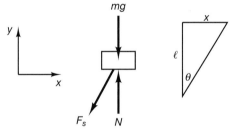

$$\sin \theta = \frac{x}{\sqrt{\ell^2 + x^2}}$$

$$\cos \theta = \frac{\ell}{\sqrt{\ell^2 + x^2}}$$

The magnitude of the spring force $F_s = k(l - l_0)$ and at any position is

$$\mathbf{F}_s = k\left[\sqrt{l^2 + x^2} - l\right]\left(-\frac{x}{\sqrt{l^2 + x^2}}\hat{\mathbf{i}} - \frac{l}{\sqrt{l^2 + x^2}}\hat{\mathbf{j}}\right)$$

The other forces are

$$\mathbf{N} = N\hat{\mathbf{j}}$$

$$\mathbf{W} = -mg\hat{\mathbf{j}}$$

The equations of motion are

$$m\frac{d^2x}{dt^2} = -k\left[\sqrt{l^2 + x^2} - l\right]\frac{x}{\sqrt{l^2 + x^2}}$$

$$m\frac{d^2y}{dt^2} = N - mg - k\left[\sqrt{l^2 + x^2} - l\right]\frac{l}{\sqrt{l^2 + x^2}}$$

The motion is constrained in the *y*-direction, and the acceleration in that direction is zero. The first equation of motion forms a nonlinear differential equation for the position *x*. This equation can be solved using numerical integration, and the details are shown in the computational supplement. A numerical example of the spring-slider mechanism is shown in the accompanying plot with the values $m = 1$ kg, $l = 1$ m, $x_0 = 0.2$, and $k = 200$ N/m. The dashed line in the plot represents a sinusoidal oscillation. The spring slider closely approximates sinusoidal motion.

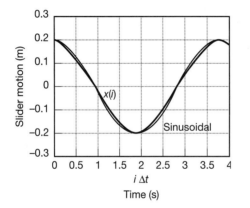

2.2.3 DETERMINATION OF THE DIRECTION OF THE NORMAL AND FRICTION FORCES

When a body is supported by contact with inclined surfaces, the direction of the normal force can be obtained from the geometry of the situation or by use of the cross product. An example of the use of trigonometry to obtain a normal unit vector for a two-dimensional problem is shown in Figure 2.2. If the surface is horizontal (in the *x*-direction), the normal is in the positive *y*-direction. Therefore, if the surface makes an angle of θ with the horizontal *x*-axis, the unit normal vector makes an angle θ with the vertical *z*-axis. The unit normal vector is

$$\hat{\mathbf{n}} = -\sin(\theta)\hat{\mathbf{i}} + \cos(\theta)\hat{\mathbf{j}} \qquad (2.12)$$

In this case, the normal was found by examination of the drawing.

An alternative approach to determine the unit vector normal to the surface—which can be easily extended to three-dimensional problems—uses the fact that the cross product is perpendicular to the plane formed by the two vectors in the vector multiplication. Consider a surface, shown in Figure 2.3, that makes an angle of θ degrees with the horizontal

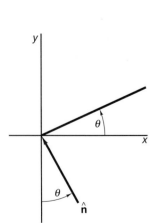

Figure 2.2

xz-plane. Note that this surface is the same as that shown in Figure 2.2, only drawn in three-dimensional perspective. The unit normal vector is

$$\hat{\mathbf{n}} = \frac{\hat{\mathbf{T}} \times \hat{\mathbf{t}}}{|\hat{\mathbf{T}} \times \hat{\mathbf{t}}|} \qquad (2.13)$$

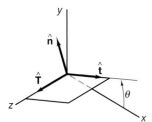

Figure 2.3

The unit vectors tangent to the plane are

$$\hat{\mathbf{t}} = \cos(\theta)\hat{\mathbf{i}} + \sin(\theta)\hat{\mathbf{j}}$$
$$\hat{\mathbf{T}} = \hat{\mathbf{k}} \qquad (2.14)$$

In this case, the two vectors tangential to the plane are orthogonal, and there is no need to divide by the magnitude of the cross product in Eq. (2.13). The unit normal vector is

$$\hat{\mathbf{n}} = \hat{\mathbf{T}} \times \hat{\mathbf{t}} = \hat{\mathbf{k}} \times [\cos(\theta\hat{\mathbf{i}} + \sin(\theta)\hat{\mathbf{j}}]$$
$$\hat{\mathbf{n}} = -\sin(\theta)\hat{\mathbf{i}} + \cos(\theta)\hat{\mathbf{j}}$$

This equation agrees with the result obtained in Eq. (2.12) and is an alternative approach to trigonometry.

If the contact is on an inclined surface in space, Eq. (2.13) can be used to determine a unit normal vector to the surface, and the normal force can be expressed in vector form as $\mathbf{N} = N\mathbf{n}$. In this regard, consider the inclined surface shown in Figure 2.4. The unit tangent vectors $\hat{\mathbf{t}}$ and $\hat{\mathbf{T}}$ lie in the *xz*- and *yz*-planes, respectively, and are

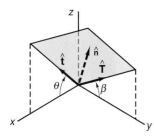

Figure 2.4

$$\hat{\mathbf{t}} = \cos(\theta)\hat{\mathbf{i}} + \sin(\theta)\hat{\mathbf{k}}$$
$$\hat{\mathbf{T}} = \cos(\beta)\hat{\mathbf{j}} + \sin(\beta)\hat{\mathbf{k}} \qquad (2.15)$$

The scalar product between the two unit tangent vectors is not zero, and therefore, the two vectors are not orthogonal. From Eq. (2.13), the unit normal vector is

$$\hat{\mathbf{n}} = \frac{\hat{\mathbf{t}} \times \hat{\mathbf{T}}}{|\hat{\mathbf{t}} \times \hat{\mathbf{T}}|} = \frac{-\sin(\theta)\cos(\beta)\hat{\mathbf{i}} - \cos(\theta)\sin(\beta)\hat{\mathbf{j}} + \cos(\theta)\cos(\beta)\hat{\mathbf{k}}}{\sqrt{\sin^2\theta\cos^2\beta + \cos^2\theta}} \qquad (2.16)$$

Note that the surface partially constrains only one degree of freedom—that is, normal into the surface of contact. The body is free to separate from the surface if the applied force on the particle has a component normally outward from the inclined surface.

If friction exists between the particle and the inclined surface, the friction always opposes the direction of motion. The particle is constrained to move on the surface of the plane, and the friction force will be tangent to the plane and opposite in direction to the tangential component of the sum of all the other forces applied to the particle. Let $\mathbf{P}$ be the vector sum of all the forces that act on a particle except the friction and normal force—that is, the force of gravitational attraction and other applied forces. The unit vector of this force can be expressed in a component normal to the plane and a component tangential to the plane. First, determine a unit vector in the $\mathbf{P}$-direction:

$$\hat{\mathbf{p}} = \frac{\mathbf{P}}{|\mathbf{P}|}$$

The component of this unit vector in the normal direction is

$$\mathbf{P}_n = (\hat{\mathbf{P}} \cdot \hat{\mathbf{n}})\hat{\mathbf{n}}$$

Therefore, the component of the unit vector in the **P**-direction tangential to the plane is

$$\mathbf{P}_t = \hat{\mathbf{P}} - \mathbf{P}_n \tag{2.17}$$

A unit vector tangent to the inclined plane and in the direction of the friction force is

$$\hat{\mathbf{f}} = \frac{\mathbf{P}_t}{|\mathbf{P}_t|} \tag{2.18}$$

Sample Problem 2.7

A constant force $\mathbf{P} = -500\hat{\mathbf{i}} - 300\hat{\mathbf{j}}$ is applied to a 20-kg block to push it up an inclined surface. The coefficient of kinetic friction between the block and the surface is 0.3. If the block starts at rest from point D, as shown in the accompanying diagram, determine the position of the block after 2 s.

Solution   The following forces act on the block:

$$\mathbf{P} = -500\hat{\mathbf{i}} - 300\hat{\mathbf{j}}$$
$$\mathbf{W} = -20(9.81)\hat{\mathbf{k}}$$
$$\mathbf{N} = N\hat{\mathbf{n}}$$
$$\mathbf{F} = F\hat{\mathbf{f}}$$

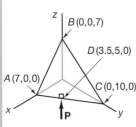

We will determine the unit vectors in the normal direction and in the direction of the friction force using Eqs. (2.16) through (2.18). Two vectors lying in the inclined plane can be obtained from the inclined plane's intercepts with the coordinate planes. These vectors are

$$\mathbf{AB} = -7\hat{\mathbf{i}} + 7\hat{\mathbf{k}}$$
$$\mathbf{AC} = -7\hat{\mathbf{i}} + 10\hat{\mathbf{j}}$$

The unit vector normal to the inclined plane is

$$\hat{\mathbf{n}} = \frac{\mathbf{AC} \times \mathbf{AB}}{|\mathbf{AC} \times \mathbf{AB}|}$$

$$\hat{\mathbf{n}} = 0.634\hat{\mathbf{i}} + 0.444\hat{\mathbf{j}} + 0.634\hat{\mathbf{k}}$$

The unit vector in the $(\mathbf{P} + \mathbf{W})$ direction is

$$\hat{\mathbf{p}} = \frac{\mathbf{P} + \mathbf{W}}{|\mathbf{P} + \mathbf{W}|} = -0.813\hat{\mathbf{i}} - 0.488\hat{\mathbf{j}} - 0.319\hat{\mathbf{k}}$$

This vector may be expressed in components normal and tangential to the inclined surface, using Eq. (2.17). The components are

$$\mathbf{P}_n = (\hat{\mathbf{p}} \cdot \hat{\mathbf{n}})\hat{\mathbf{n}} = -0.592\hat{\mathbf{i}} - 0.414\hat{\mathbf{j}} - 0.592\hat{\mathbf{k}}$$
$$\mathbf{P}_t = \hat{\mathbf{p}} - \mathbf{P}_n = -0.221\hat{\mathbf{i}} - 0.074\hat{\mathbf{j}} + 0.273\hat{\mathbf{k}}$$

The unit vector for the friction force (opposite to the direction of motion) is

$$\hat{\mathbf{f}} = -\frac{\mathbf{P}_t}{|\mathbf{P}_t|} = 0.617\hat{\mathbf{i}} + 0.205\hat{\mathbf{j}} - 0.760\hat{\mathbf{k}}$$

Since the acceleration will be in the opposite direction of the friction force unit vector, the vector equation of motion can be written as

$$\mathbf{P} + \mathbf{W} + \mathbf{N} + \mathbf{F} = ma(-\mathbf{f})$$

Using the fact that the magnitude of the friction force is $|\mathbf{F}| = 0.3$ N, we obtain, for the three scalar equations of motion in the x-, y-, and z-directions, respectively,

$$-500 + 0.634\ N + (0.617)(0.3)\ N = -0.617(20)a$$
$$-300 + 0.444\ N + (0.205)(0.3)\ N = -0.205(20)a$$
$$-20(9.81) + 0.634\ N - (0.760)(0.3)\ N = 0.760(20)a$$

Although this system appears to be three equations in three unknowns, the equations are not linearly independent, as the motion is restricted to move in a plane normal to the surface. Any two may be solved for the magnitudes of the normal force and the acceleration:

$$N = 573.7\ \text{N} \qquad\qquad a = 2.438\ \text{m/s}^2$$

The acceleration vector is

$$\mathbf{a} = -a\hat{\mathbf{f}} = 2.438\,(-0.617\hat{\mathbf{i}} - 0.205\hat{\mathbf{j}} + 0.760\hat{\mathbf{k}})$$

The acceleration vector is constant, and the vector differential equation of motion can be integrated. The initial velocity is zero, and the initial position is

$$\mathbf{r}_0 = 3.5\hat{\mathbf{i}} + 5\hat{\mathbf{j}}\ \text{m}$$

The position at any time is obtained by integration of the acceleration vector, yielding

$$\mathbf{r}(t) = \mathbf{a}t^2/2 + \mathbf{r}_0$$

The position at time equal to 2 s is

$$\mathbf{r}(2) = [3.5 - 2(2.438)(0.617)]\hat{\mathbf{i}} + [5 - 2(2.438)(0.207)]\hat{\mathbf{j}}$$
$$+ 2(2.438)(0.76)\hat{\mathbf{k}}$$
$$= [0.492\hat{\mathbf{i}} + 3.991\hat{\mathbf{j}} + 3.706\hat{\mathbf{k}}]\ \text{m}$$

| **Sample Problem 2.8** |

If the pulley system in the accompanying illustration is released from rest, determine the velocity of the 100-kg block C after it has moved 0.5 m. Neglect friction and the weight of the pulleys.

$$m_A = 25\ \text{kg} \quad m_B = 40\ \text{kg} \quad m_C = 100\ \text{kg}$$

Solution The motions of the three masses and the pulley P are constrained by the fixed length of the two ropes. The two constraint equations are

$$(x_A - x_P) + (x_B - x_P) = l_1$$
$$x_P + x_C = l_2$$

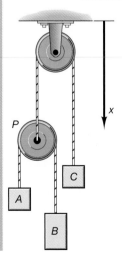

where x is taken to be positive in the downward direction. The position of the pulley can be eliminated from the two constraint equations, yielding a single constraint equation for the three masses:

$$x_A + x_B + 2x_C = l_1 + 2l_2$$

The constraint equation for the accelerations is obtained by differentiating this equation twice. The result is

$$a_A + a_B + 2a_C = 0$$

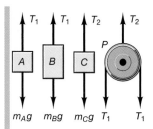

We will now construct free-body diagrams of each mass and the pulley P:
The equation of motion for each block is

$$m_A g - T_1 = m_A a_A$$

$$m_B g - T_1 = m_B a_B$$

$$m_C g - T_2 = m_C a_C$$

$$2T_1 - T_2 = 0$$

We now have five equations in five unknowns—the three accelerations and the two tensions. The tension in the second cord is eliminated from the equation of motion for mass C, using the relationship $T_2 = 2T_1$. The four remaining equations can be written in matrix notation as follows:

$$\begin{bmatrix} 1 & 1 & 2 & 0 \\ m_A & 0 & 0 & 1 \\ 0 & m_B & 0 & 1 \\ 0 & 0 & m_C & 2 \end{bmatrix} \begin{bmatrix} a_A \\ a_B \\ a_C \\ T_1 \end{bmatrix} = \begin{bmatrix} 0 \\ m_A g \\ m_B g \\ m_C g \end{bmatrix}$$

This system of equations may be solved symbolically or numerically, using the given values of the masses. A symbolic solution yields

$$\begin{bmatrix} a_A \\ a_B \\ a_C \\ T_1 \end{bmatrix} = \begin{bmatrix} \dfrac{m_A m_C + 4m_A m_B - 3m_B m_C}{m_A m_C + 4m_A m_B + m_B m_C} g \\ \dfrac{-3m_A m_C + 4m_A m_B + m_B m_C}{m_A m_C + 4m_A m_B + m_B m_C} g \\ \dfrac{-m_A m_C + 4m_A m_B - m_B m_C}{m_A m_C + 4m_A m_B + m_B m_C} g \\ \dfrac{4m_A m_B m_C}{m_A m_C + 4m_A m_B + m_B m_C} g \end{bmatrix}$$

The numerical values are

$$a_A = -5.139 \text{ m/s}^2 \qquad a_B = 0.467 \text{ m/s}^2$$
$$a_C = 2.336 \text{ m/s}^2 \qquad T_1 = 373.7 \text{ N}$$

Analyzing the motion of mass C that is under constant acceleration downward:

$$a_C = 2.336 \text{ m/s}^2$$

$$v_C = 2.336\, t + v_0 \text{ m/s}$$
$$v_0 = 0$$

$$x_C = 1.168\, t^2 + x_0 \text{ m}$$
$$x_0 = 0$$

For $x_c = 0.5$,

$$t^2 = \frac{0.5}{1.168}$$

$$t = 0.654 \text{ s}$$

$$v_C = 2.336 \cdot 0.654 = 1.528 \text{ m/s}$$

Problems

2.1 A person pushes a 200-lb box across a floor with a force of 50 lb, as shown in Figure P2.1. If the kinetic coefficient of friction between the box and the floor is 0.1, determine the time taken to move the box 100 ft.

Figure P2.1

2.2 A dog team attempts to pull a 150-kg sled up a 1% grade, as depicted in Figure P2.2. If the coefficient of static friction between the sled runners and the snow is 0.4, determine the tension on the sled harness required to break the sled free. Once the sled starts moving, the coefficient of kinetic friction is 0.3. If this tension is maintained by the dogs, determine the time required for the team to reach a velocity of 5 m/s.

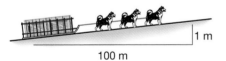

100 m   1 m

Figure P2.2

2.3 In Figure P2.3, a truck starts from a stopped position with an acceleration of 10 ft/s². The coefficients of friction between the 60-lb crate and the bed of the truck are $\mu_s = 0.2$ and $\mu_k = 0.15$. Determine whether the crate slips and, if it slips, the acceleration of the crate relative to the truck.

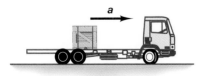

Figure P2.3

2.4 Determine the acceleration of block A for each of the cases shown in Figure P2.4. Neglect friction and the mass of the pulley. All weights in lbs.

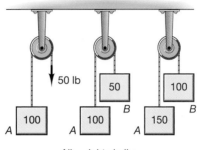

All weights in lbs.

Figure P2.4

2.5 A 100-kg man stands in a freight elevator, as shown in Figure P2.5. Determine the force between the man's feet and the elevator floor if (a) the elevator accelerates upward at 2 m/s² and (b) the elevator accelerates downward at 2.5 m/s².

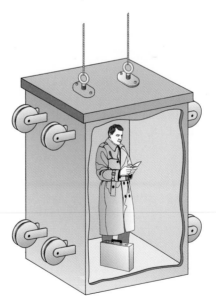

Figure P2.5

2.6 A 10-kg box is thrown on the midpoint of a 20-m-long conveyor belt moving up at a constant speed of 0.2 m/s, as illustrated in Figure P2.6. If the coefficient of static friction between the box and the conveyor belt is 0.4 and the coefficient of kinetic friction is 0.3, determine the amount of time before the box leaves the conveyor belt. Is this a properly designed conveyor system and if not, to what angle should the 30° angle be reduced?

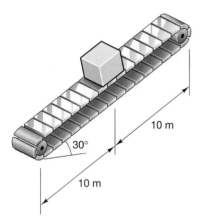

Figure P2.6

2.7 In Figure P2.7, a 100-N force is applied to a 20-kg block *A* that is attached by a cable-pulley system to a 40-kg block *B*. Determine the acceleration of the two blocks and the tension in the cable.

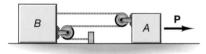

Figure P2.7

2.8 Determine the acceleration of block *B* in the pulley system shown in Figure P2.8. Neglect the weight of the pulleys and friction.

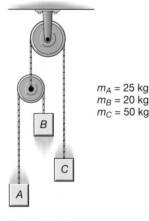

$m_A = 25$ kg
$m_B = 20$ kg
$m_C = 50$ kg

Figure P2.8

2.9 If the pulley system in Figure P2.9 is released from rest, determine the velocity of the 100-kg block after it has moved 0.5 m. Neglect friction and the weight of the pulleys.

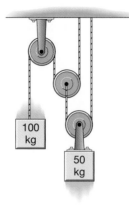

Figure P2.9

2.10 A mass of 2 kg is attached to a spring with a spring constant of 100 N/m, as shown in Figure P2.10. If the mass is released from rest with the spring stretched 0.3 m, write the differential equation of motion and determine the displacement–time relationship. Neglect friction between the mass and the surface.

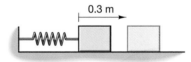

Figure P2.10

2.11 In Figure P2.11, three identical cars are pulled up an incline of slope α by an applied force **P**. Determine a general expression for the acceleration of the cars and the force in the connectors between cars *A* and *B* and between cars *B* and *C*.

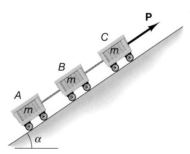

Figure P2.11

2.12 The coefficient of kinetic friction between the two blocks and the 20° inclined plane in Figure P2.12 is 0.2. If the blocks are released from rest, determine the time required for the 50-lb block to move 3 ft. What is the tension in the cable between the blocks?

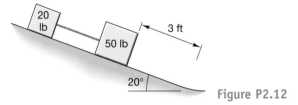

Figure P2.12

2.13 In Figure P2.13, a 200-kg rocket sled is at rest when it is given a timed burn for 6 s, producing a thrust of $F(t) = 500 [t^2 + 2t]$ N. If the coefficient of static friction between the sled and the rails is 0.5, and the coefficient of kinetic friction is 0.4, determine (a) the time before the sled begins to move; (b) the position and velocity of the sled at the end of the burn; and (c) the total distance the sled will travel before coming to rest.

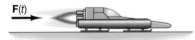

Figure P2.13

2.14 If the coefficient of static friction is 0.3 and the coefficient of kinetic friction is 0.25 between all surfaces in Figure P2.14, determine the acceleration of block A.

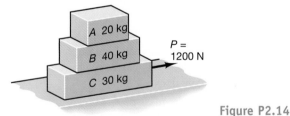

Figure P2.14

2.15 Solve Problem 2.14 if the force P is 400 N.

2.16 In Figure P2.16, the mass of block A is 50 kg, and the mass of block B is 10 kg. The coefficient of static friction is 0.2, and the coefficient of kinetic friction is 0.15 between all surfaces. Determine the distance that block B will slide to the right.

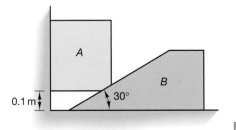

Figure P2.16

2.17 The spring–mass system shown in Figure P2.17 is released from rest, and the mass m slides vertically on a smooth rod. If the spring has a spring constant k and is unstretched at the position shown in the figure, write the differential equation of motion for the mass.

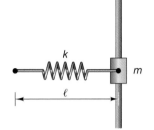

Figure P2.17

2.18 Develop the differential equation for the spring–mass system of Problem 2.17 if the coefficient of kinetic friction between the mass and the rod is μ_k. (*Note: Friction always opposes motion so the friction term should be multiplied by* $sign(v) = \dfrac{v}{|v|}$; *that is, if the velocity is positive, the friction is in the negative direction and this will reverse when the velocity is negative. You must check your computational software to determine how the program handles 0/0. You wish the value to be zero for this case.*)

2.19 If the mass is 2 kg, the length l is 0.3 m, and the spring constant k is 300 N/m, numerically solve the equation of motion of Problem 2.17. Plot the resulting motion versus time.

2.20 If the system in Figure P2.20 is released from rest, determine the time required for block B to hit the floor; $m_A = 20$ kg, $m_B = 30$ kg, and $d = 0.5/$m.

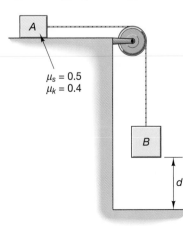

Figure P2.20

2.21 After walking 100 ft up the 30° incline (*yz*-plane) in Figure P2.21, a 130-lb climber attempts to cross an ice face with crampons and ice axes. As she attempts to advance, she loosens an ice ax, reducing the effective coefficient of static friction to 1.0. If she slips, the effective coefficient of kinetic friction between her crampons and the ice surface is 0.8.
 a) Does she slip?
 b) If she does slip, what is her velocity when she slides to the base of the ice wall? (Will she be seriously injured?)

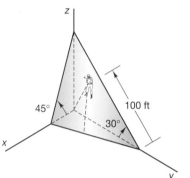

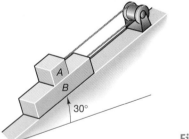

y   **Figure P2.21**

2.22 A 20-kg block *A* rests on the 60-kg plate *B* in the position shown in Figure P2.22. Neglecting the mass of the rope and pulley, and knowing that the coefficient of kinetic friction between *A* and *B* is 0.2 and the coefficient of kinetic friction between *B* and the plane is 0.1, determine the acceleration of the block relative to the plate.

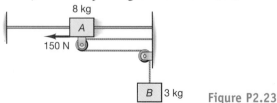

Figure P2.22

2.23 The system shown in Figure P2.23 is at rest when a constant 150-N force is applied to the collar *B*. Neglecting the effect of friction, determine
 a) the time at which the velocity of *B* will be 2.5 m/s to the left, and
 b) the corresponding tension in the cable.

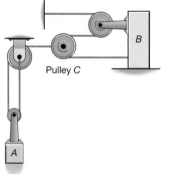

Figure P2.23

2.24 Determine the constant acceleration of the system shown in Figure P2.24 in terms of the masses, the incline angles, and the coefficient of kinetic friction. Neglect the friction between the rope and the pulley, and consider the pulley to be massless. The coefficient of kinetic friction between the masses and the inclined planes is μ_k.

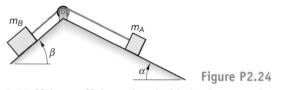

Figure P2.24

2.25 If the coefficient of static friction between the crate and the platform in Figure P2.25 is μ_s, determine the maximum acceleration that the elevator platform can have up the incline so that the crate does not slip.

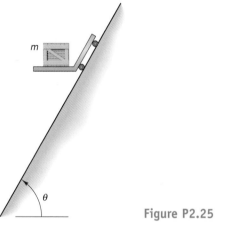

Figure P2.25

2.26 Determine the maximum acceleration that the elevator platform in Problem 2.25 can have *down* the incline so that the crate does not slip.

2.27 In Figure P2.27, the mass of block *A* is 20 kg, and the mass of block *B* is 10 kg. Neglect friction between block *B* and the plane and the mass of the pulleys. Determine the time necessary for block *A* to descend 1.5 m, and find the tension in the cables.

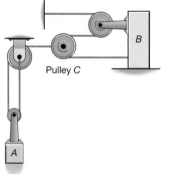

Figure P2.27

2.28 The counterweighted pulley system in Figure P2.28 is used to lower mass B to the floor. Determine the time required for the mass to reach the floor if it is released from rest from the position shown. Neglect friction and the masses of the pulleys. Use $m_A = 10$ kg, $m_B = 80$ kg, and $m_C = 20$ kg.

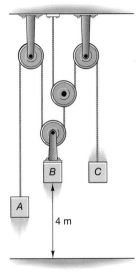

Figure P2.28

2.29 A 10-kg block A and a 20-kg block B are pulled along a horizontal plane by a force $P = 100$ N, as shown in Figure P2.29. If the coefficient of static friction is 0.2 and the coefficient of kinetic friction is 0.15 between all surfaces, determine the acceleration of each block.

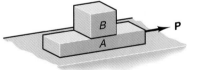

Figure P2.29

2.30 Determine the acceleration of each block in Problem 2.29 if the coefficient of static friction is only 0.17.

2.31 A 30-kg block B rests on a 50-kg block A, and the system is released from rest at the position shown in Figure P2.31. Neglect friction between all surfaces, and determine the time for block B to slide off block A.

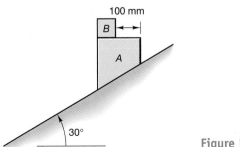

Figure P2.31

2.32 A spring-slider mechanism makes an angle α with the vertical, as depicted in Figure P2.32. The system is released from rest at the position shown, where the spring is unstretched. If the coefficient of friction is μ_k, write the general equations of motion of the mass m. The spring has a spring constant k and an unstretched length l. (*See note on Problem 2.18.*)

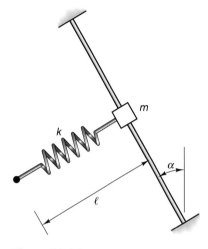

Figure P2.32

2.33 If the mass in Problem 2.32 is 5 kg and the angle α is 30°, design the mechanism so that it will damp out to rest after two oscillations. (*Hint:* Choose appropriate values for the unstretched length of the spring, the spring constant, and the coefficient of kinetic friction.)

2.34 The spring constant of a bungee cord is inversely proportional to its length $k = \dfrac{c}{l}$, where l is the original length of the cord and c is a constant related to the material and cross section of the cord. A 110-lb female bungee jumper leaps from a 150-ft platform using a 60-ft cord with the constant $c = 300$ lb. Determine the height above the ground at which the jumper's velocity is zero and the force on her legs at that point. *Show that this force is independent of the amount of free fall (length of the cord).*

2.3 DISCONTINUITY AND SINGULARITY FUNCTIONS

In Section 2.2, the forces that we hypothesized to act on the particles were continuous function of time, position, or velocity. However, in many application, these forces act in a discontinuous manner. Some examples are the intermittent firing of an engine and hitting a shock pad, a spring, or a padded dashboard. When these type of forces act on a particle, discontinuities and singularities occur in the differential equations, and the calculation of the velocity and the position are usually handled by sectioning the time of the event into several separate parts. A useful alternative approach is to use functions that are discontinuous to describe the forces. In most cases, the differential equation containing these functions is integrated numerically. In this section, singularity functions will be introduced, including the *unit doublet function* and the *Dirac delta function*. Nonsingular discontitutity functions, such as the *Heaviside step function* or the *unit step function,* will be developed from the unit doublet function and the Dirac delta function. Singularity functions were first suggested for use in the mechanics of solids to determine the deflection of beams by W. H. Macauley ("Note on the Deflection of Beams," *Messenger of Mathematics*, Vol. 48, pp. 129–130, 1919) and also appeared in Crandal and Dahl's text, *An Introduction to the Mechanics of Solids,* in 1959.

The *impulse* of an impact force can be considered to act over a time interval of zero. The impulse of a force is equal to the integral of the force over time and will be discussed in detail in Chapter 3. An impulse of this nature can be treated as force over zero time by use of the *Dirac delta function.* Using two different notations, we may write this as

$$\langle t - a \rangle_{-1} = \delta(t - a) = 0 \qquad t \neq a$$

$$\int_{-\infty}^{t} \langle \zeta - a \rangle_{-1} d\zeta = \int_{-\infty}^{t} \delta(\zeta - a)d\zeta = \begin{cases} 0, & t < a \\ 1, & t > a \end{cases} \qquad (2.19)$$

where a is the time when the impact force occurred. Regarding the two notations, for the Dirac delta function, the brackets subscripted by (-1) commonly are used in texts on the mechanics of materials, while the use of the delta to indicate this function is more common in other areas of mathematics. The function is shown in Figure 2.5.

Examination of the figure indicates that the Dirac delta function is of infinite height at point a and of zero width, but with an area under the function such that $\infty \cdot 0 = 1$. Although this expression appears strange at first, this type of product between infinity and zero appears elsewhere in mathematics and usually is encountered in a beginning calculus course. Consider, for example, the following function evaluated at $x = 0$:

$$\frac{1}{x} \cdot \sin x \to \frac{0}{0} \text{ at } x = 0$$

This function is determinate and can be evaluated by the use of l'Hôpital's rule:

$$\lim_{x \to 0} \frac{f(x)}{g(x)} = \lim_{x \to 0} \frac{f'(x)}{g'(x)}$$

$$\lim_{x \to 0} \frac{\sin x}{x} = \lim_{x \to 0} \frac{\cos x}{1} = 1$$

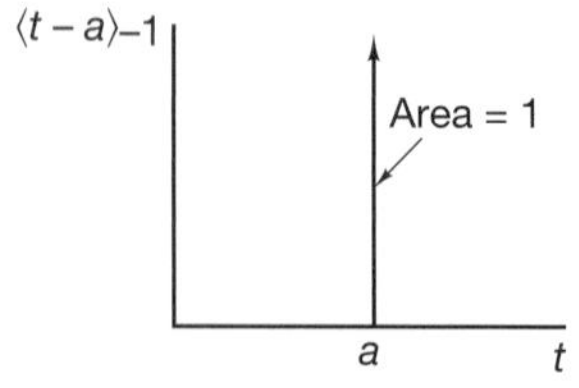

Paul A. M. Dirac defined a function that is zero everywhere, except at $t = a$, and that has an area of unity. An important property of the Dirac delta function is that

$$\langle x - a \rangle_{-1} = \delta(x - a) = 0 \qquad x \neq a$$

$$\int_{-\infty}^{x} f(\zeta)\langle \zeta - a \rangle_{-1} d\zeta = \int_{-\infty}^{x} f(\zeta)\delta(\zeta - a)d\zeta = \begin{cases} 0 & x < a \\ f(a), & x > a \end{cases} \qquad (2.20)$$

Figure 2.5

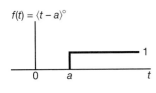

$f(t) = \langle t - a \rangle^0$

Figure 2.6

Therefore, the integral of the product of a function with the Dirac delta function that includes the point a is equal to the value of the function at a.

A superscripted type of bracket will be used to identify discontinuity functions, and their differentials and integrals are defined. Consider the function shown in Figure 2.6. This function is the *Heaviside step function,* named after the English physicist and electrical engineer, Oliver Heaviside (1850–1925). The step function has the value zero for all values of t less than a and has the value unity for all values of t greater than a. It is discontinuous precisely at a, stepping from zero to unity at that point. The definition of the function can be written mathematically as

$$\langle t - a \rangle^0 = \begin{cases} 0 & t < a \\ (t - a)^0 = 1 & t > a \end{cases} \tag{2.21}$$

The Heaviside step function is denoted by $H(t - a)$ or $\phi(t - a)$ in many applications, but the notation here will be used for consistency with other texts on the mechanics of materials. Note that for values of t greater than a, the step function is equal to $(t - a)$ raised to the zeroth power, that is, unity. The Heaviside step function can be generalized to form a group of polynomials having the value zero until it reaches the point a:

$$\langle t - a \rangle^n = \begin{cases} 0 & t < a \\ (t - a)^n & t > a \end{cases} \tag{2.22}$$

Each of these polynomial functions has the value zero before the point a, and each takes on a nonzero value at that point. These functions comprise a class of functions called *singularity functions of order* n. The Heaviside step function may be used to start any function at the point a as follows:

$$\langle t - a \rangle^0 F(t) = \begin{cases} 0 & t < a \\ F(t) & t > a \end{cases} \tag{2.23}$$

Here, the step function has been used as a constant, multiplying the function $F(t)$ and causing the function to start at the desired point a. The derivative of a discontinuity function of order n is defined as

$$\frac{d}{dt} \langle t - a \rangle^n = n \langle t - a \rangle^{n-1} \quad \text{for } n \geq 1 \tag{2.24}$$

Note that Eq. (2.24) does not apply to the Heaviside step function. The derivative of that function is more complex, and the derivative for values less than a is zero, as well as for values greater than a for which the slope of the curve is zero. At the point a, the function is discontinuous and the derivative is not defined. However, if the curve goes from zero slope to infinite slope and then back to zero slope, the derivative of the step function can be seen to be the Dirac delta function. Thus,

$$\frac{d}{dx} \langle t - a \rangle^0 = \langle t - a \rangle_{-1} \tag{2.25}$$

The Dirac delta function is defined to have an area of unity at the point a. If the Dirac delta function is integrated,

$$\int_0^t \langle u - a \rangle_{-1} \, du = \begin{cases} 0 & \text{for } t < a \\ 1 & \text{for } t > a \end{cases} = \langle t - a \rangle^0 \tag{2.26}$$

Table 2.1
Derivatives and Integrals of Discontinuity

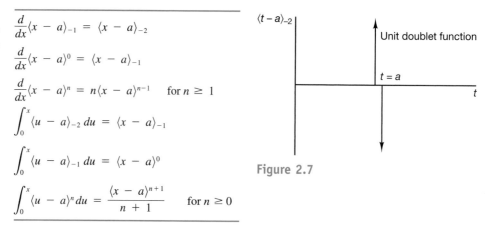

$$\frac{d}{dx}\langle x - a \rangle_{-1} = \langle x - a \rangle_{-2}$$

$$\frac{d}{dx}\langle x - a \rangle^{0} = \langle x - a \rangle_{-1}$$

$$\frac{d}{dx}\langle x - a \rangle^{n} = n\langle x - a \rangle^{n-1} \quad \text{for } n \geq 1$$

$$\int_{0}^{x}\langle u - a \rangle_{-2}\, du = \langle x - a \rangle_{-1}$$

$$\int_{0}^{x}\langle u - a \rangle_{-1}\, du = \langle x - a \rangle^{0}$$

$$\int_{0}^{x}\langle u - a \rangle^{n}\, du = \frac{\langle x - a \rangle^{n+1}}{n+1} \quad \text{for } n \geq 0$$

Figure 2.7

This solution is consistent with Eq. (2.25), as the integral of the Dirac delta function is equal to the Heaviside step function. As indicated earlier, the Dirac delta function is also written using the following notation:

$$\delta(t - a) = \langle t - a \rangle_{-1} \tag{2.27}$$

The bracket notation is consistent with that used for other singularity functions, but should not be confused to mean the inverse of $\langle x - a \rangle$. The derivative of the Dirac delta function is defined, and it also has physical applications. Examination of the Dirac delta function shows that it has zero slope up to the point a, at which the slope becomes positive infinity, followed by a negative infinity slope, and then slope zero for all values of t greater than a. A function having these characteristics is called a *unit doublet function* and is shown in Figure 2.7.

The notation used for the unit doublet function is $\langle t - a \rangle_{-2}$ and is consistent with the notation used for other singularity functions. The unit doublet function can be considered to be a unit concentrated moment, or couple. Graphically, it appears as a positive Dirac delta function and a negative Dirac delta function separated by a distance of zero. Note that the unit doublet function is equivalent to a unit couple in a clockwise direction.

A list of derivatives and integrals for the discontinuity functions is given in Table 2.1. The Heaviside step function is used as a constant multiplier to initiate a function that is not a polynomial at some point. Integration of this function may be accomplished by parts:

$$\int_{a}^{b} u\, dv = uv\Big|_{a}^{b} - \int_{a}^{b} v\, du$$

$$u = \langle \xi - a \rangle^{0} \qquad dv = F(\xi)d\xi$$

$$du = \langle \xi - a \rangle_{-1}d\xi \qquad v = \int F(\xi)d\xi = G(\xi) \tag{2.28}$$

$$\int_{0}^{x} \langle \xi - a \rangle^{0} F(\xi)d\xi = \big[\langle \xi - a \rangle^{0} G(\xi)\big]\Big|_{0}^{x} - \int_{0}^{x} \langle \xi - a \rangle_{-1} G(\xi)d\xi$$

$$= \langle x - a \rangle^{0} G(x) - \langle x - a \rangle^{0} G(a)$$

The last term is integrated using the fact that the Dirac delta function is zero everywhere, except at the point a. If $x < a$, this integral is zero, and if $x > a$, the integral is the value of the function $G(x)$ at a.

A simple example of use of the Heaviside step function occurs when a mass m is subjected to an impact force for an instant at time t_1. The force during impact will be modeled using the Dirac delta function;

$$F(t) = P\langle t - t_1\rangle_{-1}$$

where P is the magnitude of the impact force. The dynamic equation of motion is

$$m\ddot{x} = P\langle t - t_1\rangle_{-1}$$

This is an equation of rectilinear motion where the acceleration is a function of time. The velocity is

$$m\dot{x} = P\langle t - t_1\rangle^0 + v_0$$

The position is

$$mx = P\langle t - t_1\rangle^1 + v_0 t + x_0$$

For the values $m = 100$ kg, $P = 1000$ N, $t_1 = 3$ s, $v_0 = 0$ m/s and $x_0 = 2$ m, the velocity–time and position–time curves are shown here.

$$m := 100 \qquad P := 1000 \qquad t_1 := 3$$
$$x_0 := 2 \qquad v_0 := 0$$

$$v(t) := \frac{1}{m} \cdot (P \cdot \phi(t - t_1) + m \cdot v_0)$$

$$x(t) := \frac{1}{m} \cdot [P \cdot \phi(t - t_1) \cdot (t - t_1) + m \cdot v_0 \cdot t + m)x_0]$$

$$t := 0, 0.01 \ldots 6$$

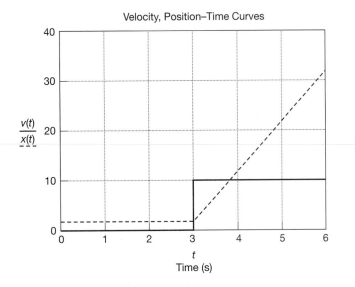

Velocity, Position–Time Curves

Note, that the velocity is zero until the time is 3 s, at which time the velocity at that instant becomes a little more than 10 m/s. The position increases from the initial position of 2 m at a constant rate from this time.

Sample Problem 2.9

A particle of mass m falls under the influence of gravitational attraction for 5 s when it collides with a rigid body and undergoes a impact force equivalent to 9 g's. Determine the resulting motion of the particle.

Solution

The force acting on the particle can be written as

$$F(t) = -mg + m(9)g\langle t - 5\rangle_{-1}$$

The coordinate x is taken positive in the upward direction, and the differential equation of motion is

$$m\ddot{x} = -mg + m(9)g\langle t - 5\rangle_{-1}$$

The mass can be canceled in the equation, and integrating yields the velocity and displacement as

$$\dot{x} = -gt + 9g\langle t - 5\rangle^0 + C_1$$

$$x = -g\frac{t^2}{2} + 9g\langle t - 5\rangle^1 + C_1t + C_2$$

where C_1 and C_2 are constants of integration. If the initial velocity and displacement are zero, the constants of integration C_1 and C_2 are zero. The displacement–time curve for the first 10 seconds of motion is

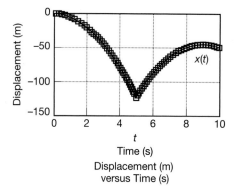

Time (s)
Displacement (m)
versus Time (s)

Sample Problem 2.10

A 200-lb stuntman jumps from a 20-ft platform and lands on a mat to break his fall, as depicted in the accompanying illustration. The mat has a spring constant of 100 lb/ft and damping linearly proportional to the velocity of 40 lb s/ft. Note that the damping occurs only while the mat is being compressed. Write the equation of motion for the fall of the stuntman, and plot his motion with time. Determine the force exerted during contact with the mat.

Solution

We will select the coordinate origin at the platform and the coordinate x to be positive in the downward direction. The force acting on the stuntman is

$$F(x) = mg - \langle x - 20\rangle^0[k(x - 20) + \langle\dot{x}\rangle^0c\dot{x}]$$

where k is the spring constant and c is the damping coefficient. Note that the mat does not exert a force on the stuntman until he has fallen 20 ft, and if he separates from the mat, it

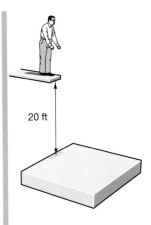

20 ft

has no influence. Note also that the damping term acts only when he is compressing the mat—that is, when the velocity is positive. The equation of motion is

$$\frac{d^2x}{dt^2} = g - \langle x - 20 \rangle^0 \left[\frac{k}{m}(x - 20) + \left\langle \frac{dx}{dt} \right\rangle^0 \frac{c}{m} \frac{dx}{dt} \right]$$

This equation is solved numerically, and the position–time graph is shown in the accompanying plot. (For details, see the Computational Supplement.)

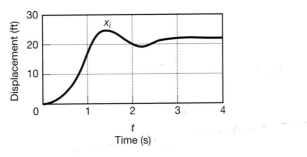

Note that, since the force depends upon the velocity and displacement, it cannot be determined until the differential equation is solved. The force that the mat exerts on the stuntman is

$$F(t) = \langle x - 20 \rangle^0 [k(x - 20) + \langle v \rangle^0 cv]$$

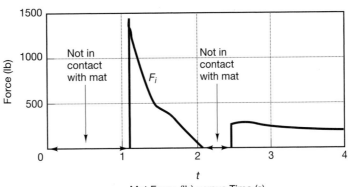

Mat Force (lb) versus Time (s)

On contact, the stunt man experiences a force of 1400 to 1500 pounds, or 7 to 7.5 BW (body weight). Although this figure may seem high, it is consistent with forces encountered in basketball, dismounts in gymnastics, and high-impact aerobics. The force settles to 200 lb, or 1 BW, supporting the stuntman's weight.

2.4 SPECIAL PROBLEM

At the end of some sections a special problem will be introduced to show how the methods of dynamics can be used to gain understanding of research questions.

This special problem arose when a biomedical engineer was studying the effects of a knee impact during an automobile accident. In the event of a front-end collision, frequently the passenger in the front seat will "tunnel" under the seat belt such that his/her knee will impact the dash. This can result in femur fracture and/or kneecap damage. The bioengineer had developed an experiment where the lower limb from a fresh cadaver was firmly mounted in a large mass and then impacted at the front of the knee by a force pulse lasting about 20 ms. The magnitude of the impact force was 7000 N applied as a sine wave for the 20 ms. The femur was assumed to have a stiffness of 1.3×10^6 N/m and although the damping of bone is not known, the damping c was assumed to be equal to 200 N $\cdot$ s/m.

The difficulty the research engineer was having was that the femur fractured at this impact load where the limb from the cadaver in a car seat did not fracture. The biomedical engineer approached a dynamics student for help with the problem. The dynamics student constructed a general model of the problem as shown here.

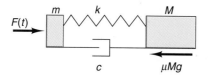

Here m is the mass of the femur and M is the mass of the mount or the upper body. To formulate the equations of motion for the two masses, we will use the δ notation for the Heaviside step function and note that if the argument of this function is negative, the step function is zero. (In the manner, we can stop the impact force after 20 ms.) The impact force can now be written as

$$F(t) = P \cdot \delta(\alpha - t) \sin \frac{\pi t}{\alpha}$$

where α is the duration time of the sine impulse and P is the maximum impact force. The equations of motion for the two masses are coupled through their velocities and positions and may be written as

$$a_1(v_1, x_1, v_2, x_2, t) = \frac{1}{m}\left[F(t) - k(x_1 - x_2) - c(v_1 - v_2)\right]$$

$$a_2(v_1, x_1, v_2, x_2, t) = \frac{1}{m}\left[k(x_1 - x_2) + c(v_1 - v_2) - \mu Mg\frac{v_2}{|v_2|}\right]$$

Now these two equations are coupled and can be solved only numerically. The following numerical values were selected:

$m = 16$ kg, $M = 660{,}000$ kg (mounting device), $k = 13{,}00{,}000$ N/m, $c = 200$N $\cdot$ s/m and a coefficient of kinetic friction between the mount and the base of $\mu = 0.1$. From the solution, the following curves were examined. First the spring compression $(x_1 - x_2)$ and the position of m (x_1) were plotted against time for the first 22 ms. This is shown in Figure 2.8.

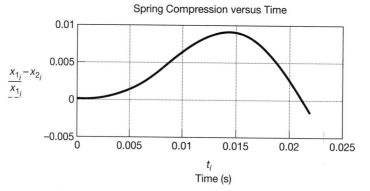

Figure 2.8

We can see that the compression of the spring and the position of the mass *m* are identical, indicating that the mounting mass *M* did not move. The spring force (bone force) versus time is shown in Figure 2.9 and the velocity of mass *m* is shown in Figure 2.10.

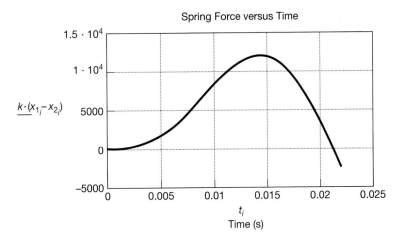

Figure 2.9

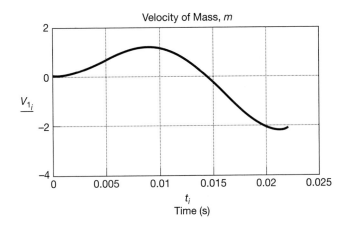

Figure 2.10

Note that the impulsive force had a maximum value of 7000 N and a duration of 20 ms, but due to *reflective waves* from the large mass at the end, the maximum force on the femur shown as the spring force is 12,000 N. This is because the wave time of the force moving down the femur is less than the 20 ms, allowing the wave to reflect and amplify the spring force with the maximum force occurring at 14 ms.

When the resisting mass is dropped to 66 kg (mass of the upper body), that mass moves and the spring force decreases, as shown in Figures 2.11, 2.12, and 2.13. Note that Figure 2.11 shows the spring compression and the movement of the first mass. For the case of the large fixture mass, the second mass does not move, and the spring compression is equal to the movement of the first mass. When the resisting mass in the model is reduced to the mass of the upper body, this mass moves and the spring compression is less. Therefore the spring force (bone force) is less.

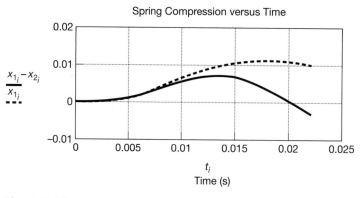

Figure 2.11

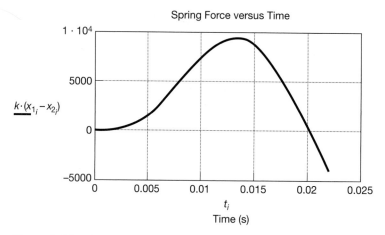

Figure 2.12

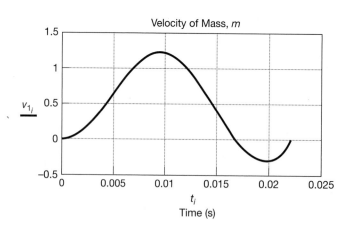

Figure 2.13

This "simple" dynamic analysis explains why the two experiments gave different results and which data were closer to actual, human impact conditions.

Problems

2.35 A 60-kg skydiver jumps at an altitude of 3 km. She free-falls for 1000 m and then deploys her chute. The chute produces a drag proportional to the velocity, that is, $\mathbf{F}_d = cv$ N, where c is related to the area of the chute by $c = 4.0$ A. Design the chute's area so that the terminal velocity of the skydiver is 9 m/s.

2.36 A 70-kg diver dives from a 10-m platform. Upon entering 3 m into the water, a buoyancy force equal to his body weight acts on him, and the water offers a drag force $F_d = cv$, where $c = 500$ Ns/m. Determine the impact force when the diver hits the water and the depth beneath the surface that he will go before stopping. (*Note: The initial impact force is large, which is why trained divers "punch a hole in the water"—so that the elbows will not buckle and drive their hands back into their faces.*)

2.37 A 1200-kg automobile accelerates at a linearly increasing rate of 0.7 m/s²/s for 5 seconds, and then decelerates by a constant rate of 2 m/s² until the car comes to a stop. Determine the distance that the car travels from start to stop.

2.38 A stuntwoman must make a jump from a height of 26 ft and land on a 6-ft thick mat (free-fall would be 20 ft). If she weighs 120 lb, design the mat such that the maximum total force acting on her body (gravity and contact force) is only about five times her body weight. Assume that the mat has a damping that is linearly proportional to the velocity and a spring force that is proportional to the square of the mat compression. (*Hint: Using trial-and-error methods, adjust the spring constant and the viscous damping coefficient. If she compresses the mat by 6 ft, she will make contact with the ground.*)

2.39 An experimental rocket engine delivers five constant bursts of propulsion of 50,000 N each for 3-s burns at intervals of 3 s. If the rocket has a mass of 500 kg, determine the final velocity of the rocket and the distance traveled at the end of the fifth burn. The initial velocity of the rocket is 100 m/s.

2.40 A 5-oz baseball crosses home plate at a velocity of 120 ft/s horizontally when the batter hits it with a force of **F** lb directed upward at a 20° angle with the field. If the impact with the bat lasts 15 ms, determine the force required to hit the ball so that it lands over 350 ft from home plate.

2.41 In Figure P2.41, a 2-kg box starts from rest and slides 4 m down a 45° incline, striking a spring with a spring constant of $k = 500$ N/m. If the coefficient of kinetic friction between the box and the incline is 0.6,

write the differential equation of motion, and plot the motion until the box comes to rest.

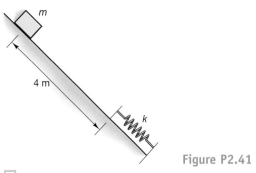

Figure P2.41

2.42 In Problem 2.41, add a damper parallel with the spring with a damper force proportional to the velocity. Choose a damper coefficient so that the box will not lose contact with the spring.

2.43 Consider a child sitting in the passenger seat of a car that is equipped with dual air bags. The mass of the child's head is 3.5 kg, and the damping force of the air bag is $F_d = cv$, where $c = 5.0$ N s/m. Here v is the relative velocity between the child's head and the air bag upon deployment of the latter during a collision. If the maximum impact force is to be less than 500 N when the child's head (moving forward at 30 m/s) meets the air bag, determine the maximum deployment velocity of the air bag.

2.44 A 5-kg particle is dropped from rest under the influence of gravity through a medium with viscous resistance proportional to the square of the particle's velocity; that is, $|\mathbf{F}_d| = -cv|v|$. If the damping constant is $c = 0.5$ N s²/m² and the particle falls 5 m before coming in contact with a spring with linear spring constant $k = 10$ N/m, determine the position of the particle as a function of time. (See Figure P 2.44.)

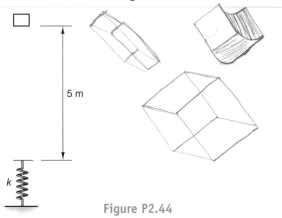

Figure P2.44

2.5   NORMAL AND TANGENTIAL COORDINATES

As shown in Chapter 1, tangential and normal coordinates are very useful in analyzing the curvilinear motion of a particle. The equations of motion in normal and tangential coordinates were given in Eq. (2.11) and are repeated here in a different form:

$$\sum F_t = m\frac{d^2s}{dt^2}$$

$$\sum F_n = \frac{m}{\rho}\left(\frac{ds}{dt}\right)^2$$

$$(2.29)$$

In these equations, s is the distance along the path of motion, and ρ is the radius of curvature of the path of motion. If the radius of curvature is constant, we can relate the differential distance along the path to the change in angle as

$$ds = \rho\, d\theta$$

Eq. (2.29) can then be written as follows:

$$\sum F_t = m\rho\frac{d^2\theta}{dt^2}$$

$$\sum F_n = m\rho\left(\frac{d\theta}{dt}\right)^2$$

$$(2.30)$$

The forces involved in a dynamics problem may easily be determined if the tangential acceleration and the magnitude of the velocity are known for a particular time or a particular position on the path of motion. This case is a form of the inverse dynamics problem. Conversely, if the forces are given as functions of time, then finding a solution to the direct dynamics problem is difficult, due to the nonlinear nature of the equation of motion in the normal direction.

Sample Problem 2.11   A car makes a left turn from a stopped position, increasing its speed at a rate of 5 ft/s². If a book is on the dashboard of the car, at what time will the book begin to slide if the coefficient of static friction between the book and the dashboard is 0.18? The radius of the curve of motion is 20 ft.

Solution   The tangential acceleration and the magnitude of the velocity at any time are

$$a_t = 5$$
$$v = 5t$$

The magnitude of the acceleration vector at any time is

$$|\mathbf{a}| = \sqrt{a_t^2 + a_n^2} = \sqrt{5^2 + \frac{(5t)^2}{20}} = \sqrt{5^2 + 1.25t^2}$$

The friction force must provide both components of the acceleration and will be in the direction of the acceleration vector. Therefore, the friction force is

$$f_{\text{max}} = \mu\, mg$$
$$\mu\, mg = m|\mathbf{a}|$$
$$0.18(32.2) = \sqrt{5^2 + 1.25t^2}$$

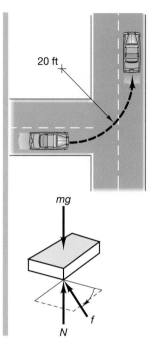

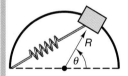

Solving for time yields

$$t = 2.622 \text{ s}$$

The angular distance traveled when the book slips is

$$s = \frac{5t^2}{2}$$

$$\rho\theta = s$$

$$\theta = \frac{5(2.622)^2}{2(20)} = 0.859 \text{ radians} = 49.2°$$

Thus, the car makes a little more than half of its left-hand turn before the book begins to slip.

Sample Problem 2.12

Determine the differential equation of motion for the spring-slider mechanism shown in the accompanying diagram. Neglect friction between the slider and the curved bar, and neglect the mass of the spring. The slider of mass m is released from rest when $\theta = 30°$, and the unstretched length of the spring with spring constant k is R.

Solution   First, we will draw a free-body diagram of the slider:
The length of the spring is

$$l_s = 2R \cos \beta$$

The magnitude of the spring force is

$$F_s = kR(2 \cos \beta - 1)$$

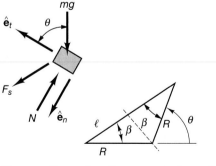

The equations of motion can now be written in the tangential and normal directions as

$$\hat{\mathbf{e}}_t : -mg \cos \theta + F_s \sin \beta = mR \frac{d^2\theta}{dt^2}$$

$$\hat{\mathbf{e}}_n : -N + mg \sin \theta + F_s \cos \beta = mR \left(\frac{d\theta}{dt}\right)^2$$

The differential equation of motion can be obtained from the tangential equation:

$$mR \frac{d^2\theta}{dt^2} = -mg \cos\theta + kR\left(2\cos\frac{\theta}{2} - 1\right)\sin\frac{\theta}{2}$$

This expression is the nonlinear differential equation of motion for the slider. The equation can be solved using computational software, and the details are shown in the Computational Supplement. A plot of the angle versus time is shown, where the initial angle $\theta_0 = \pi/6$, the initial angular velocity is zero, $R = 0.2$ m, $m = 1$ kg, and $k = 600$ N/m.

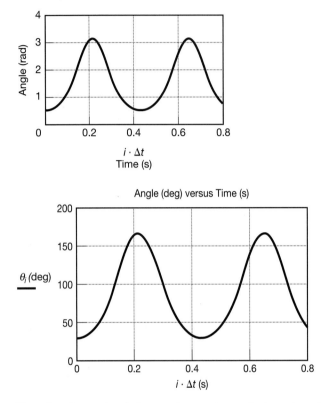

Note that the motion is harmonic, as expected, since there is no damping in the system.

Sample Problem 2.13

Determine the equations of motion for a simple, plane pendulum composed of a mass m at the end of a cord of length L. The pendulum is released from rest at an initial angle θ_0. Solve the linearized and nonlinear equations.

Solution   The problem is best formulated in normal and tangential coordinates. The free-body diagram of the pendulum is as follows:

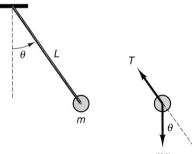

The radius of the pendulum's swing is a constant equal to L, so the equations of motion are

$$-T + mg \cos \theta = -mL\dot{\theta}^2$$
$$-mg \sin \theta = mL\ddot{\theta}$$

For small angles, the second equation becomes

$$\ddot{\theta} + \frac{g}{L}\theta = 0$$

with the initial condition $\theta(0) = \theta_0$. The linear second-order differential equation with constant coefficients can be solved, yielding

$$\theta(t) = \theta_0 \cos \sqrt{\frac{g}{L}} \, t$$

This nonlinear differential equation is solved numerically. (See the Computational Supplement.) The solution is

$$\theta(t + \Delta t) = \theta(t) + \omega(t)\Delta t$$
$$\omega(t + \Delta t) = \omega(t) - \frac{g}{L}[\sin \theta(t)] \, \Delta t$$

where $\omega = d\theta/dt$.

Problems

2.45 Determine the normal force between the seat and a 130-lb passenger in the car shown in Figure P2.45 as the car reaches the bottom of the hill if the speed of the car is constant at 60 mph.

Figure P2.45

2.46 The police investigate an accident that they suspect occurred because some high school students went on a ride to "fly a hill." If the radius of curvature of the hill is 500 ft, determine the minimum speed the car was traveling in order to become airborne at the top of the hill.

2.47 Use the illustration in Figure P2.47 to develop a general formula to determine the maximum speed that a car can go around a banked curve in terms of the bank angle, the radius of the curve, and the coefficient of static friction between the tires and the pavement. Formulas of this nature are necessary for highway design and accident reconstruction.

Figure P2.47

2.48 If an exit ramp on an expressway is designed with a radius of curvature of 100 ft and maximum traffic speed of 25 mph, determine the minimum bank angle. In bad weather, the coefficient of static friction could be as low as 0.3.

2.49 Jet fighter pilots can black out if they pull out of a high speed dive too sharply. This loss of consciousness is caused by loss of blood to the brain. If the maximum safe acceleration is 5 g's, determine the minimum radius of curvature at the bottom of a 700-mph dive.

2.50 Determine the angle β at which a particle of mass m leaves the smooth curved surface if the particle's initial velocity is v_0.

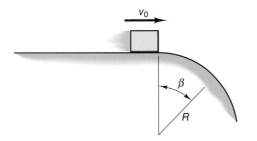

Figure P2.50

2.51 A block of mass m slides down the surface of a smooth bowl of radius R, as shown in Figure P2.51. Write the differential equation of the block's motion. If the block is released from rest at $\theta = \theta_0$, determine the velocity for any position of the block.

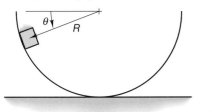

Figúre P2.51

2.52 If the coefficient of kinetic friction between the block and the bowl in Problem 2.51 is μ_k, write the differential equation of motion for the block. (*Note: Remember that friction always opposes the motion. See note on Problem 2.18.*)

2.53 For the conditions given in Problem 2.52, select a mass, bowl radius, and coefficient of kinetic friction, and numerically design a system such that the block stops after two oscillations.

2.54 If the coefficient of kinetic friction between the slider and the bar in Sample Problem 2.12 is μ_k write the equation of motion for the slider. Using the values $m = 1$ kg, $k = 600$ N/m, $R = 0.2$ m, and $\mu_k = 0.18$ (a lightly greased rod), study the motion for the first 1.5 s and determine the final equilibrium position. (*Hint: the normal force between the slider and the bar may be directed in the either positive or negative normal direction, and friction always opposes motion, such that $f = -\mu_k|N|\omega/|\omega|$ where ω is the angular velocity of the slider.*)

2.55 To determine the coefficient of static friction, a block is placed on a rotating disk, as illustrated in Figure P2.55. The disk accelerates from rest at a constant angular acceleration α, and the time between the start of rotation and the instant that the block slips is measured by a digital timer. Develop the general formula to determine the coefficient of static friction in terms of the radius r, the angular acceleration, g, and the time it takes for the block to slip.

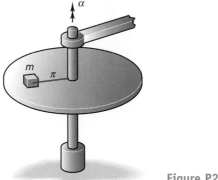

Figure P2.55

2.56 Two blocks are connected by a rope of length l, as shown in Figure P2.56, and placed on a disk rotating at a constant angular velocity ω. If $m_A > m_B$ and friction is negligible, determine the position of the two blocks, in terms of their masses and the length of the rope.

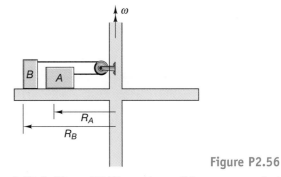

Figure P2.56

2.57 In Figure P2.57, a mass m slides on a smooth ring of radius R, which is rotated at a constant angular velocity ω. Determine the angle θ at which the mass will remain while the ring rotates at the constant angular velocity.

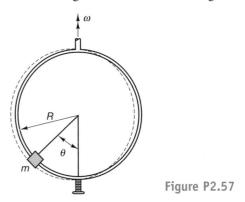

Figure P2.57

2.58 A roller-coaster car is to go around a hoop of radius R, as illustrated in Figure P2.58. For safety reasons, the force between the occupants and their seats should be one-half the occupant's weight at the top of the hoop. Determine the minimum velocity of the car as it enters the bottom of the hoop. Neglect friction between the wheels and the track.

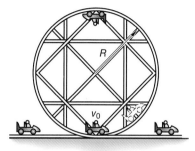

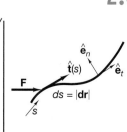

Figure P2.58

2.59 For the roller-coaster car in Problem 2.58, what is the normal force between the seat and the occupants at the bottom of the hoop? Could this force cause spinal compression problems?

2.60 In Figure P2.60, determine the acceleration of a 3-kg block A and the acceleration of a 2-kg block B at the instant that they are released when $\theta = 30°$. Neglect friction between block A and the plane.

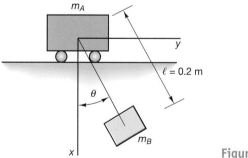

Figure P2.60

2.61 Write the equation of motion for the angle $\theta(t)$ in Problem 2.60. (*Hint: Create the free-body diagram with block B at a general position, and choose the coordinate system as shown in the figure.*)

2.62 Determine the acceleration of the 3-kg block A and the acceleration of the 2-kg block B in Figure P2.62 at the instant that they are released on a plane inclined at $\beta = 30°$ to the horizontal. Neglect friction between block A and the plane.

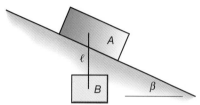

Figure P2.62

2.63 Write the equation of motion for the angle $\theta(t)$ that block B makes with the vertical in Problem 2.62.

2.64 Numerically solve the differential equation in Problem 2.61 using the following numerical values: $m_A = 3$ kg, $m_B = 2$ kg, $l = 0.5$ m, $\theta(0) = 20°$, and $\omega(0) = 0$.

2.65 Numerically solve the differential equation in Problem 2.63 using the following numerical values: $m_A = 3$ kg, $m_B = 2$ kg, $l = 0.5$ m, $\beta = 30°$, $\theta(0) = 20°$, and $\omega(0) = 0$.

2.66 At the Indianapolis 500, the lateral acceleration on a curve is measured as 3 g's. If the average speed of a car is 200 mph, determine the radius of the curve. The coefficient of static friction between the tires and the surface is 0.85. Determine the bank angle of the curve so that the cars will not slip.

2.6 TWO-DIMENSIONAL PARAMETRIC EQUATION OF DYNAMICS

In Section 1.13, the kinematics of a particle was described in terms of a parameter s—that is, the distance along the trajectory of the particle. If a particle is constrained to move in the xy-plane, a possible trajectory is shown in Figure 2.14. The unit vector $\hat{\mathbf{t}}(s)$ is the tangential base vector $\hat{\mathbf{e}}_t$. The change of the position vector is

$$\mathbf{dr} = \hat{\mathbf{t}}(s)ds \tag{2.31}$$

The parameter s is a function of time, and its first derivative is the magnitude of the velocity of the particle. The second derivative of s is the magnitude of the tangential

Figure 2.14

component of the acceleration. The unit tangent vector can be related to the rectangular coordinates as

$$\hat{\mathbf{t}}(s) = \cos \theta(s)\hat{\mathbf{i}} + \sin \theta(s)\hat{\mathbf{j}} \tag{2.32}$$

The angle $\theta(s)$ is the angle between the unit tangent vector and the x-axis and is a function of position s and, therefore, a function of time. The position vector to the particle is obtained by integrating Eq. (2.31):

$$\mathbf{r}(s) = \int_0^s \hat{\mathbf{t}}(\eta)d\eta \tag{2.33}$$

If $\theta(s)$ is specified, the trajectory of the particle can be obtained from the scalar components of Eq. (2.33) as

$$x(s) = x_0 + \int_0^s \cos \theta(\eta)d\eta$$
$$\tag{2.34}$$
$$y(s) = y_0 + \int_0^s \sin \theta(\eta)d\eta$$

where x_0 and y_0 are the initial coordinates of the particle.

The velocity of the particle in terms of the parameter s is

$$\mathbf{v}(s) = \frac{ds}{dt}\left\{\cos \theta(s)\hat{\mathbf{i}} + \sin \theta(s)\hat{\mathbf{j}}\right\} \tag{2.35}$$

Differentiating Eq. (2.35) with respect to time yields

$$\mathbf{a}(s) = \frac{d^2s}{dt^2}\left\{\cos \theta(s)\hat{\mathbf{i}} + \sin \theta(s)\hat{\mathbf{j}}\right\} + \frac{ds}{dt}\left\{-\sin \theta(s)\hat{\mathbf{i}} + \cos \theta(s)\hat{\mathbf{j}}\right\}\frac{d\theta(s)}{ds}\frac{ds}{dt}$$

$$\mathbf{a}(s) = \frac{d^2s}{dt^2}\left\{\cos \theta(s)\hat{\mathbf{i}} + \sin \theta(s)\hat{\mathbf{j}}\right\} + \left(\frac{ds}{dt}\right)^2 \frac{d\theta(s)}{ds}\left\{-\sin \theta(s)\hat{\mathbf{i}} + \cos \theta(s)\hat{\mathbf{j}}\right\}$$
$$\tag{2.36}$$

The first term is the tangential acceleration, and the second term is the normal acceleration. The equations of motion are now written in a similar manner as those shown in Section 2.4. The equations of motion can be written in terms of the parameter s:

$$\text{Tangential: } \sum \mathbf{F} \cdot [\cos \theta(s)\hat{\mathbf{i}} + \sin \theta(s)\hat{\mathbf{j}}] = m\frac{d^2s}{dt^2} \tag{2.37}$$

$$\text{Normal: } \sum \mathbf{F} \cdot [-\sin \theta(s)\hat{\mathbf{i}} + \cos \theta(s)\hat{\mathbf{j}}] = m\left(\frac{ds}{dt}\right)^2 \frac{d\theta(s)}{ds} \tag{2.38}$$

| **Sample Problem 2.14** | A driver wants to test-drive a vehicle in a tightening spiral path at a constant speed on the salt flats of Utah. If the coefficient of static friction between the tires and the ground is 0.5 and the test driver is driving at 60 ft/s, determine the point at which the car will slide out of the curve if the curve is given by |

$$\theta(s) = 1 + \left(\frac{s}{1000}\right)^2$$

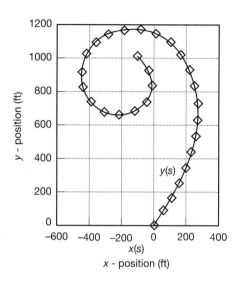

Solution The normal acceleration at any point on the curve is

$$\theta(s) = 1 + \left(\frac{s}{1000}\right)^2$$

$$\frac{d\theta}{ds} = \frac{2s}{(1000)^2}$$

$$a_n = (60)^2 \frac{2s}{(1000)^2} = \frac{7.2s}{1000}$$

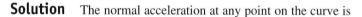

The maximum friction force for the car to hold to the flats is

$$F_n = 0.5\, mg$$

This force is needed to prevent sliding; therefore, the required normal acceleration is

$$a_n = 7.2s/1000$$

The equation of motion is

$$F_n = ma_n$$

and the car will slide when

$$0.5\,(32.2) = 7.2s/1000$$
$$s = 2236 \text{ ft into the path}$$

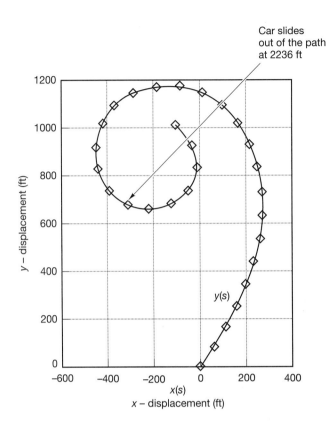

The path of motion is shown in the Computational Supplement.

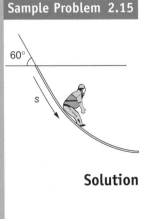

Sample Problem 2.15

A child with a mass of 20 kg slides down a 3-m long slide in the playground. The curve of the slide can be expressed parametrically as

$$\theta(s) = -60° \left[1 - \left(\frac{s}{3} \right)^2 \right]$$

If the slide lenth is 1 m and the child starts with zero velocity, determine the velocity of the child at the bottom of the slide. The coefficient of kinetic friction between clothing and the slide surface is 0.2.

Solution   This problem is best formulated in normal and tangential coordinates using Eq. (2.30). A free-body diagram of the child modeled as a particle is shown in the accompanying diagram for a positive angle θ. *The slide goes down to the right, but the free-body diagram has been drawn for a positive angle to avoid sign errors in formulating the mathematics of the problem.* Friction is assumed to oppose motion.

The equation of motion in the normal direction is

$$N - W \cos \theta(s) = m \left(\frac{ds}{dt} \right)^2 \frac{d\theta}{ds}$$

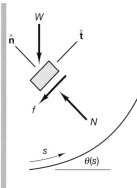

The equation of motion in the tangential direction is

$$-W \sin \theta(s) - f = m \frac{d^2 s}{dt^2}$$

Substituting the normal force into the equation of motion in the tangential direction yields

$$\frac{d}{dt}\left(\frac{ds}{dt}\right) = -g \sin \theta(s) - \mu \left\{ g \cos \theta(s) + \left(\frac{ds}{dt}\right)^2 \frac{d\theta}{ds} \right\}$$

Note that the differential equation of motion is independent of the mass of the child. This expression is a first-order nonlinear differential equation for the magnitude of the velocity vector. Note that s is a function of time, and therefore, θ is a function of time. The problem may be solved using modern computational tools or by creating special computer programs to integrate the equation numerically. The numerical coupled equations for Euler's method are

$$s(t + \Delta t) = s(t) + v(t)\Delta t$$

$$\theta(t + \Delta t) = -60\left[1 - \left(\frac{s(t)}{3}\right)^2\right] \deg$$

$$\frac{d\theta}{ds}(t + \Delta t) = \frac{120}{9} s(t)$$

$$v(t + \Delta t) = v(t) + \left[-g \sin \theta(t) - \mu\left\{g \cos \theta(t) + (v(t))^2 \frac{d\theta(t)}{ds}\right\}\right]\Delta t$$

Initial values for each of these functions are selected and used to predict the next value of the variable. (The numerical solution is shown in the Computational Supplement.)

Problems

All problems in this section require the use of computational software.

2.67 A safety engineer is to analyze a 3-m-long slide in a playground. The contour of the slide is given by the parametric equation $\theta(s) = -70°[1-(s/L)^3]$, where s is measured along the length of the slide and L is the length of the slide. Note that the initial angle with the horizontal is 70° in the downward direction and the angle at the end of the slide is 0°. If the coefficient of kinetic friction for the slide is 0.3, determine the velocity of a child after she slides down the slide. Compare this velocity to the velocity if friction is ignored. How long does it take the child to reach the bottom of the slide?

2.68 For the slide in Problem 2.67, graph the slide contour, and determine the velocity at the ground of a child

who has fallen from the top of the slide. Assume that the bottom of the slide is 0.6 m above the ground.

2.69 Numerically design a 20-m-high water slide for an amusement park. The slide should have a parametric curve $\theta(s) = -\alpha[1 - (s/L)^2]$, where α is the initial downward angle and L is the length of the slide. The equivalent coefficient of kinetic friction can be adjusted by the water velocity on the slide. You want to design the slide to get the maximum amount of time on the slide, but still have a positive velocity at the bottom.

2.70 The vertical contour of a country road is given by the parametric curve $\theta(s) = (s/25{,}000) \cos(s/500)$. This curve is valid for the first 3000 m of the roadway. Plot the contour of the road, and determine the normal force between the roadway and a car traveling along the road at a constant speed of 100 km/hr. (The force can be expressed in *g*'s.)

2.71 A car in an amusement park ride runs on a horizontal plane along a parametric curve for the first 7000 ft. The equation that describes this curve is $\theta(s) = (s/4000)$ $\cos(s/400) - (s/4000)$. If the coefficient of static friction between a person's clothing and the car's seat is 0.5, will a passenger in the car slip if the car runs at an average speed of 66 ft/s?

2.72 A snowmobile wants to execute a tight turn along a 20-m path with a parametric curve $\theta(s) = 5\sin^{-1}$ $[0.2 - 0.5\cos(s/10)]$. Determine the minimum coefficient of static friction between the snowmobile's runners and the snow in order for the snowmobile to execute the turn at a speed of 10 m/s.

2.7 POLAR COORDINATES

When a particle is constrained to move in a curvilinear manner, it is usually easier to select curvilinear coordinates as the base coordinate system. If the particle's path is circular in nature, then polar or cylindrical coordinates are best for modeling the particle's path. For general curvilinear motion, the normal and tangential coordinates are most appropriate. The scalar equations of motion in polar coordinates are

$$\sum F_r = m(\ddot{r} - r\dot{\theta}^2)$$

$$\sum F_\theta = m(2\dot{r}\dot{\theta} + r\ddot{\theta})$$

(2.39)

The equations for the velocity and acceleration are given in Section 1.8.

If the external forces acting on the particle are known, the direct dynamics problem requires the solution of the two coupled nonlinear differential equations. Solution of these equations is possible in a closed form in some limited cases, but, in general, will require some numerical techniques. The inverse dynamics problem—that is, when *r* and *θ* are given as functions of time and the forces required to sustain the particle's motion are to be determined—is not difficult. As discussed earlier, if the kinematic data are obtained by a motion analysis system, noise will increase with differentiation, and care must be taken to use digital filters and other numerical techniques.

Sample Problem 2.16

A 2-kg mass moves in a horizontal plane such that

$$r(t) = 3 - 2t + 3t^2$$

$$\theta(t) = 1 + t^2$$

Plot the path that the particle makes in the horizontal plane, and determine the external force acting on the particle during the first 2 s if the angle is expressed in radians and the radius is expressed in meters.

Solution   This example is an inverse dynamics problem, and the acceleration is obtained by differentiation of the polar coordinates as follows:

$$\dot{r}(t) = -2 + 6t$$
$$\ddot{r}(t) = 6$$
$$\dot{\theta}(t) = 2t$$
$$\ddot{\theta}(t) = 2$$

The force required to sustain this motion is found using Eq. (2.39):

$$F_r(t) = 2[6 - (3 - 2t + 3t^2)(4t^2)] = 12 - 24t^2 + 16t^3 - 24t^4 \text{ N}$$

$$F_\theta(t) = 2[2(-2 + 6t)(2t) + (3 - 2t + 3t^2(2))] = 12 - 24t + 60t^2$$

The path of motion is shown in the accompanying *xy*-plot (in meters) for the first 4 s.

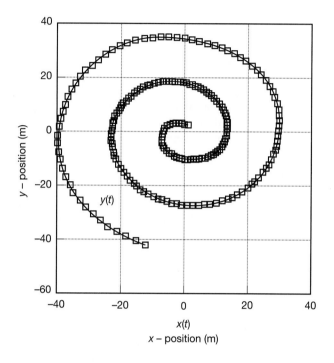

The forces needed to produce this motion are graphed for the first 3 seconds in the following plot:

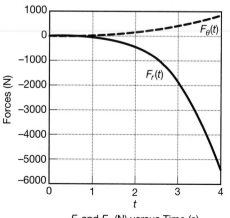

F_r and F_θ (N) versus Time (s)

Sample Problem 2.17

A 2-kg mass is attached to a spring and released from rest at an angle $\theta = 30°$ and with the spring unstretched. If the spring's unstretched length is 300 mm and the spring constant is 1600 N/m, determine the motion of the mass.

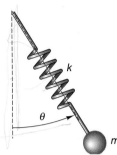

Solution The free-body diagram for the mass shows the two forces acting on the mass: The spring force is $F_s = k(r - 0.3)$ N.

The equations of motion are

$$-k(r - 0.3) + mg \cos \theta = m(\ddot{r} - r\dot{\theta}^2)$$

$$-mg \sin \theta = m(2\dot{r}\dot{\theta} + r\ddot{\theta})$$

The two coupled differential equations are nonlinear and are solved numerically in the Computational Supplement.

The results of the numerical solution for one complete oscillation is shown in the accompanying graph. (Details of the solution are shown in the Computational Supplement.) The angle of the spring–pendulum is plotted against time for the first 2 s:

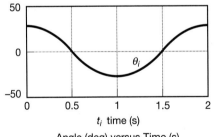

Angle (deg) versus Time (s)

We can plot the *xy*-position of the mass for one oscillation.

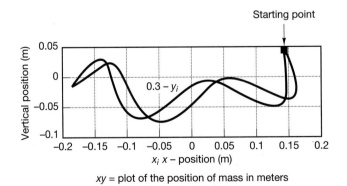

xy = plot of the position of mass in meters

Problems

2.73 A particle moves such that its path (in polar coordinates) as a function of time is given by

$$r(t) = 0.300 + 0.100 \cos(\pi t)$$

$$\theta(t) = \frac{\pi}{6} \sin(\pi t)$$

Determine the velocity and acceleration of the particle as a function of time, and determine the forces required to act on the particle in order for it to maintain the motion.

2.74 A particle of mass 1 kg is subjected to a force $\mathbf{F}(t) = (3t^2 - 1)\, \hat{\mathbf{e}}_r + \cos[(\pi/6)t)]\hat{\mathbf{e}}_\theta$ N. Determine the trajectory of the particle if the particle starts from rest, with an initial position of $r = 2$ m, and $\theta = 0$.

2.75 A 4-kg mass attached to a spring with a spring constant of 40 N/m rotates on a smooth table, as depicted in Figure P2.75. The unstretched length of the spring is 0.2 m. If the particle rotates such that its angular position is given by $\theta(t) = 0.1t^2$ rad, and the particle starts from rest with $r = 0.2$ m, determine the position of the particle at any time.

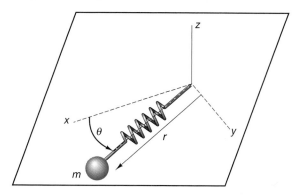

Figure P2.75

2.76 Determine the position of the particle in Problem 2.75 at any time if the coefficient of kinetic friction between the mass and the table is 0.2 and the spring constant is 20 N/m. (*Hint: Friction will resist the velocity in both the radial and transverse directions.*) (*See note on Problem 2.18.*)

2.77 The radius of a cam is given by $r(\theta) = 0.200 [2 - \cos\theta]$ m. The spring-loaded 4-kg cam follower is rotated in a horizontal plane by a fixed bar at a constant angular velocity of 1.5 rad/s, as shown in Figure P2.77. Determine the radial component of the normal force as a function of position if the spring constant is 650 N/m and the unstretched length of the spring is 0.100 m.

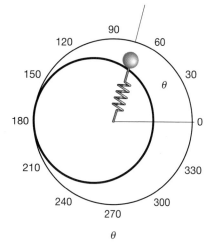

Figure P2.77

2.78 Determine the radial component of the normal force in Problem 2.77 if the angular velocity of the cam follower is $\omega(\theta) = 2 - \cos(\theta)$ rad/s.

2.79 Adjust the mass in Sample Problem 2.17 so that the oscillation will be symmetric about the vertical position. Oscillation should appear as shown in Figure P2.79.

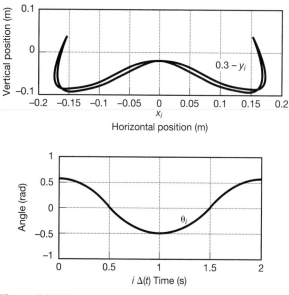

Figure P2.79

2.80 A pendulum hangs in a vertical plane as shown in Figure P2.80. The mass m is attached to the pivot point by a spring with unstretched length L and spring constant k.

a) Write the equations of motion for the system if the system is released from rest at an angle θ_0.

b) Numerically solve for the motion if $\theta_0 = 30$ degrees, $m = 2$ kg, $k = 60$ N/m, and $L = 2$ m.

c) Increase the stiffness of the spring to 600 N/m and show that the solution agrees with the solution of Sample Problem 2.13.

(*Hint: If the equations are solved using the Euler method, the solution for $r(t)$ will be unstable unless the time increment is very small. For nonlinear problems of this nature, it may be necessary to use a Runge-Kutta method.*)

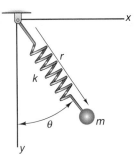

Figure P2.80

2.81 In Figure P2.81, a particle is given an initial velocity at the bottom of a vertical circular ring of radius R. Determine the minimum initial angular velocity required for the particle to move around the ring without losing contact with the ring's inner surface.

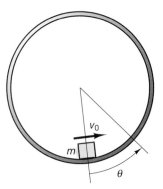

Figure P2.81

⌨ **2.82** A 4-kg mass slides between two rings, as shown in Figure P2.82. If the mass is given an initial velocity of 5 m/s when it is at its lowest point, and the

coefficient of kinetic friction between the rings and the mass is 0.2, determine the motion of the mass. (*Hint: The normal force can act in either the positive or negative radial direction, so the friction force must be written as:* $f = -\mu_k |N| (\dot{\theta}/|\dot{\theta}|)$)

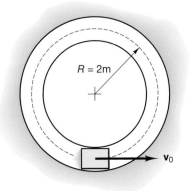

Figure P2.82

2.83 A mass m is constrained to the surface of a smooth disk of radius R by a spring with spring constant k and unstretched length $R/8$, as illustrated in Figure P2.83. Write the equation of motion of the mass if the disk is in a horizontal plane and the mass is given an initial velocity v_0 when $\theta = 0$.

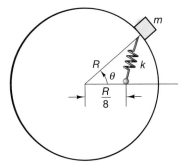

Figure P2.83

2.84 Write the equation of motion for the disk in Problem 2.83 if the disk is in the vertical plane.

2.85 Write the equation of motion for the disk in Problem 2.83 if the disk is in a horizontal plane and the coefficient of kinetic friction between the disk and the mass is μ_k. (*See note on Problem 2.18.*)

⌨ **2.86** Place the disk of Problem 2.85 in the vertical plane, and pick values for the mass, spring constant, disk radius, initial angle, and coefficient of kinetic friction such that the motion damps out after two oscillations.

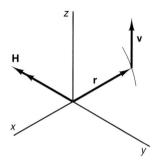

Figure 2.15

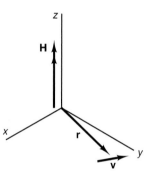

Figure 2.16

2.7.1 ANGULAR MOMENTUM OF A PARTICLE

It is useful in our discussion of polar coordinates to introduce the concept of the *angular momentum of a particle about a point*. Consider a particle of mass m moving in space, as illustrated in Figure 2.15. The position of the particle at any time is specified by the position vector $\mathbf{r}(t)$, starting from the origin of a Newtonian reference coordinate system. A quantity called the angular momentum is defined as the moment of the linear momentum $m\mathbf{v}$ and is designated by $\mathbf{H}_0$. We have

$$\mathbf{L} = m\mathbf{v}$$
$$\mathbf{H}_0 = \mathbf{r} \times \mathbf{L} = \mathbf{r} \times (m\mathbf{v}) \tag{2.40}$$

The units of the angular momentum vector in the S.I. system are $\mathrm{m \cdot kg \cdot m/s = kg \cdot m^2/s}$ and in the U.S. customary system are $\mathrm{slugs \cdot ft^2/s}$. Note that the angular momentum vector has a subscript zero to indicate that the angular momentum is taken about the origin. The angular momentum of a particle can be computed about any point in space. The scalar components of the angular momentum, expressed in rectangular coordinates, are

$$\mathbf{H}_0 = (x\hat{\mathbf{i}} + y\hat{\mathbf{j}} + z\hat{\mathbf{k}}) \times (mv_x\hat{\mathbf{i}} + mv_y\hat{\mathbf{j}} + mv_z\hat{\mathbf{k}})$$
$$H_{0x} = myv_z - mzv_y$$
$$H_{0y} = mzv_x - mxv_z \tag{2.41}$$
$$H_{0z} = mxv_y - myv_x$$

If the particle is constrained to move in a plane—for example, the xy-plane—the angular momentum is in the z-direction, as illustrated in Figure 2.16. The angular momentum about the origin for this planar motion is

$$\mathbf{H}_0 = m(xv_y - yv_x)\hat{\mathbf{k}} \tag{2.42}$$

The time rate of change of the angular momentum is obtained by differentiation of Eq. (2.40):

$$\dot{\mathbf{H}}_0 = \frac{d\mathbf{H}_0}{dt} = \dot{\mathbf{r}} \times m\mathbf{v} + \mathbf{r} \times \frac{d}{dt}(m\mathbf{v})$$
$$\dot{\mathbf{r}} \times m\mathbf{v} = \mathbf{v} \times m\mathbf{v} = m(\mathbf{v} \times \mathbf{v}) = 0$$

Therefore,

$$\dot{\mathbf{H}}_0 = \mathbf{r} \times \frac{d}{dt}(m\mathbf{v}) \tag{2.43}$$

Now, recall that Newton's second law is

$$\sum \mathbf{F} = \frac{d}{dt}(m\mathbf{v}) \tag{2.44}$$

If the vector product of a position vector $\mathbf{r}$ is taken with both sides of Eq. (2.44) (in other words, cross Eq. (2.44) with $\mathbf{r}$), the following relationship is obtained:

$$\mathbf{r} \times \sum \mathbf{F} = \mathbf{r} \times \frac{d}{dt}(m\mathbf{v})$$
$$\sum \mathbf{M} = \frac{d}{dt}\mathbf{H}_0 \tag{2.45}$$

The left-hand side of Eq. (2.45) is the moment of the forces about the origin, and from Eq. (2.41), the right-hand side is the time derivative of the angular momentum about the origin. Therefore, Eqs. (2.44) and (2.45) relate the forces to the change of linear momentum and the moment to the change of angular momentum, thereby completely describing the dynamics of the particle in space.

For specific types of plane motion, it is convenient to express Eqs. (2.44) and (2.45) in polar coordinates. The equation for the angular momentum is then

$$\begin{aligned} \mathbf{r} &= r\hat{\mathbf{e}}_r \\ \mathbf{v} &= \dot{r}\hat{\mathbf{e}}_r + r\dot{\theta}\hat{\mathbf{e}} \\ \mathbf{H}_0 &= \mathbf{r} \times m\mathbf{v} = mr^2\dot{\theta}\hat{\mathbf{e}}_z \end{aligned} \tag{2.46}$$

Equations (2.42) and (2.43) become

$$\begin{aligned} \sum F_r &= m(\ddot{r} - r\dot{\theta}^2) \\ \sum F_\theta &= m(r\ddot{\theta} + 2\dot{r}\dot{\theta}) \\ \sum M_0 &= \sum rF_\theta = m(2r\dot{r}\dot{\theta} + r^2\ddot{\theta}) \end{aligned} \tag{2.47}$$

Note that even though the motion is planar, the laws of dynamics have scalar equations in all three dimensions.

2.7.2 CENTRAL-FORCE MOTION

If the only force acting upon a particle is directed toward a fixed point in space, the particle undergoes what is called *central-force motion*. This motion has many applications in celestial mechanics. It also is of great historical importance, as will be shown later in the text. Problems involving this type of motion are expressed in polar coordinates, and the origin of each system is chosen to be the fixed point in space toward which the central force is directed. The equations of motion for such a system are

$$\begin{aligned} F_r &= m(\ddot{r} - r\dot{\theta}^2) \\ F_\theta &= m(r\ddot{\theta} + 2\dot{r}\dot{\theta}) \end{aligned} \tag{2.48}$$

Figure 2.17 shows the motion for the case in which a particle is subjected to a single central force. The force exerted upon the particle acts inward toward the origin and can be written as

$$\mathbf{F} = -F_r\hat{\mathbf{e}}_r \tag{2.49}$$

The equations of motion become

$$\begin{aligned} -F_r &= m(\ddot{r} - r\dot{\theta}^2) \\ 0 &= m(r\ddot{\theta} + 2\dot{r}\dot{\theta}) \end{aligned} \tag{2.50}$$

The second equation can be written as

$$\frac{1}{r}\left[\frac{d}{dt}\left(r^2\frac{d\theta}{dt}\right)\right] = 0 \tag{2.51}$$

Integrating this equation yields

$$r^2\dot{\theta} = h \tag{2.52}$$

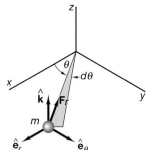

Figure 2.17

where h is the constant of integration. The constant h is the angular momentum of the particle divided by the mass, or the angular momentum per unit mass:

$$h = \frac{H_0}{m} \tag{2.53}$$

Examining Figure 2.17, we can see that the shaded triangle has an area of

$$A = \tfrac{1}{2}r\,(r\,d\theta) \tag{2.54}$$

and the change in this area with time is

$$\frac{dA}{dt} = \tfrac{1}{2}r^2\,\dot{\theta} \tag{2.55}$$

This change in area with time is called the *areal velocity*. Note that, for a particle under the influence of central force motion, the areal velocity is a constant. In the case of the solar system, where central-force motion is due to the sun's influence on a planet, the planet's position vector from the sun sweeps out equal areas in equal time intervals. This phenomenon was discovered by the German astronomer Johannes Kepler and became his second law of planetary motion. More will be said of his work later.

The concept of constant angular momentum per unit mass can be used to rewrite the first equation of motion as

$$r^2\,\dot{\theta} = h$$

$$\dot{\theta}^2 = \frac{h^2}{r^4} \tag{2.56}$$

and

$$\frac{dr}{dt} = \frac{dr}{d\theta}\dot{\theta} = \frac{dr}{d\theta}\frac{h}{r^2} = -h\frac{d}{d\theta}\left(\frac{1}{r}\right) \tag{2.57}$$

Differentiating again, we obtain

$$\frac{d^2r}{dt^2} = \frac{d}{dt}\left[-h\frac{d}{d\theta}\left(\frac{1}{r}\right)\right] = \frac{d}{d\theta}\left[-h\frac{d}{d\theta}\left(\frac{1}{r}\right)\right]\dot{\theta}$$

But

$$\dot{\theta} = \frac{h}{r^2}$$

$$\frac{d^2r}{dt^2} = -\frac{h^2}{r^2}\frac{d^2}{d\theta^2}\left(\frac{1}{r}\right) \tag{2.58}$$

Using Eq. (2.58), we can write the first equation of motion as

$$-F_r = m\left[-\frac{h^2}{r^2}\frac{d^2}{d\theta^2}\left(\frac{1}{r}\right) - \frac{h^2}{r^3}\right] \tag{2.59}$$

The form of this differential equation suggests a change of variables. We introduce a new variable, $u = 1/r$ and the equation becomes

$$h^2u^2\frac{d^2u}{d\theta^2} + h^2u^3 = \frac{F_u}{m}$$

or

$$\frac{d^2u}{d\theta^2} + u = \frac{F_u}{mh^2u^2} \tag{2.60}$$

This expression is the equation that defines the path of motion of the particle under the influence of a central force, expressed in terms of the new variable u. $F(u)$ has been taken as positive directed toward the fixed origin.

Consider an example of a particle (a space vehicle) subject only to the force of the gravitational attraction of the earth. By Newton's law of gravitational attraction, the force that the earth exerts on the particle is

$$F_r = \frac{GMm}{r^2} \tag{2.61}$$

where G is the universal gravitational constant, M is the mass of the earth, m is the mass of the particle, and r is the distance of the particle from the center of the earth. Changing variables from r to u and substituting into the differential equation for the path of motion yields

$$\frac{d^2u}{d\theta^2} + u = \frac{GM}{h^2} \tag{2.62}$$

The homogeneous part of the solution of this differential equation can be written

$$u_h = C \cos(\theta - \theta_0) \tag{2.63}$$

and the particular part of the solution is

$$u_p = \frac{GM}{h^2} \tag{2.64}$$

Therefore, the full solution of the differential equation becomes

$$u = C \cos(\theta - \theta_0) + \frac{GM}{h^2} \tag{2.65}$$

Examining Figure 2.18, we choose polar coordinates such that $\theta_0 = 0$. The differential equation for the path of motion of the space vehicle becomes

$$\frac{1}{r} = u = C \cos\theta + \frac{GM}{h^2} \tag{2.66}$$

This expression is the equation of a class of curves known as conic sections (circles, ellipses, parabolas, or hyperbolas) that are expressed in analytic geometry as

$$\frac{1}{r} = A(1 + e \cos\theta) \tag{2.67}$$

where A and e are constants. The equation is written in the following form, with the choices constants as shown:

$$e = \frac{Ch^2}{GM} \quad \text{and} \quad A = \frac{GM}{h^2}$$

$$\frac{1}{r} = \frac{GM}{h^2}[1 + e \cos\theta] \tag{2.68}$$

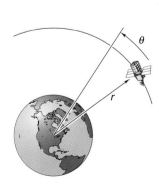

Figure 2.18

Note that the constant A is determined by the initial angular momentum per unit mass h of the particle. If the motion begins at $r = r_0$ and $\theta = \theta_0$, then

$$\frac{1}{r_0} = \frac{GM}{h^2}[1 + e] \quad \text{and} \quad e = \frac{h^2}{GMr_0} - 1 \qquad (2.69)$$

If the initial position and the initial angular momentum are given, the constants are known in terms of these initial conditions.

The effects of different initial conditions can be examined by looking at the constant e and determining the path of motion of the particle (in this case, the space vehicle):

1. $e < 1$, or $h^2 < 2GMr_0$. The path of the particle forms an ellipse because the radius always remains finite since for any value of θ, $(1 + e\cos\theta) > 0$.

2. $e = 0$, or $h^2 = GMr_0$. The path of motion is a circle for this case because the radius is a constant.

3. $e = 1$, or $h^2 = 2GMr_0$. The path of motion becomes a parabola, and the radius becomes infinite when $\theta = 180°$.

4. $e > 1$, or $h^2 > 2GMr_0$. The path becomes a hyperbola, and the radius becomes infinite at some angle between $90°$ and $180°$. The radius becomes infinite as θ approaches

$$\theta_1 = \cos^{-1}\left(-\frac{1}{e}\right)$$

These different paths of motion are shown in Figure 2.19. The free flight of the space vehicle is determined by the position and velocity at the end of the last power stage, or when burnout occurs. (See Figure 2.20.) Hence, the angular momentum per unit mass of the space vehicle is $h = r_0 v_0$, and since, from this point onward, the vehicle is subject to the central-force attraction of the earth, the equations for central-force motion hold. If the vehicle has sufficient velocity at this point, it can escape the earth's gravitational field and travel in a hyperbolic path of motion. The lowest possible escape velocity is a parabolic path when $e = 1$. It follows that

$$h = \sqrt{2GMr_0}$$

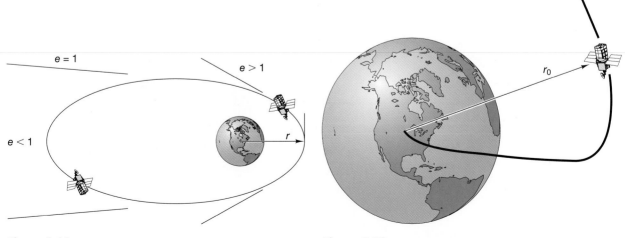

Figure 2.19   **Figure 2.20**

The escape velocity is

$$v_{esc} = v_0 = \sqrt{\frac{2GM}{r_0}}$$

The constant g is introduced and written as

$$g = \frac{GM}{R^2}$$

or

$$GM = gR^2$$

where R is the radius of the earth, which is 6.37×10^6 m, or 3960 miles. The escape velocity is thus

$$v_{esc} = \sqrt{\frac{2gR^2}{r_0}}$$

If the initial velocity of the space vehicle at this height is less than the escape velocity, the vehicle will assume an elliptical orbit around the earth. All trajectories of planets or other space satellites are elliptical in nature. For a satellite in an elliptical orbit, the minimum distance from the center of the earth to the satellite is called the *perigee* of the orbit. The maximum distance from the earth's center to the path of motion is called the *apogee*. The center of the earth is located at one of the *foci* of the ellipse. (See Figure 2.21.) The perigee occurs when $\theta = 0$, and therefore, $r_p = r_0$. The maximum distance from the earth—the apogee—occurs when the angle θ is 180° and

$$r_p = \frac{r_0}{\left(\dfrac{2\,GM}{r_0 v_0^2} - 1\right)}$$

The semimajor axis of the ellipse, a, can be written in terms of the perigee and the apogee as follows:

$$a = \frac{r_a + r_p}{2}$$

The minor axis of the ellipse, b, can be found by noting that for an ellipse, the sum of the distances from the two foci to any point on the ellipse is a constant. Designate a point A at the perigee, and let O be the center of the earth, or one of the foci. The other focus is

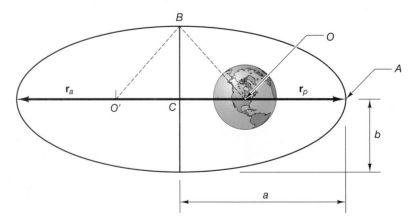

Figure 2.21

designated by O', as noted in Figure 2.21. The distance from each focus to A must equal the distance from each focus to B:

$$O'A + OA = O'B + OB$$

but

$$O'A + OA = 2a$$

Therefore,

$$OB = a$$

The minor axis b can be found by examining the right triangle COB. We have

$$a^2 = b^2 + (a - r_p)^2$$

$$\left(\frac{r_a + r_p}{2}\right)^2 = b^2 + \left(\frac{r_a - r_p}{2}\right)^2$$

$$b = \sqrt{r_a r_p}$$

The area of the ellipse is

$$A = \pi ab = \frac{\pi}{2}(r_a + r_p)\sqrt{r_a r_p}$$

The areal velocity is defined as

$$\frac{dA}{dt} = \frac{h}{2}$$

and the period, or the time to complete one full orbit, is

$$\tau = \frac{2A}{h} = \frac{\pi}{h}(r_a + r_p)\sqrt{r_a r_p}$$

As mentioned earlier, the German astronomer Johannes Kepler announced his three laws of planetary motion over the period from 1609 to 1619. These laws are as follows:

1. *The orbit of each planet about the sun is an ellipse with the sun at one of its foci.*
2. *The line joining the sun and any planet (the radius vector of the planets orbit) sweeps out equal areas in equal times as the planet travels around the sun.*
3. *The square of the period of revolution of a planet about the sun is proportional to the cube of the average distance of the planet from the sun.*

Kepler was able to determine these laws by using the ideas of Copernicus and the data of the Dutch astronomer Tycho Brahe, who spent the years from 1576 to 1596 observing the sun, moon, and planets. Kepler's work initially was slowed by his belief that the orbits of the planets were circular. This concept had been held true since ancient Greek times. Some of the data upon which Kepler based his laws are given in Table 2.2.

Over a period of 150 years, Copernicus, Galileo, and Kepler replaced the scientific theory of an earth-centered (geocentric) universe, with that of a sun-centered (heliocentric) universe. This revolution set the stage for the introduction of Newton's law of universal gravitation.

Newton knew that the acceleration of the moon, while directed toward the earth, is less than the acceleration of objects nearer to the Earth. This observation allowed him to develop the concept of gravitation. He was able to obtain the equation for an ellipse and derive Kepler's laws from theory. It is difficult to fully understand the changes in

Table 2.2
Planetary Motion about the Sun

Planet	Period (years)	Minor Elliptical Axis (1 astronomical unit = 93 × 10^6 miles)
Mercury	0.24	0.39
Venus	0.62	0.72
Earth	1.00	1.00
Mars	1.88	1.53
Jupiter	11.90	5.21
Saturn	29.50	9.55

scientific thinking that were occurring during Newton's time. Most previous work had been based upon inductive reasoning—that is, one would observe some phenomenon and then develop theories based upon the observations. The great mathematician René Descartes believed, on the other hand, that laws and theories obtained from inductive reasoning were flawed and that deductive reasoning was the essence of science. Deductive reasoning starts with theory and uses mathematics to derive laws therefrom. These laws can then be confirmed by observation. Newton was able to combine the deductive approach of Descartes with the inductive approach used by Galileo to develop his theories of motion.

Newton's gravitational law indicates that not only the sun, but also the other planets, should exert a force on any one planet during that planet's movement in the heavens. Therefore, the planets are not under pure central force motion. The orbit of any planet is then not perfectly elliptical because of perturbations by the other planets. Newton was concerned that these perturbations in a planet's orbit might make it unstable and cause the planet to fall out of orbit. A century later, the French mathematician and astronomer Pierre Laplace (1749–1827) showed that the planetary orbits were in fact, stable.

In 1781, a new planet, Uranus, was discovered, and using Newton's laws, astronomers calculated the effects of this planet on the orbits of the other planets. Some discrepancies existed between these calculations and the observed orbits. The French astronomer U. J. J. Leverrier and the English astronomer J. C. Adams suspected the existence of another planet. Using Newton's laws, they were able to compute the mass of the unknown planet, which would explain the discrepancies between the calculated and the observed orbits of the other planets. In 1846, Neptune was discovered exactly where they had predicted it to be. The orbit of Neptune then indicated that still another planet must exist in the solar system. In 1930, Pluto was discovered, and most of the discrepancies between observation and theory were reconciled. A debate remains about whether the remaining discrepancies are due to slight errors in the difficult observations of the planets' orbits or whether yet another planet exists in our solar system.

This demonstration of the power of Newton's laws led philosophers and scientists to begin to seek "the laws of nature." Some felt that Newton had proved that if all was predictable, then all was predestined. The "social sciences" were created, and scholars began to search for the laws that governed human behavior. Adam Smith tried to define the "laws" of economics. Newton had started the Age of Reason. The American and French revolutions were based upon the idea that people had natural rights based upon natural laws. In this regard, Benjamin Franklin wrote an essay entitled *On Liberty and Necessity: Man in the Newtonian Universe.*

Sample Problem 2.18

A 2-kg particle is attached to a cord that passes through a hole in a table. If the particle is pulled toward the center of the hole with a constant radial velocity of 0.1 m/s after being given an initial angular velocity $\dot{\theta} = 2$ rad/s when $r = 1$ m, determine the motion of the particle and the tension in the cord.

Solution

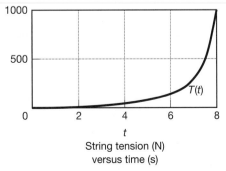

The initial conditions are $\dot{\theta} = 2$ rad/s, $r = 1$ m, and $\dot{r} = 0.1$ m/s. The differential equations of motion are

$$-T = 2(-r\dot{\theta}^2)$$
$$0 = 2(r\ddot{\theta} + 2\dot{r}\dot{\theta})$$

We have seen that the second differential equation can be solved in the form of Eq. (2.50), where $h = 2$ m^2/s, obtained from the initial values. Therefore, the angular velocity is

$$\dot{\theta} = \frac{2}{r^2} = \frac{2}{(1 - 0.1t)^2} \text{ rad/s}$$

where t is time in seconds. Integrating and using the initial condition that $\theta = 0$ when $t = 0$ gives

$$r(t) = 1 - 0.1t$$
$$\theta(t) = 20\left(\frac{1}{1 - 0.1t} - 1\right)$$

The rectangular coordinates of the particle are obtained from

$$x(t) = r(t)\cos\theta(t) \qquad y(t) = r(t)\sin\theta(t)$$

The first 8 s of motion are shown in the following *xy*-plot:

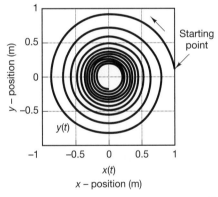

The tension in the cord is $T(r) = (8/r^3)$N. The tension as a function of time is given in the following diagram:

String tension (N) versus time (s)

The tension reaches the value of 1000 N after 8 seconds, and an infinite tension would be required to pull the particle to the center of the table.

Sample Problem 2.19

A communications satellite is launched parallel to the surface of the earth at an altitude of $r_p = 700$ km and with a initial velocity of 10,000 m/s. The radius of the earth is 6378 km, and the mass of the earth is 5.976×10^{24} kg, and the gravitational constant $G = 66.73 \times 10^{-12}$ m³/(kg · s²). Determine the maximum altitude at the apogee of the satellite.

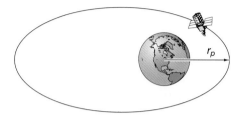

Solution

After launch, the satellite is under only the influence of gravitational attraction, and the motion is governed by Eq. (2.66); that is,

$$\frac{1}{r} = C \cos \theta + \frac{GM}{h^2} \, \text{m}^{-1}$$

where C is a constant determined by the initial conditions at launch. The initial radius is the radius of the earth added to the altitude:

$$r_0 = 6.378 \times 10^6 + 0.7 \times 10^6 = 7.078 \times 10^6 \text{ m}$$

The constant $GM = 399 \times 10^{12}$ m³/s², and the angular momentum $h = r_0 v_0 = 70.78 \times 10^9$ m²/s. At the initial position $\theta = 0$, Eq. (2.66) gives the constant C:

$$1.41 \times 10^{-7} = C + 0.80 \times 10^{-7}$$
$$C = 0.61 \times 10^{-7}$$

The equation for the radial position of the satellite from the center of the earth is given by

$$\frac{1}{r} = 0.61 \times 10^{-7} \cos \theta + 0.80 \times 10^{-7} \, \text{m}^{-1}$$

The maximum radial position is at the apogee, when $\theta = 180°$:

$$\frac{1}{r_a} = 0.19 \times 10^{-7}$$

$$r_a = 0526 \times 10^8 \text{ m} = 52{,}600 \text{ km}$$

Problems

For the following problems, assume the mass of the earth to be 5.98×10^{24} kg, the radius of the earth to be 6380 km, and $G = 66.7 \times 10^{-12}$ m³/(kg · s²).

2.87 Determine the velocity of a satellite that is in a circular orbit 500 km above the earth's surface.

2.88 Show that the period of a satellite in circular orbit around the earth is proportional to $R^{3/2}$, where R is the radius of the orbit from the earth's center.

2.89 Communication satellites are placed in geosynchronous orbit—that is, in a circular orbit such that they complete one full revolution about the earth while remaining above a fixed point on the planet. The period of a satellite in geosynchronous orbit is one sidereal day (the interval between two successive transits of the March equinox over the upper meridian of a place), which is 23 hrs and 56 min. Determine, in S.I. units, the altitude of such a satellite above the surface of the earth and its orbital velocity.

2.90 A space shuttle is in a circular orbit 400 km above the earth's surface. A satellite is to be launched from the shuttle and placed in a circular orbit 700 km above the earth's surface. The procedure to accomplish this feat is called a *Hohmann transfer* and involves two rocket boosts (See Figure P2.90). The first boost changes the satellite's velocity from v_1, the shuttle's velocity, to a velocity v_2 such that the satellite has an elliptical orbit with an apogee that is 700 km above the surface of the earth. The velocity of the satellite at this point is v_3. The second rocket boost changes the velocity at this point to the required circular orbital velocity v_4. Determine the required velocity changes during the two rocket boosts.

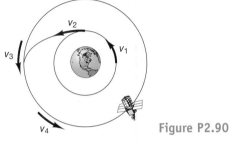

Figure P2.90

2.91 Using the satellite-launching space shuttle of Problem 2.90, determine the two velocity changes necessary to put a communication satellite in geosynchronous orbit. (See Problem 2.89.)

2.92 The elliptical path of a satellite has an eccentricity of 0.12. If the speed of the satellite at the perigee of its orbit is 4000 m/s, determine its speed at the apogee. Also, determine its altitude above the surface of the earth at the apogee of its orbit.

2.93 A meteoroid approaches the earth along a parabolic path. If it is first observed at 200,000 km from the earth's center, as shown in Figure P2.93. Determine the closest distance above the earth's surface that the meteoroid will reach.

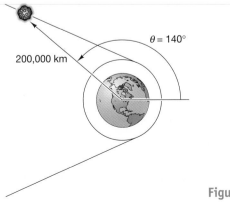

Figure P2.93

2.94 Orbital data for the *Mir* space station (shown in Figure P2.94) indicates that the altitude at the apogee of its orbit is 392.62 km and at the perigee of the orbit is 385.42 km. The period of one complete revolution around the earth is 92.34 min. Determine the eccentricity and the angular momentum of *Mir*'s path. (*Note: The orbit is decreasing about 10 km/yr as the station is taken out of service.*)

2.95 The *Atlantis* shuttle is placed in a circular orbit at an altitude of 100 km. Using the Hohmann transfer procedure (see Problem 2.90), determine the two changes in velocity that will put the shuttle in the *Mir* space station orbit (from Problem 2.94).

2.96 On June 25, 1997, the *Progress* M-34 supply vessel crashed into the *Mir* space station. Determine the velocity changes necessary to use the *Soyuz*-TM spacecraft to ferry the two cosmonauts and the American astronaut from the *Mir* orbit to a circular orbit at an altitude of 150 km. The *Soyuz* has a mass of 7100 kg, a length of 7 m, a maximum diameter of 2.7 m, and a capacity of three persons. The *Soyuz* serves as an escape lifeboat for the *Mir* space station.

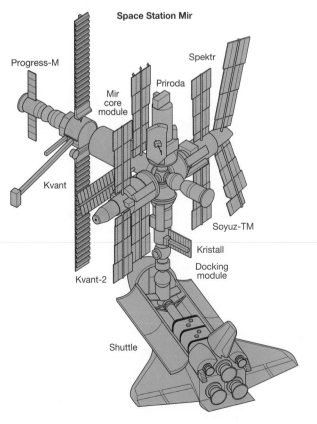

Figure P2.94

2.8 THREE-DIMENSIONAL PARTICLE DYNAMICS IN CURVILINEAR COORDINATES

The differential equations of motion are in a simple form when written in rectangular coordinates, and we have examined three-dimensional problems in those coordinates. For two-dimensional particle motion, we have shown that many problems are best formulated in curvilinear coordinate systems, such as normal–tangential or polar coordinates. The principal mathematical difficulty one finds when using these coordinate systems is that the base unit vectors are not constant. This variability adds extra nonlinear terms to the differential equations of motion. However, using computational software, we can obtain approximate numerical solutions to these nonlinear differential equations. The principal advantage of using curvilinear coordinates is that the constraints on the motion are easier to express in these coordinates. In advanced engineering courses, you will see that the boundary conditions for partial differential equations dictate the choice of the coordinate system. For example, if you were to study the heat distribution in a bar with a circular cross section, cylindrical coordinates would be the logical choice, even though the partial differential equation of heat conduction will be more complex. In this section, we will consider two three-dimensional coordinates systems— cylindrical and spherical—and show examples of problems that are best formulated in one or the other of them. There are many other curvilinear coordinates that can be used to express the equations of motion, and the vector expressions for the position, velocity, and acceleration of particles modeled in such systems are obtained by methods similar to those presented here.

2.8.1 CYLINDRICAL COORDINATES

The cylindrical coordinates and their base vectors are shown in Figure 2.22.

The components of the acceleration vector in cylindrical coordinates were developed in Section 1.9 and are given by

$$a_r = \ddot{r} - r\dot{\theta}^2$$
$$a_\theta = r\ddot{\theta} + 2\dot{r}\dot{\theta} \tag{2.70}$$
$$a_z = \ddot{z}$$

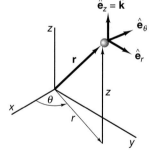

Figure 2.22

When constraints or forces dictate the use of cylindrical coordinates, the particle dynamics problem will involve the solution of three differential equations involving the accelerations given in Eq. (2.70). In general, two of these equations will be nonlinear and coupled. The equations of motion are

$$\sum F_r = ma_r = m(\ddot{r} - r\dot{\theta}^2)$$
$$\sum F_\theta = ma_\theta = m(r\ddot{\theta} + 2\dot{r}\dot{\theta}) \tag{2.71}$$
$$\sum F_z = ma_z = m\ddot{z}$$

These equations also may include further nonlinear terms or additional coupling, depending upon the forces acting on the particle. We will seek analytical solutions to the differential equations when possible and will use numerical solution methods in other cases.

Sample Problem 2.20

A particle slides down a smooth spiral channel under the influence of gravity. (See accompanying diagram.) If the particle is released from rest, the radius of the spiral is R, and the pitch of the spiral is $p = \tan \beta$, determine the velocity at any angle θ and the force on the channel wall.

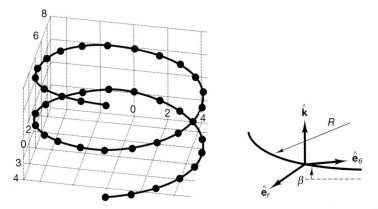

Helical Motion in Space

Solution At any point on the spiral, the free-body diagram of the particle shows the weight, the normal force $\mathbf{N}$ on the channel bottom, and the force $\mathbf{f}$ on the channel wall, as follows:

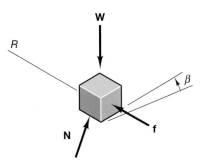

In this problem, $r = R$, a constant, so the derivatives of r are zero. Therefore, the forces acting on the particle are

$$\mathbf{W} = -mg\hat{\mathbf{k}}$$

$$\mathbf{N} = N(\sin \beta \hat{\mathbf{e}}_\theta + \cos \beta \hat{\mathbf{k}})$$

$$\mathbf{f} = -f\hat{\mathbf{e}}_r$$

The equations of motion are

$$-f = -mr\dot{\theta}^2$$

$$N \sin \beta = mr\ddot{\theta}$$

$$N \cos \beta - mg = m\ddot{z}$$

The particle is constrained to move along the spiral path until it slips. Therefore, the acceleration vector can be written as

$$\mathbf{a} = a(\cos \beta \hat{\mathbf{e}}_\theta - \sin \beta \hat{\mathbf{k}}) - r\dot{\theta}^2 \hat{\mathbf{e}}_r$$

The last two equations of motion can be written in terms of the magnitude of the constant acceleration a:

$$N \sin \beta = ma \cos \beta$$
$$N \cos \beta - mg = -ma \sin \beta$$

Eliminating the normal force N from the two equations gives

$$a = g \sin \beta \qquad N = mg \cos \beta$$

The second and third differential equations can be written as

$$\ddot{\theta} = \frac{g}{R} \sin \beta \cos \beta$$
$$\ddot{z} = -g \sin^2 \beta$$

Integrating the first equation and using the initial condition that the particle starts from rest yields

$$\frac{d\dot{\theta}}{d\theta} \dot{\theta} = \frac{g}{R} \sin \beta \cos \beta$$
$$(\dot{\theta})^2 = \left(2 \frac{g}{R} \sin \beta \cos \beta \right) \theta$$

The magnitude of the wall force at any position to prevent the particle from sliding is

$$|\mathbf{f}| = (2mg \sin \beta \cos \beta)\theta$$

The two differential equations can be directly integrated as functions of time, which yields

$$\ddot{\theta} = \frac{g}{R} \sin \beta \cos \beta \qquad\qquad \ddot{z} = -g \sin^2 \beta$$

$$\dot{\theta}(t) = \left(\frac{g}{R} \sin \beta \cos \beta \right) t + \dot{\theta}_0 \qquad\qquad \dot{z}(t) = -(g \sin^2 \beta) t + \dot{z}_0$$

$$\theta(t) = \left(\frac{g}{R} \sin \beta \cos \beta \right) \frac{t^2}{2} + \dot{\theta}_0 t + \theta_0 \qquad z(t) = -(g \sin^2 \beta) \frac{t^2}{2} + \dot{z}_0 t + z_0$$

The integration constants $\dot{\theta}_0$, θ_0, $\dot{z}_0$, and z_0 are zero if the origin of the coordinate system is set at the level of release of the particle.

Sample Problem 2.21

In Sample Problem 2.20, it is desired to design an amusement park ride such that the bank angle γ changes with the revolution angle θ in a manner so that the sliding particle would not slip even if there is no retaining walls on the channel. Determine the design specification for the banking angle in terms of position angle θ.

Solution   The normal force vector is given by

$$\mathbf{N} = N[-\sin \gamma \hat{\mathbf{e}}_r + \cos \gamma (\sin \beta \hat{\mathbf{e}}_\theta + \cos \beta \hat{\mathbf{k}})]:$$

There is no wall force and the weight vector is

$$\mathbf{W} = -mg \hat{\mathbf{k}}$$

The equations of motion are

$$N \sin \gamma = mr \dot{\theta}^2$$
$$N \cos \gamma \sin \beta = mr \ddot{\theta} = ma \cos \beta$$
$$N \cos \gamma \cos \beta - mg = m\ddot{z} = -ma \sin \beta$$

The acceleration vector is in the same direction as in Sample Problem 2.19. Solving these equations as before yields

$$a = g \sin \beta$$

$$N = mg \frac{\cos \beta}{\cos \gamma}$$

Substituting the value for N into the first equation of motion yields

$$r\dot{\theta}^2 = g \cos \beta \tan \gamma$$

The second equation of motion, after substitution for a may be written as

$$r\dot{\theta} \frac{d\dot{\theta}}{d\theta} = g \sin \beta \cos \beta$$

Integrating yields

$$r\dot{\theta}^2 = 2\theta g \sin \beta \cos \beta$$

combining these two expressions gives an equation for γ in terms of θ:

$$\gamma(\theta) = \tan^{-1}(2\theta \sin \beta)$$

In this case the acceleration is constant in magnitude and differential equations for z and θ can be solved by direct integration.

2.8.2 SPHERICAL COORDINATES

When a particle is constrained to move on a spherical surface or inside a hemisphere or a conically shaped surface, spherical coordinates are the natural choice to mathematically express these constraints. The base vectors and coordinates are shown in Figure 2.23. The components of the velocity and acceleration vectors were developed in Section 1.9, and the velocity components are

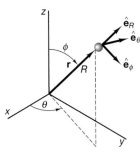

Figure 2.23

$$v_R = \dot{R}$$
$$v_\phi = R\dot{\phi} \tag{2.72}$$
$$v_\theta = R \sin \phi \dot{\theta}$$

The equations of motion in spherical coordinates are obtained using the expressions for acceleration given in Eq. (1.120):

$$\sum F_R = ma_R = m[\ddot{R} - R\dot{\phi}^2 - R\sin^2\phi\dot{\theta}^2]$$
$$\sum F_\phi = ma_\phi = m[R\ddot{\phi} + 2\dot{R}\dot{\phi} - R\sin\phi\cos\phi\dot{\theta}^2] \tag{2.73}$$
$$\sum F_\theta = ma_\theta = m[R\sin\phi\ddot{\theta} + 2\dot{R}\dot{\theta}\sin\phi + 2R\dot{\phi}\dot{\theta}\cos\phi]$$

The equations of motion are nonlinear and coupled, and, in general, can be solved only by numerical methods. It is useful to express the spherical unit base vectors in terms of the rectangular base vectors $\hat{\mathbf{i}}$, $\hat{\mathbf{j}}$, and $\hat{\mathbf{k}}$. These relationships are as follows:

$$\hat{\mathbf{e}}_R = \sin\phi\cos\theta\hat{\mathbf{i}} + \sin\phi\sin\theta\hat{\mathbf{j}} + \cos\phi\hat{\mathbf{k}}$$
$$\hat{\mathbf{e}}_\phi = \cos\phi\cos\theta\hat{\mathbf{i}} + \cos\phi\sin\theta\hat{\mathbf{j}} - \sin\phi\hat{\mathbf{k}} \tag{2.74}$$
$$\hat{\mathbf{e}}_\theta = -\sin\theta\hat{\mathbf{i}} + \cos\theta\hat{\mathbf{j}}$$

A spherical pendulum is set into motion when $\phi = 30°$, with an initial velocity in the θ direction of 0.5 m/s. If the pendulum has a length of 2 m and a mass of 3 kg, determine its motion for the first 5 s.

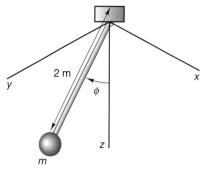

Solution
A free-body diagram of the pendulum mass is first constructed, as depicted in the following illustration:

The coordinate R is constant and equal to 2 m. The only forces acting on the mass are its weight and the tension in the cord. These forces can be expressed in spherical coordinates using Eq. (2.74):

$$\mathbf{W} = mg\hat{\mathbf{k}} = mg \cos\phi\hat{\mathbf{e}}_R - mg \sin\phi\hat{\mathbf{e}}_\phi$$

$$\mathbf{T} = -T\hat{\mathbf{e}}_R$$

Noting that R is constant, we find that the equations of motion become

$$m[-R\dot{\phi}^2 - R \sin^2\phi\dot{\theta}] = mg \cos\phi - T$$
$$R\ddot{\phi} - R \sin\phi \cos\dot{\theta}^2 = -g \sin\phi$$
$$R \sin\phi\ddot{\theta} + 2R\dot{\phi}\dot{\theta} \cos\phi = 0$$

The motion of the pendulum is described by the second and third of these differential equations, and the first equation can be used to determine the tension in the cord. The last two equations are nonlinear and coupled and must be solved numerically. However, before using numerical integration, note that the last differential equation can be written as

$$\frac{1}{R \sin\phi} \frac{d}{dt}(R^2 \sin^2\phi\dot{\theta}) = 0$$

$$R^2 \sin^2\phi\dot{\theta} = h$$

As in the case of central-force motion, the angular momentum per unit mass is conserved and equal to h. The value of h can be determined from the initial values, and θ can be eliminated from the differential equation for ϕ:

$$\dot{\theta} = \frac{h}{R^2 \sin^2\phi} = \frac{R \sin\phi_0 v_{\theta 0}}{R^2 \sin^2\phi} = \frac{\sin\phi_0 v_{\theta 0}}{R \sin^2\phi}$$

The differential equation for ϕ is

$$\ddot{\phi} = \frac{\cos\phi(\sin^2\phi_0 v_{\theta 0}^2)}{R^2 \sin^3\phi} - \frac{g}{R} \sin\phi$$

If the initial velocity in the θ-direction, $v_{\theta 0}$, is zero, then this expression is the differential equation for a plane pendulum. The equation can be solved numerically, and the variation of the angle ϕ is shown in the accompanying graph. This angle cannot be zero; that is, the pendulum cannot reach a vertical position, due to the conservation of angular momentum. The minimum angle in this case is 0.018 rad or 1°.

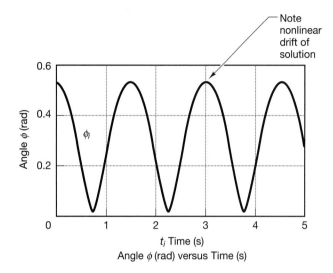

Angle ϕ (rad) versus Time (s)

The angular motion should be periodic, as there is no damping in the solution. The small drift is due to the nonlinearity of the solution and can be reduced by taking a smaller time increment in the numerical integration. The motion can be viewed from the top—that is, the *xy*-plane. As viewed from the top, the motion starts at the far right, comes into the center, circles, and goes out to the left. The path then comes back to the center, circles, goes down, comes back to the center, circles, and goes to the right to repeat the motion.

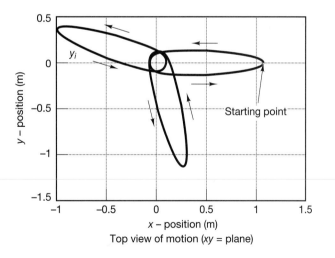

Top view of motion (*xy* = plane)

2.8.3 PARAMETRIC EQUATIONS IN TANGENTIAL, NORMAL, AND BINORMAL COORDINATES

In Section 1.13, we developed an expression for the unit tangent vector along a specified curve in space. The use of parametric equations to solve particle dynamics problems was further developed in Section 2.5, and now, we will present these equations for curves in three dimensions. A general curve with its tangent vector is shown in Figure 2.24. The unit

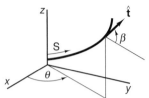

Figure 2.24

tangent vector can be written in terms of the two angles $\theta(s)$ and $\beta(s)$ that are both functions of the parameter s; that is, these angles will change along the length of the curve and completely describe the shape of the curve. The unit tangent vector is written in the form

$$\hat{\mathbf{t}}(s) = \cos\theta(s)\cos\beta(s)\hat{\mathbf{i}} + \sin\theta(s)\cos\beta(s)\hat{\mathbf{j}} + \sin\beta(s)\hat{\mathbf{k}} \qquad (2.75)$$

The velocity vector must be tangent to the curve and has a magnitude of ds/dt. Thus, we have

$$\frac{d\mathbf{r}}{dt} = \frac{ds}{dt}\hat{\mathbf{t}} = \frac{ds}{dt}[\cos\theta(s)\cos\beta(s)\hat{\mathbf{i}} + \sin\theta(s)\cos\beta(s)\hat{\mathbf{j}} + \sin\beta(s)\hat{\mathbf{k}}] \qquad (2.76)$$

Differentiation of the velocity vector yields the acceleration vector:

$$\mathbf{a} = \frac{d^2\mathbf{r}}{dt^2} = \frac{d^2s}{dt^2}[\cos\theta(s)\cos\beta(s)\hat{\mathbf{i}} + \sin\theta(s)\cos\beta(s)\hat{\mathbf{j}} + \sin\beta(s)\hat{\mathbf{k}}]$$

$$+ \left(\frac{ds}{dt}\right)^2 \left\{ \begin{array}{l} \left[-\sin\theta(s)\cos\beta(s)\dfrac{d\theta}{ds} - \cos\theta(s)\sin\beta(s)\dfrac{d\beta}{ds}\right]\hat{\mathbf{i}} \\[2mm] + \left[\cos\theta(s)\cos\beta(s)\dfrac{d\theta}{ds} - \sin\theta(s)\sin\beta(s)\dfrac{d\beta}{ds}\right]\hat{\mathbf{j}} \\[2mm] + \left[\cos\beta(s)\dfrac{d\beta}{ds}\right]\hat{\mathbf{k}} \end{array} \right\} \qquad (2.77)$$

The first term on the right-hand side is the tangential acceleration and is directed along the curve. The second term is the normal acceleration, with a magnitude equal to the square of the speed divided by the radius of curvature at that point on the curve. The bracketed vector in the second term is not a unit vector and has a magnitude of the reciprocal of the radius of curvature, as was shown in Section 1.13. Although the acceleration vector is given in terms of the rectangular unit base vectors, the problem can be formulated in both of these components or in the tangential and normal direction. There is no acceleration in the binormal direction. The unit normal vector can be obtained from Eqs. (1.151) and (1.153) and is

$$\hat{\mathbf{n}} = \frac{1}{\sqrt{\left(\dfrac{d\theta}{ds}\right)^2 \cos^2\beta + \left(\dfrac{d\beta}{ds}\right)^2}}$$

$$\left\{ \begin{array}{l} \left[-\sin\theta(s)\cos\beta(s)\dfrac{d\theta}{ds} - \cos\theta(s)\sin\beta(s)\dfrac{d\beta}{ds}\right]\hat{\mathbf{i}} \\[2mm] + \left[\cos\theta(s)\cos\beta(s)\dfrac{d\theta}{ds} - \sin\theta(s)\sin\beta(s)\dfrac{d\beta}{ds}\right]\hat{\mathbf{j}} \\[2mm] + \left[\cos\beta(s)\dfrac{d\beta}{ds}\right]\hat{\mathbf{k}} \end{array} \right\} \qquad (2.78)$$

Note that if the angle β is zero everywhere along the curve—that is, if the curve lies in the xy-plane—the unit normal vector becomes

$$\hat{\mathbf{n}} = -\sin\theta(s)\hat{\mathbf{i}} + \cos\theta(s)\hat{\mathbf{j}}$$

This equation agrees with the two-dimensional formulation given in Section 2.5.

Formulation of the dynamics problem in this manner is very useful for solving inverse dynamics problems. In these problems, the path of the particle is known, and we wish to determine the forces required to produce the accelerations needed to make the particle follow that path. If the forces are specified, we would determine the path using more conventional coordinate systems.

Sample Problem 2.23

A 1000-kg roller-coaster car moves along a circular track that varies in elevation in a sinusoidal manner as shown in figure. The 2000-m path of the roller coaster is given by

$$\theta(s) = \frac{\pi s}{1000} \qquad \beta(s) = \frac{\pi}{2} \sin\left(\frac{\pi s}{500}\right) \text{ rad}$$

Determine the normal force exerted on the riders if the roller coaster moves at a constant velocity of 15 m/s.

Solution   We can determine the rectangular coordinates of the car at any position along the curve using the coordinate relationships

$$x(s) = x_0 + \int_0^s \cos\beta(\eta)\cos\theta(\eta)d\eta$$

$$y(s) = y_0 + \int_0^s \cos\beta(\eta)\sin\theta(\eta)d\eta \text{ m}$$

$$z(s) = z_0 + \int_0^s \sin\beta(\eta)d\eta \text{ m}$$

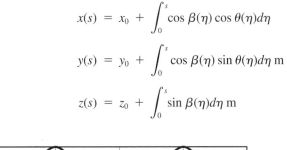

Elevation along track

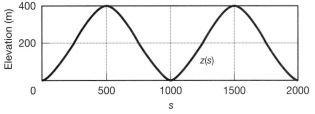

Plane view (xy) of track

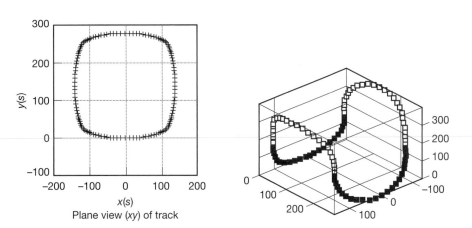

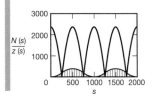

A plane view and a three-dimensional view of the track can now be shown. The magnitude of the normal force is plotted with the elevation of the track to determine when these forces are a maximum. The value of the normal force is given in newtons. The normal force is directed up for the first 250 m of track, down for the next 500 m, and then up for the next 500 m as the car goes to the low point in the track. The car weighs 9810 N, so the normal force is about 25% of the weight.

Problems

2.97 Determine the motion of the particle in Sample Problem 2.20 if the coefficient of kinetic friction between the particle and the slide (at both the base and the wall) is μ_k. *(Hint: Friction always opposes motion, and therefore, the friction force will be opposite to the tangential acceleration vector.)*

2.98 In Figure P2.98, a 3-kg particle is given an initial velocity in the θ-direction of 20 m/s and slides inside a vertical smooth tube of radius 1.5 m. The particle starts at the top of the 50-m-high tube. Determine the path of motion for the first 2 s of sliding.

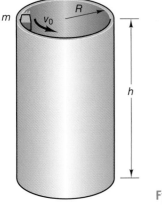

Figure P2.98

2.99 Determine the path of the particle in Problem 2.98 if the coefficient of kinetic friction between the tube wall and the particle is 0.2. *(Hint: The velocity vector will have two components—one in the z-direction and one in the θ-direction—and the friction vector will be in the opposite direction of the velocity vector.)*

2.100 The particle in Problem 2.99 has an initial velocity $\mathbf{v} = 20\hat{\mathbf{e}}_\theta + 20\hat{\mathbf{k}}$ m/s. Plot the motion for the first 4 s if the particle starts its path at a height of 30 m.

2.101 A particle of mass m is projected down a smooth, horizontal tube of radius R with an initial velocity v_0 at a pitch angle β with the radial direction, as shown in Figure P2.101. Determine the minimum initial velocity such that the mass moves in a spiral motion down the tube while the mass is in contact with the inner surface of the tube.

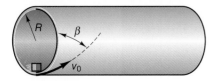

Figure P2.101

2.102 Determine the minimum initial velocity if the particle in Problem 2.101 goes through at least one revolution when the coefficient of kinetic friction between the particle and the inner surface of the tube is μ_k. $m = 3$ kg, $r = 0.6$ m, $\beta = 50°$, and $\mu_k = 0.2$

2.103 The particle in Figure P2.103 is released in a semicircular sluiceway that is inclined at an angle β with the horizontal. If the particle is released on the edge of the sluiceway, determine the path of motion for a 2-kg particle in a sluiceway of radius 1 m set at an incline angle of 20°.

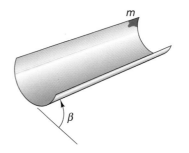

Figure P2.103

2.104 Determine the path of the particle in Problem 2.103 if the kinetic coefficient of friction between the particle and the inner surface of the sluiceway is 0.3.

2.105 In Figure P2.105, a 2-kg particle is placed at the rim of a smooth 1.5-m hemispherical bowl and released. If the particle is given an initial velocity $v = 10\hat{\mathbf{e}}_\theta$ m/s, determine the path of motion due to gravitational attraction for the first 2 s.

Figure P2.105

2.106 Determine the path of the particle in Problem 2.105 if the coefficient of kinetic friction between the particle and the bowl is 0.3.

2.107 In Figure P2.107, a 5-kg particle is given an initial velocity of $\mathbf{v} = 10\hat{\mathbf{e}}_\theta$ m/s the top of a 2-m-high inverted cone with a cone angle of 30°. Determine the path that the particle follows if the inside surface of the cone is smooth.

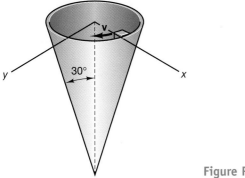

Figure P2.107

2.108 Determine the path of the particle in Problem 2.107 if the coefficient of kinetic friction between the inner surface of the cone and the particle is 0.6.

2.109 A 5-kg particle is placed on a smooth hemisphere of radius 6 m at a position $\phi = 10°$ and $\theta = 0°$ and is given an initial velocity of $\mathbf{v} = 4\hat{\mathbf{e}}_\theta - 0.5\hat{\mathbf{e}}_\phi$ m/s, as dipicted in Figure P2.109. Determine the path of the particle, and determine the particle's coordinates when it falls off the hemisphere.

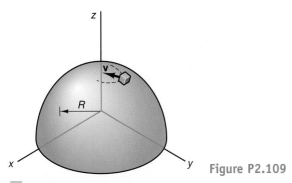

Figure P2.109

2.110 Determine the path of the particle in Problem 2.109 if the coefficient of kinetic friction between the hemisphere and the particle is 0.2.

2.111 In Sample Problem 2.22, replace the cord of the the pendulum with a spring with spring constant 200 N/m and an unstretched length of 2 m. Consider the same initial conditions, and determine the motion of the particle.

2.112 The 6000-ft path of a hilly curve is given by

$$\theta(s) = \frac{s}{2000} + e^{\frac{-s}{1000}} \text{ rad}$$

$$\beta(s) = \frac{\pi}{12} \sin\left(\frac{\pi s}{3000}\right) \text{ rad}$$

If a 3000-lb car drives on the road at a constant speed of 60 mph, determine the required normal force to keep the car on the road. The following diagrams in Figure P2.112 show a plane view of the curve and the elevation of the roadway as a function of the distance from the start of the hill:

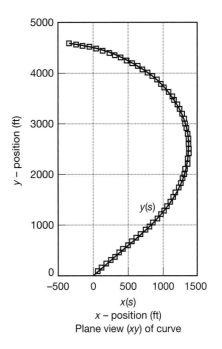

Plane view (*xy*) of curve

Hill contour

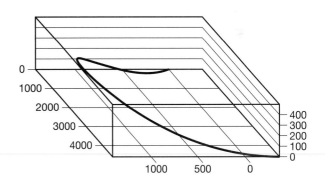

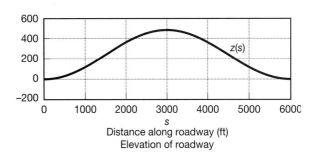

s
Distance along roadway (ft)
Elevation of roadway

Figure P2.112

2.113 Determine the required normal force along the roadway if the car of Problem 2.112 starts from rest and accelerates at a constant rate of 2 ft/s² for the first 3000 ft of the roadway and then decelerates at a rate of 1 ft/s² for the next 3000 ft.

2.114 Determine the total acceleration acting on the car of Problem 2.113 at any point along the roadway.

2.115 A 10-m slide in a playground, shown in Figure P2.115, is curved such that its path is given by

$$\theta(s) = \frac{\pi s}{20} \text{ rad}$$

$$\beta(s) = -\frac{\pi}{3}\left[1 - \left(\frac{s}{10}\right)^2\right] \text{ rad}$$

If the slide is smooth, determine the velocity of a 30-kg girl at the bottom of the slide.

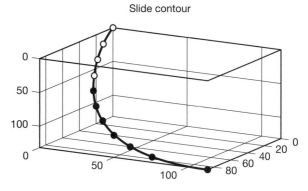

Slide contour

Figure P2.115

2.116 Determine the velocity of the child in Problem 2.115 if the coefficient of kinetic friction between the girl's clothing and the slide is 0.2.

2.9   DETECTION OF MODELING AND CONSTRAINT ERRORS IN COMPLEX PARTICLE DYNAMICS PROBLEMS

Frequently, the most difficult problem in the solution of a particle dynamics problem is modeling the mechanism and formulating the equations of motion and the constraint equations. The resulting differential equations of motion can be solved numerically, but if they are nonlinear, the solution may become unstable over time. There is no definite procedure to recognize errors in modeling or the assumptions made during this process, and at times the error will not become apparent until a solution over time is attempted.

We will take an example to show how such errors can occur and for this case, how to recognize the problem. Homework Problem 2.83 is modified for this example.

Sample Problem 2.24

A mass m is constrained to the surface of a smooth disk of radius R by a spring with spring constant k and unstretched length $R/2$, as illustrated here. Write the equation of motion of the mass if the disk is in a horizontal plane, including the affect of friction (coefficient of kinetic friction μ).

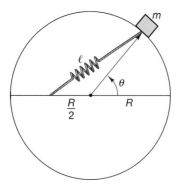

Solution   We will first write an equation for the length of the spring l at any position θ.

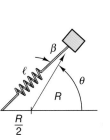

$$l^2 = \frac{R^2}{4} + R^2 - R^2 \cos(180 - \theta)$$

$$l^2 = \frac{R^2}{4} + (5 + 4\cos\theta) \qquad l(\theta) = \frac{R}{2}\sqrt{5 + 4\cos\theta}$$

We will determine the sine and cosine of the angle β using the law of cosines and the law of sines.

$$\frac{R^2}{4} = l^2 + R^2 - 2lR\cos\beta$$

$$\cos\beta = \frac{2 + \cos\theta}{\sqrt{5 + 4\cos\theta}}$$

$$\frac{\sin\beta}{R/2} = \frac{\sin\theta}{l}$$

$$\sin\beta = \frac{\sin\theta}{\sqrt{5 + 4\cos\theta}}$$

The spring force is

$$F_s(\theta) = k[l(\theta) - R/2]$$

The normal force can be obtained from the equation of motion in the R direction:

$$N(\theta, \dot{\theta}) = F_s(\theta)\cos\beta - mR\dot{\theta}^2$$

The equation of motion is

$$mR\ddot{\theta} = 2mR\dot{\theta} + F_s(\theta)\sin\beta - \mu N(\theta, \dot{\theta})\frac{\dot{\theta}}{|\dot{\theta}|}$$

Now solve the equation of motion for the case of no friction, $\mu = 0$, using the values:

$$m = 0.2 \text{ kg}, \qquad R = 0.25 \text{ m, and } k = 1000 \text{ N/m}$$

Release the system from rest when $\theta = 30°$ and plot the angular position with respect to time. The resulting plot for the first 0.4 s is shown here.

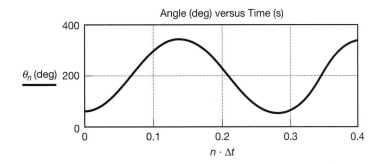

On first examination, the solution looks very good, as the mass is oscillating between 30° and 330°. Now let us examine the normal force during the same time period.

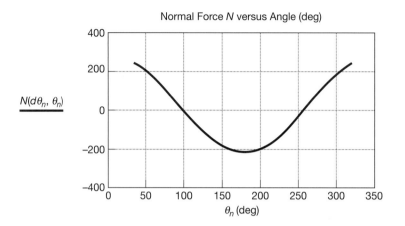

Normal Force *N* versus Angle (deg)

$N(d\theta_n, \theta_n)$

θ_n (deg)

This plot shows that the normal force becomes negative somewhere between 90° and 100°. But the normal force cannot be negative if the mass is sliding along the outside surface of the disk. The change in sign shows that the constraint that the mass slides along the surface cannot be obtained as the term proportional to the square of the velocity is greater than the spring force. We might have noticed this by seeing that the spring force is zero when the angle is 180° and the velocity would be large.

   The question now arises: Could the device be designed such that the normal force could be either positive or negative at any point? This could be obtained by having the mass slide in a circular groove in the disk instead of on the outside surface of the disk. If this was the case, the solution would be correct for the case that friction was zero. However, note that in the equation of motion, if the normal force is negative, the friction force would be positive. But friction resists motion (not producing it), so the equation of motion has to be altered. This easily can be accomplished by putting absolute magnitude designation on the normal force.

$$mR\ddot{\theta} = 2mR\dot{\theta} + F_s(\theta)\sin\beta - \mu|N(\dot{\theta}, \theta)|\frac{\dot{\theta}}{|\dot{\theta}|}$$

Now this equation is valid for the case where friction exists. If the coefficient of kinetic friction was 0.15, the angular position for the first 0.4 s would be

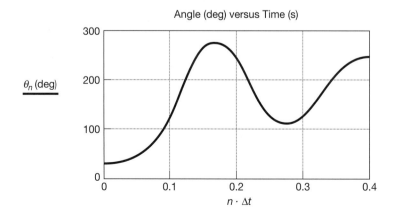

Angle (deg) versus Time (s)

θ_n (deg)

$n \cdot \Delta t$

and examination of the normal force for this period of time yields

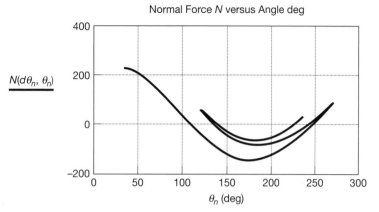

Normal Force N versus Angle deg

$N(d\theta_n, \theta_n)$

θ_n (deg)

Note that the normal force still changes sign and the friction force begins to damp out the oscillations.

Very powerful methods for the solution of particle dynamics problems have been developed in this chapter, but caution must be used to fully examine any solution. When the solution does not appear to be reasonable, the cause must be explained. This is particularly true when the equation of motion is nonlinear and the solution may not be stable. This can be overcome by the use of more accurate methods of numerical integration that can be found in most texts on differential equations or numerical analysis.

Chapter 3

WORK–ENERGY AND IMPULSE–MOMENTUM FIRST INTEGRALS OF MOTION

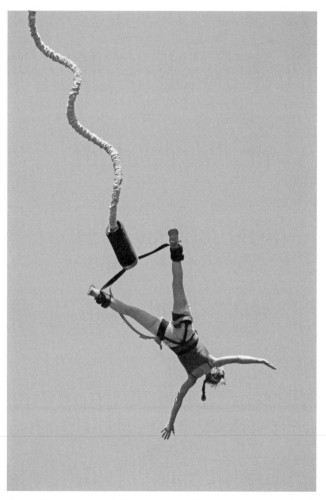

The spring constant of the bungee cord is inversely proportional to the length of the cord so that the maximum force exerted on the jumper is independent of the distance the jumper falls. This fact is important to climbers and others using safety ropes.

3.1 INTRODUCTION

Newton formulated his theories of dynamics based on linear momentum as the fundamental variable of motion, relating the forces acting on a particle to the change in its linear momentum. Vectors were used to characterize the geometry of the particle's motion—that is, the position, velocity, and acceleration of the particle. The 18th century mathematician Joseph-Louis Lagrange (1736–1813) was an analyst, not a geometer. Lagrange formulated the study of dynamics using general analysis in his masterpiece, *Mécanique analytique* (Analytical Mechanics), which he started writing as a boy of 19, and which was published when Lagrange was 52. Lagrange based his study on the concepts of work and energy, which are scalar quantities, and developed the new mathematics of calculus of variations. We will look at the scalar concepts of work–energy and also the vector concepts of impulse–momentum by examining the *first integrals of the equations of motion*. It is frequently possible to obtain useful information about the motion of a particle without solving the full differential equations of motion. It was shown in Chapters 1 and 2 that, if the acceleration of a particle is a function of only time or of only position, the second-order differential equation can be solved by separation of variables, even in cases of nonlinear differential equations. If the forces acting on the particle are known functions of position, the first integral of the particle's motion is an equation of *work–energy*. If the forces acting on the particle are known functions of time, the first integral of the particle's motion is an equation of *impulse–momentum*. It is important to realize that most of the problems that can be solved in this manner can also be solved by the methods that were introduced in Chapter 2. However, the first integrals of the particle's motion provide valuable conceptual tools.

3.2 POWER, WORK, AND ENERGY

We will begin our study of work–energy concepts by reviewing some mathematical observations. Consider the square of a vector **A,** defined as the scalar

$$A^2 = \mathbf{A} \cdot \mathbf{A} \tag{3.1}$$

If the vector is a function of time, the time derivative of the square of the vector is

$$\frac{d}{dt}(A^2) = \frac{d}{dt}(\mathbf{A} \cdot \mathbf{A}) = 2\mathbf{A} \cdot \frac{d\mathbf{A}}{dt} \tag{3.2}$$

Newton based his laws on the concept of a quantity of motion called *linear momentum,* expressed by

$$\mathbf{L} = m\mathbf{v}$$
$$\mathbf{F} = \frac{d}{dt}(\mathbf{L}) = \frac{d}{dt}(m\mathbf{v}) \tag{3.3}$$

Lagrange and other mathematicians and physicists thought that the fundamental quantity of motion was *kinetic energy,* defined as

$$T = \frac{1}{2}m(\mathbf{v} \cdot \mathbf{v}) \tag{3.4}$$

Kinetic energy is a scalar quantity such that, unlike linear momentum, it has no direction. For example, it is not possible to determine the kinetic energy in the

x-direction of a particle's path. Using Eq. (3.2), we find that the change in kinetic energy is

$$\frac{dT}{dt} = \frac{d}{dt}\left[\frac{1}{2}m(\mathbf{v} \cdot \mathbf{v})\right] = \mathbf{v} \cdot \frac{d}{dt}(m\mathbf{v}) \tag{3.5}$$

We can relate the concept of kinetic energy to Newton's equation of motion by taking the scalar product of Newton's second law with the velocity vector, yielding

$$\mathbf{F} \cdot \mathbf{v} = \mathbf{v} \cdot \frac{d}{dt}(m\mathbf{v}) \tag{3.6}$$

The term $\mathbf{F} \cdot \mathbf{v}$ is called the *power* and, from Eq. (3.5), can be determined to be equal to the time rate of change of the kinetic energy. The power P is also a scalar quantity and is equal to

$$P = \mathbf{F} \cdot \mathbf{v} = \frac{dT}{dt} \tag{3.7}$$

This scalar equation can be written as

$$\mathbf{F} \cdot d\mathbf{r} = dT \tag{3.8}$$

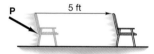

Figure 3.1

The left-hand side of this equation is the *work* done by the force $\mathbf{F}$ as it undergoes an infinitesimal displacement $d\mathbf{r}$. Work is literally defined as "to produce results" or "to perform a function." Consider the simple example illustrated in Figure 3.1. The desired result, or function, in this case, is to push a chair 5 ft across the floor. The vertical component of the force P does nothing to help produce the desired result, and therefore, does no work. The force P works only in the direction of the movement.

Consider a particle moving from position $\mathbf{r}$ to position $\mathbf{r} + d\mathbf{r}$ under the action of a force $\mathbf{F}$, as shown in Figure 3.2. The differential work done on the particle during this differential change in the position vector is

$$dU = \mathbf{F} \cdot d\mathbf{r} \tag{3.9}$$

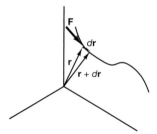

Figure 3.2

The units of work are expressed in m · N or ft · lb. The S.I. units of work are *Joules* (J) and are named after the physicist James P. Joule. Joules can be related to the other units of work by the following expressions:

$$\begin{aligned} 1\,\text{J} &= 1\,\text{m} \cdot 1\,\text{N} \\ 1\,\text{ft} \cdot 1\,\text{lb} &= 0.3048\,\text{m} \cdot 4.448\,\text{N} = 1.356\,\text{J} \end{aligned} \tag{3.10}$$

Work is a scalar quantity, and Eq. (3.9) can be written as

$$dU = F_x dx + F_y dy + F_z dz \tag{3.11}$$

The total work done by a particle as it moves from a position 1 to a position 2 is

$$U_{1\to2} = \int_{r_1}^{r_2} \mathbf{F} \cdot d\mathbf{r} \tag{3.12}$$

where the integration must be performed along the path of movement. The first integral of the particle's motion can now be written as

The work done on a particle going from position 1 to position 2 = change in kinetic energy at positions 1 and 2, or

$$U_{1\to2} = T_2 - T_1 \tag{3.13}$$

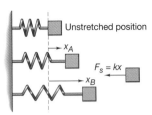

Figure 3.3

3.2.1 WORK OF A SPRING FORCE

Consider a mass attached to a fixed point by a spring with a spring constant k, as shown in Figure 3.3. The mass is moved from position A to position B, where x is measured from the unstretched length of the spring. The differential work done by the spring force is

$$dU = -F_s dx = -kx dx \tag{3.14}$$

The work done by the spring force when the mass is moved from position A to position B is

$$U_{A \to B} = -\int_{x_A}^{x_B} kx \, dx = \frac{1}{2}kx_1^2 - \frac{1}{2}kx_2^2 \tag{3.15}$$

The work done by a nonlinear spring force is computed in a similar manner. For example, consider the spring in Figure 3.3 to be a nonlinear spring such that the spring force is related to the stretch of the spring by

$$F_s = kx^n \tag{3.16}$$

Using Eqs. (3.14) and (3.15), we find that the work done by the spring force as the mass is moved from position A to B is given by

$$U_{A \to B} = -\int_{x_1}^{x_2} kx^n dx = \frac{1}{n+1}kx_1^{n+1} - \frac{1}{n+1}kx_2^{n+1} \tag{3.17}$$

Tendons in the human body (tissue that joins muscle to bone) act as nonlinear springs, and the spring force of a tendon is proportional to the square of the stretch of the tendon.

3.2.2 WORK OF THE GRAVITATIONAL ATTRACTION FORCE BETWEEN TWO MASSES

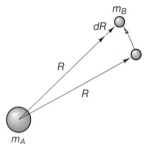

Consider two masses, as shown in Figure 3.4. We wish to determine the work done by the gravitational force on mass B as the mass moves relative to mass A. We will use a spherical coordinate system with its origin taken at the center of mass A. The gravitational force acting on mass B is directed toward mass A and has a magnitude

$$F = G\frac{m_A m_B}{R^2} \tag{3.18}$$

The differential work as mass B moves is

$$dU = -G\frac{m_A m_B}{R^2} dR \tag{3.19}$$

Figure 3.4

The work done by a finite displacement from a position 1 to a position 2 along mass B's path is equal to

$$U_{1 \to 2} = -\int_{R_1}^{R_2} G\frac{m_A m_B}{R^2} dR = \frac{Gm_A m_B}{R_2} - \frac{Gm_A m_B}{R_1} \tag{3.20}$$

The principle of work–energy extends the methods of separable first order scalar differential equations to the vector differential equation of motion by reducing that vector equation into a scalar differential equation. It is important to realize, as was stated earlier, that the principle of work–energy yields only a partial solution of the dynamics problem that gives velocity as a function of position.

3.2.3 POWER AND EFFICIENCY

We introduced work by first defining the concept of power in Eq. (3.7). Power is the time rate at which work is done and is equal to the time rate of change of the kinetic energy. We all have a layman's definition of power when we look at the horsepower of an engine, pay our electric bill, or choose an electric motor for a appliance. A small motor or a large motor may be used to do the same amount of work, but the large motor will do the work in a shorter time period. Power, like work, is a scalar quantity and may be written

$$P = \mathbf{F} \cdot \mathbf{v} = \frac{dU}{dt} = \frac{dT}{dt} \tag{3.21}$$

Power is the rate at which work is done and is measured in units of *Watts* (W), named after the engineer James Watt. The Watt may be expressed as

$$1 \text{ W} = 1 \text{ J}/1 \text{ sec} = \frac{1 \text{ meter} \cdot 1 \text{ newton}}{1 \text{ sec}} \tag{3.22}$$

Power is also measured in U.S. customary units as *horsepower,* where

$$1 \text{ hp} = 550 \text{ ft} \cdot \text{lb}/\text{s} = 746 \text{ W} \tag{3.23}$$

It is not known whether or not this relationship is based on the assumption that a horse could pull 550 pounds forward 1 foot in 1 second. Automobiles and other engines are rated in horsepower, and the owner of a car may wish to think of the car as equal to 300 horses, except that it costs less and is easier to feed.

The ratio of the work done *by* a machine to the work done *on* the machine is called the *mechanical efficiency e* of the machine. This ratio of the output work to the input work is written as

$$e = \frac{U_{\text{output}}}{U_{\text{input}}} \tag{3.24}$$

This definition is based on the assumption that the work is done at a constant rate. The mechanical efficiency of a machine at any instant of time is

$$e = \frac{P_{\text{output}}}{P_{\text{input}}} \tag{3.25}$$

The output power is always less than the input power, due to friction or electrical and thermal energy losses. Therefore, the efficiency of any machine is less than one.

Sample Problem 3.1

Consider a particle of mass 3 kg propelled with an initial velocity of 2 m/s up an inclined plane of 30°, as shown in the accompanying diagram. If the coefficient of kinetic friction is 0.4 between the particle and the incline surface, determine the distance the block will slide up the incline before coming to a stop.

Solution A free-body diagram of the particle is as follows:

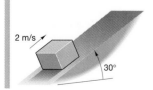

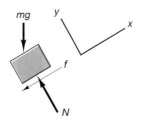

The motion of the particle is in the positive x-direction, and the only forces acting in that direction and doing work on the particle are the friction force and a component of the gravitational attraction. Both of these forces oppose the motion and, therefore, do negative work. The particle stops at the maximum distance up the incline, and the kinetic energy at that point is zero. Therefore, the work energy equation is

$$U_{1\rightarrow2} = -(mg\sin 30 + f)x = T_2 - T_1 = -\frac{1}{2}mv_0^2$$

$$f = \mu_k N = \mu_k mg\cos 30$$

$$-m9.81(\sin 30 + 0.4\cos 30)x = -\frac{1}{2}m2^2$$

$$x = 0.241 \text{ meter}$$

Note that this problem is a straightforward application of work–energy, or the first integral of the motion, as the forces are constant during the movement, and therefore, the acceleration is a constant.

Sample Problem 3.2

Using the accompanying diagram, determine the velocity at point 2 of a particle's path of motion, where the particle, starting from rest at point 1, slides down a smooth, circular path under the action of a constant force.

Solution   The path of the particle's movement is

$$x^2 + y^2 = R^2$$

$$d\bar{r} = dx\hat{\mathbf{i}} + dy\hat{\mathbf{j}}$$

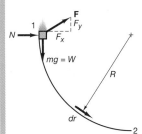

Differentiating the path of movement yields

$$2xdx + 2ydy = 0$$

Therefore,

$$dy = -\frac{x}{y}dx = -\frac{x}{\pm\sqrt{R^2 - x^2}}dx$$

The normal force between the particle and the surface does no work, and the work done by the forces **F** and **W** is

$$U_{1\rightarrow2} = \int_1^2 (\mathbf{F} + \mathbf{W})\cdot d\mathbf{r} = \int_1^2 (F_x dx + [F_y - W]dy)$$

$$U_{1\rightarrow2} = \int_{-R}^0 \left[F_x + \frac{x}{\sqrt{R^2 - x^2}}[F_y + W] \right] dx = F_x R - [F_y - W]R$$

$$U_{1\rightarrow2} = \frac{1}{2}mv_2^2$$

$$v_2 = \sqrt{\frac{2(F_x R - [F_y - W]R)}{m}}$$

If the particle slid down the path under the influence of gravitational attraction only, the force **F** is equal to zero, and

$$U_{1\rightarrow2} = mgR$$

$$v_2 = \sqrt{2gR}$$

Sample Problem 3.3

An automobile is traveling on a level road when the driver attempts to stop in order to avoid an accident. Assuming a full-wheel lockup, determine a formula for distance over which the car skids before it comes to a complete stop, in terms of the kinetic coefficient of friction and the initial velocity.

Solution

$$U_{1 \to 2} = T_2 - T_1$$

$$-fs = 0 - \frac{1}{2} mv_0^2$$

$$f = \mu_k mg$$

$$s = \frac{v_0^2}{2\mu_k g}$$

The coefficient of kinetic friction on dry pavement between the tires and road is approximately 0.7; therefore, we have the following table:

v_0 (mph)	v_0 (fps)	s (ft)
60	88	171
50	73	119
30	44	43

Police officers and accident reconstructionists use this formula to determine speeds of automobiles involved in accidents from the measurements of skid marks on the road. Stopping distances increase with the decrease in the coefficient of kinetic friction. For example, on wet roads, the coefficient is 0.5, and on snow or ice, it will be from 0.1 to 0.3.

Sample Problem 3.4

If the pulley system in the accompanying illustration is released from rest, determine the velocity of the 40-kg block *B* after it has moved 0.5 m. Neglect friction and the weight of the pulley.

Solution

$m_A = 25$ kg
$m_B = 40$ kg

The motion of the two masses is constrained by the fixed length of the rope. The constraint equations are

$$(x_A - x_P) + (x_B - x_P) = l_1$$

where x is taken as positive in the downward direction. The position of the pulley is constant, and the constraint equation can be written as

$$x_A + x_B = 2x_P + l_1$$

The constraint equations for the velocities and the changes in position are obtained by differentiating the following equation:

$$\Delta x_A + \Delta x_B = 0$$
$$v_A + v_B = 0$$

We will now construct free-body diagrams of each mass. The work–energy equations for the two masses are

$$(m_A g - T)\Delta x_A = \frac{1}{2} m_A v_A^2$$

$$(m_B g - T)\Delta x_B = \frac{1}{2} m_A v_B^2$$

We now have four equations for the four unknowns Δx_A, T, v_A, and v_B. Using the two constraint equations and the specified change in position for block B, we obtain

$$(m_B - m_A)g\Delta x_B = \frac{1}{2}(m_B + m_A)v_B^2$$

This equation could have been obtained directly for the system of the two masses, as the internal force T does no work. The velocity of block B can now be found for any displacement. For $\Delta x_A = 0.5$ m,

$$v_B = \sqrt{\frac{2(m_B - m_A)\,g\Delta x_A}{(m_A + m_B)}} \qquad\qquad v_B = 1.51 \text{ m/s}$$

Sample Problem 3.5

If the system shown in the following diagram starts from rest under the influence of a constant force $\mathbf{F}$, determine the velocity for any position x.

The coefficient of kinetic friction is the same between both blocks and the surface.

Solution The motion of the two blocks is constrained by the length of the cable joining them. The equation of constraint is

$$2(x_A - x_B) + (c - x_B) = L$$

The change in the position and velocity satisfy the constraints:

$$2\Delta x_A - 3\Delta x_B = 0$$
$$2v_A - 3v_B = 0$$

A single free-body diagram can be constructed for the system, or separate free-body diagrams can be constructed for each element, shown as follows:

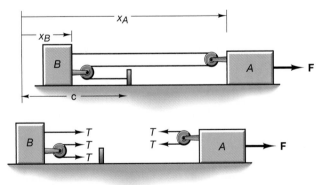

The work–energy equation for each block is

$$\text{Block } B\text{: } (3\,T - f_B)\Delta x_B = \tfrac{1}{2}m_B v_B^2$$
$$\text{Block } A\text{: } (F - 2\,T - f_A)\Delta x_A = \tfrac{1}{2}m_A v_A^2$$

If the masses A and B are considered as a single system, the work–energy relationship becomes

$$-f_B \Delta x_B + (F - f_A)\Delta x_A = \tfrac{1}{2}(m_B v_B^2 + m_A v_A^2)$$

Note that the internal forces (tension in the cable) do no work on the system. The system equation can also be obtaining by adding the equations for the two blocks as treated separately and using the relationship between the change in displacements.

Substituting for the values of the friction forces and using the constraints yields

$$-\mu m_B g\left(\frac{2}{3}\Delta x_A\right) + (F - \mu m_A g)\Delta x_A = \frac{1}{2}\left[m_B\left(\frac{2}{3}v_A\right)^2 + m_A v_A^2\right]$$

$$v_A = \sqrt{\frac{6[F - \mu g(2m_B + 3m_A)]\Delta x_A}{(4m_B + 9m_A)}}$$

Sample Problem 3.6

A block of mass m is released from rest to slide down a plane inclined at an angle α and with a coefficient of kinetic friction of μ_k. If a spring at the bottom of the incline is compressed to a maximum distance Δ, determine the distance d the block slides before it comes into contact with the spring. The spring has a spring constant of k. Discuss any limiting relationship between the incline angle and the coefficients of friction.

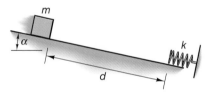

Solution

A free-body diagram of the block when it is contact with the spring is shown as follows:

Only the friction, the component of the weight parallel to the plane, and the spring force do work as the block slides down the plane. Since the block is released from rest and the velocity is zero when the spring is compressed to its maximum value, the initial and final kinetic energy are zero. Therefore, the total change in the kinetic energy is zero, and the total work done is also zero. The work done by the block as it moves from a point 1 to a point 2 is

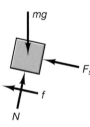

$$U_{1\to2} = (mg \sin \alpha - \mu_k mg \cos \alpha)(d + \Delta) - \frac{1}{2}k\Delta^2 = 0$$

Solving for the distance d yields

$$d = \frac{\frac{1}{2}k\Delta^2 - (mg \sin \alpha - \mu_k mg \cos \alpha)\Delta}{(mg \sin \alpha - \mu_k mg \cos \alpha)}$$

The component of the weight must be greater than the kinetic-friction force for the block to slide down the incline. Therefore, the coefficient of static friction must satisfy the relationship

$$\mu_s \leq \tan \alpha$$

This condition is necessary for the block to start to slide. Since the coefficient of kinetic friction is approximately 20 to 25% less than the coefficient of static friction, the coefficient of friction must be much less than the tangent of the incline angle for the problem to be well posed.

Problems

3.1 A 5-kg mass is dropped down a vertical shaft. If it is released from rest, determine the velocity of the particle after it has fallen 10 m.

3.2 In Figure P3.2, a man pushes a 100-lb box across a floor with a constant force of 50 lb. If the kinetic coefficient of friction between the box and the floor is 0.2, determine the velocity of the box after it has been moved 20 ft.

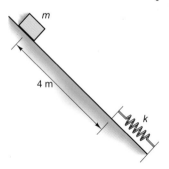

Figure P3.2

3.3 A 2-kg box starts from rest and slides 4 m down a 45° incline, where it strikes a spring with a spring constant of $k = 500$ N/m, as shown in Figure P3.3. If the coefficient of kinetic friction between the box and the incline is 0.6, determine the maximum compression of the spring.

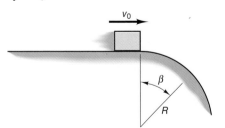

Figure P3.3 (Figure P2.41)

3.4 In Figure P3.4, determine the angle β at which a particle of mass m leaves a curved surface if its initial velocity is v_0.

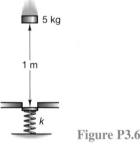

Figure P3.4 (Figure P2.50)

3.5 A block of mass m slides down the surface of a smooth bowl of radius R, as illustrated in Figure P3.5. If the block is released from rest when $\theta = \theta_0$, determine the velocity for any angle θ.

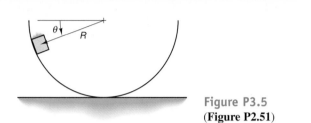

Figure P3.5
(Figure P2.51)

3.6 A 5-kg mass is dropped from a height of 1 m onto a initially compressed spring with a spring constant of 500 N/m, as depicted in Figure P3.6. If the initial compression of the spring is 50 mm, determine the maximum compression of the spring.

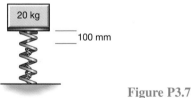

Figure P3.6

3.7 Determine the maximum distance that a 20-kg mass will fall if it is released from rest just above the nested spring system shown in Figure P3.7. The outer spring has a spring constant of 200 N/m, and the inner spring has a spring constant of 500 N/m.

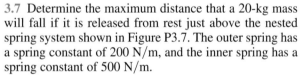

Figure P3.7

3.8 Determine the maximum distance the mass in Problem 3.7 will fall if it is released 200 mm above the outer spring.

💻 **3.9** A mass slides down a smooth bar which forms a circular guide of radius R. If a spring with a spring constant k is coiled around the bar, as shown in Figure P3.9, develop an expression to determine the angle at which the mass stops moving. Solve numerically when $m = 2$ kg, $R = 0.3$ m, and $k = 200$ N/m.

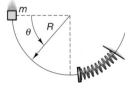

Figure P3.9

3.10 A mass attached to a rope forms a pendulum, as shown in Figure P3.10. If the mass is released from rest at a horizontal position and swings in a vertical plane, determine an expression for the tension in the rope at any position θ.

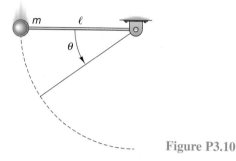

Figure P3.10

3.11 Determine the maximum angle that the mass in Problem 3.10 will attain.

3.12 A 3000-lb pickup truck tows a 2500-lb trailer at a speed of 60 mph, as depicted in Figure P3.12. If the coefficient of kinetic friction between the tires and the road is 0.7 and the driver applies the brakes to both the car and the trailer, determine (a) the distance that they travel before they stop, and (b) the horizontal force on the trailer's hitch.

Figure P3.12

3.13 Determine the distance to stop the truck and trailer in Problem 3.12 if they are decending a 10% grade.

3.14 Determine the distance to stop the truck and trailer in Problem 3.12 and the force on the trailer's hitch if the brakes are applied only to the car.

3.15 In Figure P3.15, a 6-kg block slides off the end of a smooth slide of radius 2 meters and falls 2 meters to the floor. Determine the speed of the block at the bottom of the slide and when it hits the floor.

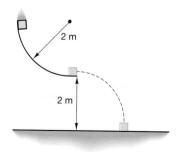

Figure P3.15

3.16 A 3-kg block slides down an incline onto a horizontal surface, as shown in Figure P3.16. If the coefficient of kinetic friction between the incline and the horizontal surface holding the block is 0.3, determine the distance D that the block will slide on the horizontal surface before coming to a stop.

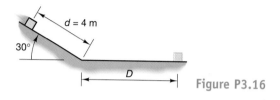

Figure P3.16

3.17 In Figure P3.17, a belt system is used to transport boxes from a container on the second floor of a building to a first-floor container. The system is locked while the boxes are placed on belt, and it can transport four boxes at a time. The system is driven by the weight of the boxes, and there is negligible friction between the belt and the drum. If the boxes are of equal mass, determine the velocity at which each box will leave the belt.

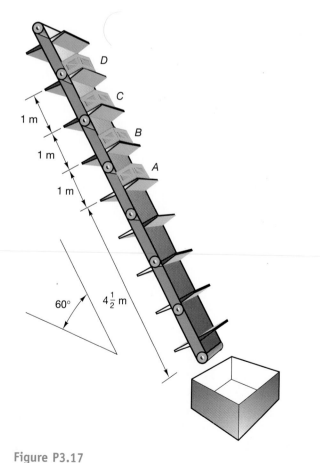

Figure P3.17

3.18 In Figure P3.18, determine the minimum power needed for a winch to haul a 100-kg crate up a 30° incline at a constant rate of 2 m/s if the coefficient of kinetic friction between the crate and the inclined surface is 0.4.

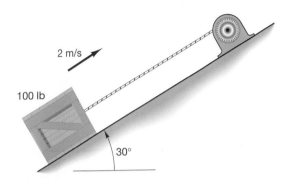

100 lb

2 m/s

30°

Figure P3.18

3.19 If a 2-kW winch was needed to haul the crate in Problem 3.18, what would be the angle of the incline?

3.20 If a 2-kW winch was needed to haul the crate up the 30° incline in Problem 3.18, what would be the coefficient of kinetic friction between the crate and the inclined surface?

 3.21 The 2-kg block in Figure P3.21 slides along a smooth rod when the 5-kg weight is released. The spring constant of the spring shown in the figure is 100 N/m, and the spring is unstretched when the system is released from rest in the position shown. Determine the maximum distance that the 2-kg block will slide to the right.

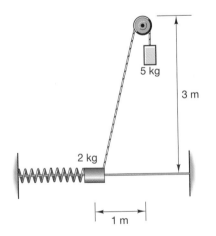

5 kg

3 m

2 kg

1 m

Figure P3.21

 3.22 Develop a general expression for the velocity of the 5-kg block in Problem 3.21 as a function of displacement of the 2-kg block.

3.23 In Figure P3.23, determine the velocity of the 3-kg block A after it has moved 200 mm on a smooth surface, starting from rest.

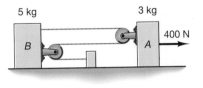

5 kg   3 kg

B   A   400 N

Figure P3.23

3.24 Determine the velocity of the 3-kg block A in Problem 3.23 if the coefficient of kinetic friction between the blocks and the surface is 0.3.

3.25 In Figure P3.25, determine the velocity of the 50-kg block after the 100-kg block has moved 500 mm. Neglect friction and the mass of the pulleys.

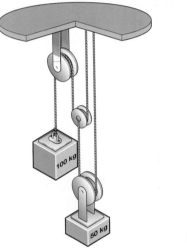

100 kg

50 kg

Figure P3.25

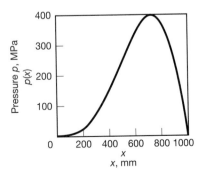 **3.26** The preassure p in a rifle barrel varies with the position of the bullet in the 1000-mm barrel by the relationship $p(x) = 10^{-3}x^2 \sin(\pi x/1000)$ MPa, as shown in the graph in Figure P3.26. If the diameter of the bore of the barrel is 10 mm and the mass of the bullet is 20 g, determine the velocity of the bullet as it leaves the barrel. Neglect friction between the bullet and the barrel.

Figure P3.26

3.27 Determine the velocity of the bullet in Problem 3.26 as it leaves the barrel if the coefficient of kinetic friction between the bullet and the barrel is 0.3.

3.28 In Figure P3.28, if the spring is stretched to a length l when θ is zero, determine the velocity of the mass A just before it strikes the horizontal guide ($\theta = 90°$). Neglect friction between the blocks and the guides. (*Hint: The velocity of block* B *is zero at this point.*)

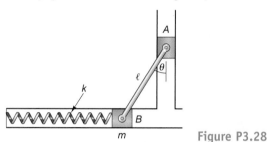

Figure P3.28

3.29 Determine the maximum compression x of the nonlinear spring ($F_s = 200x^2$N) when a 10-kg mass is released from rest 3 m above the spring, as depicted in Figure P3.29.

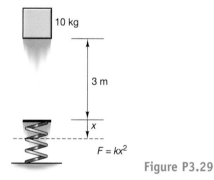

Figure P3.29

3.30 A mass is released from rest at a height h above an identical mass that is resting in equilibrium on a linear spring with a spring constant k, as shown in Figure P3.30. Determine the maximum additional compression Δ of the spring.

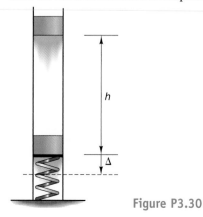

Figure P3.30

3.31 For the car in Figure P3.31, develop a formula to design the bumper stiffness k such that the car can be stopped from a speed v with a maximum deformation of the bumper Δ. Consider the bumper as a linear spring.

Figure P3.31

3.32 A cart moves at a constant velocity v and is suddenly brought to a stop, as illustrated in Figure P3.32. Determine the maximum angle that a mass which hangs from the cart will swing to if it is attached to the cart by a cord of length l.

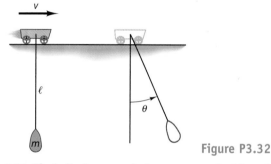

Figure P3.32

3.33 If a ball of mass m is thrown upward with an initial velocity v, determine the velocity of the ball when it returns to its initial position. Neglect air resistance.

3.34 Determine the maximum height that the ball in Problem 3.33 will attain.

3.35 In Figure P3.35, a 1-lb particle slides down a smooth helical wire of radius 5 inches when released from rest at a coordinate position A (5, 0, 8) m and reaches the end of the wire at coordinate position B (5, 0, 0) m. Determine the velocity of the particle when it reaches the end of the wire.

Helical motion in space

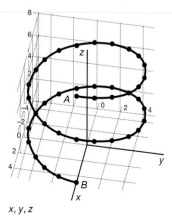

x, y, z

Figure P3.35

3.36 Determine the velocity of the particle in Problem 3.35 when it reaches the bottom of the helical wire if it starts at the top with a velocity of 2 ft/s.

3.37 A 1-kg collar slides down a smooth wire starting at a position with coordinates (5, 7, 10 m) and ending at a position with coordinates (2.5, 3.5, 0 m), as shown in Figure P3.37. If the collar has an initial velocity of 2 m/s, determine the velocity of the collar at the bottom of the wire.

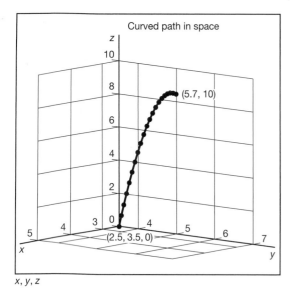

Figure P3.37

3.38 Determine the velocity of the 5-kg collar in Figure P3.38 after it starts from rest and moves 200 mm under the action of a 200-N force. Neglect friction between the collars and guide rods. (Both ends of the cable are attached to the 4-kg mass.)

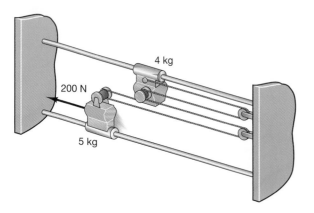

Figure P3.38

3.39 Determine the velocity of the 5-kg collar in Problem 3.38 if the collar has an initial velocity of 5 m/s.

3.40 Determine the velocity of the 4-kg collar in Problem 3.38 after this collar has moved 200 mm, starting from rest. The coefficient of kinetic friction between the collars and the rods is 0.4.

3.41 In Figure P3.41, a 110-lb female acrobat falls 50 ft onto a safety net that has an effective force–deformation relationship $F_{net} = 60\delta^3$ where δ is the net deformation in feet. Determine the minimum distance that the net should be mounted above the floor so that the acrobat will not touch the floor when she falls.

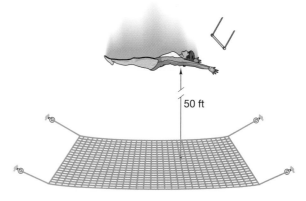

Figure P3.41

3.42 A stunt driver wants to drive a car around a vertical loop of radius R, as shown in Figure P3.42. What is the minimal initial velocity v_0 at which the car must be traveling at the bottom of the loop in order to successfully make the loop?

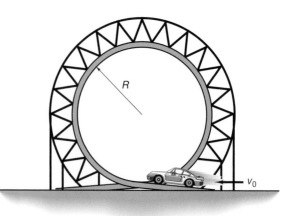

Figure P3.42

3.43 In Figure P3.43, a 5-kg piston is driven by the force displacement curve $F(x) = 500[x - (x^2/2)]$ N in a 2-m cylinder. If the piston starts from rest, determine its velocity at the end of the cylinder.

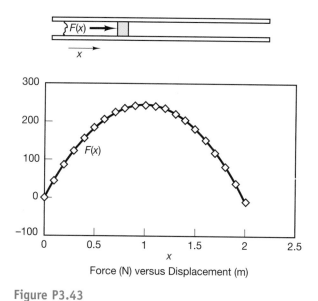

Figure P3.43

Force (N) versus Displacement (m)

3.44 Determine the velocity of the piston in Problem 3.43 at the end of the cylinder if the force–displacement is defined by $F(x) = 500(2x^2 - x^3)$ N. (See Figure P3.44.)

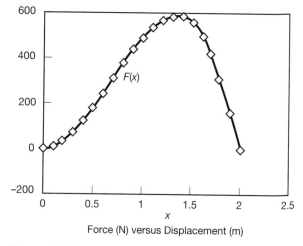

Figure P3.44

Force (N) versus Displacement (m)

3.45 A commercial airplane has a mass of 50,000 kg and can attain a takeoff speed of 55 m/s in 30 s. What is the average power that its engines deliver?

3.3 CONSERVATIVE FORCES AND POTENTIAL ENERGY

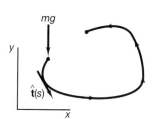

Figure 3.5

Consider the work done by the force of gravitational attraction as a particle of mass m moves in a plane perpendicular to the earth, but at a distance close to the surface of the earth. This movement is illustrated in Figure 3.5. The work done by the gravitational force is

$$U_{1 \to 2} = \int_{\mathbf{r}_1}^{\mathbf{r}_2} \mathbf{F} \cdot d\mathbf{r} = -mg \int_{y_1}^{y_2} dy = mg(y_1 - y_2) \tag{3.26}$$

The work done by the gravitational force is independent of the path of motion and depends only on the final and original positions of the particle. If the work done by a force is independent of its path of motion and depends only on the original and final positions of the particle on which it acts in space, the force is called a *conservative force*. For such forces, if the particle upon which they act is moved along a closed path (having the same initial and final positions), the work done is zero:

$$\oint \mathbf{F} \cdot d\mathbf{r} = 0 \tag{3.27}$$

The integrand in the work integral must be an exact differential of some function

$$\mathbf{F} \cdot d\mathbf{r} = -dV \tag{3.28}$$

The minus sign is introduced to simplify later equations. The function V is called a *potential function*. The potential function for a conservative force can be obtained by noting that

$$dV = \frac{\partial V}{\partial x} dx + \frac{\partial V}{\partial y} dy + \frac{\partial V}{\partial z} dz \tag{3.29}$$

The change in the position vector and the force vector can be written as

$$d\mathbf{r} = dx\,\hat{\mathbf{i}} + dy\,\hat{\mathbf{j}} + dz\,\hat{\mathbf{k}}$$

$$\mathbf{F} = F_x\hat{\mathbf{i}} + F_y\hat{\mathbf{j}} + F_z\hat{\mathbf{k}} \tag{3.30}$$

The integrand of the work integral becomes

$$\mathbf{F}\cdot d\mathbf{r} = F_x dx + F_y dy + F_z dz = -\left(\frac{\partial V}{\partial x}\,dx + \frac{\partial V}{\partial y}\,dy + \frac{\partial V}{\partial z}\,dz\right)$$

Therefore,

$$F_x = -\frac{\partial V}{\partial x}$$
$$F_y = -\frac{\partial V}{\partial y} \tag{3.31}$$
$$F_z = -\frac{\partial V}{\partial z}$$

These equations can be solved to determine the potential function V. For example, let $\mathbf{F}$ be the force of gravitational attraction. Equation (3.31) can then be written as

$$0 = -\frac{\partial V}{\partial x}$$
$$-mg = -\frac{\partial V}{\partial y}$$
$$0 = -\frac{\partial V}{\partial z}$$

The second equation yields

$$V = mgy + f(x, z)$$

The first and third equations state that V cannot be a function of either x or y; therefore,

$$V = mgy + C$$

The constant C does not effect the force and may be chosen arbitrarily. In this case, V is usually written as V_g and is called the gravitational potential. V_g is the *potential energy* of the particle with respect to the gravitational attraction force. As the body is raised, the potential energy increases, and the particle has the capability to perform positive work when released from this position. The gravitational force and, therefore, the work that it can do, is independent of the constant C. The reference datum plane or reference height that determines C can be chosen arbitrarily. The work done by a conservative force moving from a point 1 to a point 2 is

$$U_{1\to2} = V_1 - V_2 \tag{3.32}$$

The directed derivative in vector calculus is denoted by

$$\nabla = \hat{\mathbf{i}}\frac{\partial}{\partial x} + \hat{\mathbf{j}}\frac{\partial}{\partial y} + \hat{\mathbf{k}}\frac{\partial}{\partial z} \tag{3.33}$$

Eq. (3.31) can be written as

$$\mathbf{F} = -\nabla V \tag{3.34}$$

The symbol ∇ is called the *del operator* and the term ∇V is called the *gradient of* V and is sometimes written as *grad V*. It is equal to the directed derivative of V in each of the coordinate directions. Some other properties of the del operator are given in Mathematics Window 3.1.

MATHEMATICS WINDOW 3.1

Cartesian Coordinates

The del vector operator: $\nabla = \hat{\mathbf{i}}\dfrac{\partial}{\partial x} + \hat{\mathbf{j}}\dfrac{\partial}{\partial y} + \hat{\mathbf{k}}\dfrac{\partial}{\partial z}$

The gradient of a scalar function $\varphi(x, y, z)$: $\nabla\varphi = \hat{\mathbf{i}}\dfrac{\partial\varphi}{\partial x} + \hat{\mathbf{j}}\dfrac{\partial\varphi}{\partial y} + \hat{\mathbf{k}}\dfrac{\partial\varphi}{\partial z}$

The divergence of a vector: $\nabla\cdot\mathbf{u} = \dfrac{\partial u_x}{\partial x} + \dfrac{\partial u_y}{\partial y} + \dfrac{\partial u_z}{\partial z}$

The curl of a vector:

$$\nabla\times\mathbf{u} = \hat{\mathbf{i}}\left(\frac{\partial u_z}{\partial y} - \frac{\partial u_y}{\partial z}\right) + \hat{\mathbf{j}}\left(\frac{\partial u_x}{\partial z} - \frac{\partial u_z}{\partial x}\right) + \hat{\mathbf{k}}\left(\frac{\partial u_y}{\partial x} - \frac{\partial u_x}{\partial y}\right)$$

Special scalar operator, Laplacian: $\nabla^2 = \nabla\cdot\nabla = \dfrac{\partial^2}{\partial x^2} + \dfrac{\partial^2}{\partial y^2} + \dfrac{\partial^2}{\partial z^2}$

It can be shown that the *curl of the gradient* of any function is zero—that is,

$$\nabla\times\nabla\varphi = 0 \tag{3.35}$$

Therefore, another definition of a conservative force is one whose curl is zero:

$$\nabla\times\mathbf{F} = \nabla\times(-\nabla V) = 0 \tag{3.36}$$

To test whether a force is a conservative force or not, take the curl of the force and see if it is equal to zero. A potential function can be found for any conservative force, and this function is called the *potential energy*.

The properties of the del operator in cylindrical coordinates are shown in Mathematics Window 3.2 and in spherical coordinates in Mathematics Window 3.3.

MATHEMATICS WINDOW 3.2

Cylindrical Coordinates

The del vector operator: $\nabla = \hat{\mathbf{e}}_r\dfrac{\partial}{\partial r} + \hat{\mathbf{e}}_\theta\dfrac{1}{r}\dfrac{\partial}{\partial\theta} + \hat{\mathbf{k}}\dfrac{\partial}{\partial z}$

The gradient of a scalar function $\varphi(r, \theta, z)$:

$$\nabla\varphi = \hat{\mathbf{e}}_r\frac{\partial\varphi}{\partial r} + \hat{\mathbf{e}}_\theta\frac{1}{r}\frac{\partial\varphi}{\partial\theta} + \hat{\mathbf{k}}\frac{\partial\varphi}{\partial z}$$

The divergence of a vector: $\nabla\cdot\mathbf{u} = \dfrac{1}{r}\dfrac{\partial}{\partial r}(ru_r) + \dfrac{1}{r}\dfrac{\partial u_\theta}{\partial\theta} + \dfrac{\partial u_z}{\partial z}$

The curl of a vector:

$$\nabla\times\mathbf{u} = \hat{\mathbf{e}}_r\left(\frac{1}{r}\frac{\partial u_z}{\partial\theta} - \frac{\partial u_\theta}{\partial z}\right) + \hat{\mathbf{e}}_\theta\left(\frac{\partial u_r}{\partial z} - \frac{\partial u_z}{\partial r}\right) + \hat{\mathbf{k}}\left(\frac{1}{r}\frac{\partial}{\partial r}(ru_\theta) - \frac{1}{r}\frac{\partial u_r}{\partial\theta}\right)$$

Laplacian: $\nabla^2 = \dfrac{\partial^2}{\partial r^2} + \dfrac{1}{r}\dfrac{\partial}{\partial r} + \dfrac{1}{r^2}\dfrac{\partial^2}{\partial\theta^2} + \dfrac{\partial^2}{\partial z^2}$

MATHEMATICS WINDOW 3.3

Spherical Coordinates

The del vector operator: $\nabla = \hat{\mathbf{e}}_R \dfrac{\partial}{\partial R} + \hat{\mathbf{e}}_\phi \dfrac{1}{R}\dfrac{\partial}{\partial \phi} + \hat{\mathbf{e}}_\theta \dfrac{1}{R\sin\phi}\dfrac{\partial}{\partial \theta}$

The gradient of a scalar function $\psi(R, \phi, \theta)$:

$$\nabla\psi = \hat{\mathbf{e}}_R \frac{\partial \psi}{\partial R} + \hat{\mathbf{e}}_\phi \frac{1}{R}\frac{\partial \psi}{\partial \phi} + \hat{\mathbf{e}}_\theta \frac{1}{R\sin\phi}\frac{\partial \psi}{\partial \theta}$$

The divergence of a vector:

$$\nabla\cdot\mathbf{u} = \frac{1}{R^2}\frac{\partial}{\partial R}(R^2 u_R) + \frac{1}{R\sin\phi}\frac{\partial}{\partial \phi}(u_\phi \sin\phi) + \frac{1}{R\sin\phi}\frac{\partial u_\theta}{\partial \theta}$$

The curl of a vector:

$$\nabla\times\mathbf{u} = \hat{\mathbf{e}}_R \frac{1}{R^2\sin\phi}\left[\frac{\partial}{\partial \phi}(u_\theta R\sin\phi) - \frac{\partial}{\partial \theta}(Ru_\phi)\right]$$

$$+ \hat{\mathbf{e}}_\phi \frac{1}{R\sin\phi}\left[\frac{\partial u_R}{\partial \theta} - \frac{\partial}{\partial R}(u_\theta R\sin\phi)\right] + \hat{\mathbf{e}}_\theta \frac{1}{R}\left[\frac{\partial}{\partial R}(Ru_\phi) - \frac{\partial u_R}{\partial \phi}\right]$$

Laplacian:

$$\nabla^2\psi = \frac{1}{R^2}\frac{\partial}{\partial R}\left(R^2 \frac{\partial \psi}{\partial R}\right) + \frac{1}{R^2\sin\phi}\frac{\partial}{\partial \phi}\left(\sin\phi \frac{\partial \psi}{\partial \phi}\right) + \frac{1}{R^2\sin\phi}\frac{\partial^2 \psi}{\partial \theta^2}$$

It has been shown that the force of gravitational attraction is independent of the path of motion of a particle when the particle is in close proximity to the surface of the earth. It also can be shown that gravitational attraction is a conservative force when the particle's distance from the earth's center varies. The constant g is defined as

$$g = \frac{GM_E}{R_E^2}$$

where M_E and R_E are the mass are the mass and radius of the earth, respectively. Consider a particle of mass m at a distance r from the center of the earth, as shown in Figure 3.6. The force of gravitational attraction is

$$\mathbf{F} = -\frac{mgR_E^2}{r^2}\hat{\mathbf{e}}_r$$

This expression has been written in polar coordinates, and the proof of the expression will be presented in two dimensions, but still can be done in complete generality in spherical coordinates, using Eqs. (1.114) and (1.115). The del operator in polar coordinates is

$$\nabla = \hat{\mathbf{e}}_r \frac{\partial}{\partial r} + \hat{\mathbf{e}}_\theta \frac{1}{r}\frac{\partial}{\partial \theta}$$

$$\frac{d\hat{\mathbf{e}}_r}{d\theta} = \hat{\mathbf{e}}_\theta$$

$$\frac{d\hat{\mathbf{e}}_\theta}{d\theta} = -\hat{\mathbf{e}}_r$$

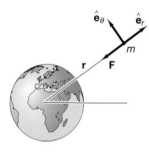

Figure 3.6

The curl of the gravitational force is

$$\nabla \times \mathbf{F} = \left(\hat{\mathbf{e}}_r \frac{\partial}{\partial r} + \hat{\mathbf{e}}_\theta \frac{1}{r} \frac{\partial}{\partial \theta}\right) \times \left(-\frac{mgR_E^2}{r^2} \hat{\mathbf{e}}_r\right)$$

$$\nabla \times \mathbf{F} = \hat{\mathbf{e}}_r \times \frac{\partial}{\partial r}\left(-\frac{mgR_E^2}{r^2}\right)\hat{\mathbf{e}}_r - \hat{\mathbf{e}}_\theta \times \frac{mgR_E^2}{r^2} \frac{1}{r} \frac{\partial}{\partial \theta}(\hat{\mathbf{e}}_r)$$

$$\nabla \times \mathbf{F} = \frac{\partial}{\partial r}\left(-\frac{mgR_E^2}{r^2}\right)(\hat{\mathbf{e}}_r \times \hat{\mathbf{e}}_r) - \frac{mgR_E^2}{r^2}\frac{1}{r}(\hat{\mathbf{e}}_\theta \times \hat{\mathbf{e}}_\theta)$$

$$\nabla \times \mathbf{F} = 0$$

Note that the cross product of any vector with itself is zero. Since the curl of the force is zero, the force is conservative and the work done by this force is independent of the path of motion. The potential energy of the particle is found from the general relationship in Eq. (3.31):

$$\mathbf{F} = -\nabla V$$

Therefore,

$$\frac{\partial V}{\partial r} = \frac{mgR_E^2}{r^2}$$

$$\frac{1}{r}\frac{\partial V}{\partial \theta} = 0$$

$$V = -\frac{mgR_E^2}{r} + C$$

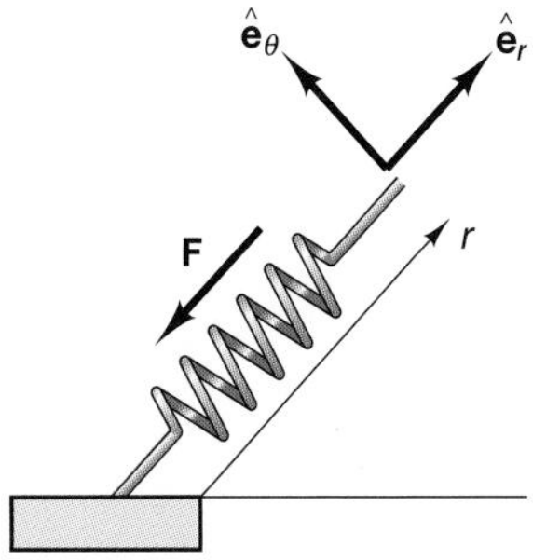

Figure 3.7

The constant of integration is usually taken to be zero.

Another conservative force is the force exerted by a spring during stretching or compressing of the spring. A spring is shown as oriented anywhere in the plane in Figure 3.7, using polar coordinates. Although the development will be done in two dimensions, it can be extended to three dimensions by use of spherical coordinates.

The force in the spring is

$$\mathbf{F} = f(r - r_0)\hat{\mathbf{e}}_r$$

where r_0 is the unstretched length of the spring. All materials can be modeled as springs, but many, such as ropes, tendons, etc., cannot support compression, so that the spring force previously given is only valid if $r > r_0$. The ligaments and tendons of the body are springs of this nature, and the spring force is related to the stretch by nonlinear relationships. The spring force is dependent only on the change in r and the curl of this force is zero, so any spring force is a conservative force.

Consider a linear spring which can support either compression or tension with a spring constant of k:

$$\mathbf{F} = -k(r - r_0)\hat{\mathbf{e}}_r = -ks\hat{\mathbf{e}}_r$$

where

$$s = (r - r_0)$$

The change of length, s, is a measure of the stretch of the spring and can have both positive and negative values. If s is positive, the spring force is in the negative r-direction—that is pulling to restore the spring to its unstretched length—and if s is negative, the spring force is in the positive r-direction. The potential energy in the spring at any position is

$$\frac{dV}{dr} = k(r - r_0)$$

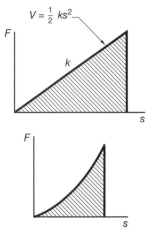

Figure 3.8

Since $ds = dr$, this expression can be written as

$$\frac{dV}{ds} = ks$$

$$V_s = \frac{ks^2}{2}$$

The potential energy in the spring is proportional to the stretch squared. Beams used as diving boards and leaf springs can be modeled as linear springs, and the spring force is proportional to the deflection, and the potential energy is proportional to the deflection squared. Torsion bars can also be modeled as springs, and the spring moment or torque is proportional to the twist angle. If the spring force is plotted as a function of the stretch, the potential energy can be seen to be the area under this curve, as illustrated in Figure 3.8. In Figure 3.8, the energy is shown for a linear and nonlinear spring as the area under the curve. The work done by a conservative force is therefore equal to

$$U_{1\rightarrow 2} = V_1 - V_2 = -\Delta V \tag{3.37}$$

If positive work is done, the potential energy decreases, or some of the potential energy is used to do the positive work.

3.4 CONSERVATION OF ENERGY

If all of the forces acting on a particle are conservative forces, the work done during any movement of the particle can be written as the change in the total potential energy:

$$U_{1\rightarrow 2} = V_1 - V_2 = -\Delta V \tag{3.38}$$

The work done during this movement is also equal to the change in the kinetic energy. Thus,

$$U_{1\rightarrow 2} = T_2 - T_1 = \Delta T \tag{3.39}$$

When only conservative forces are present, Eqs. (3.38) and (3.39) can be combined to give the principle of *conservation of the total energy*.

$$E = V + T \tag{3.40}$$

where E is the total energy of the system. For a conservative system, the change in the total energy is

$$\Delta E = 0 \tag{3.41}$$

The total energy of the system is conserved in this case, and the energy may change from potential energy to kinetic energy, but no energy is lost. Note that forces like friction and drag are not conservative forces and always dissipate energy from the system. Since these forces are always present, there is no such thing as a conservative system. The energy dissipated by friction or drag is converted to heat energy, which was not considered in this case. Conservation of energy can be written as

$$\frac{1}{2}mv^2 + V = \text{constant} \tag{3.42}$$

Sample Problem 3.7

If a particle of mass m is dropped from rest from a height h above the ground, determine the velocity of the particle as it hits the ground, neglecting air resistance.

m ○

h

Solution

The total energy of the particle when it is dropped is its gravitational potential energy:

$$E_1 = V_1 = mgh$$

Just before it hits the ground, the particle's potential energy is zero, and it has only kinetic energy:

$$E_2 = T_2 = \tfrac{1}{2}mv^2$$

Since the system is conservative, $E_1 = E_2$, and

$$v = \sqrt{2gh}$$

Sample Problem 3.8

A particle of mass m is dropped from a height h onto an uncompressed linear spring with a spring constant k. Develop an equation relating the maximum compression of the spring to the height and mass of the particle.

Solution

Only conservative forces act on the particle, and therefore, conservation of energy can be used. The velocity at release is zero, as is the velocity when the particle compresses the spring to its maximum. Let Δ be the maximum compression of the spring, and take the datum plane for the gravitational energy to be this maximum compressed position. The energy at release is

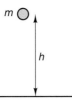

$$E_1 = mg(h + \Delta)$$

When the mass fully compresses the spring, the energy is

$$E_2 = \tfrac{1}{2}k\Delta^2$$

The energy is constant, so

$$mg(h + \Delta) = \tfrac{1}{2}k\Delta^2$$

Solving the quadratic equation for s yields

$$\Delta = \frac{1}{k}\left(mg + \sqrt{(mg)^2 + 2mghk}\right)$$

Note that even if h is zero, the compression of the spring will be double the compression when the weight (mg) is applied statically.

Problems

3.46 In Figure P3.46, a 2-kg box slides 4 meters down a 30° incline, where it strikes a linear spring with a spring constant of 200 N/m. Determine the maximum compression of the spring.

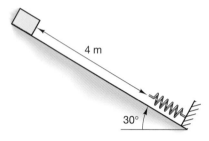

Figure P3.46

3.47 A particle of mass m is dropped from a height h and allowed either to slide down a smooth inclined plane or to slide on a smooth, curved chute as shown in Figure P3.47. In which case will the velocity of the particle be the largest, if the particle is released from rest in all cases?

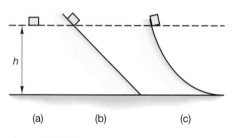

(a)   (b)   (c)

Figure P3.47

3.48 Determine the maximum distance a 20-kg mass will fall if it is released from rest just above the nested spring system shown in Figure P3.48. The outer spring has a spring constant of 200 N/m, and the inner spring has a spring constant of 500 N/m.

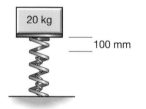

Figure P3.48 (Figure P3.7)

3.49 Determine the maximum distance that the mass in Problem 3.48 will fall if it is released 200 mm above the outer spring.

3.50 A mass slides down a smooth circular bar of radius R and contacts a spring with a spring constant k, as shown in Figure P3.50. Determine the maximum angle θ that the mass will attain. $m = 2$ kg, $R = 2$ m, and $k = 20$ N/m.

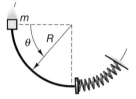

Figure P3.50 (Figure P3.9)

3.51 In Figure P3.51, a 2-kg mass is released from rest when the spring A is compressed 200 mm. Determine the maximum distance that the mass will travel up the smooth bar.

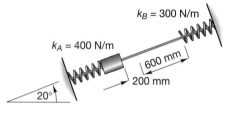

Figure P3.51

3.52 A 3-kg mass is released from rest at an angle $\theta = 60°$ and slides along a smooth, semicircular bar of radius 0.5 m, as shown in Figure P3.52. If the initial length of the spring is 0.5 m and the spring constant is 800 N/m, determine the maximum angle that the mass will attain.

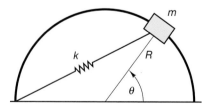

Figure P3.52

3.53 A nonlinear spring force is given by $F_s = kx^n$, where x is the stretch of the spring. Show that the spring force is a conservative force, and determine the potential energy function for the spring.

3.54 Prove that the force $\mathbf{F} = x^2\mathbf{i} + yz^2\mathbf{j} + y^2\hat{z}\mathbf{k}$ is a conservative force, and determine the corresponding potential function V.

3.55 Show that the maximum force on a bungee jumper's leg is independent of the lengh of the bungee cord if the cord stiffness is inversely proportional to the length of the cord. Refer to Figure P3.55.

Figure P3.55

3.56 Determine the launch velocity of a rocket if the rocket is to escape the gravitational field of the earth—that is, the *escape velocity*. (*Hint: As* r *approaches infinity, the velocity of the rocket will approach zero; therefore, the energy will be zero.*)

3.57 A force $\mathbf{F}(x, y) = (x\hat{\mathbf{i}} + y\hat{\mathbf{j}})/(x^2 + y^2)$ acts on a particle. Prove that the force is a conservative force, and determine the potential function associated with $\mathbf{F}$.

3.58 An automobile can develop a propulsive force on its rear tires equal to $F = \mu_s(\%mg)$—that is, the coefficient of static friction multiplied by the percent of the weight acting on the driving wheels. If the weight distribution of the automobile is 60% on the front tires and 40% on the rear tires, determine the distance for the car to go from 0 to 60 mph if the car has (a) rear-wheel drive (b) front-wheel drive, and (c) four-wheel drive. $\mu_s = 0.7$.

3.59 The mass m in Figure P3.59 is attached to a nondeformable cord and released from rest when $\theta = 0$. Use conservation of energy to determine the velocity at any angle.

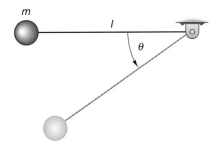

Figure P3.59

3.60 In Figure P3.60, a 6-kg block is released from rest, slides off the end of a smooth slide of radius 2 m, and then falls 2 m to the floor. Determine the magnitude of the velocity just before it hits the floor.

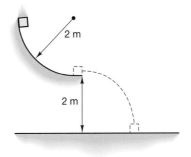

Figure P3.60 (Figure P3.15)

3.61 A 2-kg block slides along a smooth rod and is at rest when a 5-kg block is released, as illustrated in Figure P3.61. The spring constant of the spring shown in the Figure is 100 N/m, and the spring is unstretched when the system is released from rest in the position shown. Determine the maximum distance that the 2-kg block will slide to the right. The spring is attached to the 2-kg mass.

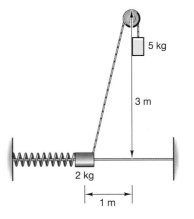

Figure P3.61 (Figure P3.21)

3.62 Determine the velocity of the 3-kg block A in Figure P3.62 after it has moved 200 mm on a smooth surface, starting from rest.

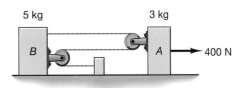

Figure P3.62 (Figure P3.23)

3.63 In Figure P3.63, determine the velocity of the 50-kg block after the 100-kg block has moved a distance of 500 mm. Neglect friction and the mass of the pulleys.

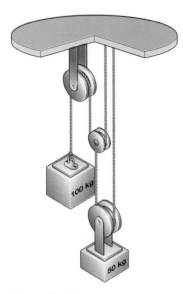

Figure P3.63 (**Figure P3.25**)

3.64 In Figure P3.64, determine the maximum compression x of the nonlinear spring ($F_x = 200\, x^2$ N) when a 10-kg mass is released at a height of 3 m above the spring.

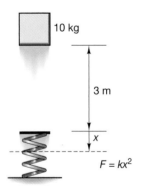

Figure P3.64 (**Figure P3.29**)

3.65 A mass is released from rest at a height h above an identical mass that is resting in equilibrium on a linear spring with a spring constant k, as depicted in the Figure P3.65. Determine the maximum additional compression Δ of the spring.

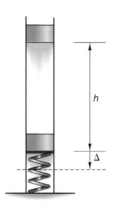

Figure P3.65 (**Figure P3.30**)

3.66 If a ball is thrown upward with an initial velocity v, show by conservation of energy that, neglecting the effects of air resistance, the velocity of the ball will be equal to its initial velocity when it returns to its initial position.

3.67 A particle slides down a smooth curved path from point A to point B, as shown in Figure P3.67. Show that the magnitude of the particle's velocity at point B is independent of the curvature of the path; that is, the curvature of the path will only affect the direction of the velocity at point B.

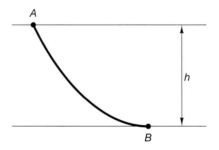

Figure P3.67

3.68 Determine if the force: $\mathbf{F}(x, y, z) = x^3 y^2 z^2 \hat{\mathbf{i}} + 1/2\, x^4 yz^2 \hat{\mathbf{j}} + 1/2\, x^4 y^2 z \hat{\mathbf{k}}$ is a conservative force, and if it is, determine the potential function associated with it.

3.69 Prove that the force $\mathbf{F}(r, \theta, z) = (r^2 \sin \theta + z^2)\hat{\mathbf{e}}_r + [(r^2/3) \cos \theta]\hat{\mathbf{e}}_\theta + 2rz\hat{\mathbf{k}}$ is a conservative force, and determine the potential function associated with it.

3.70 Prove that the force $\mathbf{F}(R, \phi, \theta) = (R^2 \sin \phi \cos \theta)\hat{\mathbf{e}}_R + [(R^2/3)\cos \phi \cos \theta]\hat{\mathbf{e}}_\theta - [(R^2/3) \sin \theta]\hat{\mathbf{e}}_\theta$ is a coservative force, and determine the potential function associated with it.

3.5 PRINCIPLE OF IMPULSE AND MOMENTUM

The principle of work and energy was obtained by taking the first integral of the equation of the motion of a particle with respect to position. Since this expression was a scalar equation, only the magnitude of the velocity could be obtained from this integration. Another useful first integral of the motion is obtained by integrating the equation of motion with respect to time. This integration, of course, can be done only if the forces acting on the particle are known functions of time. Newton's second law is

$$\sum \mathbf{F} = \frac{d\mathbf{L}}{dt} \tag{3.43}$$

where $\mathbf{L} = m\mathbf{v}$. The forces acting on the particle change the vector quantity $\mathbf{L}$, the linear momentum. If the forces acting on the particle are functions of time, integrating Eq. (3.43) with respect to time yields

$$\int_{t_1}^{t_2} \sum \mathbf{F}\,dt = \mathbf{L}_2 - \mathbf{L}_1 = m\mathbf{v}_2 - m\mathbf{v}_1 \tag{3.44}$$

It is clear that this integral is just the first integral of the equation of motion. This equation is different from the work–energy equation, as this expression is a vector equation, and the forces acting in one coordinate direction change only the linear momentum in that direction. The term on the left of this equation is called the *linear impulse,* and the term on the right is the *linear momentum.* This equation establishes the *principle of impulse and momentum.* The units of both impulse and momentum are the units of force and time, and therefore, impulse and momentum are expressed in N · s or lb · s.

The impulsive force is a function of time and, in general, varies during its period of application. The average of the total force acting on a particle over a period of time is

$$\sum \mathbf{F}_{av} = \frac{1}{t_2 - t_1} \int_{t_1}^{t_2} \sum \mathbf{F}\,dt \tag{3.45}$$

The relationship between the average impulsive force acting on the particle to the change in linear momentum is

$$(t_2 - t_1) \sum \mathbf{F}_{av} = m\mathbf{v}_2 - m\mathbf{v}_1 \tag{3.46}$$

As an example, consider the vertical ground reaction force between a human foot and the ground during walking, as shown in Figure 3.9. The force is given in percent of body

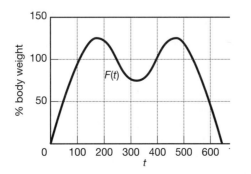

Figure 3.9

weight, and time is measured in milliseconds. The linear impulse from the vertical ground reaction force is

$$\int_0^{650} \mathbf{F}(t)dt = 5.718 \times 10^4 \ \%\text{BW} \cdot ms$$

The average impulsive force is obtained from Eq. (3.45):

$$\mathbf{F}_{av} = 87.97 \ \%\text{BW}$$

If these numbers were actual data from a walking test, the change in momentum in the vertical direction could be computed.

A large force that acts over a short period of time is called an *impulsive force* and occurs during phenomena such as the impact of a bat with a ball, collision of cars, etc. This type of force will be considered in detail in Section 3.6. If the average impulse force is zero, the linear momentum does not change during that interval of time.

3.5.1 IMPULSE AND MOMENTUM OF SEVERAL PARTICLES

If two or more particles are interacting, the particles may be treated independently, and Eq. (3.44) can be written for each particle. These vector equations, one for each particle, may be added to form a single vector impulse and momentum equation for the system of particles. The value of adding the equations for the individual particles is not initially apparent, as one would suppose that information would be lost in such a combination of equations. However, internal forces in the system—that is, forces due to the interaction of one particle on another particle—always occur in pairs, according to Newton's third law. When the equations are added, the impulse of internal forces cancel, and only the impulse from external forces produces a change in the total momentum of the system. The treatment of systems of particles will be developed fully in Chapter 4, and only systems containing a few particles will be considered in this section. Consider the two particles connected by a spring resting on a smooth, horizontal surface, as shown in Figure 3.10. If the spring is initially stretched and the particles are released from rest at that time, the equations of impulse and momentum for each particle are

m_A m_B

Figure 3.10

$$\int_{t_1}^{t_2} \mathbf{F}_A(t)dt = m_A\mathbf{v}_{A2} - m_A\mathbf{v}_{A1}$$

$$\int_{t_1}^{t_2} \mathbf{F}_B(t)dt = m_B\mathbf{v}_{B2} - m_B\mathbf{v}_{B1} \tag{3.47}$$

The only force acting on the particles is the internal spring force, and if the two equations are added, the impulse from this force cancels, yielding

$$m_A\mathbf{v}_{A2} - m_A\mathbf{v}_{A1} + m_B\mathbf{v}_{B2} - m_B\mathbf{v}_{B1} = 0$$

$$m_A\mathbf{v}_{A2} + m_B\mathbf{v}_{B2} = m_A\mathbf{v}_{A1} + m_B\mathbf{v}_{B1} \tag{3.48}$$

The total momentum of the system is the same at any instant of time when no external forces act on the system. This phenomenon is known as the *principle of conservation of linear momentum*. This principle states that in the absence of external forces, the linear momentum of a system of particles is a constant. In the example shown in Figure 3.10, since the initial momentum is zero, the final momentum of the system is zero, and

$$m_A\mathbf{v}_{A2} + m_B\mathbf{v}_{B2} = 0$$

$$m_A\mathbf{v}_{A2} = -m_B\mathbf{v}_{B2} \tag{3.49}$$

This equation establishes a relationship between the velocities of particle A and particle B at any time, but does not allow the velocities to be determined. To determine the velocity of either particle, the equation of motion for that particle must be written and the differential equation solved. Conservation of linear momentum shows that the particles move toward each other with velocities that are inversely proportional to their masses. In Chapter 4, the linear momentum of the system will be related to the velocity of the center of mass of the system.

Sample Problem 3.9

A 3-kg package slides down a smooth 30° incline and lands on a 10-kg cart that is at rest, as shown in the accompanying illustration. Determine the velocity of the cart and package immediately after the box lands on the cart. The package is released from rest 5 m up the incline.

Solution

We will first determine the velocity of the package as it leaves the incline. This calculation can be done using either conservation of energy or by determining the constant acceleration of the package on the incline and integrating the differential equation of motion. Using conservation of energy,

$$mg(5 \sin 30°) = \frac{1}{2}mv^2$$

$$v = \sqrt{2(9.81)(5 \sin 30°)}$$

$$v = 7.00 \text{ m/s}$$

The velocity of the package is directed parallel to the incline plane. Using conservation of linear momentum in the horizontal direction of the system that is composed of the package and the cart yields

$$m_p v_{px} = (m_p + m_c)v_f$$

$$3(7) \cos 30° = (3 + 10)v_f$$

$$v_f = 1.4 \text{ m/s}$$

Sample Problem 3.10

A 50-kg girl jumps from the rear of a 30-kg boat that is moving forward at a rate of 1 m/s. If the girl's velocity relative to the boat is 2 m/s when she jumps, what is the final velocity of the girl and the boat?

Solution

The initial momentum of the system (girl and boat) is

$$L_1 = (m_B + m_G)v_b = (50 + 30)(1) = 80 \text{ kg} \cdot \text{m/s}$$

Let v_{Bf} be the velocity of the boat after the girl has jumped, and let v_G be the absolute velocity of the girl. The velocity of the girl can be written as

$$v_G = v_{Bf} + v_{G/Bf}$$

The final momentum of the system is

$$L_2 = m_B v_{Bf} + m_G(v_{Bf} + v_{G/Bf}) = 30v_{Bf} + 50(v_{Bf} - 2)$$

The momentum of the system is conserved; therefore,

$$L_2 = L_1$$

$$80v_{Bf} - 100 = 80$$

$$v_{Bf} = 2.25 \text{ m/s}$$

The absolute velocity of the girl is

$$v_G = 2.25 - 2.00 = 0.25 \text{ m/s}$$

Note that both the girl and the boat are moving to the right after the jump.

Problems

3.71 In Figure P3.71, a 1500-kg car is driving down a 10% incline at 100 km/hr when a constant braking force of 7000 N is applied. Determine how long it will take to stop the car.

Figure P3.71

3.72 A 5-oz ball is thrown horizontally with a velocity of 90 mph and is hit by a batter, as shown in Figure P3.72. If the ball leaves the bat with a velocity of 110 mph at an angle of 30° with the horizontal, determine the average impulsive force if the contact time between the ball and the bat is 20 ms.

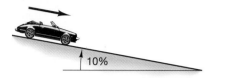

Figure P3.72

3.73 A railroad car of mass m is moving on a level rail at a velocity v when it hits and couples with two identical coupled railroad cars, as depicted in Figure P3.73. Determine the final velocity of the three coupled railroad cars.

Figure P3.73

3.74 In Figure P3.74, a 70-lb boy runs at a speed of 20 ft/s and jumps on a 20-lb snow sled. If the coefficient of kinetic friction between the sled runners and the snow is 0.2, how far will the sled slide on level, snow-covered ground?

Figure P3.74

3.75 A man throws a 5-kg package into a 3-kg box, as shown in Figure P3.75. Will the package land in the box? If it lands in the box, what is the velocity of the box and package after the box lands?

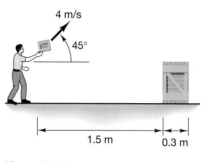

Figure P3.75

3.76 What maximum angle will the box and package in Problem 3.75 reach if the box is suspended from the ceiling by a 3-m rope?

3.77 A 10,000-kg jet airplane must reach a speed of 60 m/s for takeoff. If the jet engine increases its thrust at a rate of 20,000 N/s, determine the time necessary for the airplane to reach takeoff speed. Neglect air resistance and other forces.

3.78 The two weights in Figure P3.78 are released from rest. Use the principle of impulse and momentum to determine the velocity of the weights after 0.75 s.

Figure P3.78

3.79 A 3000-lb car is driven at a speed of 20 ft/s relative to a 200,000-lb ferry, as shown in Figure P3.79. Initially, the car and ferry are at rest in the water, and water resistance can be neglected. Determine the speed of the ferry in the water.

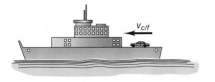

Figure P3.79

3.80 In Figure P3.80, the 4000-lb car B runs a stop sign and hits the 3500-lb car A. After the collision, the cars become entangled and slide together for 60 ft in the direction shown in the figure. If the coefficient of kinetic friction between the entangled cars and the pavement is 0.7, determine the initial speed of each car at the time of collision.

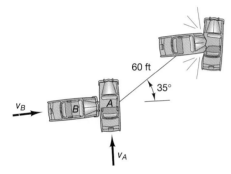

Figure P3.80

3.81 From the scenario in Problem 3.80, if the velocity of car A is known to be 25 mph and the cars slide 60 ft, determine the initial velocity of car B and the angle of the slide of the wreckage to the direction of travel of car B.

3.82 From the scenario in Problem 3.80, if the initial velocity of car A is 30 mph and the initial velocity of car B is 40 mph, determine the distance that the wreckage will slide before coming to a stop.

3.83 A 2-kg block slides down a smooth circular chute of radius 4 m and lands on a 5-kg sled, as shown in Figure P3.83. If the coefficient of kinetic friction between the sled and the ground is 0.3, determine the distance that the sled and the block will slide.

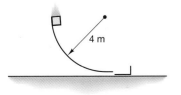

Figure P3.83

3.84 A car collides head-on with a bus, and the wreckage becomes entangled after the collision, as depicted in Figure P3.84. If the car weighs 2000 kg and was traveling at an initial speed of 100 km/hr, and the bus weighs 35,000 kg and was traveling at an initial speed of 80 km/hr, determine the velocity of the wreckage after the collision.

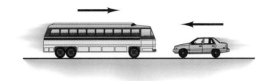

Figure P3.84

3.85 If the coefficient of kinetic friction between the wreckage and the road in Problem 3.84 is 0.6, determine the final position of the wreckage, as measured from the point of impact.

3.86 Determine the amount of energy lost in the collision in Problem 3.84.

3.87 If the collision in Problem 3.84 lasts 200 ms, determine the average force between the car and the bus.

3.88 When a 60-g tennis ball is served, it is impacted by the racket when the ball is at the maximum height of the toss. If the velocity of the serve is 190 km/hr and the time of impact with the racket is 50 ms, determine the average impulsive force acting on the ball.

3.89 In Figure P3.89, a collar B of mass m_B slides on a smooth rod when the mass m_A is released from a horizontal position when the system is at rest. Determine the velocity of the collar when $\theta = 90°$.

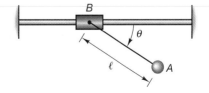

Figure P3.89

3.90 Determine a general expression for the velocity of collar B in Problem 3.89 when $\theta = 90°$ if the mass A is released from a angle $\theta = \beta$ when the system is at rest.

3.91 In Figure P3.91, a field-goal kicker kicks a 0.4-kg football over a goal post 40 meters downfield such that the ball is 4 meters above the ground as it crosses the crossbar. If the time that the kicker's foot is in contact with the ball is 0.4 s, determine the average impulsive force between the foot and the ball.

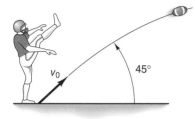

Figure P3.91

3.92 A force **P** with a force–time curve, as shown in Figure P3.92, is applied to a 40-kg box. If the coefficients of friction between the box and the floor are $\mu_s = 0.6$ and $\mu_k = 0.5$, determine (a) the time when the box will start to move, (b) the time when it stops moving, and (c) the maximum velocity attained by the box.

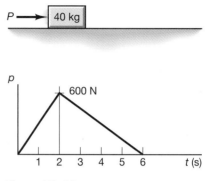

Figure P3.92

3.93 Solve Problem 3.92 if the force **P** is directed downward at an angle 30° with the horizontal.

3.6 IMPACT

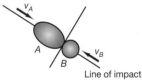

Figure 3.11

When two or more particles collide such that the internal forces produced during that collision are large compared to the external forces, the collision is called an *impact*. Impacts may occur due to planned events such as hitting a golf ball or connecting rail cars together, or impacts may occur accidentally, as in automobile accidents. An impact between two objects, modeled as particles, is shown in Figure 3.11.

The *line of impact* is the common normal to the surfaces at the point of impact. If the line of impact passes through the center of mass of each particle, the impact is called a *central impact*. When an object is modeled as a particle, the object occupies a point in space and all impacts are central impacts. If the objects are given size and shape, the common normal to the surfaces at the point of impact may not pass through the center of mass of both the objects and the impact is said to be *eccentric impact*. Eccentric impacts will be considered when dynamics of rigid bodies are examined. If the velocities of the two particles before impact are along the line of impact, the impact is called *direct central impact*. If the velocity of one or both of the particles is not along the line of impact, *oblique central impact* occurs. If the effects of external forces are neglected during the impact, the momentum of the system is conserved and

$$m_A\mathbf{v}_A + m_B\mathbf{v}_B = m_A\mathbf{v}'_A + m_B\mathbf{v}'_B \tag{3.50}$$

where the unprimed velocities are the velocities before impact and the primed velocities are the velocities after impact. If the particles adhere and remain together after the impact, the impact is a *perfectly plastic impact,* and the postimpact velocities of the two particles are equal. This postimpact velocity can be wholly determined in this case from Eq. (3.50) without any consideration of the impacting process. If the impact is such that no kinetic energy is lost during the impact, the impact is said to be perfectly *elastic impact*. Using the definition of kinetic energy, conservation of kinetic energy during impact may be written as

$$\frac{1}{2}m_A v_A'^2 + \frac{1}{2}m_B v_B'^2 = \frac{1}{2}m_A v_A^2 + \frac{1}{2}m_B v_B^2 \tag{3.51}$$

Multiplying Eq. (3.51) by two and rearranging terms yields

$$m_A(v_A'^2 - v_A^2) = m_B(v_B^2 - v_B'^2)$$

Factoring both sides of the equation gives

$$m_A(v_A' - v_A)(v_A' + v_A) = m_B(v_B - v_B')(v_B + v_B')$$

From the conservation of momentum, we obtain

$$m_A(v_A' - v_A) = -m_B(v_B' - v_B)$$

Therefore, conservation of energy becomes

$$(v_A' + v_A) = (v_B' + v_B) \tag{3.52}$$

This relationship yields a second equation to determine the final velocities. Therefore, in the two extreme cases—that is, plastic or elastic impact—the final velocities after impact can be obtained.

3.6.1 DIRECT CENTRAL IMPACT

If the particles travel along the line of impact with velocities v_A and v_B, direct central impact will occur, and the objects collide as shown in Figure 3.12.

At the point of maximum deformation of the impacting particles, the velocities of both particles will be equal. This velocity is designated as **u** in Figure 3.12. During impact, the contacting force **R** acts on the particles, as shown in Figure 3.13.

Let the impact begin at time t_1, the maximum deformation occur at time t_u, and the end of impact occur at time t_2. The impulse momentum equation for particle A is

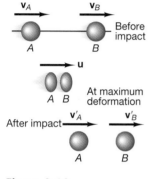

v_A v_B

Before impact

A B

u

At maximum deformation

A B

After impact —

v_A' v_B'

A B

Figure 3.12

A

$R(t)$

B

$R(t)$

Figure 3.13

a) $\quad \displaystyle\int_{t_1}^{t_u} -R(t)dt = m_A u - m_A v_A$

b) $\quad \displaystyle\int_{t_u}^{t_2} -R(t)dt = m_A v_A' - m_A u$

$$\tag{3.53}$$

Similar equations can be written for particle B, yielding

a) $\quad \displaystyle\int_{t_1}^{t_u} R(t)dt = m_B u - m_B v_B$

b) $\quad \displaystyle\int_{t_u}^{t_2} R(t)dt = m_B v_B' - m_B u$

$$\tag{3.54}$$

In general, during an impact, kinetic energy is dissipated due to permanent deformation of the particles, generation of heat or sound, or internal friction. As a result, the impulse during the time period from t_u to t_2 is less than the impulse during the initial period of contact. The first period is called the *period of deformation,* and the second period is called the *period of restitution.* The ratio of the impulse during the period of restitution to the impulse during the period of deformation is called the *coefficient of restitution* and is designated by

$$e = \frac{\displaystyle\int_{t_u}^{t_2} R(t)dt}{\displaystyle\int_{t_1}^{t_u} R(t)dt} \tag{3.55}$$

If Eq. (3.53) b is divided by Eq. (3.53) a, and Eq. (3.54) b is divided by Eq. (3.54) a, the following relationships are obtained:

$$(u - v_A)e = v'_A - u$$

$$(u - v_B)e = v'_B - u$$

Subtracting the second equation from the first yields

$$e = \frac{v'_B - v'_A}{v_A - v_B} \tag{3.56}$$

The coefficient of restitution is the *ratio of the relative separation velocity to the relative approach velocity*. If the particles are to collide, the initial velocity of *A* must be greater than the velocity of *B*, and after collision, if they are to separate, the velocity of *B* must be greater than the velocity of *A*. If the coefficient of restitution is zero, the relative separation velocity is zero and the impact is perfectly plastic. If the coefficient of restitution is unity, the separation velocity equals the approach velocity and the kinetic energy of the system is conserved. This situation is the case of perfectly elastic impact.

If the value of the coefficient of restitution is known or assumed, the final velocity of both particles can be determined from Eq. (3.50), conservation of momentum, and Eq. (3.56) for any initial velocities. Substitution of Eq. (3.56) into Eq. (3.50) yields

$$v'_A = \frac{1}{m_A + m_B}[(m_A - em_B)v_A + m_B(1 + e)v_B]$$

$$v'_B = \frac{1}{m_A + m_B}[(m_B - em_A)v_B + m_A(1 + e)v_A] \tag{3.57}$$

The initial energy in the system is

$$E = \frac{1}{2}\left(m_A v_A^2 + m_B v_B^2\right) \tag{3.58}$$

Using Eq. (3.57), we may write the final energy in the system as

$$E' = \frac{1}{2}\left\{\frac{1}{m_A + m_B}\left[m_A(m_A + e^2 m_B)v_A^2 + 2m_A m_B(1 - e^2)v_A v_B\right.\right.$$

$$+ \left.\left. m_B(m_B + e^2 m_A)v_B^2\right]\right\} \tag{3.59}$$

For a coefficient of restitution equal to 1, Eq. (3.59) becomes

$$E' = E \tag{3.60}$$

For a coefficient of restitution equal to zero, the final energy becomes

$$E' = \frac{1}{2(m_A + m_B)}(m_A v_A + m_B v_B)^2 = \frac{1}{2}(m_A + m_B)(v')^2 \tag{3.61}$$

where v' is the velocity of the two masses after the plastic collision.

Relating the final energy and, therefore, the energy lost during impact to the coefficient of restitution shows that, although this coefficient is clearly an empirical value, making it specific is equivalent, to specifying the energy lost in the collision. In many commercial traffic accident reconstruction programs, the energy lost is estimated by measuring the crush and using a spring constant to estimate the energy required to crush the involved vehicles. This method is subject to an error to the same degree as the use of an estimate of the coefficient of restitution.

Sample Problem 3.11

Police investigate an accident that involved a head-on collision between two cars. The point of impact is identified, and skid marks precollision and post-collision are measured for each vehicle. The police diagram of the scene is shown in the accompanying diagram. The weight of vehicle *A* is 4000 lb, and the weight of vehicle *B* is 3200 lb. Tests on the pavement show that the coefficient of kinetic friction is 0.7 for both vehicles, and examination of the damage leads to an estimate of a coefficient of restitution as 0.5. As an accident reconstructionist, determine the initial velocities of the two vehicles.

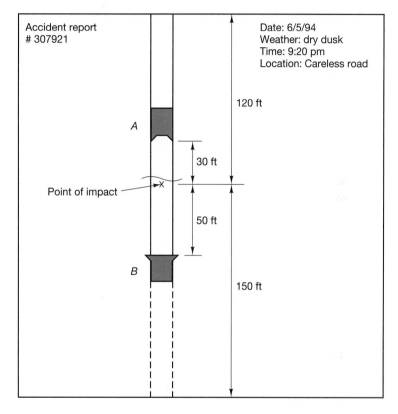

Solution

Problems of this nature are solved by starting at the end positions of the vehicles, where the velocity is known to be zero, and working back through the accident until the initial velocities of each vehicle can be determined with a reasonable amount of certainty.

The 4000 lb vehicle *A* slides 30 ft after impact before coming to rest. Therefore, using a work–energy approach,

$$-\mu mgs = 0 - \frac{1}{2}m(v')^2$$

where *s* is the slide distance and v' is the post-impact velocity. Notice that the mass of the vehicle does not enter into this calculation. The post-impact speeds are

$$v' = \sqrt{2\mu gs}$$
$$v'_A = 36.8 \text{ fps}$$
$$v'_B = 47.5 \text{ fps}$$

The speeds of the vehicles at impact are obtained using conservation of momentum and the coefficient of restitution. The initial direction of vehicle *A* is assumed to be positive. Therefore, initially, *A* has a positive speed and *B* a negative speed; after impact, *A* is negative and B is positive. Using this information, we obtain

$$m_A v_A + m_B v_B = m_A v'_A + m_B v'_B$$

$$4000 v_A - 3600 v_B = -4000(36.8) + 3600(47.5) = 23,800$$

The momentum is positive after impact, and since it is conserved, the momentum of A before impact was greater than the momentum of B. The separation velocity is related to the approach velocity by the coefficient of restitution as

$$(v'_B - v'_A) = e(v_A - v_B)$$

$$[36.8 - (-47.5)] = 0.5[v_A - (-v_B)]$$

$$v_A + v_B = 168.6$$

There are two simultaneous linear equations for the two pre-impact velocities that may be solved as follows:

$$\begin{pmatrix} v_A \\ v_B \end{pmatrix} := \begin{pmatrix} 4 & -3.6 \\ 1 & 1 \end{pmatrix}^{-1} \cdot \begin{pmatrix} 23.8 \\ 168.6 \end{pmatrix}$$

$$v_A = 82.995 \qquad\qquad v_B = 85.605$$

The initial speeds of the vehicles are obtained by work–energy relations, using the skid distances:

$$-\mu m g s = \frac{1}{2} m (v^2 - v_i^2)$$

$$v_i = \sqrt{2 \mu g s + v^2}$$

$$v_{Ai} = 110.9 \text{ fps, or } 75.6 \text{ mph}$$

$$v_{Bi} = 118.7 \text{ fps, or } 80.9 \text{ mph}$$

If this example were an actual accident, the conclusion would be that both drivers were speeding (assuming a 55 mph speed limit) and both were driving over the center line, leading to the accident. Assuming 1.75 s for perception and reaction, the drivers were separated by a distance of

$$S = 110.9\,(1.75) + 120 + 118.7\,(1.75) + 150 = 671.8 \text{ ft}$$

when they became aware of the impending collision. If they survived, both drivers would be held equally at fault for the accident.

    An accident analysis program is shown in the Computational Supplement for direct central impact.

3.6.2 OBLIQUE CENTRAL IMPACT

Now, consider the case in which the preimpact velocities of two particles are not directed along the line of impact, as shown in Figure 3.14. An axis system has been established at the point of impact designated in normal and tangential directions. If friction is ignored in the tangential direction, no forces act in that direction and the component of the momentum of each particle in the tangential direction is conserved. Therefore, the component of the velocity in that direction before and after impact is the same:

$$v'_t = v_t \tag{3.62}$$

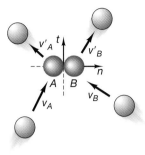

Figure 3.14

If a rectangular coordinate system is established such that the *x*-axis is along the line of impact, the components of the velocity in both the *y*- and *z*-directions are unaffected by the impact. Therefore, Eq. (3.62) is applicable in both of these directions. In the normal direction, the equations of direct central impact are applicable and linear momentum is conserved:

$$m_A(v_A)_n + m_B(v_B)_n = m_A(v'_A)_n + m_B(v'_B)_n \tag{3.63}$$

The normal components of the velocities before and after impact satisfy the relationship

$$e = \frac{(v'_B)_n - (v'_A)_n}{(v_A)_n - (v_B)_n} \tag{3.64}$$

Oblique central impacts are analyzed by separating the velocities into components and applying Eqs. (3.63) and (3.64) to the normal components and Eq. (3.62) to the tangential components.

3.6.3 IMPACT WITH A STATIONARY OBJECT

Many impacts occur between particles and stationary objects, such as a wall, and can be analyzed using the coefficient of restitution. Conservation of linear momentum in the normal direction is not applicable, as the stationary object is considered to have infinite mass and zero velocity, and therefore, the momentum of the stationary object is undefined. If friction is neglected, the tangential components of the velocity are equal before and after impact, and the normal component after impact is related to the normal component before impact by

$$(v'_A)_n = -e(v_A)_n \tag{3.65}$$

If the impact is perfectly elastic, the postimpact velocity in the normal direction is the negative of the preimpact normal component of the velocity. In this case, the angle that the velocity vector makes with the wall before and after impact is the same.

Impact with a Golf Club An interesting coefficient of restitution occurs with the new metal drivers. As the technology of developing golf clubs evolved using titanium faces and other metals, professional golfers (and amateurs) were able to get greater distances on their drives. Pros regularly were hitting drives well over 300 yards and some close to 350 yards. The Professional Golf Association (PGA) began to worry that this was reducing par 5 holes to par 4 holes. There was only so far that they could move back the tee, so they placed a limit on the coefficient of restitution for the driver. The maximum value of this coefficient was set at $e = 0.83$. The actual coefficient at impact with a golf ball would depend upon the compression of the winding of the ball (80, 90, or 100), and the PGA also is watching the development of new golf balls. To establish a feel for the velocities involved when a professional golfer hits a drive, the club head speed is around 120 mph and the golf ball leaves the face of the club at a speed of 180 mph.

  (This information was provided by Dr. Tom Mase of Cal Poly, San Luis Obispo, who has worked on the development and analysis of golf equipment.)

Sample Problem 3.12

Consider a corner shot of a racketball which makes contact with the side wall and then the front wall of the racketball court, as shown in the accompanying diagram. Friction and the change in the vertical component of the velocity due to gravitational attraction are neglected. The coefficient of restitution for the ball and wall is 0.9. Determine the velocity of the ball as it comes off the front wall $\mathbf{v}''$ if the initial velocity of the ball is $\mathbf{v} = -12\,\mathbf{i} - 4\,\mathbf{j} - 3\,\mathbf{k}$ m/s.

Solution   The *y*-direction is the normal direction for the first impact, so the *x*- and *z*-components of the velocity are unchanged and the *y*-component is the negative of the preimpact component times the coefficient of restitution:

$$\mathbf{v}' = -12\,\mathbf{i} + (0.9)4\,\mathbf{j} - 3\,\mathbf{k} = -12\,\mathbf{i} + 3.6\,\mathbf{j} - 3\,\mathbf{k}$$

During the impact with the front wall, the *x*-direction is the normal direction and the components of the velocity in the *y*- and *z*-directions are unchanged by the impact. The velocity of the ball as it comes off the front wall is

$$\mathbf{v}'' = (0.9)12\,\mathbf{i} + 3.6\,\mathbf{j} - 3\,\mathbf{k} = 10.8\,\mathbf{i} + 3.6\,\mathbf{j} - 3\,\mathbf{k}$$

The magnitude of the initial velocity is

$$|\mathbf{v}| = 13 \text{ m/s}$$

The magnitude of the final velocity is

$$|\mathbf{v}''| = 12.1 \text{ m/s}$$

Sample Problem 3.13   A pool player wishes to make a bank shot off of the far wall in order to put the 13-ball in the corner pocket, as shown in the accompanying illustration. If the coefficient of restitution between the 13-ball and the wall is 0.8 and the coefficient of restitution between the cue ball and the 13-ball is 1.0, determine the initial angle of the line of impact between the cue ball and the 13-ball neccessary to make the shot. Will the player scratch (the cue ball goes in the pocket)?

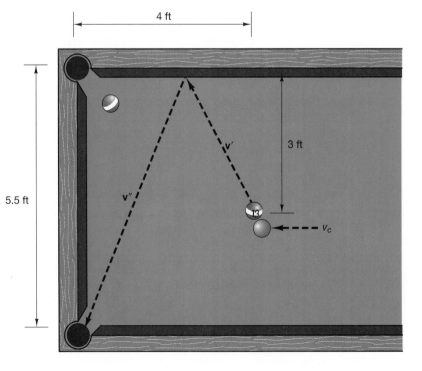

Solution   The 13-ball is at rest when it is struck by the cue ball, so its tangential velocity is zero before and after impact, and it will move along the line of impact at the angle α, such that

$$\tan \alpha = \frac{v'_y}{v'_x} = \frac{3}{d}$$

After the ball makes an impact with the side of the pool table, the ball's velocity in the y-direction is

$$v''_y = -ev'_y$$

$$\tan \alpha' = \frac{ev'_y}{v'_x} = \frac{5.5}{d'}$$

$$\tan \alpha' = e \tan \alpha$$

$$d = 3/\tan \alpha$$

$$d' = 5.5/\tan \alpha' = (0.8)5.5/\tan \alpha$$

$$d + d' = 4$$

$$\tan \alpha = \frac{3 + (0.8)5.5}{4}$$

$$\alpha = 61.61°$$

Examining the initial impact between the cue ball and the 13-ball, we get

$$\mathbf{V}_c = V_c \sin \alpha \hat{\mathbf{e}}_t + V_c \cos \alpha \hat{\mathbf{e}}_n$$

In the normal direction, the conservation of momentum gives us

$$V_c \cos \alpha = V'_c + v'$$

With a coefficient of restitution of unity, we get

$$v' - V'_c = (1)V_c \cos \alpha$$

Therefore,

$$v' = V_c \cos \alpha$$

and

$$V'_c = 0$$

The cue ball will move in the tangential direction. The point at which it will strike the side of the table is $X_c = 2.5 \tan \alpha = 4.625$, but there is only 4 ft of table left, so it will strike the end of the table and not the side. As it travels the 4 ft in the x-direction, it will move

$$Y_c = 4/\tan \alpha = 2.16 \text{ ft}$$

Therefore, the player will not scratch.

Problems

3.94 A car A of mass 2000 kg moving at a velocity of 20 km/hr rear-ends a stopped car B of mass 1500 kg, as shown in Figure P3.94. If the coefficient of restitution between the cars is $e = 0.8$, determine (a) the velocities of the cars after impact, (b) the energy lost during impact, and (c) the average impulsive force if the impact took 350 ms.

Figure P3.94

3.95 If car A in Problem 3.94 was traveling 100 km/hr when its driver applied the brakes in a four-wheel lockup 10 m behind car B before the collision, determine (a) the velocities of the cars after impact and (b) the amount of kinetic energy lost during the skid and the amount lost in the collision. Assume that the coefficient of kinetic friction between the tires and the pavement is 0.05.

3.96 A ball is dropped on a hard surface from a height of 3 m, as illustrated in Figure P3.96. If it rebounds to a height of 2 m, determine the coefficient of restitution between the ball and the surface.

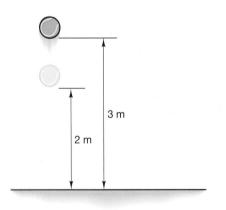

3 m

2 m

Figure P3.96

3.97 In Figure P3.97, three cars of identical mass are free to roll on a horizontal surface. (Ignore friction.) If cars B and C are at rest when they are struck by car A moving at a velocity V, determine the final velocity of car C in terms of the coefficient of restitution e and the velocity V.

(Hint: There will be multiple impacts; treat each impact as an impact between only two of the cars.)

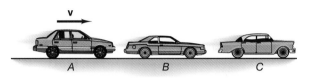

Figure P3.97

3.98 Determine the final velocity of each of the cars in Problem 3.97 if the initial velocity of car A is 20 ft/s and $e = 1$.

3.99 In Figure P3.99, two identical objects A and B undergo direct central impacts. The initial velocity of A is V, and B starts from rest. Determine the velocities of A and B after impact if (a) $e = 0$ (plastic impact), (b) $e = 1$ (elastic impact), and (c) $e = 0.5$.

Figure P3.99

3.100 Determine the coefficient of restitution if the velocity of object B in Problem 3.99 after impact is 0.7 V.

3.101 Determine the percent of energy lost for each of the three coefficients of restitution derived from Problem 3.99.

3.102 A ball is dropped from a height h onto a surface, where it bounces, as depicted by Figure P3.102. If the coefficient of restitution between the ball and the surface is e, determine a general expression for the height of the nth bounce.

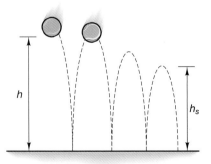

h

h_s

Figure P3.102

3.103 If the sphere in Figure P3.103 hits a surface at an approach angle of θ, determine the rebound angle β in terms of the coefficient of restitution e and the approach angle.

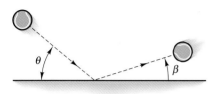

Figure P3.103

3.104 In Figure P3.104 a ball is thrown against a padded inclined plane with a velocity $\mathbf{v} = 5\hat{\mathbf{i}} - 6\hat{\mathbf{j}}$ m/s. If the coefficient of restitution is 0.7, determine the velocity after the impact.

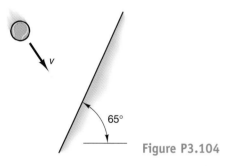

Figure P3.104

3.105 A cue ball is shot at an angle of 30° with a velocity of $v_1 = 0.5$ m/s, and it strikes the 2-ball at the angle shown in Figure P3.105. If the coefficient of restitution between the balls is 0.95 and between the 2-ball and the padded wall is 0.6, determine the velocity vectors $\mathbf{v}_1'$ and $\mathbf{v}_2''$.

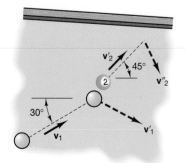

Figure P3.105

3.106 In Figure P3.106, sphere *A* has a mass of 15 kg and a radius of 100 mm, and sphere *B* has a mass of 10 kg

and a radius of 60 mm and impact as shown in the figure. If the coefficient of restitution is 0.6, determine the velocities of the two spheres after impact.

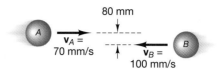

Figure P3.106

3.107 A sphere is thrown against a padded surface with a velocity $\mathbf{v} = -0.5\hat{\mathbf{i}} - 0.1\hat{\mathbf{j}}$ m/s, as shown in Figure P3.107. If the coefficient of restitution is 0.8, determine the velocity of the sphere after impact. (*Hint: Use Eq. (2.13) to determine the normal to the surface.*)

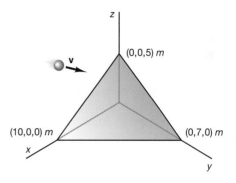

Figure P3.107

3.108 Determine the velocity of the sphere in Figure P3.107 after impact if the initial velocity is $\mathbf{v} = -2\hat{\mathbf{i}} - 0.3\hat{\mathbf{j}} - \hat{\mathbf{k}}$ m/s and the coefficient of restitution is 0.3.

3.109 What is the velocity of the sphere in Figure P3.107 after impact if the coefficient of restitution is (a) $e = 1$ and (b) $e = 0$?

3.110 A ball is thrown against an inclined surface, as shown in Figure P3.110, with a velocity $\mathbf{v} = -2\hat{\mathbf{i}} - 10\hat{\mathbf{j}} - 5\hat{\mathbf{k}}$ ft/s. Determine the velocity of the ball after impact if (a) $e = 1$, (b) $e = 0$, and (c) $e = 0.5$.

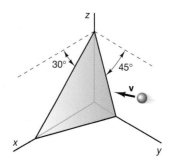

Figure P3.110

3.111 A 2-kg sphere is dropped on a wedge resting on a smooth plane, impacting the wedge with a velocity of 3 m/s, as depicted in Figure P3.111. If the wedge has a mass of 5 kg and the coefficient of restitution between the wedge and the sphere is 0.7, determine the velocity of the sphere after the impact.

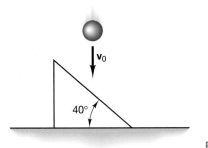

Figure P3.111

3.112 A bioengineer wishes to explain the mechanism of a closed head injury to a physician by modeling the skull as a hollow sphere of mass m_s and the brain as a smaller sphere of mass m_b, as illustrated in Figure P3.112. Initially, the skull and the brain are moving with a velocity v_0, the impact of the skull with the wall has a coefficient of restitution e_1, and the secondary impact of the brain with the skull has a coefficient of restitution of e_2. Determine the velocity of the brain and the skull after the second impact. Does the brain impact the front of the skull?

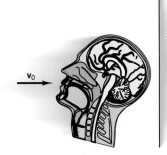

Figure P3.112

3.113 A 25-g bullet is fired into 5-kg block of wood with a velocity of 800 m/s, as shown in Figure P3.113. If the coefficient of kinetic friction between the block and the floor is 0.4, determine the distance that the block will slide.

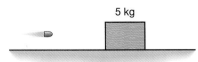

Figure P3.113

3.114 In Figure P3.114, the bank angle of a cue ball leaving the wall must be β if the 8-ball is to go in the pocket. If the coefficient of restitution of the wall is e, determine the angle θ of the cue ball with the wall.

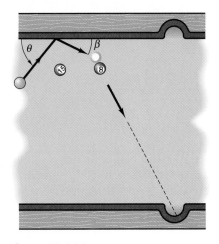

Figure P3.114

3.115 For the system in Figure P3.115, determine an expression for the velocity of the wedge in terms of the velocity of the sphere V, the coefficient of restitution e, the mass of the sphere m_s, the mass of the wedge m_w, and the angles θ and β.

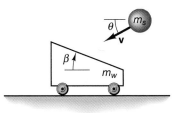

Figure P3.115

3.116 A ball is thrown into a square corner, as shown in Figure P3.116. If the coefficient of restitution is the same for both impacts, show that the exit path from the corner is parallel to the approach path.

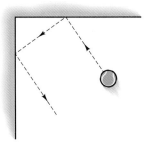

Figure P3.116

SYSTEM OF PARTICLES

Chapter **4**

The mass flow from the fireman's hose results in a large force pushing back on the fireman.

4.1   INTRODUCTION

The dynamics of single particles or a small number of particles were been discussed in Chapters 1 through 3. It is important to consider a group of particles as a system and study the dynamics of that system. Initially, the system of particles will consist of a constant number of particles, and the equations for the system will be developed. In the last part of this chapter, particles will leave or enter the system so that the total mass of the system will vary with its dynamics. Problems of this nature occur in rocket propulsion systems, where fuel particles are discharged from each system. These systems will be variable mass systems, and the center of mass will vary with the dynamics.

4.2   GENERAL EQUATIONS FOR A SYSTEM OF PARTICLES

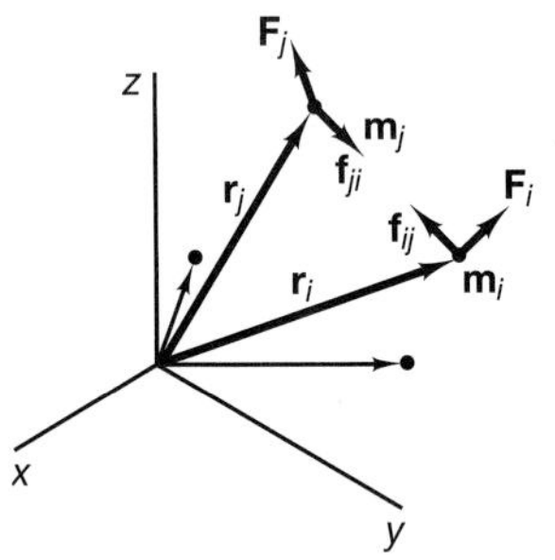

Figure 4.1

Consider the system of particles shown in Figure 4.1. Each particle is identified with a subscript, and two arbitrary particles are designated as the ith and the jth particles. The ith particle is subjected to force external to the system, $\mathbf{F}_i$ and internal forces from each of the other particles, $\sum_j \mathbf{f}_{ij}$. Newton's second law can be written for this particle as

$$\mathbf{F}_i + \sum_j \mathbf{f}_{ij} = \frac{d(m_i \mathbf{v}_i)}{dt} = \dot{\mathbf{L}}_i \tag{4.1}$$

where $\mathbf{L}_i$ is the linear momentum of the ith particle. The internal forces occur in pairs and are equal in magnitude, opposite in direction, and collinear.

All of the particles will be considered as part of a system. Information about any particular particle will be lost when the system equations are written, but the general dyanamic equations for the system will be obtained. For each particle, Newton's equation of motion can be written in the same manner as for the ith particle. To consider the system as a whole, add all of the equations of motion for the individual particles. This addition yields

$$\sum_i \mathbf{F}_i + \sum_i \sum_j \mathbf{f}_{ij} = \sum_i \frac{d}{dt}(m_i \mathbf{v}_i) \tag{4.2}$$

For each internal force, there is an equal and opposite internal force, or

$$\mathbf{f}_{ij} = -\mathbf{f}_{ji}$$

So, the sum of the internal forces equals zero. Therefore,

$$\sum_i \sum_i \mathbf{f}_{ij} = 0 \tag{4.3}$$

The system is affected only by the external forces that act upon it, so

$$\sum_i \mathbf{F}_i = \frac{d\mathbf{L}}{dt} = \sum_i \frac{d}{dt}(m_i \mathbf{v}_i) \tag{4.4}$$

where $\mathbf{L}$ is the linear momentum of the system defined by the sum of the linear momentum of each of the particles.

In a similar manner, the effect that the moment of all the forces (internal and external) acting on the system has on the change in the angular momentum of the system can be determined. The angular momentum of the system about the origin o is defined as the sum of the angular momenta of all of the particles in the system:

$$\mathbf{H}_o = \sum_i (\mathbf{r}_i \times m_i \mathbf{v}_i) \tag{4.5}$$

The sum of the moments about the origin is

$$\mathbf{M}_o = \sum_i \mathbf{M}_{oi} = \sum_i \mathbf{r}_i \times \mathbf{F}_i + \sum_i \sum_j \mathbf{r}_i \times \mathbf{f}_{ij} = \sum_i \mathbf{r}_i \times \frac{d}{dt}(m_i \mathbf{v}_i) \tag{4.6}$$

Examination of the rate of change of the angular momentum in Eq. (4.5) yields

$$\frac{d\mathbf{H}_o}{dt} = \frac{d}{dt}\sum_i \mathbf{r}_i \times m_i \mathbf{v}_i = \sum_i \left(\frac{d\mathbf{r}_i}{dt} \times m_i \mathbf{v}_i + \mathbf{r}_i \times \frac{d}{dt}(m_i \mathbf{v}_i) \right) \tag{4.7}$$

The first term in the differentiation of the product is zero, as it is the cross product between two parallel vectors:

$$\frac{d\mathbf{r}_i}{dt} \times m_i \mathbf{v}_i = \mathbf{v}_i \times m_i \mathbf{v}_i = 0 \tag{4.8}$$

Therefore, the rate of change of the angular momentum is

$$\frac{d\mathbf{H}_o}{dt} = \sum_i \left(\mathbf{r}_i \times \frac{d}{dt}(m_i \mathbf{v}_i) \right) \tag{4.9}$$

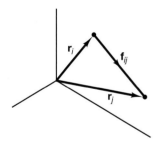

The moment of an internal force can be examined using the system shown in Figure 4.2. Since the line of action of the internal force lies on a path between the ith and the jth particles, it can be shown that the sum of the moments of the internal forces is zero:

$$\mathbf{r}_i \times \mathbf{f}_{ij} = \mathbf{r}_j \times \mathbf{f}_{ij}$$

$$\mathbf{f}_{ij} = -\mathbf{f}_{ji}$$

Figure 4.2

Therefore,

$$\sum_i \sum_i \mathbf{r}_i \times \mathbf{f}_{ij} = 0 \tag{4.10}$$

Therefore, the moment equation for the system of particles does not involve the moment of the internal forces. The rate of change of the total angular momentum of the system of particles is dependent only on the moment of the external forces acting on the system:

$$\mathbf{M}_o = \frac{d\mathbf{H}_o}{dt}$$

$$\sum_i \mathbf{r}_i \times \mathbf{F}_i = \sum_i \mathbf{r}_i \times \frac{d}{dt}(m_i \mathbf{v}_i) \tag{4.11}$$

So far, we find that the only advantage to dealing with all of the particles as a system is that the internal forces affect neither the change of the linear momentum nor the angular momentum of the system.

4.3   CENTER OF MASS OF A SYSTEM OF PARTICLES

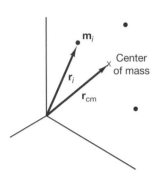

Center of mass

Figure 4.3

The center of mass is one of the most important concepts in the development of rigid body dynamics. Since, in the earth's gravitational field, the center of weight (or the center of gravity) coincides with the center of mass if the particles in question are not widely distributed in height above the earth, the terms are frequently used interchangeably. Consider the system of particles shown in Figure 4.3. The point in space that is the center of mass is defined as follows:

$$\mathbf{r}_{cm} = \frac{\sum_i m_i \mathbf{r}_i}{\sum_i m_i} \tag{4.12}$$

Refer to Chapter 5 of *Statics* for information regarding this definition. Note that, in general, this point will not coincide with any particle, and as the particles move, the center of mass will also move. These terms define the position vector from the point of reference—the origin—to the center of mass. Since this expression is a vector equation, it may be separated into its scalar components:

$$x_{cm} = \frac{\sum_i m_i x_i}{M}$$

$$y_{cm} = \frac{\sum_i m_i y_i}{M} \tag{4.13}$$

$$z_{cm} = \frac{\sum_i m_i z_i}{M}$$

$$M = \sum_i m_i$$

The velocity of the center of mass as it moves through space can be obtained by differentiation of Eq. (4.12) with respect to time, yielding

$$\mathbf{v}_{cm} = \frac{d\mathbf{r}_{cm}}{dt} = \frac{\sum_i m_i \mathbf{v}_i}{M} \tag{4.14}$$

This equation assumes that the mass of the particles and the total mass of the system remain constant over time. In a similar manner, the acceleration of the center of mass is obtained by differentiating Eq. (4.14), yielding

$$\mathbf{a}_{cm} = \frac{d\mathbf{v}_{cm}}{dt} = \frac{\sum_i m_i \mathbf{a}_i}{M} \tag{4.15}$$

Note that the system of particles acts linearly as if all of its mass is concentrated at the center of mass and the sum of all of the external forces governs the movement of this mathematically defined point. Therefore, the movement of that point can be determined without the detailed knowledge of the movement of each particle by the expression

$$\sum_i \mathbf{F}_i = M\mathbf{a}_{cm} \tag{4.16}$$

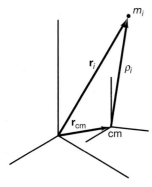

Figure 4.4

This equation relates the total external forces acting on the system to the rate of change of the linear momentum of the point that is defined as the center of mass. The system may then be treated as a single particle has been treated earlier.

By tracking the movement of the center of mass, we do not obtain information about the movement of the individual particles in space. Information concerning these movements is obtained by examining their movement relative to the center of mass. In order to do so, establish a coordinate system that moves with the center of mass, as shown in Figure 4.4. The position vector to any particle and the velocity of the particle can now be written, respectively, as

$$\mathbf{r}_i = \mathbf{r}_{cm} + \boldsymbol{\rho}_i$$
$$\mathbf{v}_i = \mathbf{v}_{cm} + \dot{\boldsymbol{\rho}}_i \tag{4.17}$$

A quantity called the *angular momentum of the system about the moving center of mass* is defined in a similar manner as the angular momentum of the system about the fixed origin. At this point, we will define the angular momentum about the center of mass as

$$\mathbf{H}_{cm} = \sum_i \boldsymbol{\rho}_i \times m_i \mathbf{v}_i \tag{4.18}$$

The velocity of each particle is the absolute velocity in Eq. (4.18), and we can use Eq. (4.17) to define the angular momentum about the center of mass in terms of the velocity relative to the center of mass:

$$\mathbf{H}_{cm} = \left(\sum_i m_i \boldsymbol{\rho}_i \right) \times \mathbf{v}_{cm} + \sum_i \boldsymbol{\rho}_i \times m_i \dot{\boldsymbol{\rho}}_i \tag{4.19}$$

The sum of the position vectors relative to the center of mass defines the distance from the center of mass to the center of mass—that is, zero. Therefore, Eq. (4.19) can be written as

$$\mathbf{H}_{cm} = \sum_i \boldsymbol{\rho}_i \times m_i \dot{\boldsymbol{\rho}}_i \tag{4.20}$$

This expression yields the surprising fact that the angular momentum about the center of mass defined using the absolute velocity of the particle is equal to the angular momentum using the relative velocity to the center of mass. This relationship will hold true *only* about the point in the system defined as the center of mass. The importance of the center of mass is now obvious, and it also should be clear why *all dynamic equations will be related to the center of mass*.

Previously, by examining the angular momentum about the fixed origin of our coordinate system, we have shown that the sum of the moments about this fixed origin equals the rate of change of the angular momentum about the origin. The angular momentum about the center of mass can be related to the angular momentum about the fixed origin as follows:

$$\mathbf{H}_o = \sum_i \mathbf{r}_i \times m_i \mathbf{v}_i \tag{4.21}$$

Using Eq. (4.17), we can write the angular momentum about the origin as

$$\mathbf{H}_o = \sum_i (\mathbf{r}_{cm} + \boldsymbol{\rho}_i) \times m_i (\mathbf{v}_{cm} + \dot{\boldsymbol{\rho}}_i)$$

Expanding the cross product and grouping terms, we get

$$\mathbf{H}_o = \mathbf{r}_{cm} \times M\mathbf{v}_{cm} + \mathbf{r}_{cm} \times \sum_i m_i \dot{\boldsymbol{\rho}}_i + \sum_i m_i \boldsymbol{\rho}_i \times \mathbf{v}_{cm} + \sum_i \boldsymbol{\rho}_i \times m_i \dot{\boldsymbol{\rho}}_i$$

The second and third terms are zero, as they define the relative velocity and position of the center of mass to itself, and the last term is the angular momentum about the center of mass. Therefore, Eq. (4.21) becomes

$$\mathbf{H}_o = \mathbf{r}_{cm} \times M\mathbf{v}_{cm} + \mathbf{H}_{cm} \tag{4.22}$$

Now, we will examine Eq. (4.11) and relate the moment about the origin to the moment about the center of mass:

$$\sum \mathbf{M}_o = \sum_i \mathbf{r}_i \times \mathbf{F}_i = \dot{\mathbf{H}}_o$$
$$\mathbf{r}_{cm} \times \sum_i \mathbf{F}_i + \sum_i (\boldsymbol{\rho}_i \times \mathbf{F}_i) = \mathbf{r}_{cm} \times M\mathbf{a}_{cm} + \mathbf{H}_{cm} \tag{4.23}$$

The first terms on either side of the equation are equal and cancel. Therefore, recognizing that the second term on the left side is the moment about the center of mass, we may write Eq. (4.23) as

$$\sum \mathbf{M}_{cm} = \dot{\mathbf{H}}_{cm} \tag{4.24}$$

Therefore, the rotational dynamics of the system can be considered as relative either to a fixed origin or to the moving center of mass of the system. This concept is fundamental to the development of rigid body dynamics.

If there are no net external forces or couples acting upon the system, the linear momentum of the system is conserved (a constant). The angular momentum about the fixed origin and the angular momentum about the moving center of mass are also conserved; that is:

$$\dot{\mathbf{L}} = 0 \quad \text{and} \quad \mathbf{L} = \text{constant}$$
$$\dot{\mathbf{H}}_o = 0 \quad \text{and} \quad \mathbf{H}_o = \text{constant} \tag{4.25}$$
$$\dot{\mathbf{H}}_{cm} = 0 \quad \text{and} \quad \mathbf{H}_{cm} = \text{constant}$$

It is then said that the momentum of the system is *conserved*.

Sample Problem 4.1

Three unattended, identical cars are parked on a level surface. None of the cars have their emergency brakes on. If car *A* rolls into the other two cars with a velocity *v*, determine the final velocity of each car after all of the impacts are completed if the coefficient of restitution for each impact is *e*.

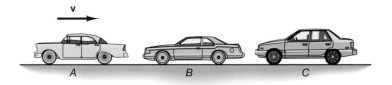

Solution   Impact between *A* and *B*:

$$mv = mv_A' + mv_B' \quad \text{(conservation of momentum)}$$
$$v_B' - v_A' = ev \quad \text{(coefficient of restitution)}$$

Therefore, after the first impact,

$$v'_B = \frac{(1 + e)}{2} v$$

and

$$v'_A = \frac{(1 - e)}{2} v$$

Impact between B and C:

$$m \frac{(1 + e)}{2} v = mv''_B + mv''_C \quad \text{(conservation of momentum)}$$

$$v''_C - v''_B = m \frac{(1 + e)}{2} v \quad \text{(impact coefficient of restitution)}$$

Therefore, after the second impact,

$$v'_A = \frac{(1 - e)}{2} v$$

$$v''_B = \frac{(1 - e)}{2} \frac{(1 + e)}{2} v$$

$$v''_C = \frac{(1 + e)}{2} \frac{(1 + e)}{2} v$$

For a perfectly elastic impact ($e = 1$), A and B will be at rest, and C will be moving with velocity v. For any value of e less than unity, the velocity of B after the second impact will be less than the velocity of A, so an additional impact between A and B will occur. For this third impact,

$$mv'_A + mv''_B = mv'''_A + m v'''_B$$

$$v'''_B - v'''_A = e(v'_A - v''_B)$$

Therefore,

$$v'''_A = \frac{(1 - e)}{2} v'_A + \frac{(1 + e)}{2} v''_B = \left[\frac{(1 - e)(1 - e)}{2} \frac{}{2} + \frac{(1 + e)}{2} \frac{(1 + e)}{2} \frac{(1 - e)}{2} \right] v$$

$$v'''_B = \frac{(1 + e)}{2} v'_A + \frac{(1 - e)}{2} v''_B = \left[\frac{(1 + e)}{2} \frac{(1 - e)}{2} + \frac{(1 - e)}{2} \frac{(1 + e)}{2} \frac{(1 - e)}{2} \right] v$$

For example, if the initial velocity of A was 10 mph and $e = 0.9$, the final velocities of the cars are

$$v'''_A = 0.476 \text{ mph} \qquad v'''_B = 0.499 \text{ mph} \qquad v'''_C = 9.025 \text{ mph}$$

The total momentum of the system is still $10m$ and is conserved during these impacts. The original energy in the system was $50m$, and the final energy is $40.96m$ kg · m²/s². The energy is not conserved, and about 20% of the kinetic energy is lost due to impacts.

Sample Problem 4.2

A 2 kg particle explodes and breaks into three equal fragments when it has a velocity of $\mathbf{V} = 10\hat{\mathbf{i}}$ m/s. If the fragments are seen to travel in the following directions, relative to the particle after the explosion, determine the velocity of each fragment and the energy lost in the explosion:

$$\mathbf{V}_a = V_a\,(0.577\hat{\mathbf{i}} + 0.577\hat{\mathbf{j}} + 0.577\hat{\mathbf{k}})$$

$$\mathbf{V}_b = V_b\,(0.333\hat{\mathbf{i}} + 0.667\hat{\mathbf{j}} - 0.667\hat{\mathbf{k}})$$

$$\mathbf{V}_c = V_c\,(-0.667\hat{\mathbf{i}} + 0.667\hat{\mathbf{j}} + 0.333\hat{\mathbf{k}})$$

Solution Each fragment has a mass of 2/3 kg, and momentum must be conserved in each direction. Thus, the following three equations of conservation of linear momentum can be written to determine the velocities:

$$\frac{2}{3}\,0.577V_a + \frac{2}{3}\,0.333V_b - \frac{2}{3}\,0.667V_c = 2(10)$$

$$\frac{2}{3}\,0.577V_a + \frac{2}{3}\,0.667V_b + \frac{2}{3}\,0.667V_c = 0$$

$$\frac{2}{3}\,0.577V_a - \frac{2}{3}\,0.667V_b + \frac{2}{3}\,0.333V_c = 0$$

Solving the system of linear equations yields

$$V_a = 20.791 \text{ m/s} \qquad V_b = 6.007 \text{ m/s} \qquad V_c = -23.993 \text{ m/s}$$

The energy in the system before the explosion was 100 J, and after the explosion, the kinetic energy of the system is 348 J. The increase in energy is due to the energy added to the system in order to fragment the particle.

Sample Problem 4.3

Three particles of equal mass in a closed system—that is, linear and angular momentum are conserved—have measured position and velocity vectors, in meters and meters/second, respectively, at time t_1:

$$\mathbf{r}_A = 3\hat{\mathbf{i}} - 4\hat{\mathbf{j}} + 2\hat{\mathbf{k}} \qquad \mathbf{v}_A = 20\hat{\mathbf{i}} + 15\hat{\mathbf{j}} + 20\hat{\mathbf{k}}$$

$$\mathbf{r}_B = -\hat{\mathbf{i}} + 3\hat{\mathbf{j}} - 3\hat{\mathbf{k}} \qquad \mathbf{v}_B = -5\hat{\mathbf{i}} - 10\hat{\mathbf{j}} - 10\hat{\mathbf{k}}$$

$$\mathbf{r}_C = -2\hat{\mathbf{i}} + \hat{\mathbf{j}} + 5\hat{\mathbf{k}} \qquad \mathbf{v}_C = -10\hat{\mathbf{i}} + 5\hat{\mathbf{j}} + 10\hat{\mathbf{k}}$$

At time t_2, the position and velocity of only two of the masses can be obtained:

$$\mathbf{r}_A = 5\hat{\mathbf{i}} + 2\hat{\mathbf{j}} + 3\hat{\mathbf{k}} \qquad \mathbf{v}_A = 15\hat{\mathbf{i}} + 20\hat{\mathbf{j}} + 25\hat{\mathbf{k}}$$

$$\mathbf{r}_B = \hat{\mathbf{i}} - 2\hat{\mathbf{j}} + 2\hat{\mathbf{k}} \qquad \mathbf{v}_B = 5\hat{\mathbf{i}} + 10\hat{\mathbf{j}} + 5\hat{\mathbf{k}}$$

Determine the velocity of particle C and a point on the tangent to the trajectory at time t_2.

Solution   This problem can be solved directly using a direct vector solution. To make our notation conform with that of most computational software packages, we will write the vectors as column matrices:

$$r_{A1} = \begin{pmatrix} 3 \\ -4 \\ 2 \end{pmatrix} \quad r_{B1} = \begin{pmatrix} -1 \\ 3 \\ -3 \end{pmatrix} \quad r_{C1} = \begin{pmatrix} -2 \\ 1 \\ 5 \end{pmatrix} \quad v_{A1} = \begin{pmatrix} 20 \\ 15 \\ 20 \end{pmatrix} \quad v_{B1} = \begin{pmatrix} -5 \\ -10 \\ -10 \end{pmatrix} \quad v_{C1} = \begin{pmatrix} -10 \\ 5 \\ 10 \end{pmatrix}$$

$$r_{A2} = \begin{pmatrix} 5 \\ 2 \\ 3 \end{pmatrix} \quad r_{B2} = \begin{pmatrix} 1 \\ -2 \\ 2 \end{pmatrix} \quad v_{A2} = \begin{pmatrix} 15 \\ 20 \\ 25 \end{pmatrix} \quad v_{B2} = \begin{pmatrix} 5 \\ 10 \\ 5 \end{pmatrix}$$

The linear momentum of the system is

$$\mathbf{L} = m(\mathbf{v}_{A1} + \mathbf{v}_{B1} + \mathbf{v}_{C1}) = m \begin{pmatrix} 5 \\ 10 \\ 20 \end{pmatrix} \text{ kg} \cdot \text{m/s}$$

Since the linear momentum of the system is conserved, the velocity of particle C is

$$\mathbf{v}_{C2} = \frac{\mathbf{L}}{m} - \mathbf{v}_{A2} - \mathbf{v}_{B2} = \begin{pmatrix} -15 \\ -20 \\ -10 \end{pmatrix} \text{ m/s}$$

The angular momentum is also conserved and is

$$\mathbf{H} = m(\mathbf{r}_{A1} \times \mathbf{v}_{A1} + \mathbf{r}_{B1} \times \mathbf{v}_{B1} + \mathbf{r}_{C1} \times \mathbf{v}_{C1}) = m \begin{pmatrix} -185 \\ -45 \\ 150 \end{pmatrix} \text{ kg} \cdot \text{m}^2/\text{s}$$

At first thought, it might appear that we can find the position of particle C in a similar manner. However, since the velocity is tangent to the trajectory and the angular momentum is independent of the choice of a point on this tangent line, there are an infinite number of positions of C which will satisfy the conservation of angular momentum. We will seek a position such that the position vector is perpendicular to the velocity vector at that point. This choice will allow us to use a direct vector method—that is, to solve for the position vector without expanding the equations into scalar notation. Let $\mathbf{p}_{C2}$ be the perpendicular position vector at time 2 s:

$$\mathbf{p}_{C2} \times \mathbf{v}_{C2} = \frac{\mathbf{H}}{m} - \mathbf{r}_{A2} \times \mathbf{v}_{A2} - \mathbf{r}_{B2} \times \mathbf{v}_{B2} = \mathbf{H}_{C2}$$

The three vectors $\mathbf{p}_{C2}$, $\mathbf{v}_{C2}$, and $\mathbf{H}_{C2}$ are mutually perpendicular. The cross product of $\mathbf{v}_{C2}$ with this equation is

$$\mathbf{v}_{C2} \times (\mathbf{p}_{C2} \times \mathbf{v}_{C2}) = \mathbf{v}_{C2} \times \mathbf{H}_{C2}$$

Using the vector identity for the triple vector product, we obtain

$$\mathbf{p}_{C2}(\mathbf{v}_{C2} \cdot \mathbf{v}_{C2}) - \mathbf{v}_{C2}(\mathbf{v}_{C2} \cdot \mathbf{p}_{C2}) = \mathbf{v}_{C2} \times \mathbf{H}_{C2}$$

The second term on the left side of the preceeding is zero, since the dot product between any two perpendicular vectors is zero. The dot product in the first term is a scalar, and the perpendicular position vector is

$$\mathbf{p}_{C2} = \frac{\mathbf{v}_{C2} \times \mathbf{H}_{C2}}{\mathbf{v}_{C2} \cdot \mathbf{v}_{C2}} = \begin{pmatrix} -1.241 \\ 3.241 \\ -4.621 \end{pmatrix} \text{ m}$$

Details of this solution can be found in the Computational Supplement. Many other examples of the direct vector solution are shown in Chapter 5.

Problems

4.1 Five identical spheres are hung from strings of equal length, as shown in Figure P4.1. If one ball is released from an angle θ, determine the response of the system. All impacts are elastic. Ignore air resistance.

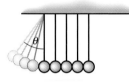

Figure P4.1

4.2 Determine the response of the system in Problem 4.1 if two balls are released.

4.3 Write a general expression for the response of the system in Problem 4.1 if n balls are released, with n varying from 1 to 5.

4.4 In Figure P4.4, a 200-g bullet is fired horizontally so that it passes through a 2-kg block A and becomes imbedded in a 4-kg block B. If the initial velocity of the bullet is 500 m/s and the velocity of block A is 20 m/s after the bullet passes through, determine the velocity of block B and the velocity of the bullet as it travels from A to B.

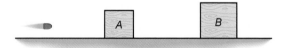

Figure P4.4

4.5 In Figure P4.5, car A weighs 3000 lb and has an initial velocity of 25 mph. If car B and car C each weigh 2500 lb and are at rest when car A strikes car B, determine the final velocities of the cars if the coefficient of restitution for all impacts is 0.8.

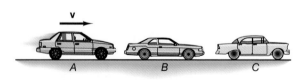

Figure P4.5

4.6 Determine the final velocities of the cars in Problem 4.5 if all of the impacts are elastic ($e = 1$).

4.7 Determine the final velocity of the cars in Problem 4.5 if all impacts are plastic.

4.8 Three particles with respective masses $m_A = 1$ kg, $m_B = 2$ kg, and $m_C = 4$ kg constitute a closed system, as in Figure P4.8. At the beginning of an experiment, the positions and the velocities of the masses at time t_1 are

$$\mathbf{r}_A = \begin{pmatrix} 3 \\ -2 \\ -1 \end{pmatrix} \text{m} \quad \mathbf{v}_A = \begin{pmatrix} 5 \\ 4 \\ -2 \end{pmatrix} \text{m/s}$$

$$\mathbf{r}_B = \begin{pmatrix} 3 \\ 5 \\ 2 \end{pmatrix} \text{m} \quad \mathbf{v}_B = \begin{pmatrix} 5 \\ -5 \\ 4 \end{pmatrix} \text{m/s}$$

$$\mathbf{r}_C = \begin{pmatrix} 8 \\ 2 \\ 0 \end{pmatrix} \text{m} \quad \mathbf{v}_C = \begin{pmatrix} -4 \\ 0 \\ -2 \end{pmatrix} \text{m/s}$$

At the conclusion of the experiment and after many collisions, only the two largest particles, B and C, are observed to have respective positions and velocities of

$$\mathbf{r}_B = \begin{pmatrix} 7 \\ -2 \\ -1 \end{pmatrix} \text{m} \quad \mathbf{v}_B = \begin{pmatrix} -1 \\ -1 \\ 2 \end{pmatrix} \text{m/s}$$

$$\mathbf{r}_C = \begin{pmatrix} 2 \\ -1 \\ -2 \end{pmatrix} \text{m} \quad \mathbf{v}_C = \begin{pmatrix} -1 \\ 0.5 \\ -1 \end{pmatrix} \text{m/s}$$

Determine the velocity of particle A and a point on its final trajectory.

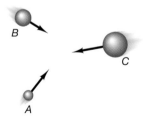

Figure P4.8

4.9 Determine the initial center of mass and the angular momentum about the center of mass in Problem 4.8. Determine the velocity of the center of mass of the system.

4.10 A system of five identical particles has the following dynamic state:

$$\mathbf{r}_A = 300\hat{\mathbf{i}} + 250\hat{\mathbf{j}} - 100\hat{\mathbf{k}} \text{ mm}$$

$$\mathbf{v}_A = -500\hat{\mathbf{i}} + 1500\hat{\mathbf{j}} - 200\hat{\mathbf{k}} \text{ mm/s}$$

$$\mathbf{r}_B = -100\hat{\mathbf{i}} - 150\hat{\mathbf{j}} - 50\hat{\mathbf{k}} \text{ mm}$$

$$\mathbf{v}_B = 1200\hat{\mathbf{i}} - 500\hat{\mathbf{j}} - 1200\hat{\mathbf{k}} \text{ mm/s}$$

$$\mathbf{r}_C = 400\hat{\mathbf{i}} + 350\hat{\mathbf{j}} + 200\hat{\mathbf{k}} \text{ mm}$$

$$\mathbf{v}_C = -800\hat{\mathbf{i}} + 1000\hat{\mathbf{j}} + 600\hat{\mathbf{k}} \text{ mm/s}$$

$$\mathbf{r}_D = 1100\hat{\mathbf{i}} - 450\hat{\mathbf{j}} + 500\hat{\mathbf{k}} \text{ mm}$$

$$\mathbf{v}_D = 1300\hat{\mathbf{i}} - 900\hat{\mathbf{j}} + 1200\hat{\mathbf{k}} \text{ mm/s}$$

$$\mathbf{r}_E = -500\hat{\mathbf{i}} + 50\hat{\mathbf{j}} + 400\hat{\mathbf{k}} \text{ mm}$$

$$\mathbf{v}_E = -900\hat{\mathbf{i}} - 1500\hat{\mathbf{j}} - 700\hat{\mathbf{k}} \text{ mm/s}$$

Determine the position of the center of mass of the system, the linear momentum per unit of mass of the system, and the angular momentum per unit of mass of the system about the origin and about the center of mass.

4.11 A 4-kg particle is given an initial velocity $\mathbf{v} = 3\hat{\mathbf{i}} - 2\hat{\mathbf{j}} + 3\hat{\mathbf{k}}$ m/s. Determine a starting position for the particle if the angular momentum about the origin is

$$\mathbf{H}_o = -10\hat{\mathbf{i}} + 30\hat{\mathbf{j}} + 30\hat{\mathbf{k}} \text{ kg} \cdot \text{m}^2/\text{s}$$

Note that the angular-momentum vector must be perpendicular to the velocity vector for this problem to have a solution. Solve by a direct vector method. (See Sample Problem 4.3.)

4.12 A 2-kg particle is given an initial position $\mathbf{r} = -1.5\hat{\mathbf{i}} + \hat{\mathbf{j}} - 1.5\hat{\mathbf{k}}$ m. Determine a starting velocity for the particle if the angular momentum about the origin is

$$\mathbf{H}_o = -20\hat{\mathbf{i}} + 60\hat{\mathbf{j}} + 60\hat{\mathbf{k}} \text{ kg} \cdot \text{m}^2/\text{s}$$

Note that the position vector must be perpendicular to the angular momentum vector for this problem to have a solution. The initial velocity may have an arbitrary component parallel to the position vector, as this component will not alter the angular momentum. Therefore, only the component of the velocity perpendicular to the position vector can be uniquely determined.

4.13 Two 10,000-kg railroad cars are rolling toward each other with the velocities shown in Figure P4.13. When the cars collide, they couple, but before coupling occurs,

a 1000-kg crate slides down a shute onto railroad car A. Determine the final velocity of the cars.

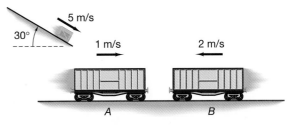

Figure P4.13

4.14 In Figure P4.14, a cart of mass M slides along a smooth channel at a velocity v when the rotating mass m attached to the car is at an angle $\theta = 0$. If the rotating mass is driven by a motor at a constant angular velocity ω, determine a general expression for the velocity of the cart for any angular position of the rotating mass.

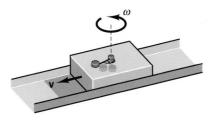

Figure P4.14

4.15 Show that if a particle has a velocity of $\mathbf{v} = 3\hat{\mathbf{i}} - 2\hat{\mathbf{j}} + 4\hat{\mathbf{k}}$ m/s, there is no position in space that it can occupy such that the angular momentum vector about the origin is in the x-direction.

4.16 A 2000-kg rocket sled launcher is moving with a velocity of 10 m/s when a 200-kg rocket is launched with a velocity relative to the sled of 200 m/s, as shown in Figure P4.16. Determine the velocity of the sled after launch.

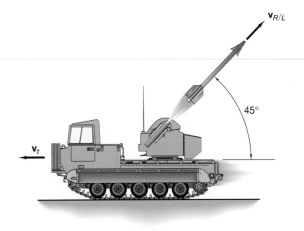

Figure P4.16

4.17 In Figure P4.17, a 10-ft, 200-lb rowboat sits in still water with a 200-lb man at the stern. If the man walks to the bow of the boat, will the center of mass of the rowboat move, and if so, how far? Ignore the resistance of the water.

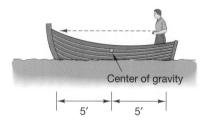

Figure P4.17

4.18 Three masses rest on a horizontal surface and are subjected to external forces, as shown in Figure P4.18. The masses are connected by springs with spring constants 500 N/m and of unstretched lengths 0.5 m. The mass, position, velocity, and external force for each mass are

$$m_A = 3 \text{ kg} \qquad \mathbf{r}_A = 2\hat{\mathbf{i}} + 3\hat{\mathbf{j}} \text{ m}$$

$$\mathbf{v}_A = 3\hat{\mathbf{i}} - \hat{\mathbf{j}} \text{ m/s} \qquad \mathbf{F}_A = 200\hat{\mathbf{i}} + 100\hat{\mathbf{j}} \text{ N}$$

$$m_B = 1 \text{ kg} \qquad \mathbf{r}_B = 4\hat{\mathbf{i}} \text{ m}$$

$$\mathbf{v}_B = 5\hat{\mathbf{i}} \text{ m/s} \qquad \mathbf{F}_B = -300\hat{\mathbf{i}} - 60\hat{\mathbf{j}} \text{ N}$$

$$m_C = 2 \text{ kg} \qquad \mathbf{r}_C = 0 \text{ m}$$

$$\mathbf{v}_C = -3\hat{\mathbf{i}} - 2\hat{\mathbf{j}} \text{ m/s} \qquad \mathbf{F}_C = 300\hat{\mathbf{i}} - 200\hat{\mathbf{j}} \text{ N}$$

Determine the linear momentum of the system and the angular momentum about the origin and about the center of mass.

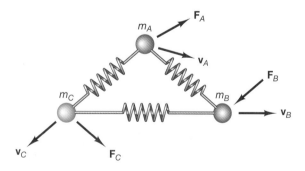

Figure P4.18

4.19 For the system in Problem 4.18, determine the position of the center of mass, the acceleration of the

center of mass, and the rate of change of the angular momentum about the center of mass.

4.20 Safety engineers are called in to investigate a midair collision between a commercial airliner and a private plane, as depicted in Figure P4.20. The estimated total weight of the airliner, its fuel, and passengers at the time of the crash is 200,000 lb, and the private plane and its pilot weigh 3000 lb. The air traffic controller estimates that the airliner was flying due east at 20,000 ft and at a speed of 200 mph, while the private plane was flying at the same altitude South 30° West at a speed of 150 mph. Considering the two planes as a system of particles, determine the velocity of the center of mass of the wreckage immediately after the collision, and determine where the center of mass of the wreckage will hit the ground, relative to the point of collision. This information is necessary to direct the search parties.

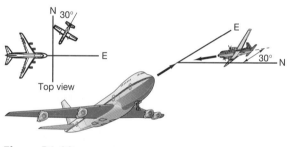

Figure P4.20

4.21 In Figure P4.21, a mass of 5 kg breaks into two pieces with respective weights 3 kg and 2 kg at a height of 30 m and at a horizontal velocity of $\mathbf{v} = 5\hat{\mathbf{i}}$ m/s. If the 3-kg fragment hits the ground at coordinates (7.013, 1.000, 0.000) m, at which coordinates will the 2-kg fragment hit the ground?

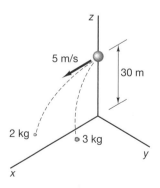

Figure P4.21

4.4   KINETIC ENERGY OF A SYSTEM OF PARTICLES

The kinetic energy of a system of particles is defined as the sum of the kinetic energy of the individual particles:

$$T = \frac{1}{2}\sum m_i \mathbf{v}_i \cdot \mathbf{v}_i \tag{4.26}$$

It is advantageous to express the kinetic energy in terms of a reference system with the origin at the center of mass. Note that this coordinate system will, in general, be moving with respect to an inertial coordinate reference. The position vector to any particle and the velocity of the particle are given in Eq. (4.17) and are repeated here as

$$\mathbf{r}_i = \mathbf{r}_{cm} + \boldsymbol{\rho}_i$$
$$\mathbf{v}_i = v_{cm} + \dot{\boldsymbol{\rho}}_i \tag{4.27}$$

The kinetic energy can be written in terms of the velocity of the particles relative to the center of mass:

$$T = \frac{1}{2}\sum m_i [v_{cm}^2 + 2\mathbf{v}_{cm} \cdot \dot{\boldsymbol{\rho}}_i + \dot{\boldsymbol{\rho}}_i \cdot \dot{\boldsymbol{\rho}}_i] \tag{4.28}$$

The middle term in Eq. (4.27) is zero by the definition of the center of mass, and the kinetic energy of the system is

$$T = \frac{1}{2}\sum m_i [\dot{\boldsymbol{\rho}}_i \cdot \dot{\boldsymbol{\rho}}_i] + \frac{1}{2} M v_{cm}^2 \tag{4.29}$$

Therefore, the kinetic energy of the system is equal to the sum of the kinetic energy of the total mass considered to be concentrated at the center of mass and the kinetic energy of the particles due to motion relative to the center of mass. Eqs. (4.26) and (4.29) are difficult to apply to a system of particles, as the velocity of each of the particles must be determined in order to apply these equations, and those values are usually not known. These equations, however, will be the foundations of the development of energy theorems for rigid bodies.

4.5   WORK–ENERGY AND CONSERVATION OF ENERGY OF A SYSTEM OF PARTICLES

Consider the system of particles shown in Figure 4.5. The work done by the forces acting on the ith particle is

$$U_{1\to2}^i = \int_{\bar{r}_{i1}}^{\bar{r}_{i2}} \left(\mathbf{F}_i + \sum \mathbf{f}_{ij}\right) \cdot d\mathbf{r}_i \tag{4.30}$$

If the work done by each particle is summed over the number of particles in the system, the work done by the internal forces does not cancel out. Even though the internal forces occur in pairs, the displacements of the particles will differ and the dot product between the internal forces and the different displacements will not be equal. The total work done by the system is

$$U_{1\to2} = \sum_i \int_{\bar{r}_{i1}}^{\bar{r}_{i2}} \left(\mathbf{F}_i + \sum_j \mathbf{f}_{ij}\right) \cdot d\mathbf{r}_i \tag{4.31}$$

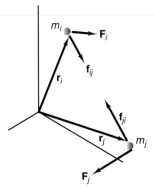

Figure 4.5

Having defined the kinetic energy of the system to be the sum of the kinetic energy of all the particles, we can write a work–energy equation as

$$U_{1 \to 2} = T_2 - T_1 \tag{4.32}$$

If all of the forces, both internal and external, are conservative, then the total energy of the system is conserved:

$$V_1 + T_1 = V_2 + T_2 \tag{4.33}$$

Again, these equations are not particularly useful to the analysis of a system of particles, as the internal forces must be determined, as well as the motion of each particle.

4.6 IMPULSE AND MOMENTUM OF A SYSTEM OF PARTICLES

The other first integral of the motion of a system of particles leads to an impulse–momentum principle. The equations of motion of the system of particles are

$$\sum_i \mathbf{F}_i = \frac{d}{dt} \mathbf{L}, \quad \text{where} \quad \mathbf{L} = \sum_i m_i \mathbf{v}_i = M\mathbf{v}_{cm}$$

$$\sum \mathbf{M}_{cm} = \frac{d}{dt} \mathbf{H}_{cm} \quad \text{or} \quad \sum \mathbf{M}_0 = \frac{d}{dt} \mathbf{H}_0 \tag{4.34}$$

The linear impulse to the system of particles is

$$\sum_i \int_{t_1}^{t_2} \mathbf{F}_i(t)dt = \mathbf{L}_2 - \mathbf{L}_1 \tag{4.35}$$

If the total linear impulse to the system is zero, the linear momentum of the system will be *conserved*. A similar relationship can be written for the angular impulse and the angular momentum of the system. This relationship is

$$\sum \int_{t_1}^{t_2} \mathbf{M}_{cm}dt = \mathbf{H}_{cm_2} - \mathbf{H}_{cm_1}$$

or

$$\sum \int_{t_1}^{t_2} \mathbf{M}_0 dt = \mathbf{H}_{0_2} - \mathbf{H}_{0_1} \tag{4.36}$$

If there is no impulsive moment about the center of mass or about the fixed origin of the coordinate system, the angular momentum of the system will be conserved. Note that the internal forces do not appear in the linear or angular impulse, as has been shown in Section 4.2.

Sample Problem 4.4

A cart supports a simple pendulum of the same mass, as shown in the accompanying diagram. If, at this position, the velocity of the mass B is V_0 and the velocity of the cart A is zero, determine the velocity of the masses when the pendulum reaches its, maximum height and when it returns to the vertical position again.

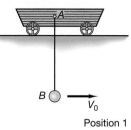

Position 1

Solution   Consider the system when the pendulum reaches its maximum angle, as depicted in the accompanying figure. There is no impulse to the system in the x-direction, so

$$L_{1x} = L_{2x}$$

$$mV_0 = m(v_A + v_B)$$

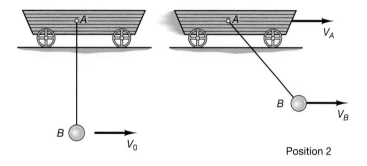

Position 2

When the pendulum reaches its maximum (or minimum) angle, the relative motion between the cart and the pendulum is zero:

$$v_{B/A} = v_B - v_A = 0$$

Therefore,

$$v_B = v_A = \frac{V_0}{2}$$

This condition is the state when the pendulum is at its minimum position. The pendulum then swings back to its position below the cart, as shown in the accompanying figure. Again, the linear momentum of the system is conserved in the x-direction, and the linear momentum at position 1 must equal the linear momentum at position 3. So,

$$V_0 = v_A + v_B$$

The energy of the system is conserved, so the kinetic energy at positions 1 and 3 must be equal:

$$\frac{1}{2} mV_0^2 = \frac{1}{2} m(v_A^2 + v_B^2)$$

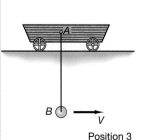

Position 3

Squaring the linear momentum equation and comparing the square to the equation for the conservation of energy yields

$$2v_A v_B = 0$$

Therefore, either v_A or $v_B = 0$. But the relative velocity of B must be negative, so

$$v_{B/A} = v_B - v_A$$
$$v_A = V_0 \text{ and } v_B = 0$$

The system "walks" forward, starting with all the momentum with B, and then, as the pendulum swings forward, the momentum is equally divided between the cart and the pendulum. As the pendulum swings back, the momentum is given to the cart, and when the pendulum reaches the furthermost position back, the momentum is equal between the cart and pendulum. When the pendulum swings forward it acquires all of the momentum, and the initial dynamic state is returned. The motion repeats in the next cycle. If there were no friction, the motion would continue indefinitely.

Problems

4.22 A 6-kg fireworks is fired upward at an 80° angle at an initial velocity of 100 m/s, as shown in Figure P4.22. At its maximum altitude, the fireworks explodes into three equal pieces. The velocities of two of the pieces are

$$\mathbf{v}_1 = 15\hat{\mathbf{i}} - 20\hat{\mathbf{j}} + 30\hat{\mathbf{k}}$$
$$\mathbf{v}_2 = -20\hat{\mathbf{i}} + 10\hat{\mathbf{j}} - 20\hat{\mathbf{k}} \text{ m/s}$$

Determine the velocity of the third piece and the energy change due to the explosion.

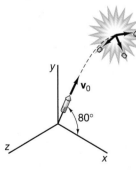

Figure P4.22

4.23 Determine the loss of energy of the system in Problem 4.5.

4.24 Determine the energy lost in the system in Problem 4.7.

4.25 Determine the change in kinetic energy in the system in Problem 4.8.

4.26 Two identical spheres are connected by a rigid bar of negligible mass and rest on a smooth horizontal surface, as shown in Figure P4.26. If a third sphere of the same mass, moving at an initial velocity V, has an elastic impact with sphere C, determine the velocities of spheres B and C as a function of the angle θ.

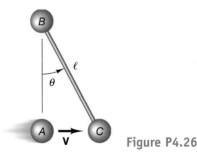

Figure P4.26

4.27 For the system in Problem 4.26, determine the velocities of spheres B and C if the mass of B is twice the mass of A and C.

4.28 In a game of billiards, as shown in Figure P4.28, ball A has an initial velocity as indicated in the figure and strikes balls B and C, which are adjacent to each other, in the direct center of where the balls touch. If friction is neglected and all impacts are elastic, determine the final velocities of the three balls of equal mass.

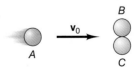

Figure P4.28

4.29 In a game of pool, shown in Figure P4.29, the cue ball has an initial velocity of $\mathbf{v}_0 = 3\hat{\mathbf{i}}$ m/s. After impact, the three balls of equal mass move in the directions shown in the figure. If all impacts are elastic, determine the magnitude of the velocity of each ball after impact.

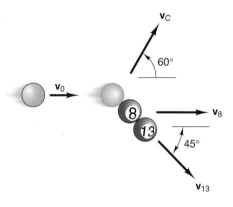

Figure P4.29

4.30 For the pool game in Problem 4.29, determine the magnitude of the velocities of the ball if the initial velocity of the cue ball is 5 m/s.

4.31 In Figure P4.31, sphere C has an initial velocity of 4 m/s and impacts spheres A and B of identical mass, which are initially at rest. All impacts are elastic, and the direction of the final velocities of the spheres is shown in the figure. Neglect friction, and determine the magnitude of the velocity of each sphere after all of the impacts have occurred. (There are two solutions; justify your choice of solution.)

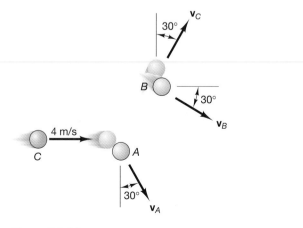

Figure P4.31

4.32 For the system in Problem 4.31, use conservation of angular momentum to determine the equation of a line between sphere A and sphere B before impact.

4.33 Determine the magnitude of the velocity of each sphere in Problem 4.31 if the initial velocity of sphere C is 7 m/s.

4.34 Two identical masses are connected by a rigid bar of negligible mass and rotate in a vertical plane, pinned as shown in Figure P4.34. If the system is released from rest in a horizontal position, determine the angular equation of motion.

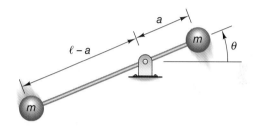

Figure P4.34

4.35 Add damping to the pin connection in Problem 4.34 so that a damping moment is generated proportional to the velocity $M_d = -c\omega$ N m, and develop the angular equation of motion.

4.36 Solve the equation of motion in Problem 4.35 by choosing a damping coefficient and mass, length, and pin offset so that the motion will damp out after two oscillations.

4.37 Three 5-kg masses are connected by rigid, massless bars, as shown in Figure P4.37. The system is in a horizontal plane and is driven by a constant torque motor. Determine the torque required to accelerate the system from rest to 10 rad/s in 6 s.

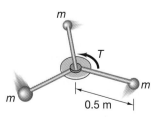

Figure P4.37

4.38 In Figure P4.38, block *A* of mass 3 kg slides on a smooth plane and supports a mass *B* of 1 kg on the end of a 200 mm rod that is pinned with a frictionless connection. If block *A* is given an initial velocity of 0.5 m/s at the position shown in the figure, determine the maximum angle that the mass *B* will attain.

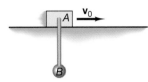

Figure P4.38

4.39 A pendulum is modeled as a mass at the end of a rigid bar of negligible mass and pinned as shown in Figure P4.39. Develop the angular equation of motion for the pendulum, and solve the nonlinear differential equation to determine the velocity at any position.

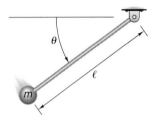

Figure P4.39

4.40 Select a value for the mass and the length of the pendulum in Problem 4.39 and solve the nonlinear differential equation to obtain the angle as a function of time, assuming the pendulum is released from a horizontal position.

4.41 A detail of the pin connection for the pendulum in Problem 4.39 is shown in Figure P4.41. Friction at this pin connection has a damping moment $M_d = -\mu_k N r(\omega/|\omega|)$, where N is the normal force at the pin, ω is the angular velocity of the pendulum, μ_k is the coefficient of kinetic friction, and r is the radius of the pin. Develop the differential equation of motion for the pendulum. Note that friction always opposes the motion. $\dfrac{\omega}{|\omega|}$ is defined to be equal to 0, when $\omega = 0$.

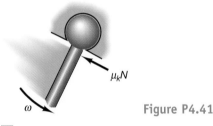

Figure P4.41

4.42 Select values for the parameters in Problem 4.41, and solve the nonlinear differential equation of motion. Choose different values of the coefficient of kinetic friction to examine the dependency of the motion on friction.

*4.7 MASS FLOWS

In the preceding sections, the system of particles that was examined was a closed system; that is, there were no particles entering the system or leaving the system, so the total mass of the system was conserved. There are many engineering applications where this is not the case, and particles are flowing in and out of a fixed, spatial *control volume*. Jet engines, rockets, turbines, fans, nozzles, pumps, etc. are examples of this situation. For a detailed understanding of these systems, the subject of fluid mechanics should be studied, and the presentation here is not meant to be a replacement for such a course. The principle of conservation of linear momentum will be used to determine the force exerted on an object due to a steady mass flow through that object or a variable mass flow through the control volume.

4.7.1 STEADY MASS FLOW

A steady mass flow is a condition in which the rate of change of the mass entering the control volume equals the rate of change of the mass leaving the control volume. The control volume may be a section of pipe or vessel through which the system of particles (modeled

Figure 4.6

as a fluid) flows, as shown in Figure 4.6. If the mass flow is steady, the rate of mass flow at A_1 must equal the rate of mass flow at A_2 for continuity of flow, or

$$\frac{dm}{dt} = \dot{m} = \rho_1 A_1 v_1 = \rho_2 A_2 v_2 \tag{4.37}$$

where ρ is the mass density of the fluid stream at points 1 and 2, v is the fluid speed at those points, and A is the cross-sectional area of the stream at those points.

To examine the forces acting on the fluid as it passes through the control volume, it is necessary to define the *auxiliary constant system of particles,* as the system contained in the control volume is a *variable system of particles.*

The auxiliary system of particles will contain two parts: (1) the mass in the control volume at time t, denoted by m, and (2) the mass that will enter the control volume in a time interval Δt, denoted by Δm. The auxiliary constant system of particles is shown in Figure 4.7. If the mass flow is steady, the momentum within the control volume will be constant, and the impulse–momentum principle for time t to $(t + \Delta t)$ is

$$(\Delta m)\mathbf{v}_1 + m\mathbf{v} + \sum \mathbf{F}\Delta t = m\mathbf{v} + (\Delta m)\mathbf{v}_2$$

$$\sum \mathbf{F} = \frac{dm}{dt}(\mathbf{v}_2 - \mathbf{v}_1) \tag{4.38}$$

where the rate of mass flow is given for most steady mass-flow systems. The inflow and outflow velocities $\mathbf{v}_1$ and $\mathbf{v}_2$ are velocities to the control volume velocity, and these velocities are assumed to be constant across the inlet and outlet cross-sectional areas. Therefore, the effects of fluid friction are neglected. For more details of fluid flow, you should consult a text in that field. The forces acting on the system must include all external forces such as gravitational attraction, support of any structures, etc.

Eq. (4.38) is a powerful application of the principle of impulse–momentum when we can define the boundary of the system within which the mass is a constant. A free-body diagram must be constructed, and all external forces are considered. A similar development can be made for the angular momentum in a steady-flow system. The moment about any point 0 in or out of the boundaries of the system equals the rate of change of the angular momentum of the system about 0. The angular momentum equation may be obtained by taking the cross product of Eq. (4.38), yielding

$$\sum \mathbf{r}_i \times \mathbf{F}_i = \sum \mathbf{M}_0 = \frac{dm}{dt}(\mathbf{r}_2 \times \mathbf{v}_2 - \mathbf{r}_1 \times \mathbf{v}_1) \tag{4.39}$$

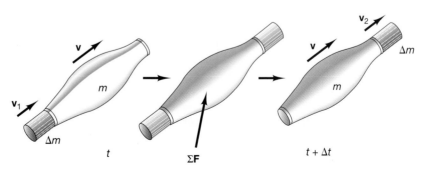

Figure 4.7

Sample Problem 4.5

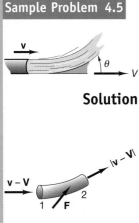

A stream of water discharged with a velocity of v is deflected by a single smooth vane moving at a velocity V, as shown in the accompanying diagram. If the cross-sectional area of the stream is A, with density ρ, determine the force the stream exerts on the vane, neglecting the weight of the fluid.

Solution   The relative velocity at position 1 is

$$\mathbf{v}_1 = (v - V)\hat{\mathbf{i}}$$

The relative velocity at position 2 is

$$\mathbf{v}_2 = (v - V)(\cos\theta\,\hat{\mathbf{i}} + \sin\theta\,\hat{\mathbf{j}})$$

The rate of mass flow is

$$\frac{dm}{dt} = A\rho(v - V)$$

Therefore, the force that the vane exerts on the fluid stream is

$$\mathbf{F} = A\rho(v - V)^2[(\cos\theta - 1)\hat{\mathbf{i}} + \sin\theta\,\hat{\mathbf{j}}]$$

4.7.2 VARIABLE MASS FLOW

Suppose that an object of mass m and velocity $\mathbf{v}$ ejects an element of mass Δm at a velocity $\mathbf{u}$ relative to the object during a period of time Δt while under the influence of external forces $\mathbf{F}$, as shown in Figure 4.8. Applying the principle of impulse and momentum, we get

$$\mathbf{F}\Delta t = (m - \Delta m)(\mathbf{v} + \Delta\mathbf{v}) + \Delta m(\mathbf{v} + \mathbf{u}) - m\mathbf{v}$$

$$= m\Delta\mathbf{v} + \Delta m\mathbf{u} - \Delta m\Delta\mathbf{v} \qquad (4.40)$$

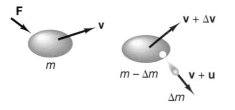

Figure 4.8

Dropping the higher order term, we may write Eq. (4.40)

$$\mathbf{F} = m\frac{d\mathbf{v}}{dt} + \frac{dm}{dt}\mathbf{u} \qquad (4.41)$$

The last term can be considered to be a psuedo-force if it is moved to the left side of Eq. (4.41). If the system absorbed mass, the impulse–momentum equation is

$$\mathbf{F} = m\frac{d\mathbf{v}}{dt} - \frac{dm}{dt}\mathbf{u} \qquad (4.42)$$

The relative velocity of the mass flow must be such that the mass is either ejected or absorbed by the object. Eqs. (4.41) and (4.42) can be written for the case when mass is both absorbed and ejected at the same time:

$$\mathbf{F} = m\frac{d\mathbf{v}}{dt} + \frac{dm_e}{dt}\mathbf{u}_e - \frac{dm_a}{dt}\mathbf{u}_a \qquad (4.43)$$

The mass flow in (absorbed) and the mass flow out (ejected) may be different.

Sample Problem 4.6

A rocket is launched vertically from the north pole with an initial mass m_0 and a burnout mass of m_{bo}. If fuel is consumed at a constant rate of q and the variation of gravitational attraction and air resistance are neglected, develop a relationship between the velocity of the rocket and the fuel consumed by the rocket. The relative exhaust velocity of the fuel is $\mathbf{u}$. Under these assumptions, determine the maximum velocity of the rocket.

Solution If the upward direction is taken as positive,

$$-mg = \frac{dm_e}{dt}(-u) + m\frac{dv}{dt}$$

The mass of the rocket decreases by the amount of fuel decrease; therefore,

$$\frac{dm}{dt} = -\frac{dm_e}{dt}$$

$$dv = -u\frac{dm}{m} - gdt$$

$$\int_0^v dv = -u\int_{m_0}^m \frac{dm}{m} - g\int_0^t dt$$

$$v = u\ln\frac{m_0}{m} - gt$$

Since the fuel is burning at a constant rate, let this rate be q, and

$$m(t) = m_0 - qt$$

The time to burnout is

$$t = \frac{m_0 - m_{bo}}{q}$$

Therefore,

$$v_{max} = u\ln\frac{m_0}{m_{bo}} - \frac{g}{q}(m_0 - m_{bo})$$

Problems

4.43 Water comes out of a nozzle with a velocity v m/s at a volume rate of V m³/s and is split into two equal streams by a fixed vane and deflected through an angle θ, as shown in Figure P4.43. If the density of the vane ρ kg/m³, determine the force required to hold the vane in place.

Figure P4.43

4.44 Determine the force acting on the vane in Problem 4.43 if the volume rate is 0.06 m³/s, the density of water is 1000 kg/m³, the velocity of the water at the nozzle is 20 m/s, and the water is deflected at a 30° angle.

4.45 Determine the thrust developed by a turbojet engine being tested in a lab. The input air velocity is zero, and the mass-flow rate of the air is 15 kg/s. The mass-flow rate of the fuel is 0.2 kg/s, and the exhaust velocity is 600 m/s.

4.46 For the airplane in Figure P4.46, the lift force **L** balances the component of the gravitational force in the y-direction so that the acceleration in that direction is zero. Develop an expression for the acceleration of a jet aircraft of mass M moving with a velocity **v**. (This value is the velocity of the air intake.) The drag force **D** is 15% of the lift force. The rate of air intake is dm/dt and the air is discharged at a velocity $4\mathbf{v}$.

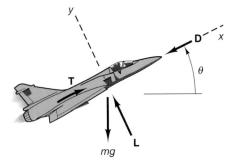

Figure P4.46

4.47 A turbojet engine's thrust reverser causes the exhaust to exit at an angle to the engine centerline, as shown in Figure P4.47. The touchdown velocity of the commercial airliner is 200 ft/s, and the exhaust velocity is 1000 ft/s. If the mass-flow of the air entering the compressor is 4 slug/s and the mass-flow rate of the fuel is

0.2 slug/s, determine the braking force that the engine exerts on the airplane.

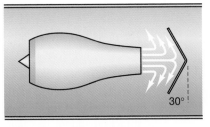

Figure P4.47

4.48 A conveyor belt system is used to load gravel into dump trucks, as depicted in Figure P4.48. The velocity of the first conveyor belt is $|\mathbf{v}_1| = 2.5$ m/s, and the velocity of the second conveyor belt is $|\mathbf{v}_2| = 3.0$ m/s. If the mass-flow rate of the gravel onto the belt is 200 kg/s, determine the reactions at the support of the second conveyor belt system. The conveyor belt assembly and the gravel it supports have a mass of 400 kg.

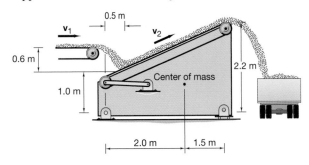

Figure P4.48

4.49 In Figure P4.49, a 20-kg snowblower is pushed forward at a rate of 0.3 m/s and takes in snow at a mass-flow rate of 5 kg/s. The snow is blown out the side of the snowblower at an angle of 45° to the horizontal at a velocity of 1 m/s. Determine the force that the woman running the snowblower must exert to push the snowblower along a level sidewalk.

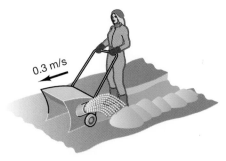

Figure P4.49

4.50 In Figure P4.50, a 2000-kg helicopter generates a 6-m-diameter slipstream with a downward air speed of 25 m/s. Using $\rho = 1.2$ kg/m³ for air, determine whether the helicopter is accelerating up or down.

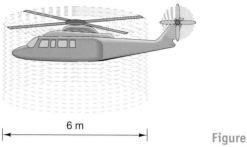

6 m

Figure P4.50

4.51 Determine the acceleration of the helicopter in Problem 4.50 that increases the prop rotation such that the downward air speed of the slipstream is 30 m/s.

4.52 A lawn sprinkler emits water with a mass-flow rate of 0.3 kg/s at each nozzle with a velocity of 6 m/s, as shown in Figure P4.52. The nozzle A is located at $(0.15, -0.03, 0.03)$ m, and the water exits so that it makes equal angles with each coordinate axis. Determine the moment about the z-axis exerted on the sprinkler by the flows from the four nozzles.

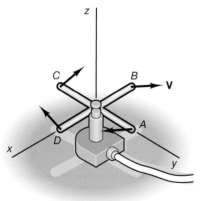

Figure P4.52

4.53 A fireboat pumps water from two pumps at a rate of 3000 kg/s at a velocity of 40 m/s, as shown in Figure P4.53. Determine the force that the pumps exert on the boat.

Figure P4.53

4.54 A chain with mass ρ per unit length slides through a hole in a horizontal table, as illustrated in Figure P4.54. Neglecting friction, use the principle of impulse and momentum to develop the differential equation to express the velocity as a function of the length of chain x.

Figure P4.54

4.55 A chain of length L and mass ρ per unit length slides off smooth table, as shown in Figure P4.55. Develop a relation for the velocity of the chain v in terms of the distance x. Determine the velocity of the chain as the last link slides off the table.

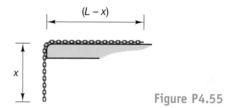

(L − x)

x

Figure P4.55

4.56 In Figure P4.56, grain exits a chute onto a conveyor belt at a flow rate of 40 kg/s and at a velocity of 4 m/s. If the conveyor belt moves at a constant speed of 1.5 m/s, determine the force that the grain exerts on the belt.

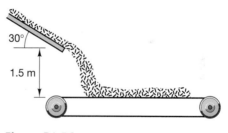

30°

1.5 m

Figure P4.56

4.57 In Figure P4.57, a fireman holds a hose at an angle θ to the horizontal. If the water velocity at the nozzle is v m/s, the volume rate of the water emitted is V m³/s, and the mass density of the water is ρ kg/m³, determine the force that the hose exerts on the fireman.

θ

Figure P4.57

Chapter 5 KINEMATICS OF RIGID BODIES

The full extension of the tennis player's arm allows high velocity of the tennis racket resulting in serves at speeds over 120 mph.

5.1 INTRODUCTION

Kinematics, as previously stated is the study of motion without consideration of the cause of that motion. As noted in Chapters 1 through 3, particles have no orientation in space, only position. In Chapter 4, systems of particles were examined, and, in general, each of the particles in a system was free to move independently. The principles of kinematics will now be applied to a system of particles that are rigidly connected, called a *rigid body*. The assumption that an object is rigid means that the deformations of the object are assumed to be negligible and that the distance between any two points on the body remains constant. The body may translate in a direction with components along the three coordinate axes and may rotate about any three axes. Therefore, the body is said to have *six degrees of freedom*, and all six degrees must be specified in order to properly locate and orient the body in space and to describe its motion. In Sections 5.12 and 5.13, it will be shown that the general motion of a rigid body in space is difficult to describe.

However, to simplify the discussion, the general motion of the rigid body will be grouped into a subset of five motions that are easy to conceptualize. Each of the five motions will be handled separately, and a mathematical description of each motion will be obtained.

1. Translation   A motion of a rigid body is said to be pure translation if every straight line between any two points on the body maintains the same orientation during movement. Figure 5.1 shows a rigid body moving in pure translation. The path of motion of point A to point A' will be the same as the path of motion of point B to B'. These paths are called the *trajectories* of the points A and B, respectively. If these paths are straight lines, the motion is said to be *rectilinear translation*. If the paths are curved lines in space, the motion is termed *curvilinear translation*. Note that the rigid body behaves essentially as a particle, as its orientation does not change during movement, so only the three coordinates of a point on the rigid body are required to locate the body in space.

In Chapter 1, we studied the motion of one particle relative to another particle. The same concept of relative motion can be used to understand the motion of a rigid body. A vector connecting two points on the body, A and B, is designated as $\mathbf{r}_{B/A}$, or the *position of* B *relative to* A. Schematically, this vector is drawn from A to B, as shown in Figure 5.1. This vector is called the *relative position vector* of B to A. Think of this vector as being glued to the body, running from one point to another. This vector can be formally defined as

$$\mathbf{r}_{B/A} = \mathbf{r}_B - \mathbf{r}_A \tag{5.1}$$

where $\mathbf{r}_A$ and $\mathbf{r}_B$ are the position vectors of points A and B, respectively, in the inertial coordinate system. In the new postion of the rigid body, these position vectors have new values, and the relative position vector is

$$\mathbf{r}'_{B/A} = \mathbf{r}'_B - \mathbf{r}'_A \tag{5.2}$$

For translational motion, either rectilinear or curvilinear, the relative position vector is constant:

$$\mathbf{r}'_{B/A} = \mathbf{r}_{B/A} \tag{5.3}$$

Any relative position vector between two points on a rigid body in pure translation remains constant with time. Note that the position vector of B relative to A moves with

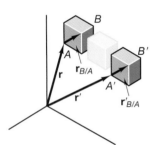

Figure 5.1

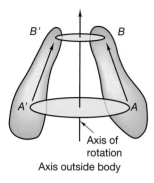

Figure 5.2

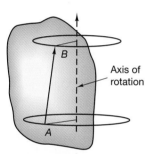

Figure 5.3

the body during translation but does not change in magnitude or direction and, thus, is a constant. It is important to remember that the magnitude of the relative position vector must always remain constant as it joins two points on a rigid body. This fact is true regardless of the type of motion of the rigid body. Since the orientation of the rigid body does not change when the body is in pure translation, a position vector to any point on the body may be used to describe the position of the entire body in space. In general, the position vector to the center of mass of the body is chosen to describe its translation. In Section 5.2, we will show that the body can indeed be treated like a particle if it only translates in space, and therefore, the kinematics equations for a particle are applicable. The body is said to have *three degrees of freedom* in pure translation, one degree of freedom for each coordinate direction.

2. Rotation about a Fixed Axis   The second motion to be examined is one of pure rotation of a rigid body about a fixed axis in space. The axis may pass through the body or be outside the body. Conceptually, the boundaries of the body can be extended so that the axis may be said to be on the body, or the body *extended*. The points on the rigid body move in parallel planes along paths that form circles centered on the axis of rotation during this motion. This motion is illustrated in Figures 5.2 and 5.3. In Figure 5.2, the fixed axis of rotation lies outside of the body, and the body is shown in two positions. The relative position vector $\mathbf{r}_{B/A}$, the position of B relative to A, although constant in magnitude, changes in orientation as it rotates around the axis. If the relative position vector were parallel to the axis of rotation, it would not change in direction, but would translate about the axis of rotation. Figure 5.3 shows a rigid body rotating about an axis that passes through the body. Again, the relative position vector rotates around the axis, changing direction as it rotates. The body may be located in space by specifying the location of a point on the body along its circular path of motion. Therefore, the body is said to have only *one degree of freedom*. In Section 5.3, this change in orientation will be used to describe the rotation of the body. The concepts of *angular velocity* and *angular acceleration* will be introduced in that section.

3. General Plane Motion   If the motion of the body is such that the paths of movement, or the trajectories, of every point on the body lie in parallel planes, the movement is called *general plane motion*. This movement has only *three degrees of freedom*, as the position and orientation of the body may be described by its translation in these parallel planes and its rotation about an axis that is perpendicular to these planes. These planes are called the *planes of motion*. Examples of this type of motion are shown in Figure 5.4. The body will be modeled as two dimensional and having a single plane of motion. Note that the relative position vector between two points on a rigid body, shown by the arrows in Figure 5.4, will change in orientation if there is rotation of the body.

Note that the change in the relative position vector between any two points on the body describes the rotation of the body. This factor will be the basis of the equations of plane motion developed in Sections 5.5 to 5.11.

4. Motion about a Fixed Point   If the motion of the rigid body is such that one point on the body or the body extended remains fixed in space, the body is said to be moving about this fixed point. Since that point on the body remains fixed in space and has no translation, only the three-dimensional orientation of the body needs to be specified, and the body is said to have *three degrees of freedom*. The spinning top, shown in Figure 5.5, has its tip, or base, fixed in space while it moves and is an example of this type of motion. Note that points on the body remain a constant distance from the fixed point and move on

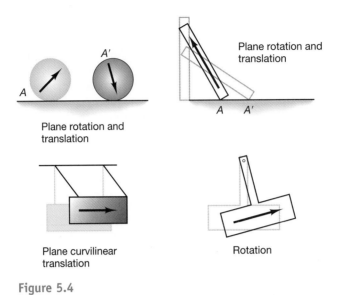

Plane rotation and
translation

Plane rotation and
translation

Plane curvilinear
translation

Rotation

Figure 5.4

Figure 5.5

Figure 5.6

the surface of spheres of radii equal to these constant distances. A relative position vector from this fixed point to another point on the body would change orientation during movement, as may be seen for the vector from the base to point *A* in Figure 5.5. This type of motion will be examined in detail in Sections 5.12 and 5.13.

5. General Motion of a Rigid Body

The most general motion that a rigid body can execute does not fall into any of the previous descriptions. For this kind of motion, the body can freely translate in any direction and assume any orientation. An example of such motion is shown in Figure 5.6. Note that any relative position vector between any two points on the body would remain constant in magnitude and can only change in orientation. The vector shown in back of the wings of the airplane in Figure 5.6 seems shorter in the second position, but this appearance is only due to the orientation of the plane. Going from the first position to the second position, the airplane has translated in three directions and has rotated in pitch (tipped down), roll (rotation about the long axis of the plane), and yaw (rotation about an axis perpendicular to the horizontal plane of the airplane). The rigid body is said to have *six degrees of freedom*. This general motion will be examined in Sections 5.12 and 5.13.

The decision as to whether a body will be treated as a particle or a rigid body depends upon the information desired. For example, imagine the short chip shot hit by a professional golfer, shown in Figure 5.7. The trajectory of the ball may be determined by modeling the ball as a particle, but if the effect of "backspin" is to be understood, the ball must be treated as a rigid body. The *backspin* is very important as the velocity of the air passing over the top is increased and over the bottom is decreased. From the *Bernoulli principle* (fluid mechanics), when the speed increases the pressure decreases. The pressure differential results in a lifting force on the ball, increasing height and distance. The aerodynamics of the golf ball has been extensively studied from the featherie, a leather pouch filled with goose feathers, to the modern golf ball. By 1930, the current golf ball with a rubber thread wound around a rubber core and coated with dimpled enamel was accepted. The dimpled surface decreases the drag on the ball by changing laminar flow

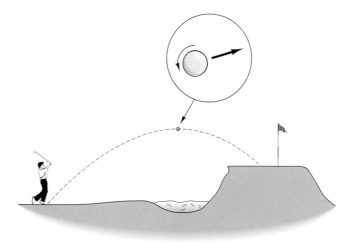

Figure 5.7

over the sphere to turbulent flow. A complete aerodynamic study of the golf ball, including the slice and hook, is beyond the scope of this text.

When plane motion is discussed, the equations of the motion will be applied to linkages, gears, or wheels that are considered to move in a plane. The concept of an *instantaneous center of rotation* will be introduced later in this chapter, and it will be shown how this concept can be used in many cases to simplify the analysis. Certain new topics will be introduced that have applications in experimental motion analysis and biomechanics. The *instantaneous center of relative rotation* between two rigid bodies will be described, and this location can be considered as an instantaneous hinge point.

Finally, Section 5.14 will consider the kinematics of general motion of a rigid body. *Coriolis acceleration* will be discussed, and the motion of a body relative to a rotating frame of reference will be examined. This material may be used to introduce many of the concepts of advanced dynamics for those interested in machine design, flight dynamics, navigation, dynamic systems, biodynamics, and other applications of dynamics. Throughout the chapter, vector mathematics will be emphasized to facilitate solution of problems.

5.2 TRANSLATION OF A RIGID BODY

The simplest motion that a rigid body can have is pure translation, and in this case, the rigid body behaves as a particle. The rigid body shown in Figure 5.8 is moving in pure translation. Note that the trajectories of all points on the body form parallel curves in space. The position of the body in space at any instant of time may be described by specifying a position vector to any point on the body, such as, point A or point B in the figure. If position vectors to both of these points are specified, then the relative position vector from point A to point B can be obtained as described in Section 5.1:

$$\mathbf{r}_B = \mathbf{r}_A + \mathbf{r}_{B/A} \tag{5.4}$$

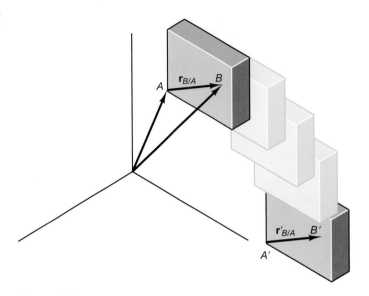

Figure 5.8

For translation, the orientation of the position vector of B relative to A remains constant, as does its magnitude. Therefore, the differentiation of the relative position vector with time is zero, and differentiation of $\mathbf{r}_B$ shows that the velocity and acceleration are

$$\begin{aligned} \mathbf{v}_B &= \mathbf{v}_A \\ \mathbf{a}_B &= \mathbf{a}_A \end{aligned} \tag{5.5}$$

Thus, the velocities and accelerations of point A and point B must be equal. Since A and B may be any points on the rigid body, then, *in pure translation, the velocities and accelerations of all points on the rigid body are equal.* To fully describe the dynamic state of the rigid body, the three components of the position vector, the velocity vector, and the acceleration vector are needed. We previously studied this motion when we examined the dynamics of a particle. As in that case, the object has three degrees of freedom—that is, translation in the three coordinate directions.

If the body moves in rectilinear translation, all points in the body move in parallel straight lines and the direction of the velocity and the acceleration vectors will remain constant, but their magnitudes will, in general, change. In the case of curvilinear translation, both the magnitude and the direction of the velocity and acceleration vectors will change.

5.3 ROTATION ABOUT A FIXED AXIS

Consider a rigid body that rotates about a fixed axis in space, as illustrated in Figure 5.9. The z-axis may be chosen to be the fixed axis about which the body rotates. This axis is called the axis of rotation. Consider the case shown in Figure 5.9, in which the rotation is positive (the right-hand rule). The coordinate system is chosen as shown in the figure, and point B on the body moves in a circular motion around the axis of rotation. Point A is the projection of the vector $\mathbf{r}$ onto the axis of rotation. Point A is then the center of the circle described by the motion of B. The position of B at any time may be described by specifying an angle θ from some initial position. The change in θ, $d\theta$, is shown in Figure 5.9. If the body moves through one complete revolution, the change in this angle would be

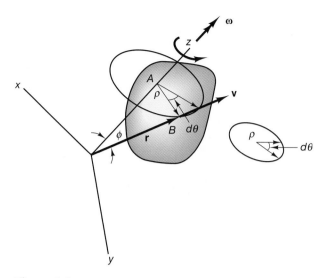

Figure 5.9

described as 1 revolution, or 360 degrees. Recall that the scientific measurement for one complete revolution is 2π radians, so 1 revolution = 2π rad = $360°$.

The change in position of point B may be seen as it moves along the circle centered at A with radius $\rho = |\mathbf{r}_B| \sin \phi$, where $\mathbf{r}_B$ is a vector from the origin on the axis of rotation to point B. The change in position of B is written as

$$ds = \rho d\theta \qquad (5.6)$$

where ds is the arc length through which B moves with the change $d\theta$. Since ρ does not change as point B moves, the magnitude of the velocity is

$$|\mathbf{v}| = \frac{ds}{dt} = \rho \frac{d\theta}{dt} \qquad (5.7)$$

The direction of the velocity vector is tangent to the path of motion, or tangent to the circle of radius ρ.

If we introduce an angular velocity vector $\boldsymbol{\omega}$ having magnitude of $\dot{\theta}$ and direction along the axis of rotation, the magnitude of the velocity is written as

$$|\mathbf{v}| = |\boldsymbol{\omega}||\mathbf{r}| \sin \phi \qquad (5.8)$$

The sense of the angular velocity vector is determined by the right-hand rule. For example, the angular velocity might appear as shown in Figure 5.10. Note that the velocity vector is perpendicular to the plane formed by the angular velocity vector and the position vector. The angular velocity and the position vectors may be used to define a plane in space, and the vector product between them is the linear velocity vector.

The following relations may now be written:

$$\boldsymbol{\omega} = \dot{\theta}\hat{\mathbf{k}}$$
$$\mathbf{v} = \boldsymbol{\omega} \times \mathbf{r} \qquad (5.9)$$

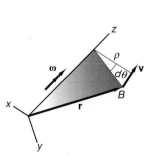

Figure 5.10

Therefore, the velocity of the point B on the body is defined as the vector product, or the cross product, of the angular velocity vector with a vector from any point on the axis of rotation to the point B. Next, the acceleration of the point B may be obtained by differentiation of the velocity vector. The product on the right side of the velocity vector equation must be formally differentiated:

$$\mathbf{a} = \frac{d}{dt}(\boldsymbol{\omega} \times \mathbf{r}) = \dot{\boldsymbol{\omega}} \times \mathbf{r} + \boldsymbol{\omega} \times \dot{\mathbf{r}}$$

But,

$$\mathbf{v} = \dot{\mathbf{r}} = \boldsymbol{\omega} \times \mathbf{r} \tag{5.10}$$

and if an *angular acceleration* vector $\boldsymbol{\alpha} = \dot{\boldsymbol{\omega}}$ is introduced, the linear acceleration of the point B may be written as

$$\mathbf{a} = \boldsymbol{\alpha} \times \mathbf{r} + \boldsymbol{\omega} \times \mathbf{v}$$
$$\mathbf{a} = \boldsymbol{\alpha} \times \mathbf{r} + \boldsymbol{\omega} \times (\boldsymbol{\omega} \times \mathbf{r}) \tag{5.11}$$

Note that for the case shown, the axis of rotation is fixed in space and coincides with the z-axis, so the angular acceleration vector is parallel to the angular velocity vector, and both vectors are in the z-direction. Therefore, only the magnitudes of $\boldsymbol{\omega}$ and $\boldsymbol{\alpha}$ change with time.

Now consider the directions of the terms of the various components of the velocity and acceleration vectors. By examining the vectors shown in Figure 5.11, we can see that the velocity vector is perpendicular to the plane formed by the position vector and the angular velocity vector. The first term in the acceleration equation, $\boldsymbol{\alpha} \times \mathbf{r}$, is parallel to the velocity vector and is tangent to the circular path of motion. This term is called the *tangential acceleration*, is denoted by

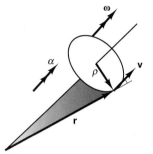

Figure 5.11

$$\mathbf{a}_t = \boldsymbol{\alpha} \times \mathbf{r} \tag{5.12}$$

and is parallel to the velocity vector. Note that this acceleration only changes the magnitude of the velocity, and not its direction. This acceleration is similar to the tangential acceleration discussed Chapter 1 and, therefore, is called the tangential acceleration. Note that if there is no angular acceleration, there will be no tangential acceleration and the magnitude of the velocity will remain constant and the velocity vector will change in direction only. Now consider the second term in the acceleration equation:

$$\mathbf{a}_n = \boldsymbol{\omega} \times \mathbf{v} = \boldsymbol{\omega} \times (\boldsymbol{\omega} \times \mathbf{r}) \tag{5.13}$$

This acceleration vector is perpendicular to both the angular velocity and the linear velocity vectors and, therefore, is normal to the circular path of motion directed inward along the vector $-\boldsymbol{\rho}$. This term is called the *normal acceleration*. The normal acceleration produces the change in direction of the velocity vector necessary for the point on the body to move in a circular motion about the fixed axis of rotation. Therefore, even if the body is rotating at a constant angular velocity about a fixed axis of rotation, every point on the body that does not lie on the axis of rotation will have an acceleration. Note that the normal acceleration vector has been written as a triple vector product, and the triple vector product identity can be used to reduce it to the sum of two vectors:

— in the $\boldsymbol{\omega}$-direction

$$\boldsymbol{\omega} \times (\boldsymbol{\omega} \times \mathbf{r}) = \boldsymbol{\omega}(\boldsymbol{\omega} \cdot \mathbf{r}) - \mathbf{r}(\boldsymbol{\omega} \cdot \boldsymbol{\omega}) \tag{5.14}$$

—in the $-\mathbf{r}$-direction

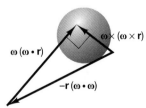

Figure 5.12

The sum of these two vectors lies in the $-\boldsymbol{\rho}$ direction, as shown in Figure 5.12. The normal acceleration may also be written as

$$\boldsymbol{\omega} \times (\boldsymbol{\omega} \times \mathbf{r}) = -(\boldsymbol{\omega} \cdot \boldsymbol{\omega})\boldsymbol{\rho} \tag{5.15}$$

A unit vector along $\boldsymbol{\omega}$ will be designated as

$$\hat{\mathbf{e}}_\omega = \frac{\boldsymbol{\omega}}{|\boldsymbol{\omega}|} \tag{5.16}$$

The vector $\boldsymbol{\rho}$ is the component of $\mathbf{r}$ which is perpendicular to $\boldsymbol{\omega}$ and may be written as:

$$\boldsymbol{\rho} = \mathbf{r} - (\hat{\mathbf{e}}_\omega \cdot \mathbf{r})\hat{\mathbf{e}}_\omega = \mathbf{r} - \frac{(\boldsymbol{\omega} \cdot \mathbf{r})\boldsymbol{\omega}}{\boldsymbol{\omega} \cdot \boldsymbol{\omega}} \tag{5.17}$$

Multiplying Eq. (5.17) by $(-\omega^2)$ yields

$$-(\boldsymbol{\omega} \cdot \boldsymbol{\omega})\boldsymbol{\rho} = (\boldsymbol{\omega} \cdot \mathbf{r})\boldsymbol{\omega} - \mathbf{r}(\boldsymbol{\omega} \cdot \boldsymbol{\omega}) \tag{5.18}$$

where $\boldsymbol{\omega} \cdot \boldsymbol{\omega} = \omega^2$.

Comparison of the right sides of Eqs. (5.14) and (5.18) completes the proof of the identity in Eq. (5.15).

This form of the normal acceleration will be useful when plane motion is examined in detail.

Sample Problem 5.1

Consider a ball that is spinning about the z-axis at a constant angular velocity of 2 rad/s, as shown in the accompanying diagram. Determine the linear velocity and acceleration of point A on the ball, and determine the angle that a position vector to A makes with the z-axis.

Solution

A coordinate system has been suggested in the figure, and the axis of rotation is the z-axis. Note that the angular velocity is constant, so the angular acceleration α, is zero. A position vector from the origin to point A is

$$\mathbf{r} = 4\hat{\mathbf{i}} + 3\hat{\mathbf{j}} + 5\hat{\mathbf{k}} \text{ in}$$

The linear velocity of point A is

$$\mathbf{v} = \boldsymbol{\omega} \times \mathbf{r}$$

where

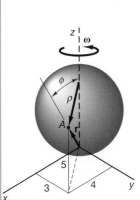

$$\boldsymbol{\omega} = 2\hat{\mathbf{k}}$$
$$\mathbf{v} = 2\hat{\mathbf{k}} \times (4\hat{\mathbf{i}} + 3\hat{\mathbf{j}} + 5\hat{\mathbf{k}})$$
$$= -6\hat{\mathbf{i}} + 8\hat{\mathbf{j}} \text{ in/s}$$

The magnitude of $\mathbf{v}$ is $|\mathbf{v}| = \sqrt{6^2 + 8^2} = 10$ in/s.

Since $\boldsymbol{\alpha} = 0$, the linear acceleration reduces to the normal component—that is,

$$\mathbf{a} = \mathbf{a}_n = \boldsymbol{\omega} \times \mathbf{v}$$
$$= 2\hat{\mathbf{k}} \times (-6\hat{\mathbf{i}} + 8\hat{\mathbf{j}}) = -(16\hat{\mathbf{i}} + 12\hat{\mathbf{j}}) \text{ in/s}^2$$

The magnitude of the acceleration is $|\mathbf{a}| = \sqrt{16^2 + 12^2} = 20$ in/s². The vector $\boldsymbol{\rho}$ from the z-axis to point A is

$$\boldsymbol{\rho} = 4\hat{\mathbf{i}} + 3\hat{\mathbf{j}}$$

and

$$\mathbf{a}_n = -(\boldsymbol{\omega} \cdot \boldsymbol{\omega})\boldsymbol{\rho} = -4(4\hat{\mathbf{i}} + 3\hat{\mathbf{j}}) = -(16\hat{\mathbf{i}} + 12\hat{\mathbf{j}}) \text{ in/s}^2$$

This answer agrees with the previous solution.

The angle that **r** makes with the axis of rotation can be found by taking the dot product of **r** with a unit vector in the direction of the axis of rotation—$\hat{\mathbf{k}}$ in this case—and dividing by the magnitude of **r**:

$$\cos \varphi = \frac{\mathbf{r} \cdot \hat{\mathbf{k}}}{|\mathbf{r}|} = \frac{5}{\sqrt{4^2 + 3^2 + 5^2}} = 0.707$$

$$\varphi = 45°$$

5.4 PLANAR PURE ROTATION ABOUT AN AXIS PERPENDICULAR TO THE PLANE OF MOTION

Consider the body shown in Figure 5.13, rotating about an axis in the *z*-direction and passing through the origin. This figure may be considered to be a slab taken through the body that was shown in Section 5.3 or a thin body with origin taken to coincide with the axis of rotation. If the angular velocity of the body is **ω**, the linear velocity of a point on the body located by the position vector **r** may be written using Eq. (5.9):

$$\mathbf{v} = \boldsymbol{\omega} \times \mathbf{r} \tag{5.19}$$

For plane motion of this type, the position vector for any point on the slab is such that **r** = **ρ** and may be written as

$$\mathbf{r} = x\hat{\mathbf{i}} + y\hat{\mathbf{j}} \tag{5.20}$$

The angular velocity has only a component in the *z*-direction and may be written as

$$\boldsymbol{\omega} = \omega\hat{\mathbf{k}} \tag{5.21}$$

Therefore, **r** and **ω** are perpendicular to each other, and the magnitude of **v** is

$$|\mathbf{v}| = |\boldsymbol{\omega} \times \mathbf{r}| = r\dot{\theta} = r\omega \tag{5.22}$$

For a rigid body rotating about a fixed axis, the acceleration can be written as

$$\mathbf{a} = \boldsymbol{\alpha} \times \mathbf{r} + \boldsymbol{\omega} \times (\boldsymbol{\omega} \times \mathbf{r}) \tag{5.23}$$

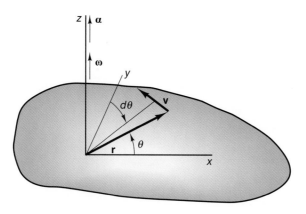

Figure 5.13

The angular acceleration has only a z-component and is written as

$$\boldsymbol{\alpha} = \alpha\hat{\mathbf{k}} \tag{5.24}$$

Using the triple vector product, the linear acceleration becomes

$$\mathbf{a} = \alpha\hat{\mathbf{k}} \times \mathbf{r} + \omega\hat{\mathbf{k}}(\omega\hat{\mathbf{k}} \cdot \mathbf{r}) - \mathbf{r}(\omega\hat{\mathbf{k}} \cdot \omega\hat{\mathbf{k}}) \tag{5.25}$$

Since $\mathbf{r}$ is in the xy-plane and $\boldsymbol{\omega}$ is in the z-direction, these two vectors are perpendicular, and the second term in Eq. (5.25) is zero. Therefore, the acceleration is

$$\mathbf{a} = \alpha\hat{\mathbf{k}} \times \mathbf{r} - \omega^2\mathbf{r} \tag{5.26}$$

As before, the first term in the acceleration equation, the tangential acceleration, is parallel to the velocity vector and is equal to the change in magnitude of the velocity vector. The second term is directed inward, parallel to the position vector, and is the normal acceleration, changing only the direction of the velocity vector. Therefore, for rotation in plane motion about a fixed axis perpendicular to the plane,

$$\begin{aligned}\mathbf{a}_t &= \alpha\hat{\mathbf{k}} \times \mathbf{r} \\ \mathbf{a}_n &= -\omega^2\mathbf{r}\end{aligned} \tag{5.27}$$

This expression could be proven directly, as $\mathbf{r} = \boldsymbol{\rho}$ in this case. Note that since the magnitude of the normal acceleration is $r\omega^2$, the magnitude can also be written as v^2/r. The magnitude of the total acceleration is the square root of the sum of the squares of its components and is

$$|\mathbf{a}| = \sqrt{a_t^2 + a_n^2} \tag{5.28}$$

These accelerations are shown in the plane in Figure 5.14.

The angular velocity and the angular acceleration can be written as the time rate of change of the angle between the position vector and the x-axis. If the angle θ is a function of time, the complete motion of the body may be described in terms of this angle and its derivatives:

$$\begin{aligned}\omega &= \frac{d\theta}{dt} \\ \alpha &= \frac{d\omega}{dt} = \frac{d^2\theta}{dt^2}\end{aligned} \tag{5.29}$$

In the most general case of pure planar rotation, the angular acceleration would be determined as a function of angular position, angular velocity, and time. This case is analogous to the rectilinear motion of a particle discussed in Section 1.3, and the angular acceleration is written as

$$\begin{aligned}\alpha &= f(\omega, \theta, t) \\ \ddot{\theta} &= f(\dot{\theta}, \theta, t)\end{aligned} \tag{5.30}$$

The equation for the acceleration is a second-order ordinary differential equation that may be linear or nonlinear. If the equation is linear and has constant coefficients, the homogeneous part can be obtained by use of an integrating factor:

$$a\frac{d^2\theta}{dt^2} + b\frac{d\theta}{dt} + c\theta = 0 \tag{5.31}$$

The integrating factor is

$$\theta = e^{\lambda t} \tag{5.32}$$

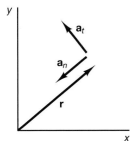

Figure 5.14

yielding the characteristic equation for λ as

$$a\lambda^2 + b\lambda + c = 0$$

$$\lambda = \frac{-b \pm \sqrt{b^2 - 4ac}}{2a} \tag{5.33}$$

The two values of λ may be real or complex, leading to solutions in terms of exponentials or trigonometric functions.

Eq. (5.30) can be nonlinear, but computational tools permit numerical integration of this nonlinear equation to obtain angular position as a function of time. Solution of this type of problem was impossible before the invention of computers, but now these problems commonly are solved through use of computer applications, as shown in Chapter 1.

If the angular acceleration is a function of time only, the angular velocity and the angular position can be obtained by direct integration:

$$\frac{d^2\theta}{dt^2} = f(t)$$

$$\frac{d\theta}{dt}(t) = \int f(t)dt + C_1 \tag{5.34}$$

$$\theta(t) = \int \left[\int f(t)dt + C_1 \right] dt + C_2$$

The two constants of integration are related to the initial angular velocity and the initial angular position. Eq. (5.34) is analogous to Eqs. (1.13) to (1.16) for rectilinear motion of a particle. There is also an analogy to the two special cases of rectilinear translation using Eqs. (1.17) to (1.23).

1. In the case when the angular acceleration is zero, the motion is called uniform rotation, and

$$\alpha = \frac{d^2\theta}{dt^2} = 0$$

In this case, the angular velocity must be constant and is denoted by ω_0:

$$\frac{d\theta}{dt} = \omega_0$$
$$\theta = \omega_0 t + \theta_0 \tag{5.35}$$

where θ_0 is the initial angle.

2. If the angular acceleration is a constant, the motion is called uniformly accelerated rotation. The constant angular acceleration is denoted by α_0:

$$\frac{d^2\theta}{dt^2} = \alpha_0$$

Integrating with respect to time yields

$$\omega = \alpha_0 t + \omega_0 \tag{5.36a}$$

where ω_0 is the initial angular velocity. Integrating again yields an expression for the angular position at any time. θ_0 is the initial angular position at time zero, so

$$\theta = \alpha_0 \frac{t^2}{2} + \omega_0 t + \theta_0 \tag{5.36b}$$

Note that these equations are applicable for a limited and restricted type of motion and thus should be used carefully. These special types of motion do occur in many problems, but they still represent special cases. For most planar pure rotation, the general Eq. (5.30) should be used.

If the angular acceleration is given as a function of the angular position only, the differential equation is separable. Using the chain rule of differentiation, we may write the angular acceleration in terms of ω and θ only, eliminating explicit dependency upon time.

$$\alpha = \frac{d\omega}{dt} = \frac{d\omega}{d\theta}\frac{d\theta}{dt} = \omega\frac{d\omega}{d\theta} \tag{5.37}$$

Separation of variables yields

$$\alpha(\theta)d\theta = \omega d\omega$$
$$\int_{\theta_0}^{\theta} \alpha(\theta)d\theta = \int_{\omega_0}^{\omega} \omega d\omega = \frac{1}{2}(\omega^2 - \omega_0^2) \tag{5.38}$$

The position–time relation can be obtained from

$$\omega = \frac{d\theta}{dt}$$
$$dt = \frac{d\theta}{\omega(\theta)} \tag{5.39}$$
$$\int_0^t dt = t = \int_{\theta_0}^{\theta} \frac{d\theta}{\omega(\theta)}$$

These integrals can sometimes be evaluated in closed form, but if necessary, they can be obtained by numerical integration using available commercial software.

For example, if the acceleration is a constant, it can be considered a function of angular position, and we may use Eq. (5.38):

$$\int_{\theta_0}^{\theta} \alpha_0 d\theta = \int_{\omega_0}^{\omega} \omega d\omega$$

Integrating both sides, where ω_0 and θ_0 are the initial angular velocity and position, respectively,

$$\alpha_0(\theta - \theta_0) = \frac{1}{2}(\omega^2 - \omega_0^2) \tag{5.40}$$

This equation may be rearranged to yield

$$\omega^2 = \omega_0^2 + 2\alpha_0(\theta - \theta_0)$$

This expression is another form of the equation of uniformly accelerated rotation given in Eq. (5.36).

The second case for which Eq. (5.30) is separable occurs when the angular acceleration is given as a function of angular velocity only. Again, this situation is analogous to the rectilinear motion of a particle in Eqs. (1.21) to (1.23):

$$\alpha(\omega) = \omega\frac{d\omega}{d\theta}$$
$$d\theta = \frac{\omega d\omega}{\alpha(\omega)} \tag{5.41}$$
$$\int_{\theta_0}^{\theta} d\theta = \int_{\omega_0}^{\omega} \frac{\omega d\omega}{\alpha(\omega)}$$

The relationship between time and the angular velocity is

$$\alpha(\omega) = \frac{d\omega}{dt}$$

$$dt = \frac{d\omega}{\alpha(\omega)} \tag{5.42}$$

$$\int_0^t dt = t = \int_{\omega_0}^{\omega} \frac{d\omega}{\alpha(\omega)}$$

In this case, angular acceleration, angular position and time are expressed as functions of angular velocity, but these algebraic expressions may be inverted to obtain angular position as a function of time.

5.4.1 VECTOR RELATIONS FOR ROTATION IN A PLANE

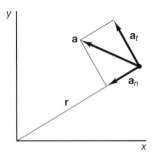

Figure 5.15 Normal and Tangential Accelerations

The acceleration vector can be considered to be composed of two components: the normal acceleration and the tangential acceleration. The relationship between these two components and the position vector is shown in Figure 5.15. Noting the directions of these two components, we may separate the acceleration vector into the normal and tangential components by use of a *triad* of unit vectors. A unit vector may be formed in the **r**-direction by dividing **r** by its magnitude:

$$\hat{\mathbf{e}}_r = \frac{\mathbf{r}}{|\mathbf{r}|} \tag{5.43}$$

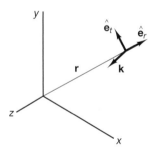

Figure 5.16

The axis of rotation has been selected to be the *z*-axis, so the unit vector in the direction perpendicular to the plane of rotation is $\hat{\mathbf{k}}$. The unit vector in the tangential direction is easily formed by the cross product:

$$\hat{\mathbf{e}}_t = \hat{\mathbf{k}} \times \hat{\mathbf{e}}_r \tag{5.44}$$

The spatial relationship of these three unit vectors is shown in Figure 5.16. These unit vectors are similar to those used when normal and tangential coordinates were discussed in Chapter 1 and may be seen to be identical to a set of unit vectors in cylindrical coordinates.

If the acceleration vector of a point on the rotating body is known, the vector may be separated into its components, and the angular velocity and acceleration may be determined from

$$\begin{aligned} a_t &= \hat{\mathbf{e}}_t \cdot \mathbf{a} = r\alpha \\ a_r &= \hat{\mathbf{e}}_r \cdot \mathbf{a} = -r\omega^2 \end{aligned} \tag{5.45}$$

The analysis can be done without consideration of the velocity in this manner. However, only the square of the angular velocity can be obtained, and not the direction of the rotation.

5.4.2 DIRECT VECTOR SOLUTION FOR ANGULAR VELOCITY AND ANGULAR ACCELERATION

Occasionally, the situation arises where the velocities and accelerations of two points on a rigid body are known and we wish to determine the angular velocity and the angular acceleration of the body. Usually, the linear velocities and accelerations are measured experimentally. In this case, the vector equations

$$\boldsymbol{\omega} \times \mathbf{r} = \mathbf{v}$$
$$\boldsymbol{\alpha} \times \mathbf{r} + \boldsymbol{\omega} \times (\boldsymbol{\omega} \times \mathbf{r}) = \mathbf{a}$$

can be solved for $\boldsymbol{\omega}$ and $\boldsymbol{\alpha}$. (*This type of solution is developed in Section 5.13 for general motion of a rigid body.*) There is no definition for vector division, and these equations appear to have no method of solution except to expand them into scalar components. However, the use of same characteristics of vector algebra leads to a *direct vector solution*. Let us formulate the problem mathematically. The following vectors are known for points A and B on a rigid body; $\mathbf{r}_A$, $\mathbf{r}_B$, $\mathbf{v}_A$, $\mathbf{v}_B$, $\mathbf{a}_A$ and $\mathbf{a}_B$. Consider first the two equations for the velocities:

$$\boldsymbol{\omega} \times \mathbf{r}_A = \mathbf{v}_A$$
$$\boldsymbol{\omega} \times \mathbf{r}_B = \mathbf{v}_B$$

Take the cross product of the first equation with the second, yielding

$$\mathbf{v}_A \times (\boldsymbol{\omega} \times \mathbf{r}_B) = \mathbf{v}_A \times \mathbf{v}_B$$

Now, consider the vector identity for the triple vector product:

$$\mathbf{A} \times (\mathbf{B} \times \mathbf{C}) = \mathbf{B}(\mathbf{C} \cdot \mathbf{A}) - \mathbf{C}(\mathbf{A} \cdot \mathbf{B})$$

Therefore,

$$\boldsymbol{\omega}(\mathbf{v}_A \cdot \mathbf{r}_B) - \mathbf{r}_B[(\boldsymbol{\omega} \times \mathbf{r}_A) \cdot \boldsymbol{\omega}] = \mathbf{v}_A \times \mathbf{v}_B$$

The second term is zero as $\boldsymbol{\omega}$ is perpendicular to $(\boldsymbol{\omega} \times \mathbf{r}_A)$ and

$$\boldsymbol{\omega} = \frac{\mathbf{v}_A \times \mathbf{v}_B}{\mathbf{v}_A \cdot \mathbf{r}_B}$$

We will rewrite the acceleration equation by identifying a vector $\mathbf{q}$ to be

$$\mathbf{q} = \mathbf{a} - \boldsymbol{\omega} \times (\boldsymbol{\omega} \times \mathbf{r})$$

The acceleration equations may now be written as

$$\boldsymbol{\alpha} \times \mathbf{r}_A = \mathbf{q}_A$$
$$\boldsymbol{\alpha} \times \mathbf{r}_B = \mathbf{q}_B$$

The solution is identical to that used for $\boldsymbol{\omega}$ and the angular acceleration is

$$\boldsymbol{\alpha} = \frac{\mathbf{q}_A \times \mathbf{q}_B}{\mathbf{q}_A \cdot \mathbf{q}_B}$$

5.4.3 CONSTRAINTS TO THE MOTION

When a body is rotating about a fixed axis, the body is obviously constrained to rotate about that axis, but additional constraints to the body also may exist and must be considered in the analysis. The two gears shown in Figure 5.17 are constrained to move about fixed axes through their centers. If the gears are not engaged with each other, as shown in Figure 5.17(a), each gear is free to rotate independently about its own axis. However, in the case shown in Figure 5.17(b), the gears are engaged with each other, and the motions are not independent, and the rotations are coupled. The linear velocity of the contacting point C must be the same when this point is considered to be on gear A and when it is considered to be on gear B. This constraint may be expressed mathematically as

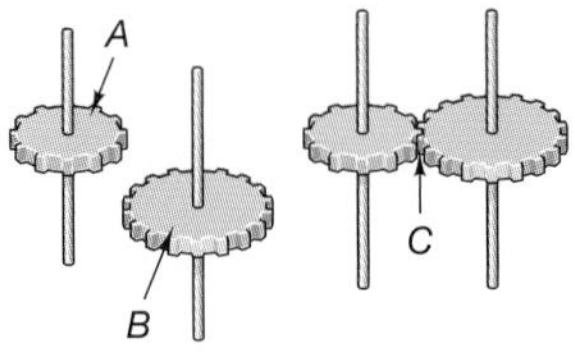

$$\mathbf{v}_C = \boldsymbol{\omega}_A \times \mathbf{r}_{C/A} = \boldsymbol{\omega}_B \times \mathbf{r}_{C/B} \tag{5.46}$$

This relationship can be written in scalar notation as

$$\omega_A r_{C/A} = -\omega_B r_{C/B} \tag{5.47}$$

Figure 5.17

Note that even if the gears have equal radii, their angular velocities are in opposite directions; that is, if gear A were rotating counterclockwise, gear B would be rotating clockwise. This situation is to be expected because the relative position vectors from the gear centers to the point of contact are collinear, but in opposite directions. If the radii of the gears are not equal, their angular velocities are not equal in magnitude. This orientation, of course, is the principal use of gears, and the *gear ratio* is equal to the ratio of the radii of the gears.

The angular accelerations of the two gears also are coupled, but this coupling is a little more difficult to formulate mathematically. If point C is considered as a point on gear A, its linear acceleration is

$$\mathbf{a}_C = \boldsymbol{\alpha}_A \times \mathbf{r}_{C/A} - (\omega_A)^2 \, \mathbf{r}_{C/A} \tag{5.48}$$

As before, the first term is the tangential acceleration and changes the magnitude of the linear velocity, while the second term (the normal acceleration) changes the direction of the linear velocity vector. If point C is considered to be a point on gear B, the tangential component of the linear acceleration must equal the tangential component of the linear acceleration when C is considered to be a point on gear A. The normal acceleration components are each directed toward their respective centers and are not equal (i.e., they point in opposite directions). The mathematical representation of this constraint is similar to that for the angular velocities and is written

$$\boldsymbol{\alpha}_A \times \mathbf{r}_{C/A} = \boldsymbol{\alpha}_B \times \mathbf{r}_{C/B} \tag{5.49}$$

Other constraints occur when belts or cables are placed over pulleys, and the belt or cable does not slip. The linear velocity of the contacting point on the pulley must equal the velocity of the belt or cable, and the tangential acceleration of this contacting point must equal the linear acceleration of the belt or cable. In this manner, the motion of different pulleys in a system may be related, and these constraints may be used to understand the motion of a variety of devices, both machines and mechanisms.

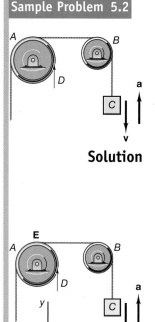

Sample Problem 5.2

In the accompanying diagram, pulley A has a radius of 600 mm, and pulley B has a radius of 400 mm. An inextensible cable rotates both pulleys without slipping as the block C is moved. If the initial velocity of C is 1000 mm/s downward and the block is accelerated at a constant rate of 100 mm/s² in the direction shown in the figure, determine (a) the initial acceleration of point D on pulley A, (b) the acceleration of point D after 8 s, and (c) the number of revolutions that pulley A will turn in 10 s. *Note: Point D is not a point fixed to the pulley but a point in space that corresponds to a point on the pulley at any instant of time.*

Solution   Establish a coordinate system with the x-axis to the right and the y-axis up, as shown in the accompanying figure.

Since the cable is inextensible, the initial linear velocity of point E on pulley A will be

$$\mathbf{v}_E^0 = 1000\hat{\mathbf{i}} \text{ mm/s}$$

The relative position vector from the center of A to E is

$$\mathbf{r}_{E/A} = 600\hat{\mathbf{j}} \text{ mm}$$

The initial angular velocity of A is

$$\boldsymbol{\omega}_A^0 = \omega_A^0 \hat{\mathbf{k}}$$

$$\boldsymbol{\omega}_A^0 \times \mathbf{r}_{E/A} = \mathbf{v}_E^0 \qquad \omega_A^0 \hat{\mathbf{k}} \times 600\hat{\mathbf{j}} = 1000\hat{\mathbf{i}}$$

$$\omega_A^0 = -1.667 \text{ rad/s}$$

The tangential acceleration of point E is

$$\mathbf{a}_{E_t} = -100\hat{\mathbf{i}} \text{ mm/s}^2$$

The angular acceleration of A is a constant and is

$$\boldsymbol{\alpha}_A = \alpha_A \hat{\mathbf{k}}$$

and

$$\boldsymbol{\alpha}_A \times \mathbf{r}_{E/A} = \mathbf{a}_{E_t}$$

$$\alpha_A = 0.167 \text{ rad/s}^2$$

a) The initial acceleration of point D is

$$\mathbf{a}_D^0 = \boldsymbol{\alpha}_A \times \mathbf{r}_{D/A} + \boldsymbol{\omega}_A^0 \times (\boldsymbol{\omega}_A^0 \times \mathbf{r}_{D/A})$$

$$= 0.167\hat{\mathbf{k}} \times 600\hat{\mathbf{i}} + (-1.667\hat{\mathbf{k}}) \times (-1.667\hat{\mathbf{k}} \times 600\hat{\mathbf{i}})$$

$$= 100\hat{\mathbf{j}} - 1667\hat{\mathbf{i}} \text{ mm/s}^2$$

The magnitude of the acceleration is

$$|\mathbf{a}_D^0| = \sqrt{100^2 + 1667^2} = 1670 \text{ mm/s}^2$$

b) The acceleration of point D after 8 s may be determined after the angular velocity of the pulley is determined after 8 s:

$$\mathbf{a}_D(t) = \boldsymbol{\alpha}_A \times \mathbf{r}_{D/A} + \boldsymbol{\omega}_A(t) \times [\boldsymbol{\omega}_A(t) \times \mathbf{r}_{D/A}]$$

For constant angular acceleration,

$$\omega_A(t) = \alpha_A t + \omega_A^0$$

$$\omega_A(8) = 0.167 \times 8 - 1.667 = -0.331 \text{ rad/s}$$

$$\mathbf{a}_D = 0.167\hat{\mathbf{k}} \times 600\hat{\mathbf{i}} - 0.331\hat{\mathbf{k}} \times (-0.331\hat{\mathbf{k}} \times 600\hat{\mathbf{i}})$$

$$= 100\hat{\mathbf{j}} - 65.77\hat{\mathbf{i}} \text{ mm/s}^2$$

The magnitude of the acceleration at 8 s is

$$119.7 \text{ mm/s}^2$$

c) The rotation of the pulley may be found by integration of the constant acceleration:

$$\theta(t) = \frac{1}{2}\alpha t^2 + \omega_0 t + \theta_0$$

$$\theta(10) = \frac{1}{2} \times 0.167 \times 10^2 - 1.667 \times 10 + 0 = -8.32 \text{ rad}$$

(the negative indicates counterclockwise rotation)
The number of revolutions is $-8.32/2\pi = -1.32$.

Sample Problem 5.3

In the accompanying diagram, an accelerometer mounted upon a block on a rotating plat-form measures an acceleration of 3 m/s² when the block is 2 m from the axis of rotation. Determine the angular velocity and acceleration of the platform at this instant.

Solution   For the position shown the position vector and the acceleration vector are

$$\mathbf{a} = -3\hat{\mathbf{i}}$$

$$\mathbf{r} = r(\cos\theta\hat{\mathbf{i}} + \sin\theta\hat{\mathbf{j}}) = 1.732\hat{\mathbf{i}} + \hat{\mathbf{j}}$$

The unit vector in the *r*-direction is

$$\hat{\mathbf{e}}_r = (0.866\hat{\mathbf{i}} + 0.5\hat{\mathbf{j}})$$

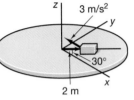

The tangential unit vector is

$$\hat{\mathbf{e}}_t = \hat{\mathbf{k}} \times \hat{\mathbf{e}}_r = -0.5\hat{\mathbf{i}} + 0.866\hat{\mathbf{j}}$$

The tangential acceleration is

$$|\mathbf{a}_t| = \mathbf{a}\cdot\hat{\mathbf{e}}_t = 1.5\ \text{m/s}^2$$

Therefore,

$$r\alpha = 1.5$$
$$\alpha = 0.75\ \text{rad/s}^2$$

The normal acceleration is

$$|\mathbf{a}_n| = \mathbf{a}\cdot\hat{\mathbf{e}}_r = -2.598\ \text{m/s}^2$$

The minus sign indicates that the acceleration is directed inward, opposite to $\hat{\mathbf{e}}_r$. Therefore,

$$r\omega^2 = 2.598$$
$$\omega = \pm1.140\ \text{rad/s}$$

Note that considering the acceleration only allows the determination of ω^2 and gives the magnitude of the angular velocity but not its direction (i.e., whether the platform is rotating in the clockwise or counterclockwise direction at that instant). The sign of the angular acceleration is not determined.

Sample Problem 5.4

The block shown in Sample Problem 5.3 is placed upon the platform when the platform is at rest. The platform is then rotated with a constant angular acceleration of $\alpha_0 = 5$ rad/s². If the block will slip when it is subjected to a total acceleration of 18 m/s², determine the time elapsed before it slips.

Solution

$$\alpha = 5$$
$$\omega = 5t$$

If $r = 2$,

$$a_t = r\alpha = 10$$
$$a_n = r\omega^2 = 50t^2$$

The total acceleration is $a = \sqrt{a_t^2 + a_n^2}$. For slipping to occur,

$$a^2 = (18)^2 = (10)^2 + (50t^2)^2$$
$$t = 0.547\ \text{s}$$

Sample Problem 5.5	A pencil is held at its point and allowed to fall to the horizontal position when released at an angle of 1° with the vertical. It will be shown in Chapter 6 that the angular acceleration is a function of the angle with the vertical and can be determined from

$$\alpha = (3g/2L)\,\sin\theta$$

where g is the gravitational acceleration and L is the length of the pencil. Calculate the angular velocity of the pencil at any position θ and the time it takes for the pencil to reach that position.

Solution   The angular velocity may be obtained from the angular acceleration at any position by

$$\omega\,d\omega = \alpha\,d\theta$$

$$\int \omega\,d\omega = \int \frac{3g}{2L}\sin\theta\,d\theta$$

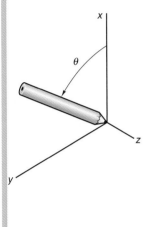

or

$$\frac{\omega^2}{2} = -\frac{3g}{2L}\cos\theta + C_1$$

where C_1 a constant of integration. To evaluate this constant, note that the pencil starts from a position $\theta = 1$ degree to the left and $\cos(1) = 0.99984$, or approximately 1.0, and $\omega = 0$ at this position. Therefore,

$$C_1 = \frac{3g}{2L}$$

Thus,

$$\omega^2 = \frac{3g}{L}(1 - \cos\theta)$$

$$\omega = \sqrt{\frac{3g}{L}(1 - \cos\theta)}$$

where the positive sign is selected, as the pencil is falling to the left. Since $\omega = d\theta/dt$,

$$\frac{d\theta}{dt} = \sqrt{\frac{3g}{L}(1 - \cos\theta)}$$

and using separation of variables and integrating to obtain θ as a function of time,

$$\int \frac{d\theta}{\sqrt{1 - \cos\theta}} = \int \sqrt{\frac{3g}{L}}\,dt$$

$$\sqrt{2}\ln\left(\tan\frac{\theta}{4}\right) = \sqrt{\frac{3g}{L}}\,t + C_2$$

where C_2 is a constant of integration. Removing the natural logarithm by taking the exponential of both sides yields

$$\tan\frac{\theta}{4} = e^{+\left(\sqrt{\frac{3g}{2L}}t + C_2\right)}$$

The initial condition of the problem is that $t = 0$ when $\theta = 1°$, so

$$\tan\left(\frac{1}{4}\right) = 0.004363 = e^{C_2}$$

$$C_2 = -5.353$$

The equation for θ becomes

$$\theta(t) = 4\tan^{-1}\left(e^{\left(\sqrt{\frac{3g}{2L}}t - 5.353\right)}\right)$$

The pencil reaches the horizontal when $\theta = 90° = \frac{\pi}{2}$, so $t = 0.839$ s.

Sample Problem 5.6

A door on the overhead compartment of a passenger airplane is modeled as shown in the accompanying illustration. When the door is released, a spring produces a constant angular acceleration of 2 rad/s². The door is completely open at $\theta = 45°$. Determine the time required to open the door and the angular velocity of the door just before it reaches the full open position. Determine the linear velocity and acceleration of the latch A at this time.

Solution   The coordinate system is established as shown in the figure, and the vector from the origin to point A may be written

$$|\mathbf{r}_{A/0}| = \sqrt{0.6^2 + 0.5^2} = 0.78$$

$$\tan^{-1}(0.6/0.5) = 50.2°$$

$$\mathbf{r}_{A/0} = 0.78\,[\cos(50.2 + \theta)\hat{\mathbf{i}} + \sin(50.2 + \theta)\hat{\mathbf{j}}]$$

$$\boldsymbol{\alpha} = +2\hat{\mathbf{k}}$$

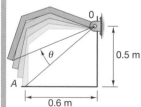

For constant angular acceleration,

$$\theta = \frac{1}{2}\alpha t^2 + \omega_0 t + \theta_0$$

$$\omega_0 = \theta_0 = 0$$

Therefore,

$$t = \sqrt{\frac{2\theta}{\alpha}} = \sqrt{\frac{2(\pi/4)}{2}} = 0.92 \text{ s}$$

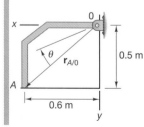

The angular velocity just before the door fully opens is determined by

$$\omega^2 = \omega_0^2 + 2\alpha(\theta - \theta_0)$$

$$\omega = \sqrt{2(2)(\pi/4)} = +1.85 \text{ rad/s}$$

The sign is chosen to agree with the coordinates. The linear velocity and acceleration of A may now be obtained:

$$\mathbf{v}_A = \boldsymbol{\omega} \times \mathbf{r}_{A/0} = (+1.85\hat{\mathbf{k}}) \times 0.78\,(\cos 95.2\hat{\mathbf{i}} + \sin 95.2\hat{\mathbf{j}})$$

$$\mathbf{v}_A = -1.44\hat{\mathbf{i}} + 0.13\hat{\mathbf{j}}\ \text{m/s}$$

$$|\mathbf{v}_A| = 1.45\ \text{m/s}$$

$$\mathbf{a} = \boldsymbol{\alpha} \times \mathbf{r}_{A/0} + \boldsymbol{\omega} \times \mathbf{v}_A =$$

$$(2\hat{\mathbf{k}}) \times 0.78(-0.09\hat{\mathbf{i}} + 0.996\hat{\mathbf{j}}) + (+1.85\hat{\mathbf{k}}) \times (1.44\hat{\mathbf{i}} + 0.13\hat{\mathbf{j}})$$

$$\mathbf{a} = -1.79\hat{\mathbf{i}} + 2.39\hat{\mathbf{j}}\ \text{m/s}^2$$

$$|\mathbf{a}_A| = 2.99\ \text{m/s}^2$$

Sample Problem 5.7

A combination of spur gears and bevel gears is used to change the direction of motion in a gear box, as shown in the accompanying figure. If gear A has an angular velocity of 4 rad/s and an angular acceleration of 2 rad/s^2, determine the output angular velocity and acceleration of shaft D.

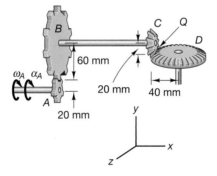

Solution Establish a coordinate system, as shown in the diagram, and designate the point of contact between gears A and B as P and the point of contact between the bevel gears C and D as Q. Considering P as a point on gear A,

$$\mathbf{r}_{P/A} = 20\hat{\mathbf{j}}$$

Considering P as a point on gear B,

$$\mathbf{r}_{P/B} = -60\hat{\mathbf{j}}$$

The constraint to the motion may be written as

$$\mathbf{v}_P = \boldsymbol{\omega}_A \times \mathbf{r}_{P/A} = \boldsymbol{\omega}_B \times \mathbf{r}_{P/B}$$

$$-4\hat{\mathbf{i}} \times 20\hat{\mathbf{j}} = \omega_B\hat{\mathbf{i}} \times (-60\hat{\mathbf{j}})$$

$$\omega_B = 1.333\ \text{rad/s}$$

The bevel gears may be considered in a similar manner. Considering point Q as part of gear C,

$$\mathbf{r}_{Q/C} = -20\hat{\mathbf{j}}$$

considering point Q as part of gear D,

$$\mathbf{r}_{Q/D} = -40\hat{\mathbf{i}}$$

Again, the velocity of point Q is equal when it is considered as part of either gear:

$$1.333\hat{\mathbf{i}} \times (-20\hat{\mathbf{j}}) = \omega_D \hat{\mathbf{j}} \times (-40\hat{\mathbf{i}})$$

$$\omega_D = -0.667\hat{\mathbf{j}} \text{ rad/s}$$

In a similar manner, the constraint on the acceleration is that at each point of contact, the tangential acceleration must be the same for either contacting gear. The angular velocities of A and B may be related as

$$\mathbf{a}_{P_t} = -2\hat{\mathbf{i}} \times 20\hat{\mathbf{j}} = \alpha_B \hat{\mathbf{i}} \times (-60\hat{\mathbf{j}})$$

$$\alpha_B = 0.667 \text{ rad/s}^2$$

Examination of the contact point Q on gears C and D yields

$$\mathbf{a}_{Q_t} = 0.667\hat{\mathbf{i}} \times (-20\hat{\mathbf{j}}) = \alpha_D \hat{\mathbf{j}} \times (-40\hat{\mathbf{i}})$$

$$\alpha_D = -0.333\hat{\mathbf{j}} \text{ rad/s}^2$$

This sample problem illustrates the use of constraints to solve kinematic problems involving gears or pulleys that are connected in a manner such that they roll against each other without slipping.

Sample Problem 5.8

The differential equation of motion for a disk rotating about its axis is given by

$$\alpha + 0.2\,\omega^2 + 4\theta = \sin(t)$$

Determine the angular velocity and angular position as functions of time if the initial conditions are $\theta_0 = 0$ and $\omega_0 = 0.1$.

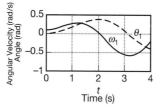

Solution   The angular acceleration is $\alpha(t) = -0.2\omega^2 - 4\theta + \sin(t)$.

This equation is nonlinear and can be solved only by numerical integration. Details of the solution are shown in the Computational Supplement, and the results are shown in the accompanying solution figure. The relationship between the angular velocity and the angle is shown in the graph. Note that the angular velocity is the slope of the angular position curve.

Problems

5.1 A disk accelerates at a constant angular acceleration of 2 rad/s². If the disk starts from rest, (a) determine the angular velocity after the disk has rotated through 2 revolutions, and (b) determine the time required for the disk to rotate through the 2 revolutions.

5.2 The angular position of a variable speed shaft is given by the relationship, $\theta = \theta_0 (1 - e^{-t} \sin \pi t)$, where θ is expressed in radians and the time t is in seconds. If $\theta_0 = 0.5$ rad, determine an expression for the angular velocity and the angular acceleration of the shaft. Plot the position, angular velocity, and angular acceleration for the first 5 s.

5.3 During startup of a prop engine, the propeller rotates through 120 revolutions in 4 s before reaching the constant operating speed, as shown in Figure P5.3. Assuming that the propeller was uniformly accelerated during startup, determine the angular acceleration during startup and the operational angular velocity.

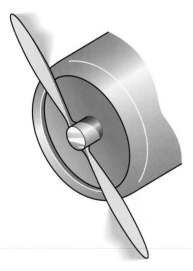

Figure P5.3

5.4 Starting from rest, a disk is rotated in an oil bath. The angular acceleration of the disk is given by $\alpha = A - c\omega$, where $A = 100$ rad/s² and the viscous resistance from the oil is $c = 10/s$ and is proportional to the angular velocity. Determine the terminal angular velocity of the disk.

5.5 A helicopter lands, and its passenger must wait until the rotor stops turning before getting out. The engine friction decelerates the rotor at a constant rate. If 8 s are

required for the rotor to stop and, initially, the rotor was rotating at 1800 rpm, determine the constant deceleration and the number of revolutions that the rotor will turn before it stops.

5.6 In Figure P5.6, a shaft rotates at a constant angular velocity of 50 rad/s counterclockwise, viewed from point B on the shaft. Determine the linear velocity and acceleration of points C and D on the trapezoidal plate welded to the shaft at the position shown in the figure.

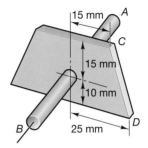

Figure P5.6

5.7 For the shaft in Problem 5.6, determine the linear velocity and acceleration of points C and D if the shaft is rotating at an angular velocity of 50 rad/s and is increasing its angular velocity at a rate of 100 rad/s².

5.8 At time zero, point A on a disk of radius 200 mm is at the position shown in Figure P5.8. The shaft is rotating at a constant angular velocity of 10 rad/s. Write in Cartesian coordinates a general expression for the linear velocity and acceleration of point A at any time. Show that the magnitude of the velocity and acceleration are constant.

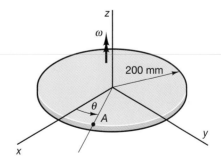

Figure P5.8

5.9 Write the general expressions for the velocity and acceleration of point A on the disk in Problem 5.8 if the shaft has an angular velocity of 10 rad/s at time zero and is accelerating at a rate of 5 rad/s².

5.10 The earth (Figure P5.10) has an equatorial radius of 6378 km and makes one complete rotation every 23.93 hours. If Cape Kennedy, Florida, is at a latitude of 25° north, what is the linear velocity of a shuttle at Cape Kennedy before launch?

Figure P5.10

5.11 Using the data on the earth from Problem 5.10, determine the angular velocity of the earth. Moscow is located at a latitude of 56° north. Determine the linear velocity of a rocket at this site before launch.

5.12 A rigid body rotates about a fixed axis. The linear velocity of point A on the body is $\mathbf{v}_A = 6\hat{\mathbf{i}} + 4\hat{\mathbf{j}} + 3\hat{\mathbf{k}}$ in/s. A position vector from a point on the axis of rotation to this point is $\mathbf{r}_A = 2\hat{\mathbf{i}} - 3\hat{\mathbf{j}}$ in. The linear velocity of a second point B on the body is $\mathbf{v}_B = -16\hat{\mathbf{i}} - 2\hat{\mathbf{j}} + 5\hat{\mathbf{k}}$ in/s, and point B is located $\mathbf{r}_B = 5\hat{\mathbf{j}} + 2\hat{\mathbf{k}}$ in from the axis of rotation. Determine the angular velocity of the body.

5.13 A rigid body rotates at a constant rate of 30 rad/s about a fixed axis through its origin that has direction cosines of (0.200, 0.420, 0.885) with the x-, y-, and z-axes, respectively. Determine the linear velocity and acceleration of a point whose position vector from the origin is $\mathbf{r} = 2\hat{\mathbf{i}} + 6\hat{\mathbf{k}}$, with units in feet.

5.14 The wrench with radius of 15 in. on the wrecker in Figure P5.14 rotates at a constant angular velocity of 4 rad/s. How long will it take to raise the front of the car to a height of 3 ft?

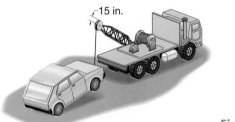

Figure P5.14

5.15 The wrench system used to raise the front end of the car in Problem 5.14 has an angular velocity of 5 rad/s and a constant angular acceleration of 2 rad/s² at the instant shown in Figure P5.14. Under these conditions, how long will it take to raise the front end of the car 3 ft?

5.16 A disk of radius 400 mm starts from rest at $\theta = 90°$ and is accelerated such that the angular acceleration is given by

$$\alpha = 5\sin\theta - 0.2\omega^2 \text{ rad/s}^2$$

Determine the velocity of a point on the edge of the disk during the first 3 s of movement.

5.17 For the disk in Problem 5.16, determine the acceleration of a point on the edge of the disk.

5.18 In Figure P5.18, a bent bar rotates at a constant angular velocity of 3 rad/s about an axis from point A to point B. Determine the magnitude of the velocity and acceleration of point C.

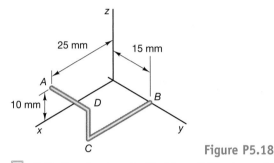

Figure P5.18

5.19 For the bar in Problem 5.18, determine the magnitude of the velocity and acceleration of point D when 3 s after starting from rest if the bar has an angular acceleration of 2 rad/s².

5.20 Pulley B drives pulley A at a constant acceleration of α_B in the speed-reduction assembly shown in Figure P5.20. Determine a general expression for the angular acceleration and the angular velocity of pulley A.

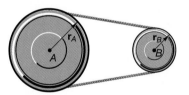

Figure P5.20

5.21 In Figure P5.21, a gear assembly consists of two equal speed reductions. Determine the angular velocity ω_B in terms of ω_A, r_1, and r_2.

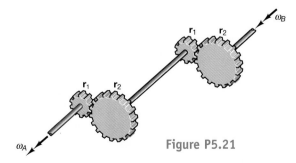

Figure P5.21

5.5 GENERAL PLANE MOTION

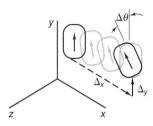

Figure 5.18 General Plane Motion

Let us examine the degrees of freedom and the movements that a rigid body can execute during plane motion. All points on the body translate in planes that are parallel, and the object is modeled as lying in a single plane, the *plane of motion*. The body can translate in two directions within this plane and rotate about an axis that, in general, is not fixed, but which will remain perpendicular to the plane of motion. The body, then, has three degrees of freedom, two translations, and one rotational, as illustrated in Figure 5.18. The general plane motion always may be considered to be the sum of a *plane curvilinear translation and a rotation*. The two different movements that can occur in a plane may be separated and may be considered to be the special cases of plane motion that were discussed previously. First, the body may move in curvilinear translation in the plane but cannot rotate and therefore, would have only two degrees of freedom. In this case, the body may be treated as a particle. The difference between this situation and general plane motion may be seen in Figure 5.19. In Figure 5.19(a), the body exhibits curvilinear translation with trajectories appearing as circles as the body moves from positions 1 to 2 to 3 to 4 and then back to position 1. The body does not rotate, as may be seen by observing the vector from *A* to *B*. This vector is a vector between two points on the rigid body and does not change in magnitude or direction during the movement. This movement, then, satisfies the condition of curvilinear translation within a plane, and the velocity of every point on the body is equal. (In particular, the relative velocity between any two points is zero.)

In Figure 5.19(b), the body appears to have a similar curvilinear motion to that of Figure 5.19(a), but it is clearly rotating, as may be seen by the change in direction of the vector from *A* to *B*. This relative position vector of *B* to *A* is not a constant, and the relative velocity is not zero. Note that the difference between rotation and no rotation depends upon whether the relative velocity vector between the two points is zero or not. If this relative velocity vector is not zero, the velocities of points *A* and *B* are not equal and the body rotates.

Another special case of plane motion occurs when the body rotates about a fixed point in the plane of motion. This situation corresponds to the body rotating about a fixed axis in space, which, in this case, would be parallel to the *z*-axis. This motion has been examined previously, but some of the ideas of the motion will be incorporated in the discussion of general plane motion. In generalized plane motion, the relative position vector between two points on the rigid body can change in direction, but not in magnitude. *The only thing that one point on the rigid body can do, relative to another point on the same rigid body, is to rotate about it.* If you were standing on point *A* and watching what point *B* was doing relative to you, point *B* would appear to be rotating about you. Similarly, if you were on point *B*, then point *A* would appear to be rotating about you.

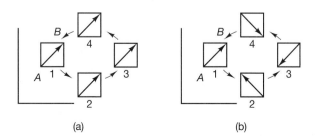

(a)   (b)

Figure 5.19 Curvilinear translation and rotation

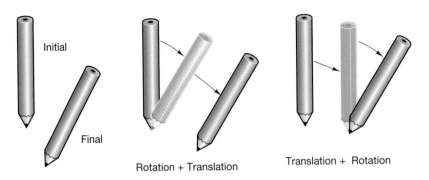

Figure 5.20

The rotation of the body initially can be considered to be completely independent of the curvilinear translation. This principle is easily demonstrated by moving a pencil along the surface of the desk, as shown in Figure 5.20. The pencil's initial orientation and the final orientation are shown in the darker shade. Notice that in its final position, the pencil has been translated and rotated, but the time of which the rotation occurred during the translation cannot be determined just by examination of the initial and final positions. For example, the pencil could have rotated and then translated, or it could be rotating during translation, or it could have translated to the final location and then rotated. The general plane motion always can be considered to be the sum of this rotation and translation, but to determine how the body moves, one must be able to couple, or synchronize, these two independent movements. It will be shown that this coupling is due to *constraints* on the movement. These constraints relate the separate degrees of freedom and are fundamental to the understanding of kinematics. Again, *the movement may always be considered to be the sum of translation and rotation*, and these two movements must be coupled to fully describe the kinematics of the rigid body.

5.5.1 ABSOLUTE AND RELATIVE VELOCITIES IN PLANE MOTION OF A RIGID BODY

The rotation of a rigid body will be obtained by examining the relative velocity between any two points on the body. Two such points are shown on the rigid body in Figure 5.21. The velocity at point B may be written in terms of the velocity at point A and the relative velocity as

$$\mathbf{v}_B = \mathbf{v}_A + \mathbf{v}_{B/A} \tag{5.50}$$

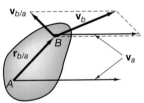

Figure 5.21

The relative velocity must be perpendicular to the relative position vector, for if it had a component parallel to the relative position vector, the length of the relative position vector would be changed. This change is not allowed, as points A and B are points on the same rigid body, and the magnitude of the relative position vector between them cannot change. The relative velocity vector now can be expressed in terms of the angular velocity vector in a manner similar to that used for describing rotation about a fixed axis because the relative motion is a pure rotation. Thus, the relative velocity

$$\mathbf{v}_{B/A} = \mathbf{v}_B - \mathbf{v}_A$$

can be written as

$$\mathbf{v}_{B/A} = \boldsymbol{\omega} \times \mathbf{r}_{B/A} \tag{5.51}$$

where $\boldsymbol{\omega}$ is the angular velocity of the body and $\mathbf{r}_{B/A}$ is the position vector of B relative to A.

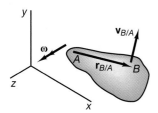

Figure 5.22

This relationship between the relative velocity of two points on a rigid body and the angular velocity of the body is fundamental to an understanding of rigid-body dynamics. Note that the linear velocities of any two points on a rigid body are used to determine the angular velocity of the body, but this angular velocity is the angular velocity of the entire rigid body. It should not be considered as the angular velocity about some point in the body. The angular velocity is a free vector and has no specified line of action. For plane motion, the angular velocity vector is in the z-direction if the plane of motion is the xy-plane. Remember that a right-handed coordinate system must be used, and the direction of the angular velocity vector is determined by the right-hand rule, as illustrated in Figure 5.22.

Before examining ways to solve for the angular velocity, note that the dot product between the relative position vector and the relative velocity vector must be zero, as these two vectors are always perpendicular, as may be seen from Eq. (5.51). This factor is again a consequence of the assumption that the body is rigid. For any rigid body,

$$\mathbf{v}_{B/A} \cdot \mathbf{r}_{B/A} = 0 \tag{5.52}$$

This relationship can be used to check vector calculations and (in many cases) eliminate errors.

Although the rotation of a rigid body in a plane has been determined by examining the velocity of one point on the body relative to another point, the angular velocity is the absolute angular velocity of the body and should not be thought of as a relative value. Another way of stating this phenomenon is that every point on the body is rotating about every other point on the body with the same angular velocity. If the xy-plane is considered to be the plane of motion, then the relative velocity and the relative position vectors lie in this plane and the angular velocity vector is in the z-direction—that is, perpendicular to the plane of motion. In order to determine the angular velocity, we write the angular velocity as

$$\boldsymbol{\omega} = \omega_z \hat{\mathbf{k}} \tag{5.53}$$

and the relative position vector as

$$\mathbf{r}_{B/A} = x_{B/A}\hat{\mathbf{i}} + y_{B/A}\hat{\mathbf{j}} \tag{5.54}$$

Since the unit vectors $\hat{\mathbf{i}}$ and $\hat{\mathbf{j}}$ are constant (that is, not functions of time), the relative velocity is obtained by differentiation of Eq. (5.54):

$$\mathbf{v}_{B/A} = \dot{x}_{B/A}\hat{\mathbf{i}} + \dot{y}_{B/A}\hat{\mathbf{j}} \tag{5.55}$$

However, the relative velocity can be expressed in terms of the angular velocity and the relative position as

$$\mathbf{v}_{B/A} = \boldsymbol{\omega} \times \mathbf{r}_{B/A} = \omega_z\hat{\mathbf{k}} \times (x_{B/A}\hat{\mathbf{i}} + y_{B/A}\hat{\mathbf{j}})$$
$$\mathbf{v}_{B/A} = -y_{B/A}\omega_z\hat{\mathbf{i}} + x_{B/A}\omega_z\hat{\mathbf{j}} \tag{5.56}$$

Equating Eq. (5.55) to Eq. (5.56) yields

$$\dot{x}_{B/A}\hat{\mathbf{i}} + \dot{y}_{B/A}\hat{\mathbf{j}} = -y_{B/A}\omega_z\hat{\mathbf{i}} + x_{B/A}\omega_z\hat{\mathbf{j}} \tag{5.57}$$

Equating components yields

$$\omega_z = \frac{\dot{y}_{B/A}}{x_{B/A}} = -\frac{\dot{x}_{B/A}}{y_{B/A}} \tag{5.58}$$

Eq. (5.58) gives two ratios from which to determine the angular velocity. The dot product between the relative velocity vector and the relative position vector is zero, and

$$\mathbf{r}_{B/A} \cdot \mathbf{v}_{B/A} = x_{B/A}\dot{x}_{B/A} + y_{B/A}\dot{y}_{B/A} = 0$$

Equating the two expressions for the angular velocity given in Eq. (5.58) is equivalent to setting the dot product of the relative position vector and the relative velocity vector equal to zero.

The solution is obtained by expanding the vectors into component form and then solving the corresponding scalar equations. If the angular velocity and the linear velocity of one point on a rigid body are known, then the linear velocity of every point on the rigid body may be determined by use of the basic equation

$$\mathbf{v}_B = \mathbf{v}_A + \boldsymbol{\omega} \times \mathbf{r}_{B/A} \tag{5.59}$$

The solution for the velocities of a rigid body always depends upon this equation.

The relative velocity equation is the basis of all velocity calculations for a rigid body and must be examined carefully to determine the number of kinematic quantities that are involved. We will write this equation in an expanded vector form:

$$\mathbf{v}_B = \mathbf{v}_A + \boldsymbol{\omega} \times (\mathbf{r}_B - \mathbf{r}_A)$$

For plane motion, the two velocity vectors and the two position vectors have components in the x- and y-directions. The angular velocity vector is in the z-direction. Therefore, there are nine scalar kinematic variables. The vector equation yields only two scalar equations and seven scalar kinematic variables must be specified to obtain a solution of the vector equations. In some cases, the condition that the relative velocity vector and the relative position vector are perpendicular to each other may be used to determine the value of a variable, but this expression is not an independent equation.

In most problems, the body will be constrained in some manner such that the rotation of the body and the translation are coupled. As was stated earlier, this coupling must be described, or the body will be free to rotate independent of the translation. This movement is a form of *constrained motion*, or a *constraint to the motion*. The body may be constrained such that the magnitude of the velocity is not known at a particular point, but the direction of the velocity is known. In another case, such as if a wheel rolls on the ground without sliding, then the point of contact between the wheel and the ground is constrained so that it cannot slide, and therefore, the velocity of that contacting point would equal that of the ground—that is, it would equal zero. If the wheel rolled on the bed of a truck without sliding, the velocity of the contacting point on the wheel with the truck bed would be equal to the velocity of the truck. A rigid body also may be constrained such that certain points on the body follow a particular path. There are no general rules for recognizing and writing the equations of constraint, and each case must be considered independently. Previously, the number of kinematic variables in the equation of relative velocity was shown to be nine. Seven of these variables must be known before the equations of relative motion can be solved. In most cases, some of these variables will be specified in the form of constraints to the motion. Examples of certain types of constraints are shown in Figure 5.23. In Figure 5.23(a), the contact point of the wheel with the plate must have a velocity equal to that of the plate if the wheel is rolling without slipping. An easier, but equally important, constraint is that the center of the wheel moves in rectilinear translation; that is, the velocity and acceleration of the center of the wheel are parallel to the plane on which the wheel rolls. In Figure 5.23(b), both ends of the bar are constrained to move on planes in space. Therefore, point A would have a velocity only in the horizontal direction and the velocity vector of point B would be along the inclined plane of $60°$. Figure 5.23(c) shows

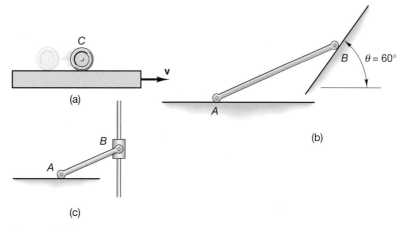

Figure 5.23

a similar constraint; point *A* must move along the horizontal plane, and point *B* is attached to a slider that must move in the vertical direction. Constraints are not difficult, but care must be taken to determine the correct mathematical formulation of them.

5.5.2 EXPERIMENTAL MOTION DATA

It was noted that for a rigid body, the relative velocity vector is always perpendicular to the relative position vector. In many cases, a body is not truly rigid, but is just assumed to be rigid. Experimentally, individuals collect data on movement by using a motion analysis system consisting of high-speed video cameras or motion picture cameras. These systems take 50 to 200 pictures per second, and the position data is digitized electronically. This position data is numerically differentiated to obtain the linear velocity and acceleration of points on the body. Numerical differentiation produces more noise in the derivatives, even if smoothing or filtering of the original data is performed, as was shown in Chapter 1. Due to the noise in the original data or noise from numerical differentiation, the relative velocity vector may not be perpendicular to the relative position vector. The component of the relative velocity vector that is parallel to the relative position vector can be computed and subtracted from the original relative velocity vector. This correction in the data is shown in the accompanying box.

ANGULAR VELOCITY FOR NOISY EXPERIMENTAL DATA

Let $\mathbf{v}_{B/A}^{D}$ and $\mathbf{r}_{B/A}^{D}$ be the relative velocity and position vectors from the raw data, and $\mathbf{v}_{B/A}^{C}$ will be the corrected relative velocity. First, form a unit vector $\hat{\mathbf{e}}_{B/A}$ in the direction of the relative position vector:

$$\hat{\mathbf{e}}_{B/A} = \frac{\mathbf{r}_{B/A}^{D}}{|\mathbf{r}_{B/A}^{D}|}$$

The component of the data-relative velocity vector along this unit vector may be determined by the projection, or scalar, product. Subtracting this component from the original data-relative velocity gives the corrected relative velocity vector $\mathbf{v}_{B/A}^{C}$ as

$$\mathbf{v}_{B/A}^{C} = \mathbf{v}_{B/A}^{D} - (\mathbf{v}_{B/A}^{D} \cdot \hat{\mathbf{e}}_{B/A})\hat{\mathbf{e}}_{B/A} = \hat{\mathbf{e}}_{B/A} \times (\mathbf{v}_{B/A}^{D} \times \hat{\mathbf{e}}_{B/A})$$

This expression is derived from the triple vector product. Note that another vector could be formed,

$$\hat{\mathbf{e}}_t = \hat{\mathbf{k}} \times \hat{\mathbf{e}}_{B/A}$$

$$\mathbf{v}_{B/A}^C = (\hat{\mathbf{e}}_t \cdot \mathbf{v}_{B/A}^D)\hat{\mathbf{e}}_t$$

This vector is now perpendicular to the relative position vector, and the angular velocity vector may be obtained as before:

$$\boldsymbol{\omega} \times \mathbf{r}_{B/A} = \mathbf{v}_{B/A}^C$$

The three vectors are now orthogonal, or mutually perpendicular.

This method is very useful in the field of biomechanics when computing the motion of body segments to determine the loading on the joints of the body. The data in these tests are always noisy due to skin movement at locations where markers are placed.

Sample Problem 5.9

A body is moving in plane motion, and in the position shown in the accompanying figure, the position vectors and linear velocities of points A and B are

$$\mathbf{r}_A = 1.60\hat{\mathbf{i}} + 1.50\hat{\mathbf{j}} \quad \text{m}$$

$$\mathbf{r}_B = 2.00\hat{\mathbf{i}} + 1.80\hat{\mathbf{j}} \quad \text{m}$$

$$\mathbf{v}_A = 3.00\hat{\mathbf{i}} \quad \text{m/s}$$

$$\mathbf{v}_B = 2.40\hat{\mathbf{i}} + 0.80\hat{\mathbf{j}} \quad \text{m/s}$$

Determine the angular velocity of the body at this instant.

Solution   Calculate the position of B relative to A and the velocity of B relative to A.

$$\mathbf{r}_{B/A} = \mathbf{r}_B - \mathbf{r}_A = 0.4\hat{\mathbf{i}} + 0.3\hat{\mathbf{j}} \quad \text{m}$$

$$\mathbf{v}_{B/A} = \mathbf{v}_B - \mathbf{v}_A = -0.6\hat{\mathbf{i}} + 0.8\hat{\mathbf{j}} \quad \text{m/s}$$

Check to be sure that $\mathbf{r}_{B/A} \cdot \mathbf{v}_{B/A} = 0$

$$0.4(-0.6) + 0.3(0.8) = 0$$

Now, the angular velocity may be written as $\boldsymbol{\omega} = \omega_z \hat{\mathbf{k}}$

$$\boldsymbol{\omega} \times \mathbf{r}_{B/A} = \mathbf{v}_{B/A}$$

$$\omega_z \hat{\mathbf{k}} \times (0.4\hat{\mathbf{i}} + 0.3\hat{\mathbf{j}}) = (-0.6\hat{\mathbf{i}} + 0.8\hat{\mathbf{j}})$$

Taking the cross product, we get

$$0.4\omega_z \hat{\mathbf{j}} - 0.3\omega_z \hat{\mathbf{i}} = -0.6\hat{\mathbf{i}} + 0.8\hat{\mathbf{j}}$$

Equating the $\hat{\mathbf{i}}$ and $\hat{\mathbf{j}}$ components yields

$$\omega_z = \frac{0.6}{0.3} = \frac{0.8}{0.4} = 2 \text{ rad/s}$$

Sample Problem 5.10	A commonly found example of a device rotating in plane motion is a wheel rolling without slipping. In designing drive trains, transmissions, transaxles, engine axles, and wheels, it is important to know the relationship between the linear velocity of a car (in this sample problem, we will examine the axle of the rear wheels of a car) and the angular velocity of the wheel–tire assembly. It is also important to know the velocity of various parts of the assembly of the wheel. Consider the wheel–tire assembly shown in the accompanying figure. The wheel has a 14-in. diameter hub, and the width of the tire adds another 5 in. to the radius. If the velocity of the car (in this case, axle) is 60 mph, calculate the velocity of a point on the rim of the wheel for the instant when $\theta = 30°$. Also, calculate the angular velocity of the wheel.

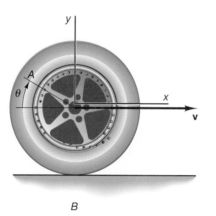

Solution   Using the coordinate system where we take x to the right and y up, with the origin at the wheel center.

$$\mathbf{r}_{B/0} = -\frac{12}{12}\hat{\mathbf{j}} \text{ ft}$$

$$\mathbf{v}_0 = 60 \text{ mph } \hat{\mathbf{i}} = 88 \text{ fps } \hat{\mathbf{i}}$$

$$\mathbf{v}_B = 0 = \mathbf{v}_0 + \boldsymbol{\omega} \times \mathbf{r}_{B/0}$$

$$\boldsymbol{\omega} = -88\hat{\mathbf{k}} \text{ rad/s}$$

This value is the angular velocity of the wheel, and the linear velocity of any point on the wheel may now be determined.

$$\mathbf{v}_A = \mathbf{v}_0 + \boldsymbol{\omega} \times \mathbf{r}_{A/0}$$

$$\mathbf{r}_{A/0} = \frac{7}{12}(-\cos 30°\hat{\mathbf{i}} + \sin 30°\hat{\mathbf{j}})$$

Thus, the linear velocity at A is

$$\mathbf{v}_A = 88\hat{\mathbf{i}} + (-88\hat{\mathbf{k}}) \times \frac{7}{12}(-\cos 30°\hat{\mathbf{i}} + \sin 30°\hat{\mathbf{j}})$$

$$\mathbf{v}_A = 113.7\hat{\mathbf{i}} + 44.5\hat{\mathbf{j}} \text{ ft/s}$$

The magnitude of the velocity at A is

$$|\mathbf{v}_A| = \sqrt{113.7^2 + 44.5^2} = 122.1 \text{ fps, or about 83 mph}$$

A boomerang is thrown so that it may be considered to be moving in plane motion. If the partial velocities are known to be the values shown in the accompanying diagram, determine the angular velocity and the unknown components of the linear velocities at points *A*, *B*, and *C* on the boomerang. All dimensions are given in inches.

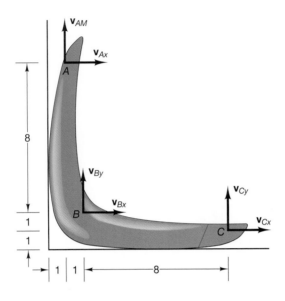

Also,

$$v_{Ax} = 10 \text{ in/s}$$
$$v_{By} = -10 \text{ in/s}$$
$$v_{Cx} = 20 \text{ in/s}$$

Solution Using the standard relative velocity expressions, we may relate the velocities at *A* and *C* to the velocity at *B* and the angular velocity as follows

$$\mathbf{v}_A = \mathbf{v}_B + \boldsymbol{\omega} \times \mathbf{r}_{A/B}$$
$$\mathbf{v}_C = \mathbf{v}_B + \boldsymbol{\omega} \times \mathbf{r}_{C/B}$$

The unknowns in the problem are the scalar components v_{Ay}, v_{Bx}, v_{Cy}, and ω. The two two-dimensional vector equations will yield four scalar equations for these unknowns. Let $\boldsymbol{\omega} = \omega\hat{\mathbf{k}}$ and for the geometry given,

$$\mathbf{r}_{A/B} = -\hat{\mathbf{i}} + 8\hat{\mathbf{j}}$$
$$\mathbf{r}_{C/B} = 8\hat{\mathbf{i}} - \hat{\mathbf{j}}$$

Forming the cross products and equating components,

$$v_{Ax} = v_{Bx} - \omega y_{A/B} \qquad 10 = v_{Bx} - 8\omega$$
$$v_{Ay} = v_{By} + \omega x_{A/B} \qquad v_{Ay} = -10 - \omega$$
$$v_{Cx} = v_{Bx} - \omega y_{C/B} \qquad 20 = v_{Bx} + \omega$$
$$v_{Cy} = v_{By} + \omega x_{C/B} \qquad v_{Cy} = -10 + 8\omega$$

Solving for each variable yields

$$\omega = 1.11 \ \text{rad/s}$$
$$v_{Bx} = 18.89 \ \text{in/s}$$
$$v_{Ay} = -11.11 \ \text{in/s}$$
$$v_{Cy} = -1.11 \ \text{in/s}$$

The system of four equations for the four unknowns can be solved using matrix methods. An alternate approach to solving the system is to use the relationship that the relative velocity must be perpendicular to the relative position vector. In this case,

$$\mathbf{v}_{A/B} \cdot \mathbf{r}_{A/B} = 0, \quad \text{or} \quad [(10 - v_{Bx})\hat{\mathbf{i}} + (v_{By} + 10)\hat{\mathbf{j}}] \cdot [-\hat{\mathbf{i}} + 8\hat{\mathbf{j}}] = 0$$
$$\mathbf{v}_{C/B} \cdot \mathbf{r}_{C/B} = 0, \quad \text{or} \quad [(20 - v_{Bx})\hat{\mathbf{i}} + (v_{Cy} + 10)\hat{\mathbf{j}}] \cdot [8\hat{\mathbf{i}} - \hat{\mathbf{j}}] = 0$$
$$\mathbf{v}_{A/C} \cdot \mathbf{r}_{A/C} = 0, \quad \text{or} \quad [(10 - 20)\hat{\mathbf{i}} + (v_{Ay} - v_{Cy})\hat{\mathbf{j}}] \cdot [-9\hat{\mathbf{i}} + 9\hat{\mathbf{j}}] = 0$$

Expanding and writing in matrix notation yields

$$\begin{bmatrix} 1 & 8 & 0 \\ -8 & 0 & -1 \\ 0 & 1 & -1 \end{bmatrix} \begin{bmatrix} v_{Bx} \\ v_{Ay} \\ v_{Cy} \end{bmatrix} = \begin{bmatrix} -70 \\ -150 \\ -10 \end{bmatrix}$$

$$v_{Bx} = 18.89$$
$$v_{Ay} = -11.11$$
$$v_{Cy} = -1.11$$

In this case, the components of the linear velocities are obtained without knowledge of the angular velocity. The angular velocity easily can be obtained by any of the relative velocity equations.

Sample Problem 5.12

In the accompanying figure, a cylinder 3 m in diameter rolls without slipping along a level surface at a velocity of 2 m/s (at the center of the cylinder). Determine the angular velocity of the cylinder, and plot the movement of point B on the surface of the cylinder for the first 9.4 s.

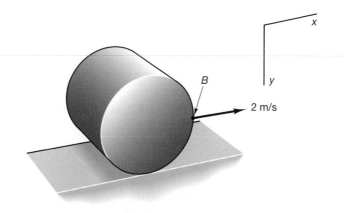

Solution The velocity of the center of the cylinder is

$$\mathbf{v}_c = 2\hat{\mathbf{i}} \ \text{m/s}$$

Let P be the point of contact of the cylinder with the surface, and noting that the velocity of this point is zero, we can write the velocity of P as

$$\mathbf{v}_P = 0 = \mathbf{v}_c + \boldsymbol{\omega} \times \mathbf{r}_{P/c}$$

$$0 = 2\hat{\mathbf{i}} + \omega\hat{\mathbf{k}} \times 1.5\hat{\mathbf{j}}$$

$$\omega = 1.333 \text{ rad/s}$$

Note that the angular velocity is positive, corresponding to clockwise rotation. The angle through which the cylinder rotates is

$$\theta = \omega t$$

The velocity of point B is

$$\mathbf{v}_B = \mathbf{v}_c + \boldsymbol{\omega} \times \mathbf{r}_{B/c}$$

At any time, the relative position vector is

$$\mathbf{r}_{B/c} = 1.5[\cos(\omega t)\hat{\mathbf{i}} + \sin(\omega t)\hat{\mathbf{j}}] \text{ m}$$

The velocity of point B is

$$\mathbf{v}_B = [2 - (1.333)(1.5)\sin(1.333t)]\hat{\mathbf{i}} + (1.333)(1.5)\cos(1.333t)\hat{\mathbf{j}} \text{ m/s}$$

The velocity can be integrated to determine the position of point B in space. The initial position will be taken as zero:

$$x(t) = [2t + 1.5\cos(1.333t)] \text{ m}$$

$$y(t) = 1.5\sin(1.333t) \text{ m}$$

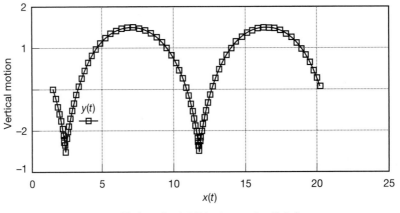

Motion of point B for two cycles (9.4 s)

5.5.3 DIRECT VECTOR METHOD TO OBTAIN THE ANGULAR VELOCITY

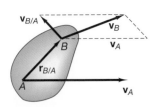

Figure 5.24

Previously, we determined the angular velocity of a rigid body by examining the relative linear velocity between two points on the rigid body. For a body in plane motion, only two points are needed to determine the angular velocity if the positions and the linear velocities of the two points are known. Consider two such points on the rigid body shown in Figure 5.24. The velocity at point B may be written in terms of the velocity at point A and the relative velocity as

$$\mathbf{v}_B = \mathbf{v}_A + \mathbf{v}_{B/A} \tag{5.60}$$

The relative velocity must be perpendicular to the relative position vector. The relative velocity vector can be expressed in terms of the angular velocity vector as

$$\mathbf{v}_{B/A} = \mathbf{v}_B - \mathbf{v}_A$$
$$\mathbf{v}_{B/A} = \boldsymbol{\omega} \times \mathbf{r}_{B/A} \tag{5.61}$$

where ω is the angular velocity of the body.

For plane motion, the angular velocity can be determined using Rodrigue's Formula (*J. Math. Pures Appl.* 5 (1840) 380–410), which is developed by use of vector identities. The relative velocity relationship is

$$\boldsymbol{\omega} \times \mathbf{r}_{B/A} = \mathbf{v}_{B/A} \tag{5.62}$$

Taking the cross product of both sides of Eq. (5.62) with the relative position vector yields

$$\mathbf{r}_{B/A} \times (\boldsymbol{\omega} \times \mathbf{r}_{B/A}) = \mathbf{r}_{B/A} \times \mathbf{v}_{B/A} \tag{5.63}$$

The left-hand side of the above expression is a triple vector product and may be written as

$$\boldsymbol{\omega}(\mathbf{r}_{B/A} \cdot \mathbf{r}_{B/A}) - \mathbf{r}_{B/A}(\mathbf{r}_{B/A} \cdot \boldsymbol{\omega}) = \mathbf{r}_{B/A} \times \mathbf{v}_{B/A} \tag{5.64}$$

But ω is perpendicular to the plane of motion, and $\mathbf{r}_{B/A} \cdot \boldsymbol{\omega} = 0$. The vector equation now can be solved by simple algebra, yielding Rodrigue's Formula for the angular velocity of a rigid body:

$$\boldsymbol{\omega} = \frac{\mathbf{r}_{B/A} \times \mathbf{v}_{B/A}}{\mathbf{r}_{B/A} \cdot \mathbf{r}_{B/A}} \tag{5.65}$$

This procedure is called a direct vector method, as we have not resolved Eq. (5.62) into components to obtain a solution. If computational software or calculators are used to compute cross products, the direct vector method is preferred to expanding into scalar components. Remember that this formula is dependent upon the relative position vector being perpendicular to the angular velocity and, therefore, is used for problems of plane motion. In this case, the two reference points can be considered to be in the plane of motion, and the angular velocity vector is perpendicular to this plane. We will modify the direct vector method to obtain the instantaneous center of rotation and the angular acceleration in later sections.

Note that the direct vector method automatically eliminates noise in the relative velocity vector parallel to the relative position vector. Therefore, there would be no need to correct the relative velocity vector as was shown in the section on Experimental Motion Data.

Problems

5.22 In Figure P5.22, a cylinder rolls to the right without slipping with a constant velocity at its center, $v_c = 4$ m/s. Determine the angular velocity of the cylinder.

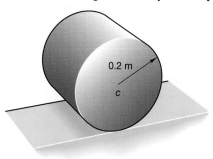

Figure P5.22

5.23 In Figure P5.23, a disk of radius r rolls to the left without slipping with an absolute velocity of v_C while the truck that carries the disk moves to the right with a velocity of v_T. Determine the angular velocity of the disk.

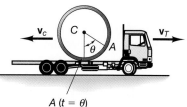

$A \; (t = \theta)$

Figure P5.23

5.24 For the disk and truck in Problem 5.23, if $r = 300$ mm, $v_C = 2$ m/s, and $v_T = 10$ m/s, determine the absolute velocity of point A on the disk as the disk rolls through one complete rotation.

5.25 Determine the angular velocity of the disk in Figure P5.23 if the velocity of the center of the disk relative to the truck is $v_{C/T}$.

5.26 A yo-yo rolls without slipping as it is pulled along, as shown in Figure P5.26. If its string is pulled at a velocity v, determine the velocity of the center of the yo-yo and the angular velocity of the yo-yo.

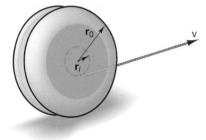

Figure P5.26

5.27 The position vectors and absolute velocity vectors of two points on a rigid body in plane motion are given by:

$$\mathbf{r}_A = \begin{pmatrix} 3 \\ 2 \\ 0 \end{pmatrix} \text{m} \qquad \mathbf{r}_B = \begin{pmatrix} 5 \\ 4 \\ 0 \end{pmatrix} \text{m}$$

$$\mathbf{v}_A = \begin{pmatrix} 3 \\ -7 \\ 0 \end{pmatrix} \text{m/s} \qquad \mathbf{v}_B = \begin{pmatrix} -5 \\ 1 \\ 0 \end{pmatrix} \text{m/s}$$

Determine the angular velocity of the body from the scalar equations.

5.28 Solve Problem 5.27 by the direct vector method.

5.29 For the rigid body in Problem 5.27, determine the linear velocity of a point C whose position is given by $\mathbf{r}_C = -2\hat{\mathbf{i}} - 3\hat{\mathbf{j}}$ m.

5.30 A body moves in general plane motion. The position vector in a fixed coordinate system and the absolute velocity of a point A are, respectively,

$$\mathbf{r}_A = -300\hat{\mathbf{i}} + 550\hat{\mathbf{j}} \text{ mm}$$

$$\mathbf{v}_A = 250\hat{\mathbf{i}} - 600\hat{\mathbf{j}} \text{ mm/s}$$

If $\boldsymbol{\omega} = 2\hat{\mathbf{k}}$ rad/s, determine the velocity of a point with position vector $\mathbf{r}_B = 240\hat{\mathbf{i}} - 200\hat{\mathbf{j}}$ mm.

5.31 Write the general expression for the velocity of the piston shown in Figure P5.31 in terms of the position of the disk drive if the disk is moving at a constant angular velocity $\boldsymbol{\omega}_d$.

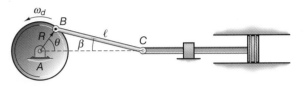

Figure P5.31

5.32 If the velocity of the disk in Figure P5.31 is 3 rad/s, the disk radius is 200 mm, and the length of the connecting bar is 500 mm, plot the velocity of the piston versus the angle θ.

5.33 A disk of radius R rolls without slipping to the right such that the constant linear velocity of its center is v. A bar of length l is attached at a point $R/2$ from the disk center, as shown in Figure P5.33. Develop a general expression for the angular velocity of the bar at any angle θ. Plot the angular velocity versus the angle θ for the case where $R = 0.5$ m, $l = 2$ m, and $v = 0.2$ m/s.

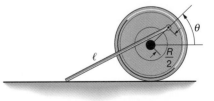

Figure P5.33

5.34 For the system in Problem 5.33, develop a general expression for the linear velocity of the end of the bar in contact with the ground. Plot the linear velocity versus the angle θ using the value given in Problem 5.33.

5.35 If the disk in Problem 5.33 is rotated with an angular velocity of ω about a fixed axis through the center of the disk, develop a general expression for the angular velocity of the bar at any angle θ. Plot the angular velocity versus the angle for the case when $R = 0.5$ m, $l = 2$ m, and $\omega = 0.4$ rad/s. Compare this result with the result in Problem 5.33.

5.36 For the system in Problem 5.35, develop a general expression for the linear velocity of the end of the bar in contact with the ground. Plot the linear velocity versus the angle θ using the values given the Problem 5.35.

5.37 Two wheels of equal radius are connected by a link, as shown in Figure P5.37. If wheel A is rotating at a constant angular velocity ω, determine the angular velocity of wheel B.

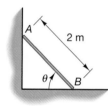

Figure P5.37

5.38 In Figure P5.38, the point B on the rigid rod AB is sliding along the horizontal surface with a constant velocity of 5 m/s. Determine a general expression for the angular velocity of the bar for any angle θ.

Figure P5.38

5.39 In Figure P5.39, gear A of radius 150 mm rotates at a constant angular velocity of 4 rad/s counterclockwise, determine the angular velocity of gear B of radius 50 mm if the angular velocity of the link AB is 2 rad/s counter clockwise.

Figure P5.39

5.40 Collar A moves with a constant velocity V along the rod shown in Figure P5.40. Determine a general expression for the velocity of point B on the rod AB.

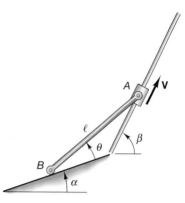

Figure P5.40

5.41 Noisy experimental data collected for two skin markers placed on a thigh are

$$\mathbf{r}_A = 200\hat{\mathbf{i}} + 300\hat{\mathbf{j}} \text{ mm} \qquad \mathbf{v}_A = 10\hat{\mathbf{i}} + 30\hat{\mathbf{j}} \text{ mm/s}$$

$$\mathbf{r}_B = 300\hat{\mathbf{i}} + 500\hat{\mathbf{j}} \text{ mm} \qquad \mathbf{v}_B = 7\hat{\mathbf{i}} + 36.5\hat{\mathbf{j}} \text{ mm/s}$$

Determine the angular velocity of the thigh.

5.42 Using the methods of plane-motion analysis, determine the angular velocity of the rod AB in Figure P5.42 and the length of the spring between A and B for the first 2 s of motion. The position vectors for A and B are, respectively,

$$\mathbf{r}_A(t) = 3t\hat{\mathbf{i}} + 5t\hat{\mathbf{j}} \text{ m}$$

$$\mathbf{r}_B(t) = (3t + \cos \pi t)\hat{\mathbf{i}} + (5t + 2 \sin \pi t)\hat{\mathbf{j}} \text{ m}.$$

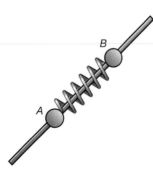

Figure P5.42

5.6 INSTANTANEOUS CENTER OF ROTATION IN PLANE MOTION

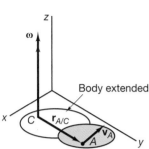

Figure 5.25 Rotation of a Rigid Body about a Fixed Point

An interesting conceptual approach can be used when a body is moving in generalized plane motion. During such motion, the body will be translating and rotating, and any point on the body will, in general, have an absolute velocity. If, instead, the body were rotating about a fixed axis in space perpendicular to the plane of motion, the intercept point of that axis of rotation with the plane of motion would have zero velocity. This intercept point could be considered to be a point on the body or the body extended— that is, the mathematical extension of the body. Such a fixed axis of rotation is shown in Figure 5.25. Now, for a rigid body, the velocity of point A can be related to the velocity of point C as

$$\mathbf{v}_A = \mathbf{v}_C + \boldsymbol{\omega} \times \mathbf{r}_{A/C} \tag{5.66}$$

The point C is chosen as the intercept of the fixed axis of rotation with the plane of motion. Therefore,

$$\mathbf{v}_C = 0 \tag{5.67}$$

The velocity of A is

$$\mathbf{v}_A = \boldsymbol{\omega} \times \mathbf{r}_{A/C} \tag{5.68}$$

The velocity vector of A is perpendicular to the relative position vector of A to C. These observations of movement about a fixed axis of rotation lead one to question whether at any instant of time, the body may be considered to be rotating about an *instantaneous axis of rotation*. The point of intercept of this instantaneous axis of rotation with the plane of motion (point C in Figure 5.25) is called the *instantaneous center of rotation*, or the *instantaneous center of zero velocity*. For this concept to be true, point C would have to have zero absolute velocity at this instant. Now, in general, point C will have an acceleration so that its velocity will not be zero an instant later. For this reason, point C is called the instantaneous center of rotation, and the point on the body or the body extended—that is the instantaneous center of rotation is—in general, constantly changing.

If the velocity of a point on the body and the angular velocity of the body are known, the instantaneous center of rotation C can be found. If the velocity of point A and the angular velocity of the body are known, then Eq. (5.66) can be written as

$$\mathbf{v}_C = \mathbf{v}_A + \boldsymbol{\omega} \times \mathbf{r}_{C/A} \tag{5.69}$$

The desired point C, the instantaneous center of rotation, has zero velocity at this instant:

$$\mathbf{v}_C = 0$$
$$\boldsymbol{\omega} \times \mathbf{r}_{C/A} = -\mathbf{v}_A \tag{5.70}$$

The location of C relative to point A may be obtained by expanding the vector equation Eq. (5.70) into scalar components:

$$\omega_z \hat{\mathbf{k}} \times (x_{C/A}\hat{\mathbf{i}} + y_{C/A}\hat{\mathbf{j}}) = -(v_{Ax}\hat{\mathbf{i}} + v_{Ay}\hat{\mathbf{j}}) \tag{5.71}$$

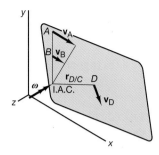

Figure 5.26

Equating components of the vector equation yields two scalar equations for the coordinates of the instantaneous center of rotation relative to point A:

$$x_{C/A} = \frac{-v_{Ay}}{\omega_z} \tag{5.72}$$

$$y_{C/A} = \frac{v_{Ax}}{\omega_z} \tag{5.73}$$

Having identified the instantaneous center of rotation, we may then determine the velocity of any point for that instant of time. This result is shown graphically in Figure 5.26. Note that the velocities are perpendicular to a line from the instantaneous center to the point in question and are equal to the angular velocity multiplied by that distance.

Historically, this method was developed to allow draftsmen to solve kinematics problems graphically. An individual could construct perpendiculars to the directions of motion and, with a scale, measure distances to this intercept (the instantaneous center of rotation). The angular velocity could be obtained by simple division of the linear velocity of a point by the distance to the instantaneous center of rotation. This method made the instantaneous center of rotation an excellent graphical tool for the solution of kinematics problems. This method then was carried forward when trigonometry (law of sines and cosines) was used to solve for sides of the resulting triangles to obtain the distances without using a scale. This routine might be viewed as a transitional method precluding the time when most texts began to use the full power of vector mathematics. Since vectors will be used in the solution of this type of problem, the value of the concept of an instantaneous center of rotation is limited. There are times when this concept can be useful for visualizing the motion at any instant. The best example of such an occasion is the case of the wheel rolling without slipping. In this case, the constraint to the motion is that the point of contact of the wheel with the ground has zero velocity and is, therefore, by definition, the instantaneous center of rotation. The relative velocity equations may be written relative to this point of contact. However, realize that these equations would be exactly the same whether the point of contact was recognized as the instantaneous center of rotation or not. In fact, if the wheel slips while it rolls, the point of contact is *not* the instantaneous center of rotation, and the linear velocity of the point of contact of the wheel is *not* zero. All kinematic problems may be solved without this concept, and one should not feel dependent upon the concept. Example problems will be presented that show the historical development from graphical solutions to trigonometric methods to the vector approach.

The instantaneous center of rotation can be determined by a *direct vector solution* if the linear velocity of one point on the body and the angular velocity of the body are known. The alternative solution for $\mathbf{r}_{C/A}$ may be obtained by use of vector identities and by crossing both sides of Eq. (5.70) with the angular velocity vector:

$$\boldsymbol{\omega} \times (\boldsymbol{\omega} \times \mathbf{r}_{C/A}) = -\boldsymbol{\omega} \times \mathbf{v}_A \tag{5.74}$$

Using the vector identity for the triple vector product, we obtain

$$\boldsymbol{\omega} \times (\boldsymbol{\omega} \times \mathbf{r}_{C/A}) = \boldsymbol{\omega}(\boldsymbol{\omega} \cdot \mathbf{r}_{C/A}) - \mathbf{r}_{C/A}(\boldsymbol{\omega} \cdot \boldsymbol{\omega}) \tag{5.75}$$

The first term on the right side is zero because in plane motion, the angular velocity vector is perpendicular to the relative position vector. Eq. (5.75) then yields

$$\mathbf{r}_{C/A} = \frac{\boldsymbol{\omega} \times \mathbf{v}_A}{\boldsymbol{\omega} \cdot \boldsymbol{\omega}} \tag{5.76}$$

At first glance, the vector solution involving cross products and dot products may look more complicated than that of expanding into scalar form, but it can be obtained quickly, and this form is compatible with computer software.

Sample Problem 5.13

A link is constrained to move as shown in the accompanying figure. If, at the instant shown, the velocity of point B is 10 in/s upward, determine the velocity of point A.

Solution   This problem will be solved by three different methods: (a) graphical (b) trigonometry and (c) vector solution.

a) Graphical Solution

To solve this problem graphically, it is necessary to lay out perpendiculars to the two velocities, one at A and one at B. The intercept of those perpendiculars is the instantaneous center of rotation. Such a layout is illustrated in the accompanying diagram.

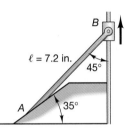

The measured distance of the instantaneous center of rotation (ICR) to B is 8.7 in., with some error in measurement. The angular velocity is

$$\omega = (10 \text{ in/s})/(8.7 \text{ in.}) = 1.15 \text{ rad/s}$$

The velocity of point A is the angular velocity times the distance from ICR to point A. This distance was measured to be 6.3 in. Therefore, the velocity of point A is

$$v_A = \omega \, d_A = 1.15(6.3) = 7.24 \text{ in/s}$$

b) Trigonometric Solution

A triangle may be formed between the ICR, point B, and point A. Since the angles are known, this triangle may be solved by the law of sines.
By the law of sines:

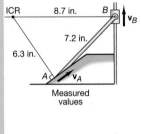

$$\frac{d_B}{\sin 80} = \frac{7.2}{\sin 55} \quad \Rightarrow \quad d_B = 8.66 \text{ in.}$$

$$\frac{d_A}{\sin 45} = \frac{7.2}{\sin 55} \quad \Rightarrow \quad d_A = 6.22 \text{ in.}$$

Now ω can be calculated as it was in the graphical solution:

$$\omega = \frac{v_B}{d_B} = \frac{10}{8.65} = 1.156 \text{ rad/s}$$

The velocity of point A is

$$v_A = \omega d_A = 1.156(6.22) = 7.19 \text{ in/s}$$

c) Vector Solution

If this problem is formulated with vectors, it is not necessary to determine the location of the instantaneous center of rotation in order to obtain the angular velocity and the velocity of point A. Instead, we can just note that

$$\mathbf{v}_A = \mathbf{v}_B + \boldsymbol{\omega} \times \mathbf{r}_{A/B}$$

Point A is constrained to move along the 35° plane:

$$\mathbf{v}_A = v_A(\cos 35° \hat{\mathbf{i}} + \sin 35° \hat{\mathbf{j}})$$

$$\mathbf{v}_B = 10\hat{\mathbf{j}}$$

$$\mathbf{r}_{A/B} = 7.2(-\cos 45° \, \hat{\mathbf{i}} - \sin 45° \hat{\mathbf{j}})$$

Substituting into the relative velocity equation, we get

$$v_A(\cos 35° \hat{\mathbf{i}} + \sin 35° \hat{\mathbf{j}}) = 10\hat{\mathbf{j}} + \omega\hat{\mathbf{k}} \times 7.2(-\cos 45° \hat{\mathbf{i}} - \sin 45° \hat{\mathbf{j}})$$

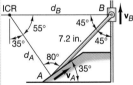

Equating components yields

$$0.819v_A = \omega(7.2)(0.707)$$

$$0.574v_A = 10 - \omega(7.2)(0.707)$$

Solving for v_A and ω, we obtain

$$v_A = 7.179 \text{ in/s}$$

$$\omega = 1.155 \text{ rad/s}$$

The location of the instantaneous center of rotation may be located now that ω is known. Locating this center relative to point B, we calculate that

$$\boldsymbol{\omega} \times \mathbf{r}_{ICR/B} = -vB$$

$$1.155\hat{\mathbf{k}} \times (x_{ICR/B}\hat{\mathbf{i}} + y_{ICR/B}\hat{\mathbf{j}}) = -10\hat{\mathbf{j}}$$

$$x_{ICR/B} = -8.65 \text{ in.}$$

$$y_{ICR/B} = 0.0$$

As may be expected, the answers from the three solutions are approximately equal. The accuracy of the graphical solution is directly related to the scale of the drawing used and the care taken in making measurements. As has been indicated earlier, the graphical and trigonometric solutions have been replaced by the vector solution in terms of common usage.

Sample Problem 5.14	Consider the automobile wheel assembly illustrated in the accompanying figure. If the wheel rolls without slipping and the velocity of the car is v to the right at the instant shown in the figure, determine the velocity of point B on the wheel. Leave the solution in terms of the radius r, the angle θ and the velocity v.

Solution If the wheel rolls without slipping, the point of contact C must have zero velocity, and therefore, C is the instantaneous center of rotation of the wheel. The angular velocity of the wheel easily can be obtained if we know the velocity of the center of the wheel (which is also the velocity of the car). To obtain this value, we calculate that

$$\boldsymbol{\omega} \times \mathbf{r}_{O/C} = \mathbf{v}_O$$

$$\mathbf{r}_{O/C} = r\hat{\mathbf{j}}$$

$$\mathbf{v}_O = v\hat{\mathbf{i}}$$

$$\omega\hat{\mathbf{k}} \times r\hat{\mathbf{j}} = v\hat{\mathbf{i}}$$

$$\omega = -\frac{v}{r}$$

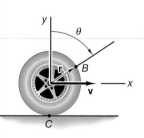

The velocity of point B may now be found:

$$\mathbf{v}_B = \boldsymbol{\omega} \times \mathbf{r}_{B/C}$$

$$\mathbf{r}_{B/C} = r\left[\sin\theta\hat{\mathbf{i}} + (1 + \cos\theta)\hat{\mathbf{j}}\right]$$

Therefore;

$$\mathbf{v}_B = v[(1 + \cos\theta)\hat{\mathbf{i}} - \sin\theta\hat{\mathbf{j}}]$$

If point B is considered to be a point fixed to the wheel, the angle θ increases as the wheel turns and has the value ωt at any time if the time t is zero when $\theta = 0$. It is easy to obtain a complete solution of the motion using any of the available software. A general solution as a function of time will be presented in Section 5.10, when parametric solutions are discussed. The velocity components and magnitude of a point D that is initially on the top of the wheel are shown in the Computational Supplement. For the numerical example, the radius of the wheel is chosen to be 10 in., and the velocity of the center of the wheel is 10 in/s to the right.

Sample Problem 5.15

The compound wheel in the accompanying figure rolls without slipping on the inner hub of the wheel, which has a radius of 20 in. If the angular velocity of the wheel is 4 rad/s counterclockwise, determine the velocity of point B.

Solution

The contact point between the inner hub and the surface has zero velocity as the wheel rolls without slipping, and therefore, this point is the instantaneous center of rotation. So,

$$\mathbf{v}_B = \mathbf{v}_C + \omega \times \mathbf{r}_{B/C}$$

$$\mathbf{v}_C = 0$$

$$\omega = 4\hat{\mathbf{k}}$$

$$\mathbf{r}_{B/C} = 20\hat{\mathbf{j}} + 40(\sin 30°\hat{\mathbf{i}} + \cos 30°\hat{\mathbf{j}})$$

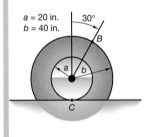

$a = 20$ in.
$b = 40$ in.
$30°$

Therefore, the velocity of point B at this instant is

$$\mathbf{v}_B = -218.6\hat{\mathbf{i}} + 80\hat{\mathbf{j}}$$

The magnitude of the velocity is

$$|v_B| = \sqrt{(218.6)^2 + (80)^2} = 232.75 \text{ in/s}$$

The numerical vector calculations can be done using a computer or a calculator as a vector calculator, as shown in the Computational Supplement.

5.7 INSTANTANEOUS CENTER OF ROTATION BETWEEN TWO RIGID BODIES

The concept of an instantaneous center of rotation was introduced in the last section, and it defined a point in space and on a body or body extended about which the body rotates at that instant. It was also noted that the use of this concept is limited, as most solutions of problems concerning such bodies do not employ graphical techniques; however, it might be useful for visualizing the problem. The concept will be extended in this section to determine the instantaneous center of relative rotation between two rigid bodies. This point might be thought of as an instantaneous hinge point between the two bodies. If the two bodies were actually hinged together, this point would maintain a fixed position on both of the bodies. This concept finds applications in the field of biodynamics (biomechanics of human motion) and is used to define the center of a joint. Consider the knee of a dancer, shown in Figure 5.27.

The knee may be considered to be a hinge between the thigh and the lower leg, or the shank. The knee joint is not an exact hinge, and the center of the joint changes as the knee

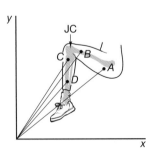

Figure 5.27

is flexed. The angular velocity of the shank relative to the thigh is called knee flexion, or extension rate, or the angular velocity of the joint. Markers are placed in such a manner that the positions and velocities of two points on the thigh and two points on the shank are measured at a particular instant, as shown in Figure 5.27. We will consider the thigh as a rigid body having markers placed at points A and B. The shank will also be considered as a rigid body, having markers at points C and D. For this example, we will assume that the thigh and the shank move in a single plane of motion (in biodynamics, called the "sagittal plane"). If a motion analysis system is used, we can determine the positions and velocities of the two points on the thigh and the two points on the shank. Therefore, $\mathbf{r}_A$, $\mathbf{r}_B$, $\mathbf{r}_C$, $\mathbf{r}_D$ and $\mathbf{v}_A$, $\mathbf{v}_B$, $\mathbf{v}_C$, $\mathbf{v}_D$ are known for any instant of time. The angular velocity of the thigh may be found for any instant of time, as follows:

$$\boldsymbol{\omega}_T \times (\mathbf{r}_B - \mathbf{r}_A) = (\mathbf{v}_B - \mathbf{v}_A) \tag{5.77}$$

The angular velocity of the shank is determined in a similar manner:

$$\boldsymbol{\omega}_S \times (\mathbf{r}_D - \mathbf{r}_C) = (\mathbf{v}_D - \mathbf{v}_C) \tag{5.78}$$

The angular velocity of the knee joint may be defined as the relative angular velocity of the shank to the thigh:

$$\boldsymbol{\omega}_J = \boldsymbol{\omega}_S - \boldsymbol{\omega}_T \tag{5.79}$$

In this case, flexion is positive and extension negative. Flexion occurs when the shank is rotating faster than the thigh.

The instantaneous center of relative rotation is defined in a manner that is consistent with that used to derive the instantaneous center of rotation in absolute plane motion. The relative instanteous center is that point on the thigh or the thigh extended that has zero relative motion to the same point considered to be on the shank or the shank extended. Therefore, we will seek a point on both bodies that has the same linear velocity. This point is called the instantaneous hinge point in biodynamics, or the joint center, or JC, and it is illustrated in Figure 5.28. Considering the joint center to be a point on the thigh, we get

$$\mathbf{v}_{JC} = \mathbf{v}_B + \boldsymbol{\omega}_T \times \mathbf{r}_{JC/B} \tag{5.80}$$

Considering the joint center also to be a point on the shank, we get

$$\mathbf{v}_{JC} = \mathbf{v}_C + \boldsymbol{\omega}_S \times \mathbf{r}_{JC/C} \tag{5.81}$$

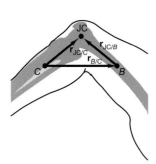

Figure 5.28

Since we are looking for a point on the thigh and a point on the shank that have the same velocity, the velocities of this point defined on both the thigh and the shank can be equated:

$$\mathbf{v}_C + \boldsymbol{\omega}_S \times \mathbf{r}_{JC/C} = \mathbf{v}_B + \boldsymbol{\omega}_T \times \mathbf{r}_{JC/B} \tag{5.82}$$

Considering the relationship of the points B, C, and JC in space, as illustrated in Figure 5.28, we obtain

$$\mathbf{r}_{JC/C} = \mathbf{r}_{B/C} + \mathbf{r}_{JC/B}$$

$$(\boldsymbol{\omega}_S - \boldsymbol{\omega}_T) \times \mathbf{r}_{JC/B} = (\mathbf{v}_B - \mathbf{v}_C - \boldsymbol{\omega}_S \times \mathbf{r}_{B/C}) \tag{5.83}$$

Note that all vectors on the right side of this equation are known, and

$$\boldsymbol{\omega}_J = \boldsymbol{\omega}_S - \boldsymbol{\omega}_T$$

and

$$\mathbf{v}_{B/C} = \mathbf{v}_B - \mathbf{v}_C$$

Therefore, the location of the joint center is

$$\boldsymbol{\omega}_J \times \mathbf{r}_{JC/B} = (\mathbf{v}_{B/C} - \boldsymbol{\omega}_S \times \mathbf{r}_{B/C}) \tag{5.84}$$

The location of the dynamically defined joint center at any instant can now be determined. The equation for the joint center may be solved by expanding the vector equation into its scalar components and solving for the x- and y-distances from point C to the joint center.

Note that the vector from C to the joint center may also be determined by a direct vector solution in a manner similar to that used by Rodrigue. We will construct a triple vector product by crossing the location of the joint center with $\boldsymbol{\omega}_J$, the angular velocity of the joint:

$$\boldsymbol{\omega}_J \times (\boldsymbol{\omega}_J \times \mathbf{r}_{JC/B}) = \boldsymbol{\omega}_J \times \mathbf{v}_{B/C} - \boldsymbol{\omega}_J \times (\boldsymbol{\omega}_S \times \mathbf{r}_{B/C}) \tag{5.85}$$

The two triple vector products can be expanded as

$$\boldsymbol{\omega}_J(\boldsymbol{\omega}_J \cdot \mathbf{r}_{JC/B}) - \mathbf{r}_{JC/B}(\boldsymbol{\omega}_J \cdot \boldsymbol{\omega}_J) = \boldsymbol{\omega}_J \times \mathbf{v}_{B/C} - \boldsymbol{\omega}_S(\boldsymbol{\omega}_J \cdot \mathbf{r}_{B/C}) + \mathbf{r}_{B/C}(\boldsymbol{\omega}_J \cdot \boldsymbol{\omega}_S) \tag{5.86}$$

Noting that the angular velocity vectors are perpendicular to the plane of motion, we can solve Eq. (5.86) directly for the position vector to the joint center from point B.

$$\mathbf{r}_{JC/B} = -\frac{\boldsymbol{\omega}_J \times \mathbf{v}_{B/C} + \mathbf{r}_{B/C}(\boldsymbol{\omega}_J \cdot \boldsymbol{\omega}_S)}{(\boldsymbol{\omega}_J \cdot \boldsymbol{\omega}_J)} \tag{5.87}$$

In biodynamic studies, however, the knee cannot be considered as a simple hinge, and the instantaneous center of relative rotation is needed to compute the moments acting at the knee joint.

Problems

5.43 A yo-yo is pulled by a string along a surface at a speed of 200 mm/s, as shown in Figure P5.43. If the yo-yo rolls without slipping, determine the angular velocity of the yo-yo and the velocity of the center of the yo-yo.

is 350 mm, determine the instantaneous center of rotation when $\theta = 0$, and determine the velocity of the collar C.

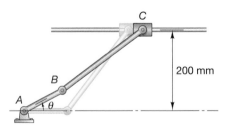

Figure P5.44

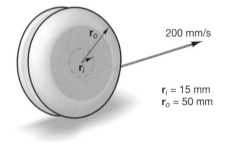

$r_i = 15$ mm
$r_o = 50$ mm

Figure P5.43

5.44 If the angular velocity of the 100-mm link AB in Figure P5.44 is 10 rad/s and the length of the link BC

5.45 Solve Problem 5.44 when $\theta = \pi/2$.

5.46 Solve Problem 5.44 when $\theta = \pi$.

5.47 Solve Problem 5.44 when $\theta = 3\pi/2$.

5.48 A roll of paper is pulled from the end at a velocity of 300 mm/s to the right, as depicted in Figure P5.48. If the center of the roll moves to the right at a velocity of 120 mm/s, determine (a) the instantaneous center of rotation, (b) the angular velocity of the roll, and (c) the length of paper in millimeters unrolled per second.

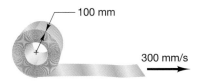

Figure P5.48

5.49 A disk rolls without slipping when pulled by a rope, as shown in Figure P5.49. If the velocity of the end of the rope is 500 mm/s, determine the angular velocity of the disk and the velocity of the center of the disk.

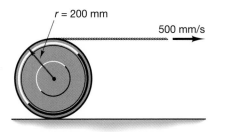

Figure P5.49

💻 **5.50** In Figure P5.50, the position vectors and velocity vectors for two points on the forearm and two points on the upper arm of a person are:

$$\mathbf{r}_A = \begin{bmatrix} 3 \\ 1.2 \\ 0 \end{bmatrix} \mathrm{m} \quad \mathbf{r}_B = \begin{bmatrix} 3.3 \\ 1.1 \\ 0 \end{bmatrix} \mathrm{m}$$

$$\mathbf{r}_C = \begin{bmatrix} 2.9 \\ 1.3 \\ 0 \end{bmatrix} \mathrm{m} \quad \mathbf{r}_D = \begin{bmatrix} 2.8 \\ 1.6 \\ 0 \end{bmatrix} \mathrm{m}$$

$$\mathbf{v}_A = \begin{bmatrix} 0.1 \\ 0.02 \\ 0 \end{bmatrix} \mathrm{m/s} \quad \mathbf{v}_B = \begin{bmatrix} 0.3 \\ 0.62 \\ 0 \end{bmatrix} \mathrm{m/s}$$

$$\mathbf{v}_C = \begin{bmatrix} 0.095 \\ 0.01 \\ 0 \end{bmatrix} \mathrm{m/s} \quad \mathbf{v}_D = \begin{bmatrix} 0.155 \\ 0.03 \\ 0 \end{bmatrix} \mathrm{m/s}$$

Determine the elbow joint center and the angular velocity of the forearm relative to the upper arm.

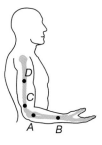

Figure P5.50

5.51 A professional tennis player serves a ball with a racket speed of 60 mph, as shown in Figure P5.51. Determine the angular velocity of the player's arm if his shoulder has a velocity of zero at the moment of contact between the racket and the ball.

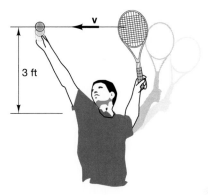

Figure P5.51

💻 **5.52** At an instant, the position and velocity of two points on a rigid body moving in plane motion shown in Figure P5.52, are:

$$\mathbf{r}_A = 2\hat{\mathbf{i}} + \hat{\mathbf{j}} \,\mathrm{m} \quad \mathbf{v}_A = 2\hat{\mathbf{i}} - 2\hat{\mathbf{j}} \,\mathrm{m/s}$$
$$\mathbf{r}_B = 5\hat{\mathbf{i}} + 4\hat{\mathbf{j}} \,\mathrm{m} \quad \mathbf{v}_B = 9\hat{\mathbf{i}} - 9\hat{\mathbf{j}} \,\mathrm{m/s}$$

Determine the angular velocity and the position of the instantaneous center of rotation of the body.

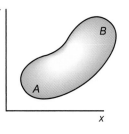

Figure P5.52

💻 **5.53** For the body in Problem 5.52, determine the velocity of a point C at a position with coordinates $(3,3,0)$.

5.54 Find the instantaneous center of rotation for the disk in Problem 5.23.

5.55 Determine the instantaneous center of rotation for the yo-yo in Problem 5.26.

5.56 Find the instantaneous center of rotation of the body in Problem 5.30, and use that point as a reference to determine the velocity of point B on the body.

5.57 Write a general expression for a position vector from A to the instantaneous center of rotation of link BC in Problem 5.31.

5.58 Determine a general expression for the position of the instantaneous center of rotation for the bar in Problem 5.38.

💻 **5.59** The thoracic cage and the pelvis are treated as rigid bodies in a biomechanics experiment to locate the instantaneous center of rotation during flexion of a person's back. Two markers A and B are placed on the pelvis, and two markers C and D are placed on the thoracic cage, as shown in Figure 5.59. If the position vectors and velocity vectors relative to the laboratory coordinate system are

$$\mathbf{r}_A = \begin{pmatrix} 1.30 \\ 1.03 \\ 0 \end{pmatrix} \text{m} \quad \mathbf{r}_B = \begin{pmatrix} 1.22 \\ 1.24 \\ 0 \end{pmatrix} \text{m}$$

$$\mathbf{r}_C = \begin{pmatrix} 1.35 \\ 1.544 \\ 0 \end{pmatrix} \text{m} \quad \mathbf{r}_D = \begin{pmatrix} 1.52 \\ 1.74 \\ 0 \end{pmatrix} \text{m}$$

$$\mathbf{v}_A = \begin{pmatrix} 0 \\ 0 \\ 0 \end{pmatrix} \text{m/s} \quad \mathbf{v}_B = \begin{pmatrix} 0.084 \\ 0.032 \\ 0 \end{pmatrix} \text{m/s}$$

$$\mathbf{v}_C = \begin{pmatrix} 0.181 \\ -0.138 \\ 0 \end{pmatrix} \text{m/s} \quad \mathbf{v}_D = \begin{pmatrix} 0.196 \\ -0.151 \\ 0 \end{pmatrix} \text{m/s}$$

then determine the location of the instantaneous center of rotation relative to marker B.

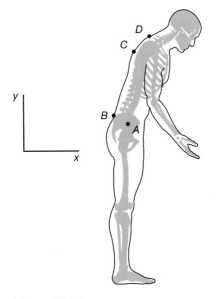

Figure P5.59

5.8 ABSOLUTE AND RELATIVE ACCELERATION OF A RIGID BODY IN PLANE MOTION

Consider a body in plane motion, as shown in Figure 5.29. We showed in Section 5.5 that the length of the (relative) vector between any two points on a rigid body is constant (i.e. the magnitude of the relative position vector is constant). In plane motion, this factor allows the determination of the angular velocity of the body if the linear velocities of two points on the body are known. This method, in fact, separates the translational motion of the body from the rotation of the body. The relative velocity is related to the angular velocity of the body by

$$\mathbf{v}_{B/A} = \boldsymbol{\omega} \times \mathbf{r}_{B/A} \tag{5.88}$$

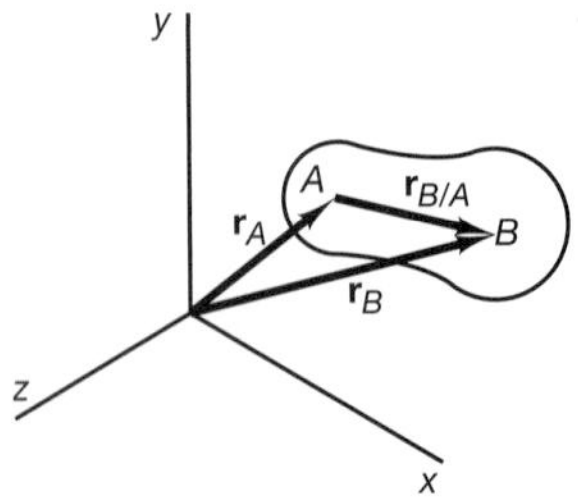

Recall that the acceleration of B relative to A is

$$\mathbf{a}_{B/A} = \frac{d\mathbf{v}_{B/A}}{dt} = \mathbf{a}_B - \mathbf{a}_A \tag{5.89}$$

Figure 5.29

Differentiation of the relative velocity vector yields

$$\frac{d\mathbf{v}_{B/A}}{dt} = \frac{d}{dt}(\boldsymbol{\omega} \times \mathbf{r}_{B/A})$$

$$\mathbf{a}_{B/A} = \boldsymbol{\alpha} \times \mathbf{r}_{B/A} + \boldsymbol{\omega} \times \mathbf{v}_{B/A} \tag{5.90}$$

where $\boldsymbol{\alpha}$ is the angular acceleration vector and must be perpendicular to the plane of motion when the body is in plane motion. The relative velocity term may be replaced by its relation to the angular velocity, yielding

$$\mathbf{a}_{B/A} = \boldsymbol{\alpha} \times \mathbf{r}_{B/A} + \boldsymbol{\omega} \times (\boldsymbol{\omega} \times \mathbf{r}_{B/A}) \tag{5.91}$$

Using the triple vector product, we may write the last term as

$$\boldsymbol{\omega} \times (\boldsymbol{\omega} \times \mathbf{r}_{B/A}) = \boldsymbol{\omega}(\mathbf{r}_{B/A} \cdot \boldsymbol{\omega}) - \mathbf{r}_{B/A}(\boldsymbol{\omega} \cdot \boldsymbol{\omega})$$

Since the angular velocity vector is perpendicular to the plane of motion, the first term is zero, and the second relative acceleration can be written as

$$\mathbf{a}_{B/A} = \boldsymbol{\alpha} \times \mathbf{r}_{B/A} - (\boldsymbol{\omega} \cdot \boldsymbol{\omega})\mathbf{r}_{B/A} \tag{5.92}$$

The first term of the expression for the relative acceleration vector is perpendicular to the relative position vector and is parallel to the relative velocity vector. This term changes the magnitude of the relative velocity vector. The second term of the expression is parallel to the relative position vector and is perpendicular to the relative velocity vector. This term changes the direction of the relative velocity vector. It is important to remember that the relative acceleration of a point on a rigid body is composed of two components that are orthogonal to each other. One component is tangential to the path of the relative motion and the other is normal to it. This observation will aid in examining constraints. Note that if a rigid body has an angular velocity and the angular acceleration is zero, every point on the body will, in general, have a linear acceleration. If the body has an angular acceleration, every point on the body will have a linear acceleration, except for a point that is fixed in space. The directions of these acceleration vectors are shown in Figure 5.30. The first diagram divides the relative acceleration vector into its two components, one normal and one tangential. In the second diagram, the linear acceleration of point A is added to the acceleration of B relative to A, yielding the linear acceleration of B. The first term in the relative acceleration equation is parallel to the relative velocity and is called the tangential relative acceleration. The second term changes the direction of the relative velocity, as it is perpendicular to the relative velocity, and is called the normal relative acceleration.

The angular acceleration of the body is obtained from the known linear velocities and accelerations of two points on the body. This method is similar to the method used to determine the angular velocity. However, the angular velocity of the body appears in the relative acceleration equation and must first be determined if the angular acceleration is to be calculated. A method to determine the angular acceleration without first determining the angular velocity will be presented later in this section. A simple rearrangement of the terms in the relative acceleration equation will show the similarity to the method used to determine the angular velocity:

$$\boldsymbol{\alpha} \times \mathbf{r}_{B/A} = \mathbf{a}_{B/A} - \boldsymbol{\omega} \times (\boldsymbol{\omega} \times \mathbf{r}_{B/A})$$

$$= (\mathbf{a}_B - \mathbf{a}_A) + (\boldsymbol{\omega} \cdot \boldsymbol{\omega})\mathbf{r}_{B/A} \tag{5.93}$$

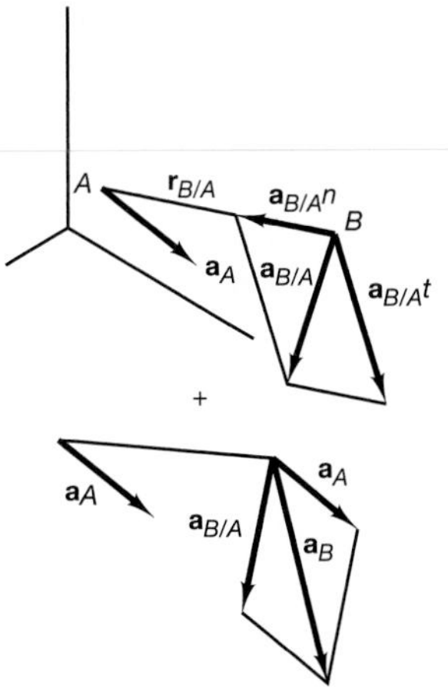

Figure 5.30

For plane motion, the vectors can be written as

$$\mathbf{r}_{B/A} = x_{B/A}\hat{\mathbf{i}} + y_{B/A}\hat{\mathbf{j}}$$

$$\boldsymbol{\alpha} = \alpha\hat{\mathbf{k}}$$

$$\mathbf{a}_A = a_{Ax}\hat{\mathbf{i}} + a_{Ay}\hat{\mathbf{j}}, \quad \mathbf{a}_B = a_{Bx}\hat{\mathbf{i}} + a_{By}\hat{\mathbf{j}}$$

Substituting these vectors into Eq. (5.93) yields

$$-y_{B/A}\alpha\hat{\mathbf{i}} + x_{B/A}\alpha\hat{\mathbf{j}} = (a_{Bx} - a_{Ax})\hat{\mathbf{i}} + (a_{By} - a_{Ay})\hat{\mathbf{j}} + x_{B/A}\omega^2\hat{\mathbf{i}} + y_{B/A}\omega^2\hat{\mathbf{j}} \quad (5.94)$$

The angular acceleration may be solved by equating the components, yielding

$$\alpha = -\frac{[a_{Bx} - a_{Ax} + x_{B/A}\omega^2]}{y_{B/A}} = \frac{[a_{By} - a_{Ay} + y_{B/A}\omega^2]}{x_{B/A}} \quad (5.95)$$

Once the angular velocity and the angular acceleration are known, the linear velocity and the linear acceleration of any two points on the body may be related as follows:

$$\mathbf{v}_B = \mathbf{v}_A + \boldsymbol{\omega} \times \mathbf{r}_{B/A}$$

$$\mathbf{a}_B = \mathbf{a}_A + \boldsymbol{\alpha} \times \mathbf{r}_{B/A} - \omega^2\mathbf{r}_{B/A} \quad (5.96)$$

These equations allow complete description of a kinematics of a rigid body moving in plane motion.

When the relative velocity equation was considered, it was advantageous to determine the number of kinematic variables involved in the relative equation so that it may be determined if the problem is well formulated. The count for the relative acceleration equation is the same as for the relative velocity equation, with the addition of the angular velocity. This equation then involves ten kinematic variables (four components of position, four components of linear acceleration, and the angular velocity and acceleration). Only two scalar equations from the relative acceleration equation can be written, so eight kinematic variables must be specified before the relative acceleration vector equation can be solved. Although this specification of variables may appear to be an unnecessary exercise in "accounting," it is absolutely necessary to be able to determine which variables are known and which are unknown before attempting a mathematical solution. In many cases, this specification will allow the determination of overlooked constraints to the motion.

*5.8.1 ALTERNATE SOLUTION OF THE ACCELERATION OF RIGID BODIES

It is sometimes advantageous to solve for the angular acceleration of a rigid body directly and to examine the velocity in detail later. Figure 5.30 shows that the normal and tangential relative acceleration components are perpendicular to each other. The normal component is parallel to the relative position vector, and the tangential component is perpendicular to the relative position vector. A unit vector may be formed along the relative position vector as follows:

$$\hat{\mathbf{e}}_{B/A} = \frac{\mathbf{r}_{B/A}}{|\mathbf{r}_{B/A}|} \quad (5.97)$$

For plane motion where the z-axis is selected as perpendicular to the plane of motion, $\boldsymbol{\omega}$ and $\boldsymbol{\alpha}$ have components only in the z-direction. A vector perpendicular to the relative position vector in the plane of motion may be formed by the cross product

$$\hat{\mathbf{e}}_t = \hat{\mathbf{k}} \times \hat{\mathbf{e}}_{B/A} \quad (5.98)$$

The relative acceleration vector may now be separated into a part that is parallel to the relative position vector and a part that is perpendicular to the relative position vector. (*Note: The component parallel to the relative position vector will be in the opposite direction from the normal relative acceleration.*) So,

$$a_r = -a_n = \hat{\mathbf{e}}_{B/A} \cdot \mathbf{a}_{B/A} \tag{5.99}$$

and

$$a_t = \hat{\mathbf{e}}_t \cdot \mathbf{a}_{B/A}$$

The angular acceleration may now be determined:

$$a_t = |\mathbf{r}_{B/A}|\alpha = \hat{\mathbf{e}}_t \cdot \mathbf{a}_{B/A}$$

Therefore,

$$\alpha = \frac{\hat{\mathbf{e}}_t \cdot \mathbf{a}_{B/A}}{|\mathbf{r}_{B/A}|} \tag{5.100}$$

The sign of the angular acceleration $\boldsymbol{\alpha}$ will determine whether the vector is in the plus or minus *z*-direction. The sign of the angular acceleration is, as always, specified by the right-hand rule. In a similar manner, we may determine the magnitude of the angular velocity:

$$a_r = -a_n = -|\mathbf{r}_{B/A}|\omega^2$$
$$a_r = \hat{\mathbf{e}}_{B/A} \cdot \mathbf{a}_{B/A} \tag{5.101}$$

Therefore,

$$\omega = \pm\sqrt{\frac{-\hat{\mathbf{e}}_{B/A} \cdot \bar{\mathbf{a}}_{B/A}}{|\bar{\mathbf{r}}_{B/A}|}}$$

At first examination, this expression would appear to involve the square root of a negative number. However, the dot product in the numerator must be negative, as the relative acceleration vector must always have a component in the negative direction of the position vector if the angular velocity is not zero. This method may be used as an alternative to or a check of the expansion of the vector equations into scalar equations, as shown previously.

A direct vector method similar to that used to obtain Rodrigue's formula may be employed to determine the angular acceleration. Crossing both sides of the acceleration equation ($\boldsymbol{\alpha} \times \mathbf{r}_{B/A}) = \mathbf{a}_{B/A} + \omega^2\mathbf{r}_{B/A}$ with the relative position vector, we obtain

$$\mathbf{r}_{B/A} \times (\boldsymbol{\alpha} \times \mathbf{r}_{B/A}) = \mathbf{r}_{B/A} \times (\mathbf{a}_{B/A} + \omega^2\mathbf{r}_{B/A}) \tag{5.102}$$

If we use the triple vector product identity on the left side of Eq. (5.102) and note that the second term on the right side is zero, as it is the cross product of two parallel vectors, Eq. (5.102) becomes

$$\boldsymbol{\alpha}(\mathbf{r}_{B/A} \cdot \mathbf{r}_{B/A}) - \mathbf{r}_{B/A}(\mathbf{r}_{B/A} \cdot \boldsymbol{\alpha}) = \mathbf{r}_{B/A} \times \mathbf{a}_{B/A} \tag{5.103}$$

The second term on the left side of Eq. (5.103) is zero for plane motion, and the angular acceleration vector may be obtained directly:

$$\boldsymbol{\alpha} = \frac{\mathbf{r}_{B/A} \times \mathbf{a}_{B/A}}{\mathbf{r}_{B/A} \cdot \mathbf{r}_{B/A}} \tag{5.104}$$

Note the similarity of this expression to Rodrigue's velocity formula, and also, realize that this expression is only true for the case of plane motion, as during its derivation, a term was eliminated using the argument that $\boldsymbol{\alpha}$ was perpendicular to the plane of motion. Note that these formulas are very simple to enter into computational software, and even some hand calculators can be programmed to perform these calculations.

Sample Problem 5.16

In Sample Problem 5.9, a rigid body moving in plane motion had the velocities and positions of two points, *A* and *B*, specified as follows:

$$\mathbf{r}_A = 1.60\hat{\mathbf{i}} + 1.50\hat{\mathbf{j}} \text{ m}$$

$$\mathbf{r}_B = 2.00\hat{\mathbf{i}} + 1.80\hat{\mathbf{j}} \text{ m}$$

$$\mathbf{v}_A = 3.00\hat{\mathbf{i}} \text{ m/s}$$

$$\mathbf{v}_B = 2.40\hat{\mathbf{i}} + 0.80\hat{\mathbf{j}} \text{ m/s}$$

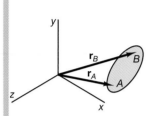

We found that the angular velocity is equal to 2 rad/s at the instant of time considered in the sample problem. If the linear accelerations of points *A* and *B* are

$$\mathbf{a}_A = 3\hat{\mathbf{i}} + 2\hat{\mathbf{j}} \text{ m/s}^2$$

$$\mathbf{a}_B = 1.1\hat{\mathbf{i}} + 1.2\hat{\mathbf{j}} \text{ m/s}^2$$

determine the angular acceleration of the rigid body at this instant.

Solution

The angular acceleration may be obtained from Eq. (5.95):

$$\alpha = -\frac{[a_{Bx} - a_{Ax} + x_{B/A}\omega^2]}{y_{B/A}} = -\frac{[1.1 - 3 + (2 - 1.6)2^2]}{(1.8 - 1.5)} = 1.0 \text{ rad/s}^2$$

$$\alpha = \hat{\mathbf{k}} \text{ rad/s}^2$$

This problem also could be solved by first decomposing the relative acceleration vector into its components. In this case, the angular velocity does not need to be determined first. The relative position vector is

$$\mathbf{r}_{B/A} = 0.4\hat{\mathbf{i}} + 0.3\hat{\mathbf{j}}$$

The magnitude of this vector is $\sqrt{(0.4)^2 + (0.3)^2} = 0.5$. Therefore, a unit vector in this direction is

$$\hat{\mathbf{e}}_{B/A} = \frac{\mathbf{r}_{B/A}}{|\mathbf{r}_{B/A}|} = 0.8\hat{\mathbf{i}} + 0.6\hat{\mathbf{j}}$$

A unit vector in the tangential direction is

$$\hat{\mathbf{e}}_t = \hat{\mathbf{k}} \times \hat{\mathbf{e}}_{B/A} = -0.6\hat{\mathbf{i}} + 0.8\hat{\mathbf{j}}$$

The relative acceleration vector is

$$\mathbf{a}_{B/A} = \mathbf{a}_B - \mathbf{a}_A = -1.9\hat{\mathbf{i}} - 0.8\hat{\mathbf{j}}$$

Now,

$$\alpha = \frac{\hat{\mathbf{e}}_t \cdot \mathbf{a}_{B/A}}{|\mathbf{r}_{B/A}|} = \frac{(-0.6\hat{\mathbf{i}} + 0.8\hat{\mathbf{j}}) \cdot (-1.9\hat{\mathbf{i}} - 0.8\hat{\mathbf{j}})}{0.5} = \frac{1.14 - 0.64}{0.5} = 1 \text{ rad/s}^2$$

The magnitude of the angular velocity also could be calculated by the above method. The angular velocity may be obtained using Rodrigue's formula:

$$\boldsymbol{\omega} = \frac{\mathbf{r}_{B/A} \times \mathbf{v}_{B/A}}{\mathbf{r}_{B/A} \cdot \mathbf{r}_{B/A}}$$

$$= \frac{(0.4\hat{\mathbf{i}} + 0.3\hat{\mathbf{j}}) \times (-0.6\hat{\mathbf{i}} + 0.8\hat{\mathbf{j}})}{(0.4\hat{\mathbf{i}} + 0.3\hat{\mathbf{j}}) \cdot (0.4\hat{\mathbf{i}} + 0.3\hat{\mathbf{j}})}$$

$$= \frac{(0.32 + 0.18)\hat{\mathbf{k}}}{(0.16 + 0.09)} = 2\hat{\mathbf{k}} \text{ rad/s}$$

Knowing the angular velocity and the velocity of point A, we may find the instantaneous center of rotation using Eq. (5.76):

$$\mathbf{r}_{C/A} = \frac{\boldsymbol{\omega} \times \mathbf{v}_A}{\boldsymbol{\omega} \cdot \boldsymbol{\omega}}$$

$$\mathbf{r}_{C/A} = \frac{2\hat{\mathbf{k}} \times 3\hat{\mathbf{i}}}{2\hat{\mathbf{k}} \cdot 2\hat{\mathbf{k}}} = 1.5\hat{\mathbf{j}} \text{ m}$$

The instantaneous center of rotation is located 1.5 m above point A and has coordinates relative to the origin of (1.6, 3.0).

Next, consider the calculation of the angular acceleration directly using Eq. (5.104).

$$\boldsymbol{\alpha} = \frac{\mathbf{r}_{B/A} \times \mathbf{a}_{B/A}}{\mathbf{r}_{B/A} \cdot \mathbf{r}_{B/A}} = \frac{(0.4\hat{\mathbf{i}} + 0.3\hat{\mathbf{j}}) \times [(1.1 - 3.0)\hat{\mathbf{i}} + (1.2 - 2.0)\hat{\mathbf{j}}]}{(0.4\hat{\mathbf{i}} + 0.3\hat{\mathbf{j}}) \cdot (0.4\hat{\mathbf{i}} + 0.3\hat{\mathbf{j}})}$$

$$\boldsymbol{\alpha} = 1\hat{\mathbf{k}} \text{ rad/s}^2$$

Note that the direct methods are solved more conveniently by mathematical software packages than are the indirect methods.

Sample Problem 5.17

For the mechanism illustrated in the accompanying diagram, point A has a constant linear velocity of 2 m/s down at the instant shown, and the bar AB makes an angle of 30° with the horizontal. Determine the angular velocity and acceleration of the rod, and determine the velocity and acceleration at point B.

Solution   The relative position vector of B to A is

$$\mathbf{r}_{B/A} = 2(-\cos 30°\hat{\mathbf{i}} - \sin 30°\hat{\mathbf{j}})$$

The velocity at B may be related to the velocity at A by

$$\mathbf{v}_B = \mathbf{v}_A + \boldsymbol{\omega} \times \mathbf{r}_{B/A}$$

$\mathbf{v}_B$ is constrained to move only in the x-direction; therefore,

$$v_B\hat{\mathbf{i}} = -2\hat{\mathbf{j}} + \omega\hat{\mathbf{k}} \times 2(-0.866\hat{\mathbf{i}} - 0.5\hat{\mathbf{j}})$$

Expanding into scalar components, we get

$$v_B = 0.5(2)\omega$$

$$0 = -2 - 0.866(2)\omega$$

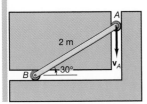

Therefore,

$$\omega = -1.155 \text{ rad/s}$$

and

$$v_B = -1.155 \text{ m/s}$$

Note that the angular velocity is clockwise and the velocity of B is in the negative direction. These are the results that would be expected. The results always should be examined to determine whether they make sense conceptually to eliminate careless errors. The angular acceleration now may be obtained by writing the relative acceleration equation:

$$\mathbf{a}_B = \mathbf{a}_A + \boldsymbol{\alpha} \times \mathbf{r}_{B/A} - \omega^2 \mathbf{r}_{B/A}$$

The acceleration of point A is zero (the velocity is constant), and the acceleration of point B is constrained to be in the x-direction. So,

$$a_B \hat{\mathbf{i}} = \alpha \hat{\mathbf{k}} \times 2(-0.866\hat{\mathbf{i}} - 0.5\hat{\mathbf{j}}) - (1.155)^2(2)(-0.866\hat{\mathbf{i}} - 0.5\hat{\mathbf{j}})$$

Expanding into scalar components, we get

$$a_B = \alpha + 2.309$$

$$0 = -1.732\alpha + 1.334$$

Therefore,

$$\alpha = 0.770 \text{ rad/s}^2 \quad \text{and} \quad a_B = 3.079 \text{ m/s}^2$$

Sample Problem 5.18

A cylinder of a 2 ft radius rolls without slipping along a horizontal surface, as shown in the accompanying figure. If the angular velocity and angular acceleration are 3 rad/s and 2 rad/s², respectively, both clockwise at the instant shown in the figure, determine the velocity and acceleration of the center of the cylinder.

Solution

The constraint of rolling without slipping restricts the velocity of point C to be equal to the velocity of the surface upon which it rolls. Therefore the velocity of point C is zero. In a similar manner, we determine that there can be no acceleration of point C parallel to the surface, so the component of the acceleration vector parallel to this surface (the x-component) is zero. The center of the wheel is constrained to move in rectilinear translation in the x-direction, so the y-component of the velocity and acceleration of point O is zero. The relative velocity relation becomes

$$\mathbf{v}_o = v_o \hat{\mathbf{i}}$$

$$\mathbf{v}_C = 0 \quad \boldsymbol{\omega} = -3\hat{\mathbf{k}}$$

$$\mathbf{r}_{o/C} = r\hat{\mathbf{j}} = 2\hat{\mathbf{j}}$$

$$\mathbf{v}_o = \mathbf{v}_C + \boldsymbol{\omega} \times \mathbf{r}_{o/C}$$

$$v_o \hat{\mathbf{i}} = \omega \hat{\mathbf{k}} \times \mathbf{r}_{o/C} = -3\hat{\mathbf{k}} \times 2\hat{\mathbf{j}} = 6\hat{\mathbf{i}}$$

Therefore,

$$\mathbf{v}_o = 6\hat{\mathbf{i}} \text{ ft/s}$$

The acceleration is handled in a similar manner:

$$\mathbf{a}_o = \mathbf{a}_C + \boldsymbol{\alpha} \times \mathbf{r}_{o/C} - \omega^2 \mathbf{r}_{o/C}$$

$$\omega = -3$$

$$\boldsymbol{\alpha} = -2\hat{\mathbf{k}}$$

$$\mathbf{a}_o = a_o\hat{\mathbf{i}} \qquad \mathbf{a}_C = a_{Cy}\hat{\mathbf{j}}$$

$$a_o\hat{\mathbf{i}} = a_{Cy}\hat{\mathbf{j}} + (-2)\hat{\mathbf{k}} \times 2\hat{\mathbf{j}} - (3)^2 2\hat{\mathbf{j}}$$

$$a_o = 4 \text{ ft/s}^2$$

$$a_{Cy} = 18 \text{ ft/s}^2$$

Sample Problem 5.19

Video data collected to determine the velocity and acceleration of a surfboard, shown in the accompanying diagram, gave the following expressions for the two ends of the board at a given instant:

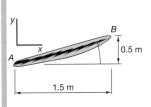

$$\mathbf{v}_A = 2.0\hat{\mathbf{i}} - 3.5\hat{\mathbf{j}} \qquad \mathbf{v}_B = 0.5\hat{\mathbf{i}} + 1.0\hat{\mathbf{j}}$$

$$\mathbf{a}_A = 3.0\hat{\mathbf{i}} + 7.0\hat{\mathbf{j}} \qquad \mathbf{a}_B = -7.5\hat{\mathbf{i}} - 6.5\hat{\mathbf{j}}$$

Determine the angular velocity and acceleration of the board at this instant. Is the surfer going up a wave, cresting a wave, or descending the front of a wave?

Solution   The angular velocity and acceleration easily are obtained from the direct vector formulae. The relative position, velocity, and acceleration are

$$\mathbf{r}_{B/A} = 1.5\hat{\mathbf{i}} + 0.5\hat{\mathbf{j}}$$

$$\mathbf{v}_{B/A} = -1.5\hat{\mathbf{i}} + 4.5\hat{\mathbf{j}}$$

$$\mathbf{a}_{B/A} = -10.5\hat{\mathbf{i}} - 13.5\hat{\mathbf{j}}$$

ω and α now may be calculated directly:

$$\boldsymbol{\omega} = \frac{\mathbf{r}_{B/A} \times \mathbf{v}_{B/A}}{\mathbf{r}_{B/A} \cdot \mathbf{r}_{B/A}} = \frac{(1.5\hat{\mathbf{i}} + 0.5\hat{\mathbf{j}}) \times (-1.5\hat{\mathbf{i}} + 4.5\hat{\mathbf{j}})}{(1.5\hat{\mathbf{i}} + 0.5\hat{\mathbf{j}}) \cdot (1.5\hat{\mathbf{i}} + 0.5\hat{\mathbf{j}})} = \frac{7.5\hat{\mathbf{k}}}{2.5} = 3\hat{\mathbf{k}} \text{ rad/s}$$

$$\boldsymbol{\alpha} = \frac{\mathbf{r}_{B/A} \times \mathbf{a}_{B/A}}{\mathbf{r}_{B/A} \cdot \mathbf{r}_{B/A}} = \frac{(1.5\hat{\mathbf{i}} + 0.5\hat{\mathbf{j}}) \times (-10.5\hat{\mathbf{i}} - 13.5\hat{\mathbf{j}})}{2.5}$$

$$= \frac{-15\hat{\mathbf{k}}}{2.5} = -6\hat{\mathbf{k}} \text{ rad/s}^2$$

The angular velocity is positive, meaning that the front of the board is rising, but the angular acceleration is negative, meaning that the angular velocity is decreasing in magnitude. The surfer is riding up the wave at this moment.

Problems

5.60 Determine the angular acceleration of the yo-yo in Problem 5.43 if the string is being accelerated at a rate of 20 mm/s² at the instant shown in Figure P5.60.

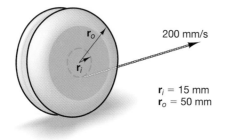

$r_i = 15$ mm
$r_o = 50$ mm

200 mm/s

Figure P5.60 (Fig. P5.43)

5.61 Determine the linear acceleration of collar C in Problem 5.44 if the rod AB has an angular velocity of 2 rad/s and a constant angular acceleration of 1 rad/s² when $\theta = 0°$. The lengths of AB and BC are 100 mm and 350 mm, respectively. (See Figure P5.61.)

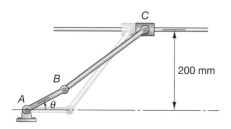

200 mm

Figure P5.61 (Fig. P5.44)

5.62 Determine the velocity and acceleration of collar C in Problem 5.61 when bar AB is in the vertical position ($\theta = 90°$). *Hint: First determine the angular velocity at this moment before beginning the acceleration analysis.*

5.63 Determine the angular acceleration of the disk in Figure P5.63 and the linear acceleration of the center of the disk if the velocity of the end of the rope is 500 mm/s to the right, the acceleration of the end of the rope is 20 mm/s² to the left, and the disk rolls without slipping.

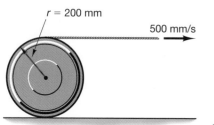

$r = 200$ mm

500 mm/s

Figure P5.63

5.64 The position, velocity, and acceleration of two points on a rigid body in plane motion are given at a instant of time as

$$\mathbf{r}_A = \begin{bmatrix} 3 \\ 1.5 \\ 0 \end{bmatrix} \text{m} \qquad \mathbf{r}_B = \begin{bmatrix} 5 \\ -2 \\ 0 \end{bmatrix} \text{m}$$

$$\mathbf{v}_A = \begin{bmatrix} 1.2 \\ 0.6 \\ 0 \end{bmatrix} \text{m/s} \qquad \mathbf{v}_B = \begin{bmatrix} 2.6 \\ 1.4 \\ 0 \end{bmatrix} \text{m/s}$$

$$\mathbf{a}_A = \begin{bmatrix} 0.2 \\ -0.4 \\ 0 \end{bmatrix} \text{m/s}^2 \qquad \mathbf{a}_B = \begin{bmatrix} -0.47 \\ -0.04 \\ 0 \end{bmatrix} \text{m/s}^2$$

Determine the angular velocity and the angular acceleration of the body.

5.65 Determine the linear velocity and acceleration of a point on the rigid body in Problem 5.64 having a position vector

$$\mathbf{r}_C = \begin{bmatrix} -2.0 \\ 2.35 \\ 0 \end{bmatrix} \text{m}$$

5.66 A 200-mm radius disk rolls without slipping on a horizontal surface. If the velocity of the center of the disk is 500 mm/s and the acceleration of the center of the disk is 200 mm/s² at a particular instant, determine the angular velocity and the angular acceleration of the disk.

5.67 Determine the velocity and acceleration of a point on the top of the disk in Problem 5.66.

5.68 The triangular plate in Figure P5.68 undergoes plane motion. If the velocity of point A and the direction of the velocity vector of point B on the plate are known, determine the angular velocity of the plate and the velocity of point C. The velocities are given by

$$\mathbf{v}_A = \begin{pmatrix} 200 \\ -200 \\ 0 \end{pmatrix} \text{mm/s} \qquad \mathbf{v}_B = v_B \begin{pmatrix} 0.275 \\ 0.962 \\ 0 \end{pmatrix} \text{mm/s}$$

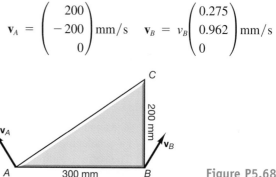

A 300 mm B

200 mm

Figure P5.68

💻 **5.69** Determine the angular acceleration of the plate in Problem 5.68 if the accelerations of point A and point B are, respectively,

$$\mathbf{a}_A = \begin{pmatrix} 250 \\ 0 \\ 0 \end{pmatrix} \text{mm/s}^2 \quad \mathbf{a}_B = \begin{pmatrix} -2450 \\ -450 \\ 0 \end{pmatrix} \text{mm/s}^2$$

💻 **5.70** A collar attached to a rod base by a spring slides along a smooth rod, as shown in Figure P5.70. The position and velocity of the collar and rod base at an instant of time are given by

$$\mathbf{r}_B = \begin{pmatrix} 1.00 \\ -1.00 \\ 0 \end{pmatrix} \text{ft} \quad \mathbf{v}_B = \begin{pmatrix} 1.500 \\ -2.000 \\ 0 \end{pmatrix} \text{ft/s}$$

$$\mathbf{r}_C = \begin{pmatrix} 2.200 \\ 1.300 \\ 0 \end{pmatrix} \text{ft} \quad \mathbf{v}_C = \begin{pmatrix} -4.475 \\ 3.373 \\ 0 \end{pmatrix} \text{ft/s}$$

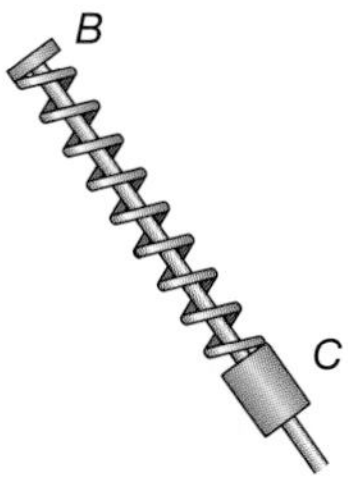

B

C

Figure P5.70

If the system moves in plane motion, determine the angular velocity of the rod and the rate at which the spring stretches. *Hint: Determine the velocity of the collar along the rod, and then treat the collar as being rigidly attached to the rod.*

5.71 Determine the angular acceleration of the rod and the acceleration of the collar along the rod in Problem 5.70 if the accelerations of the base and the collar are, respectively,

$$\mathbf{a}_B = \begin{pmatrix} 1.2 \\ 0 \\ 0 \end{pmatrix} \text{ft/s}^2 \quad \mathbf{a}_C = \begin{pmatrix} -4.762 \\ -19.840 \\ 0 \end{pmatrix} \text{ft/s}^2$$

Hint: Use the same solution method as in Problem 5.50.

5.72 Determine the angular velocity if the position and velocity of two points A and B on a rigid body are

$$\mathbf{r}_A = 3\hat{\mathbf{i}} + 4\hat{\mathbf{j}} \text{ m} \qquad \mathbf{v}_A = \hat{\mathbf{i}} - \hat{\mathbf{j}} \text{ m/s}$$
$$\mathbf{r}_B = 5\hat{\mathbf{i}} - \hat{\mathbf{j}} \text{ m} \qquad \mathbf{v}_B = 11\hat{\mathbf{i}} + 5\hat{\mathbf{j}} \text{ m/s}$$

5.73 Determine the angular acceleration of the rigid body in Problem 5.72 if the linear accelerations of A and B are

$$\mathbf{a}_A = 2\hat{\mathbf{i}} + 2\hat{\mathbf{j}} \text{ m/s}^2 \quad \mathbf{a}_B = 15\hat{\mathbf{i}} - 20\hat{\mathbf{j}} \text{ m/s}^2.$$

5.9   KINEMATICS OF A SYSTEM OF RIGID BODIES

The individual components of many machines may be modeled as rigid bodies and then combined with each other to form a system of rigid bodies that produces a desired motion. This concept is fundamental to robotics and machine design. The kinematic equations, based upon the relative motion of points on a single rigid body, are completely applicable to each rigid body within the system.

For each rigid body moving in plane motion, a single vector relationship may be written for the relative velocity and another for the relative acceleration. So, we have two scalar equations for the velocity and two scalar equations for the acceleration of each body in the system. If there were n bodies in the system in plane motion, then $2n$ scalar equations can be written for the relative velocities, and therefore, $2n$ unknown components of velocities can be determined. For each rigid body moving in plane motion, it has been shown in Section 5.5 that there are nine kinematic variables. Therefore, there will be $9n$

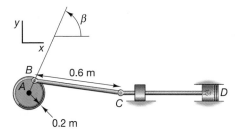

Figure 5.31

variables appearing in the velocity equations for the system. Since only $2n$ variables may be determined by the relative velocity equations, the other $7n$ variables must be specified or determined by other means. Many of these variables will be specified by the constraints to the motion. This method allows us to take an organized approach to the solution of the kinematics of such a system of rigid bodies.

For the system of rigid bodies shown in Figure 5.31, rotational motion is transformed into oscillatory rectilinear translation. The method of solution of the kinematics of a system of rigid bodies will be demonstrated through the analysis of motion of the machine in this figure.

This system may be considered to be made up of three moving rigid bodies: the rotating disk A, the link BC, and the piston CD. As the disk rotates, the piston is driven in and out of the cylinder. The disk rotates about a fixed axis, the piston CD is in rectilinear translation, and the link BC is in general plane motion. The disk and the piston have obvious constraints to their motions. There are three rigid bodies in the system, and therefore, 27 kinematic variables appear in the relative velocity equations. As discussed earlier, six scalar equations can be written for the velocities, and therefore, 21 variables must be specified in the problem. Many of these variables will appear as constraints to a motion. For example, point B on the link BC is also a point on the disk, and as such, the position, velocity, and acceleration of this point are equal when considered as part of the disk and as part of link BC. Thus, two relations are given between the position components and the velocity or acceleration components. Point C on link BC is also a point on the piston CD, and as such, the position, velocity, and acceleration components are the same. Point A is constrained such that its velocity and acceleration are zero. The piston CD is constrained to move in a horizontal direction (x-direction) only, and therefore, the velocity in the y-direction of points C and D is zero, and the angular velocity of the rod is zero. Now, the number of variables known can be counted: the position of points A and B on the disk (four known variables); the position of points B and C on the link (four known variables); the position of points C and D on the piston (four known variables); the velocity of point A (two known variables); the velocity of point B on disk equals the velocity of point B on the link (two known variables); the velocity of point C on the link equals the velocity of point C on the piston (two known variables); and the velocity of point D in the y-direction and the angular velocity of the piston (two more known variables). Thus, we have a total of 20 variables. These 20 specifications and the six scalar equations are then one variable short of the required 27 variables. Finally, it is necessary to specify the velocity of some point in the system. For example, the angular velocity of the disk could be specified or the linear velocity of the piston given. One of these velocities normally is specified in design problems, but sometimes, other design considerations can specify the angular velocity of the link or a component of the linear velocity of some other point in the system.

The count of the acceleration equations is similar to that given for the velocity equations, except there are three additional angular velocities to the 27 acceleration or position variables, bringing the total number of varibles to 30 kinematic variables in the acceleration equations. However, since these variables are known from the velocity analysis, the specification is exactly the same as for the velocity.

Consider that, at the instant shown in Figure 5.31, the disk is rotating counterclockwise at a constant rate of 10 radians per second. If the angle β is 30°, determine the velocity and acceleration of the piston at this time. First, we consider the disk as a rigid body and write the equation for the linear velocity of point B:

$$\mathbf{v}_B = \mathbf{v}_A + \boldsymbol{\omega}_A \times \mathbf{r}_{B/A}$$

where $\mathbf{v}_A = 0$, as it is the center of the disk,

$$\boldsymbol{\omega}_A = 10\hat{\mathbf{k}} \text{ rad/s, and}$$
$$\mathbf{r}_{B/A} = 0.2(\cos 30\hat{\mathbf{i}} + \sin 30\hat{\mathbf{j}}) = 0.1732\hat{\mathbf{i}} + 0.100\hat{\mathbf{j}}$$
$$\mathbf{v}_B = -1.00\hat{\mathbf{i}} + 1.732\hat{\mathbf{j}} \text{ m/s}$$

The acceleration of point B is

$$\mathbf{a}_B = \mathbf{a}_A + \boldsymbol{\alpha}_A \times \mathbf{r}_{B/A} - (\omega_A^2)\mathbf{r}_{B/A}$$

But α and $\mathbf{a}_A$ are both zero, so

$$\mathbf{a}_B = -17.32\hat{\mathbf{i}} - 10\hat{\mathbf{j}} \text{ m/s}^2$$

Now the link BC can be analyzed, knowing the velocity and acceleration of point B and that point C moves in rectilinear translation. The following analysis first considers the velocity in order to determine the angular velocity, and then it examines the acceleration, which is dependent upon the angular velocity. The link makes an angle with the horizontal θ, whose sine is $1/6$ or $\theta = 9.59°$. So, we get

$$\mathbf{v}_C = \mathbf{v}_B + \boldsymbol{\omega}_{BC} \times \mathbf{r}_{C/B}$$
$$v_C\hat{\mathbf{i}} = -1.00\hat{\mathbf{i}} + 1.732\hat{\mathbf{j}} + \omega_{BC}\hat{\mathbf{k}} \times [0.6(\cos 9.59\hat{\mathbf{i}} - 0.1\hat{\mathbf{j}})]$$
$$v_C\hat{\mathbf{i}} = -1.00\hat{\mathbf{i}} + 1.732\hat{\mathbf{j}} + 0.060\omega_{BC}\hat{\mathbf{i}} + 0.592\omega_{BC}\hat{\mathbf{j}}$$

Equating scalar components, we obtain

$$v_C = -1.00 + 0.06\omega_{BC}$$
$$0 = 1.732 + 0.592\omega_{BC}$$

Therefore,

$$\omega_{BC} = -2.926 \text{ rad/s and } v_C = -1.176 \text{ m/s}$$

The acceleration of link BC may now be analyzed, as the angular velocity of the link is known:

$$\mathbf{a}_C = \mathbf{a}_B + \alpha_{BC} \times \mathbf{r}_{C/B} - \omega_{BC}^2\mathbf{r}_{C/B}$$
$$a_C\hat{\mathbf{i}} = -17.32\hat{\mathbf{i}} - 10.00\hat{\mathbf{j}} + \alpha_{BC}\hat{\mathbf{k}} \times (0.592\hat{\mathbf{i}} - 0.100\hat{\mathbf{j}})$$
$$- (2.926)^2(0.592\hat{\mathbf{i}} - 0.100\mathbf{j})$$
$$a_C\hat{\mathbf{i}} = -17.32\hat{\mathbf{i}} - 10.00\hat{\mathbf{j}} + 0.100\alpha_{BC}\hat{\mathbf{i}} + 0.592\alpha_{BC}\hat{\mathbf{j}} - 5.068\hat{\mathbf{i}} + 0.856\hat{\mathbf{j}}$$

Equating components, we obtain

$$a_C = -17.32 + 0.100\alpha_{BC} - 5.068$$
$$0 = -10.00 + 0.592\alpha_{BC} + 0.851$$

Therefore,

$$\alpha_{BC} = 15.45 \text{ rad/s}^2 \text{ and } a_C = -20.84 \text{ m/s}^2$$

The velocity of the piston is 1.176 m/s to the left, and the acceleration of the piston is 20.84 m/s² to the left at the instant shown in Figure 5.31.

We have analyzed the machine shown in Figure 5.31 for the position where the angle β equaled 30°. Now let us consider what information we obtained by this analysis. We know the velocities and acceleration of the three components for this specified position but not for any other position. If the entire kinematics of the machine were desired, this analysis would have to be repeated over and over, maybe at 2° intervals. This would give us a step-by-step analysis, would make a nice set of homework problems, but would be useless because of the time required for any actual design analysis.

There must be a way to obtain general equations for machines (as shown in Figure 5.31) so parameters can be varied and an optimal design developed. That is the topic of Section 5.10, and this section should be considered a tedious warm-up for that section.

Sample Problem 5.20

If point C on the linkage shown in the accompanying diagram is moving to the right with a velocity of 100 mm/s and an acceleration of 10 mm/s² at the instant shown in the figure, determine the angular velocities and accelerations of links AB and BC.

Solution   First, we examine the motion of the system. It is constrained in three manners:

1. The velocity of point A is zero.
2. Point C must move in the horizontal direction.
3. Point B is a common point on links AB and BC, and its absolute velocity and acceleration are the same for both links.

Select a coordinate system and specify the relative position vectors and velocities:

$$\sin \theta = \frac{100}{200} \qquad \theta = 30°$$

$$\sin \beta = \frac{100}{250} \qquad \beta = 26.2°$$

$$\mathbf{r}_{B/A} = 200 \,(\cos 30\hat{\mathbf{i}} + \sin 30\hat{\mathbf{j}}) \quad \mathbf{r}_{B/C} = 250(-\cos 26.2\hat{\mathbf{i}} + \sin 26.2\hat{\mathbf{j}}) \text{ mm}$$

$$\mathbf{v}_A = 0 \quad \mathbf{v}_B = v_{Bx}\hat{\mathbf{i}} + v_{By}\hat{\mathbf{j}} \quad \mathbf{v}_C = 100\hat{\mathbf{i}} \text{ mm/s}$$

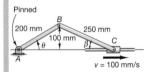
Pinned
200 mm   B   250 mm
100 mm
θ β C
A
$v = 100$ mm/s

Let ω_{AB} and ω_{BC} be the angular velocities of the two links. For each link, we can express the velocity of point B in terms of its angular velocity. Using the constraint that the two expressions for the velocity of B must be equal, we may find the angular velocities. For link AB, we determine that

$$\mathbf{v}_B = \mathbf{v}_A + \boldsymbol{\omega}_{AB} \times \mathbf{r}_{B/A}$$

$$\mathbf{v}_B = \omega_{AB}\hat{\mathbf{k}} \times 200(0.866\hat{\mathbf{i}} + 0.500\hat{\mathbf{j}})$$

$$\mathbf{v}_B = -100\omega_{AB}\hat{\mathbf{i}} + 173.2\omega_{AB}\hat{\mathbf{j}}$$

For link BC, we determine that

$$\mathbf{v}_B = \mathbf{v}_C + \boldsymbol{\omega}_{BC} \times \mathbf{r}_{B/C}$$

$$\mathbf{v}_B = +100\hat{\mathbf{i}} + \omega_{BC}\hat{\mathbf{k}} \times 250(-0.917\hat{\mathbf{i}} + 0.400\hat{\mathbf{j}})$$

$$\mathbf{v}_B = (100 - 100\omega_{BC})\hat{\mathbf{i}} - 229\omega_{BC}\hat{\mathbf{j}}$$

Equating the two expressions for the velocity at B and expanding to scalar notation yields

$$\hat{\mathbf{i}}: \quad -100\omega_{AB} = 100 - 100\omega_{BC}$$

$$\hat{\mathbf{j}}: \quad 173.2\omega_{AB} = -229\omega_{BC}$$

We can solve these two simultaneous equations:

$$\omega_{AB} = -0.57 \text{ rad/s} \qquad \omega_{BC} = 0.431 \text{ rad/s}$$

BC is rotating clockwise at this instant, and AB is rotating counterclockwise. These results agree with the conceptual examination of the problem.

For link AB, the angular acceleration may be examined by use of the relative acceleration equation:

$$\mathbf{a}_B = \mathbf{a}_A + \boldsymbol{\alpha}_{AB} \times \mathbf{r}_{B/A} - \omega_{AB}^2 \mathbf{r}_{B/A}$$

$$\mathbf{a}_B = 0 + \alpha_{AB}\hat{\mathbf{k}} \times (173.2\hat{\mathbf{i}} + 100\hat{\mathbf{j}}) - (0.57)^2(173.2\hat{\mathbf{i}} + 100\hat{\mathbf{j}})$$

$$\mathbf{a}_B = -100\alpha_{AB}\hat{\mathbf{i}} + 173.2\alpha_{AB}\hat{\mathbf{j}} - 56.3\hat{\mathbf{i}} + 32.5\hat{\mathbf{j}}$$

Link BC can be analyzed in the same manner:

$$\mathbf{a}_B = \mathbf{a}_C + \boldsymbol{\alpha}_{BC} \times \mathbf{r}_{B/C} - \omega_{BC}^2 \mathbf{r}_{B/C}$$

$$\mathbf{a}_B = 10\hat{\mathbf{i}} + \alpha_{BC}\hat{\mathbf{k}} \times (-229\hat{\mathbf{i}} + 100\hat{\mathbf{j}}) - (0.431)^2(-229\hat{\mathbf{i}} + 100\hat{\mathbf{j}})$$

$$\mathbf{a}_B = 10\hat{\mathbf{i}} - 100\alpha_{BC}\hat{\mathbf{i}} - 229\alpha_{BC}\hat{\mathbf{j}} + 42.5\hat{\mathbf{i}} - 18.6\hat{\mathbf{j}}$$

Equating the two expressions for the acceleration of point B and equating the scalar components, we get

$$-100\alpha_{AB} + 100\alpha_{BC} = 108.8$$

$$173.2\alpha_{AB} + 229\alpha_{BC} = -51.1$$

Therefore,

$$\alpha_{AB} = -0.747 \text{ rad/s}^2 \qquad \alpha_{BC} = 0.342 \text{ rad/s}^2$$

Problems

5.74 If link AB in Figure P5.74 is rotating counterclockwise at a constant angular velocity of 2 rad/s, determine the angular velocity and the angular acceleration of links BC and CD.

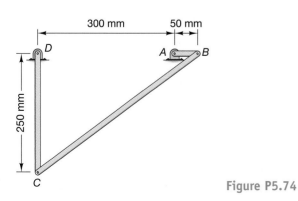

Figure P5.74

5.75 Determine the linear velocity and the linear acceleration of a point at the middle of bar BC in Problem 5.74.

5.76 Determine the angular velocity and the angular acceleration of links BC and CD in Problem 5.74 after link AB has rotated 30° from the position shown in the figure.

5.77 Determine the angular velocity and the angular acceleration of links BC and CD in Problem 5.74 after link AB has rotated 90° from the position shown in the figure.

5.78 Determine the angular velocity and the angular acceleration of links BC and CD in Problem 5.74 after link AB has rotated 180° from the position shown in the figure.

5.79 A bar of length $3r$ is attached to a disk of radius r, as shown in Figure P5.79. If the disk rolls without slipping to the left at a constant velocity of v (the velocity of the center of the disk), determine a general expression for the angular velocity of the bar.

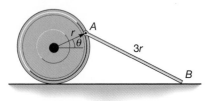

Figure P5.79

5.80 Determine a general expression for the acceleration of point B on the bar in Problem 5.79.

5.81 Determine a general expression for the linear velocity of point C in Figure P5.81 if the link AB rotates at a constant angular velocity ω.

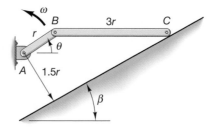

Figure P5.81

5.82 Determine a general expression for the linear acceleration of point C in Problem 5.81.

5.10 ANALYSIS OF PLANE MOTION IN TERMS OF A PARAMETER

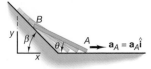

Figure 5.32

Most of the problems we have examined thus far have been analyzed for a given position or an equivalent instant in time, and we determined the velocities and accelerations of points on rigid bodies. This type of analysis is very useful, but it would have to be repeated for each new position of the body or system of bodies representing a form of *step-by-step integration.* In general, the acceleration is dependent upon the square of the current angular velocity, and this step-by-step integration is subject to difficulties. It obviously would be valuable to obtain a general expression for the motion in terms of time or position. Many of the problems involving rotation about a fixed axis assumed that the angular position, velocity, or acceleration was known, and a general analysis was conducted. A similar approach may be used for general plane motion, but, as would be expected, this approach is mathematically more difficult to formulate and solve.

As an example of this type of analysis, consider a bar whose end A is accelerated along a horizontal surface at a constant rate. The bar starts from rest against the inclined plane of angle β and slides to the floor, as illustrated in Figure 5.32. At any time, the bar makes an angle θ with the horizontal plane. Initially, $\theta = \beta$, and when the bar completes the slide, $\theta = 0$. Point A is subjected to a constant acceleration a to the right.

Point A is constrained to move in the x-direction under constant acceleration, and its motion may be written as follows:

$$\ddot{x}_A = a$$
$$\dot{x}_A = at$$
$$x_A = \frac{at^2}{2} + l\cos\beta$$

The position of point B may be expressed in terms of the parameter θ:

$$\mathbf{r}_{B/A} = -l\cos\theta\hat{\mathbf{i}} + l\sin\theta\hat{\mathbf{j}}$$

We may write the velocity of point B as $\mathbf{v}_B = \mathbf{v}_A + \boldsymbol{\omega} \times \mathbf{r}_{B/A}$

$$\mathbf{v}_B = (\dot{x}_A - l\sin\theta)\hat{\mathbf{i}} - l\cos\theta\dot{\theta}\hat{\mathbf{j}}$$

But B is constrained to move along the inclined surface, so

$$\mathbf{v}_B = v_B\cos\beta\hat{\mathbf{i}} - v_B\sin\beta\hat{\mathbf{j}}$$

Equating the two expressions and separating into scalar components, we get

$$v_B\cos\beta = at - l\sin\theta\dot{\theta}$$

$$-v_B\sin\beta = -l\cos\theta\dot{\theta}$$

Now, we multipy the first equation in the above expression by sine β and the second equation by cosine β and add the results together, yielding

$$-\dot{\theta}l(\sin\beta\sin\theta + \cos\beta\cos\theta) + at\sin\beta = 0$$

This expression may be written as $\cos(\beta - \theta)$, leaving us with

$$\cos(\beta - \theta)\dot{\theta} = \frac{at}{l}\sin\beta$$

This expression may be integrated to determine θ:

$$\int_{\beta}^{\theta}\cos(\beta - \theta)d\theta = \int_{0}^{t}\frac{at}{l}\sin\beta dt$$

$$\sin(\beta - \theta) = \frac{at^2}{2l}\sin\beta$$

Therefore,

$$\theta = \beta - \sin^{-1}\left(\frac{at^2}{2l}\sin\beta\right)$$

The angular velocity may be expressed as a function of time:

$$\dot{\theta} = -\frac{at\sin\beta}{\sqrt{l^2 - \frac{a^2t^4}{4}\sin^2\beta}}$$

Examination of the displacement of point A shows that the bar is flat to the surface when the angle $\theta = 0$: therefore,

$$\frac{at^2}{2} = l \text{ or } t = \sqrt{\frac{2l}{a}}$$

and the (maximum) angular velocity when the rod hits the surface is

$$\dot{\theta} = -\sqrt{\frac{2a}{l}}\tan\beta$$

Sample Problem 5.21

Develop a general solution for the motion of the piston shown in the accompanying diagram if the disk is rotating counterclockwise at a constant rate ω.

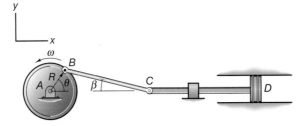

Solution   The angle θ may be written in terms of the angular velocity of the disk:

$$\theta = \omega t$$

The relative position vector of point B relative to the disk center A is

$$\mathbf{r}_{B/A} = R(\cos \theta \hat{\mathbf{i}} + \sin \theta \hat{\mathbf{j}})$$

Therefore, the velocity of point B may be written

$$\mathbf{v}_B = \omega \hat{\mathbf{k}} \times \mathbf{r}_{B/A}$$
$$\mathbf{v}_B = \omega R(-\sin \theta \hat{\mathbf{i}} + \cos \theta \hat{\mathbf{j}})$$

Now, consider the link BC. The velocity of point C may be related to the velocity of point B by

$$\mathbf{v}_C = \mathbf{v}_B + \boldsymbol{\omega}_{BC} \times \mathbf{r}_{C/B}$$
$$\mathbf{r}_{C/B} = l(\cos \beta \hat{\mathbf{i}} - \sin \beta \hat{\mathbf{j}})$$
$$v_C \hat{\mathbf{i}} = (-R\omega \sin \theta + l\omega_{BC} \sin \beta)\hat{\mathbf{i}} + (R\omega \cos \theta + l\omega_{BC} \cos \beta)\hat{\mathbf{j}}$$

Separating the above equation into scalar components, we obtain

$$v_C = -R\omega \sin \theta + l\omega_{BC} \sin \beta$$
$$0 = R\omega \cos \theta + l\omega_{BC} \cos \beta$$

Solving for the unknowns, v_C *and* ω_{BC}, we find that

$$\omega_{BC} = -\frac{R\omega \cos \theta}{l \cos \beta}$$

$$v_C = -R\omega \sin \theta - \frac{R\omega l \sin \beta \cos \theta}{l \cos \beta}$$

The angle θ has already been related to time, and we will now relate the angle β to θ and, therefore, to time. The length of line BD is equal to $R \sin \theta$ and to $l \sin \beta$, and the following relations apply:

$$R \sin \theta = l \sin \beta$$

$$\sin \beta = \frac{R}{l} \sin (\omega t)$$

$$\cos \beta = \sqrt{1 - \sin^2 \beta} = \sqrt{1 - \left(\frac{R^2}{l^2}\right) \sin^2 (\omega t)}$$

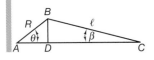

The angular velocity of the link BC and the velocity of C, (the piston) are, respectively,

$$\omega_{BC} = -\frac{R\omega \cos(\omega t)}{l\sqrt{1 - \left(\frac{R^2}{l^2}\right)\sin^2(\omega t)}}$$

$$v_C = -R\omega \sin(\omega t) - \frac{R\sin(\omega t)R\omega \cos(\omega t)}{l\sqrt{1 - \left(\frac{R^2}{l^2}\right)\sin^2(\omega t)}}$$

The acceleration of the piston can now be analyzed in a similar manner:

$$\mathbf{a}_B = -\omega^2 R(\cos\theta\hat{\mathbf{i}} + \sin\theta\hat{\mathbf{j}})$$

$$\mathbf{a}_C = \mathbf{a}_B + \boldsymbol{\alpha}_{BC} \times \mathbf{r}_{C/B} - \omega_{BC}^2\mathbf{r}_{C/B}$$

$$a_C\hat{\mathbf{i}} = -\omega^2 R\cos\theta\hat{\mathbf{i}} + \omega^2 R\sin\theta\hat{\mathbf{j}} + l\alpha_{BC}\sin\beta\hat{\mathbf{i}} + l\alpha_{BC}\cos\beta\hat{\mathbf{j}}$$
$$- \omega_{BC}^2 l\cos\beta\hat{\mathbf{i}} + \omega_{BC}^2 l\sin\beta\hat{\mathbf{j}}$$

Therefore,

$$\alpha_{BC} = -\frac{(\omega^2 + \omega_{BC}^2)R\sin\theta}{l\cos\beta}$$

$$a_C = -[\omega^2 R\cos\theta + \omega_{BC}^2 l\cos\beta] - \frac{(\omega^2 + \omega_{BC}^2)R\sin\theta}{l\cos\beta}$$

Sample Problem 5.22

Determine the time history of the position and velocity of point D on the wheel shown on the accompanying figure. The wheel is rolling to the right without slipping at a constant angular velocity ω. Consider that point D is at the top of the wheel when time is zero.

Solution

Point C, the point of contact between the wheel and the surface, has zero velocity, and the velocity of the wheel center A is in the horizontal direction; therefore,

$$\mathbf{v}_A = \mathbf{v}_C + \boldsymbol{\omega} \times \mathbf{r}_{A/C}$$
$$v_A\hat{\mathbf{i}} = 0 + (-\omega)\hat{\mathbf{k}} \times r\hat{\mathbf{j}} = \omega r\hat{\mathbf{i}}$$

Note that ω is negative in this problem, so

$$v_A = \omega r$$

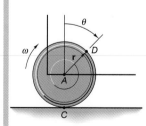

At any instant of time, the angle θ is equal to ωt, so the velocity of point D may be related to the velocity of point A as

$$\mathbf{v}_D = \mathbf{v}_A + \boldsymbol{\omega} \times \mathbf{r}_{D/A}$$

$$\boldsymbol{\omega} = -\omega\hat{\mathbf{k}} = -\frac{v_A}{r}\hat{\mathbf{k}}$$

$$\mathbf{r}_{D/A} = r(\sin\theta\hat{\mathbf{i}} + \cos\theta\hat{\mathbf{j}})$$

$$\mathbf{v}_D = v_A[(1 + \cos\theta)\hat{\mathbf{i}} - \sin\theta\hat{\mathbf{j}}]$$

Note that $v_A = \omega r$ is a constant and the velocity of the wheel center. So,

$$\mathbf{v}_D = v_A\left\{\left[1 + \cos\left(\frac{v_A t}{r}\right)\right]\hat{\mathbf{i}} - \sin\left(\frac{v_A t}{r}\right)\hat{\mathbf{j}}\right\}$$

This expression completely describes the velocity vector as a function of time. Each component may be plotted as a function of time using any of the available computer programs. The position vector of point D also may be formulated as a function of time by integration of the velocity vector:

$$\bar{\mathbf{r}}_D = \int v_D dt = \left\{ v_A\left[t + \frac{r}{v_A} \sin\left(\frac{v_A t}{r}\right) \right]\hat{\mathbf{i}} + \frac{r}{v_A}\cos\left(\frac{v_A t}{r}\right)\hat{\mathbf{j}} \right\}$$

$$\bar{\mathbf{r}}_D = \left[v_A t + r\sin\left(\frac{v_A t}{r}\right) \right]\hat{\mathbf{i}} + r\cos\left(\frac{v_A t}{r}\right)\hat{\mathbf{j}}$$

The position vector is now completely specified as a function of the parameter time.

Sample Problem 5.23

In the accompanying figure, a rod of length L is pinned to a wheel of radius r at point P. If the wheel rolls at a constant angular velocity ω without slipping, develop a general expression for the velocity of point Q with time.

Solution   Select a general parmeter θ, and formulate all variables in terms of this parameter. The velocity of point C is zero as the wheel rolls without slipping with a constant angular velocity. The relative position vector from C to P is

$$\mathbf{r}_{P/C} = r\cos\theta\hat{\mathbf{i}} + r(1 + \sin\theta)\hat{\mathbf{j}}$$

Therefore,

$$\mathbf{v}_P = \mathbf{v}_C + \boldsymbol{\omega} \times \mathbf{r}_{P/C}$$

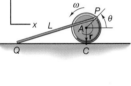

$$\mathbf{v}_P = 0 + \omega\hat{\mathbf{k}} \times r[\cos\theta\hat{\mathbf{i}} + (1 + \sin\theta)\hat{\mathbf{j}}]$$

$$\mathbf{v}_P = -r\omega[(1 + \sin\theta)\hat{\mathbf{i}} - \cos\theta\hat{\mathbf{j}}]$$

The velocity of point Q can now be related to the velocity of point P, as they are points on a rigid body. The relative velocity equation will be written in terms of the angle β, and then β will be expressed in terms of θ:

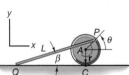

$$\mathbf{v}_Q = \mathbf{v}_P + \boldsymbol{\omega}_{PQ} \times \mathbf{r}_{Q/P}$$

$$\mathbf{r}_{Q/P} = -L(\cos\beta\hat{\mathbf{i}} + \sin\beta\hat{\mathbf{j}})$$

But $\mathbf{v}_Q$ is constrained to move in the x-direction, so

$$\mathbf{v}_Q = v_Q\hat{\mathbf{i}} = -r\omega(1 + \sin\theta)\hat{\mathbf{i}} + r\omega\cos\theta\hat{\mathbf{j}}$$

$$+ \omega_{PQ}\hat{\mathbf{k}} \times (-L)(\cos\beta\hat{\mathbf{i}} + \sin\beta\hat{\mathbf{j}})$$

$$v_Q\hat{\mathbf{i}} = [-r\omega(1 + \sin\theta) + L\omega_{PQ}\sin\beta]\hat{\mathbf{i}} + [r\omega\cos\theta - L\omega_{PQ}\cos\beta]\hat{\mathbf{j}}$$

Equating components, we get

$$\omega_{PQ} = \frac{r\omega\cos\theta}{L\cos\beta}$$

$$v_Q = -r\omega[1 + \sin\theta - \cos\theta\tan\beta]$$

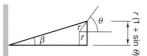

The angle β may be related to θ by consideration of the triangle in the diagram, so

$$\sin \beta = \frac{r(1 + \sin \theta)}{L}$$

$$\cos \beta = \sqrt{1 - \sin^2 \beta}$$

$$\cos \beta = \sqrt{1 - \frac{r^2}{L^2}(1 + 2 \sin \theta + \sin^2 \theta)}$$

$$\tan \beta = \frac{r/L (1 + \sin \theta)}{\sqrt{1 - \frac{r^2}{L^2}(1 + 2 \sin \theta + \sin^2 \theta)}}$$

Although this expression is algebraically complicated, it may be substituted into the expression for velocity, noting that at any time, $\theta = \omega t$. This result expresses the velocity of point Q as a function of time, as desired. A detailed example for particular values of r, L, and ω is shown in the Computational Supplement.

Problems

Note: Linkages such as the four-bar linkages shown in Figures P5.83 and P5.90 and the rig shown in Figure P5.92 present special problems in determining the relationship between the linkage angles for a genral solotuion. These relationships can be determined numerically and should always be checked to ensure that the linkages do not jam at any position.

5.83 Write a general expression for the angular velocity of link CD in Problem 5.74 in terms of the angle θ of link AB, measured counterclockwise from the position shown in Figure P5.83 if link AB is rotating counterclockwise at a constant angular velocity ω.

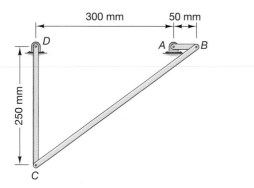

Figure P5.83 (Figure P5.74)

5.84 Determine a general expression for the angular velocity and the angular acceleration of link BC in Problem 5.83 in terms of the position and angular velocity of link AB.

5.85 Determine the velocity and acceleration of point B on the rod in Problem 5.79 and Figure P5.85, and plot the velocity and acceleration of this point for one complete revolution of the disk, assuming that $r = 200$ mm and the constant velocity of the center of the disk v is 200 mm/s. Assume disk rolls without slipping to the left.

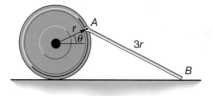

Figure P5.85 (Figure P5.79)

5.86 Determine the velocity and acceleration of point C in Problem 5.81 and Figure P5.86 for one complete revolution of link AB if $\omega = 2$ rad/s, $r = 500$ mm, and $\beta = 30°$.

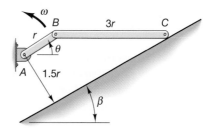

Figure P5.86 (Figure P5.81)

5.87 A rod slides down a smooth incline, as shown in Figure P5.87. Write the expression for the angular velocity of the rod in terms of the angle θ if point A on the rod moves at a constant velocity V. Plot the angular velocity if $\beta = 30°$, $L = 1.5$ m, and $V = 2$ m/s.

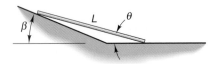

Figure P5.87

5.88 Determine a general expression for the velocity and acceleration of the collar C in Figure P5.88 if rod AB starts from rest at the position shown in the figure and has an angular acceleration of 0.2 rad/s². Plot the velocity and acceleration for two revolutions of rod AB.

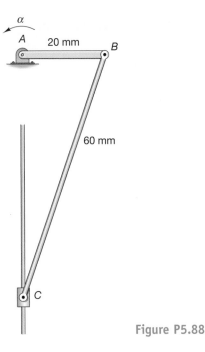

Figure P5.88

5.89 Determine a general expression for the angular velocity and the angular acceleration of rod BC in Problem 5.88.

5.90 If bar AB rotates at a constant angular velocity of 2 rad/s in a counterclockwise direction as shown in Figure P5.90, determine a general expression for the angular velocity and the angular acceleration of bar CD.

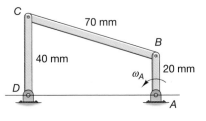

Figure P5.90

5.91 Determine a general expression for the angular velocity and the angular acceleration of bar BC in Problem 5.90.

5.92 For the oil pumping rig shown in Figure P5.92, if the motor drives the 500-mm radius disk at a constant velocity of 1.5 rad/s, determine a general expression for the velocity and acceleration of the pump rod at point D.

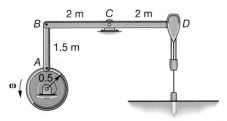

Figure P5.92

5.93 Determine a general expression for the angular velocity and the angular acceleration for bar BCD in Problem 5.92.

*5.11 GENERAL THREE-DIMENSIONAL MOTION OF A RIGID BODY

Consider the rigid body shown in two different positions in Figure 5.33. These two positions correspond to two different times t and t', respectively. Two points, A and B, have been identified on the body. Consider only the change in the relative position vector $\Delta \mathbf{r}_{B/A}$. At any given instant of time, this is equivalent to the relative velocity of B to A.

$$\mathbf{v}_{B/A} = \frac{\lim}{\Delta t \to 0} \frac{\Delta \mathbf{r}_{B/A}}{\Delta t}$$

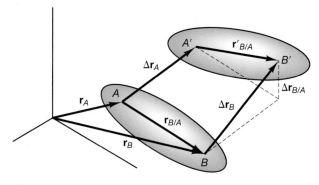

Figure 5.33

$\mathbf{v}_{B/A}$ must be perpendicular to $\mathbf{r}_{B/A}$ for a rigid body, and the general relation for any rigid body is

$$\mathbf{v}_{B/A} = \boldsymbol{\omega} \times \mathbf{r}_{B/A} \qquad (5.105)$$

where $\boldsymbol{\omega}$ is the angular velocity of the body.

Note that this relation is not restricted to plane motion, but holds for any motion of a rigid body. In plane motion, the angular velocity vector is perpendicular to the plane of motion and, therefore, perpendicular to both the relative position and the relative velocity vectors. If at any instant of time, the positions and the linear velocities of both points A and B are known, it would appear that the angular velocity could be obtained from the three scalar equations from this single vector equation. The scalar equations are written as

$$\omega_y z_{B/A} - \omega_z y_{B/A} = v_{B/Ax}$$
$$\omega_z x_{B/A} - \omega_x z_{B/A} = v_{B/Ay} \qquad (5.106)$$
$$\omega_x y_{B/A} - \omega_y x_{B/A} = v_{B/Az}$$

These expressions appear to be three equations for the three unknown components of the angular velocity. It is instructive to write these three equations in matrix notation:

$$[C][\omega] = [v_{B/A}]$$

$$\begin{bmatrix} 0 & z_{B/A} & -y_{B/A} \\ -z_{B/A} & 0 & x_{B/A} \\ y_{B/A} & -x_{B/A} & 0 \end{bmatrix} \begin{bmatrix} \omega_x \\ \omega_y \\ \omega_z \end{bmatrix} = \begin{bmatrix} v_{B/Ax} \\ v_{B/Ay} \\ v_{B/Az} \end{bmatrix}$$

The first matrix $[C]$ is called the coefficient matrix and involves the components of the relative position vector. In this case, the determinant of this matrix is zero, meaning that the equations are not linearly independent and thus cannot be solved. Students who are familiar with Kramer's Rule may note that the determinant of the coefficient matrix appears in the denominator of the solution and, in this case, would result in division by zero. The value of the determinant of the coefficient matrix is used by most computer programs for the solution of linear systems of equations to monitor for independence of the equations and for ill-conditioned sets of equations. The lack of independency of the equations is conceptually obvious if one notes that the angular velocity vector and the relative position vector have no specific orientation with each other. In plane motion, these two vectors are perpendicular. Note that in three-dimensional applications, any component of the angular

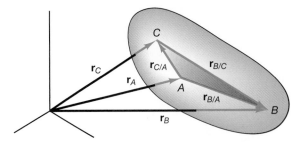

Figure 5.34

velocity that is parallel to the relative position vector would drop out of the vector product and could not be obtained.

The angular velocity of a rigid body in general three-dimensional motion can be determined by considering three points that are not collinear, as shown in Figure 5.34. Two vector equations may be written to relate the angular velocity to the relative position and velocity vectors:

$$\boldsymbol{\omega} \times \mathbf{r}_{B/A} = \mathbf{v}_{B/A} \tag{5.108}$$

$$\boldsymbol{\omega} \times \mathbf{r}_{C/A} = \mathbf{v}_{C/A} \tag{5.109}$$

A third equation also can be written to relate the velocity of C to B:

$$\boldsymbol{\omega} \times \mathbf{r}_{C/B} = \mathbf{v}_{C/B} \tag{5.110}$$

We will not use Eq. (5.110) to determine the angular velocity.

The vector equations Eqs. (5.108) and (5.109) represent an overdetermined system of equations (six scalar equations for the three unknowns) that may be expanded and solved. The solution of an overdetermined system of equations is accomplished by a least-squares algorithm. Consider the system written in matrix notation that is shown in Mathematics Window 5.1.

MATHEMATICS WINDOW 5.1

$$[C][\omega] = [v_{\text{rel}}]$$

where $[C]$ is an $(m \times n)$ matrix and $m > n$. $[\omega]$ is an $(n \times 1)$ matrix, and $[v_{\text{rel}}]$ is a $(m \times 1)$ matrix.

The least-squares solution is obtained by minimizing the squared error in

$$[C][\omega] - [v_{\text{rel}}] = 0$$

Computational software will solve the overdetermined system by different commands.

MATLAB

MATLAB will solve the system by the command:

$$[\omega] = [C]\backslash[v_{\text{rel}}]$$

Mathcad

Mathcad will solve the system by the command:

$$[\omega] = ([C]^{T}[C])^{-1}[C]^{T}[v_{\text{rel}}]$$

A direct vector solution was proposed by Youm and Yoon in 1979[*], using the cross product between the two equations:

$$(\boldsymbol{\omega} \times \mathbf{r}_{B/A}) \times (\boldsymbol{\omega} \times \mathbf{r}_{C/A}) = \mathbf{v}_{B/A} \times \mathbf{v}_{C/A} \tag{5.111}$$

A direct vector solution is possible unless $\boldsymbol{\omega}$, $\mathbf{r}_{B/A}$ and $\mathbf{r}_{C/A}$ lie in a plane. In that case, both sides of Eq. (5.111) are zero. For the general case, the triple vector product on the left side can be expanded to yield a direct vector solution. The first cross product will be viewed as a single vector making up the first vector in the triple vector product:

$$\boldsymbol{\omega}[(\boldsymbol{\omega} \times \mathbf{r}_{B/A}) \cdot \mathbf{r}_{C/A}] - \mathbf{r}_{C/A}[(\boldsymbol{\omega} \times \mathbf{r}_{B/A}) \cdot \boldsymbol{\omega}] = \mathbf{v}_{B/A} \times \mathbf{v}_{C/A} \tag{5.112}$$

The term $(\boldsymbol{\omega} \times \mathbf{r}_{B/A})$ is perpendicular to $\boldsymbol{\omega}$, and the second term is zero. Therefore, Eq. (5.112) can be written as

$$\boldsymbol{\omega}[(\boldsymbol{\omega} \times \mathbf{r}_{B/A}) \cdot \mathbf{r}_{C/A}] = \mathbf{v}_{B/A} \times \mathbf{v}_{C/A} \tag{5.113}$$

The left-hand side of Eq. (5.113) can be written in terms of $\mathbf{v}_{B/A}$ as

$$\boldsymbol{\omega}[\mathbf{v}_{B/A} \cdot \mathbf{r}_{C/A}] = \mathbf{v}_{B/A} \times \mathbf{v}_{C/A}$$
$$\boldsymbol{\omega} = \frac{\mathbf{v}_{B/A} \times \mathbf{v}_{C/A}}{\mathbf{v}_{B/A} \cdot \mathbf{r}_{C/A}} \tag{5.114}$$

We can determine if $\boldsymbol{\omega}$ lies in the plane formed by the relative position vectors (that shaded plane in Figure 5.34) by examining the cross product between their relative velocities. If the two relative velocities are parallel (as will be the case if the angular velocity is coplanar with the relative position vectors), the angular velocity can not be obtained by the direct vector method.

5.11.1 LINEAR AND ANGULAR ACCELERATION

The expressions for the angular acceleration of a body in general three-dimensional motion are obtained by differentiation of the expressions for velocities. Therefore, the relation between the linear acceleration of two points on a rigid body is

$$\mathbf{a}_B = \mathbf{a}_A + \boldsymbol{\alpha} \times \mathbf{r}_{B/A} + \boldsymbol{\omega} \times (\boldsymbol{\omega} \times \mathbf{r}_{B/A}) \tag{5.115}$$

This equation is the same expression that was used for plane motion and is equally valid for three-dimensional motion. The difference between the two conditions is that neither $\boldsymbol{\alpha}$ and $\boldsymbol{\omega}$ are needed to be parallel nor restricted to be perpendicular to the relative position vector or the linear accelerations of points A and B. The analysis of the kinematics of a body in general three-dimensional motion proceeds similarly to that for a body in plane motion. Again, note that even if the linear acceleration of two points on the body is known, it is first necessary to solve the velocity relations, as the angular velocity must be known at each instant if the accelerations are to be analyzed.

As in the analysis of velocities, the angular acceleration cannot be determined knowing only the linear acceleration of two points, and a minimum of three noncollinear points is needed. The analogy with the velocity analysis is clearer if the acceleration equation is written as follows:

$$\boldsymbol{\alpha} \times \mathbf{r}_{B/A} = \mathbf{q}_{B/A} \tag{5.116}$$
$$\boldsymbol{\alpha} \times \mathbf{r}_{C/A} = \mathbf{q}_{C/A} \tag{5.117}$$

where

$$\mathbf{q}_{B/A} = \mathbf{a}_{B/A} - \boldsymbol{\omega} \times (\boldsymbol{\omega} \times \mathbf{r}_{B/A}) \tag{5.118}$$

. . . .

[*]Y. Youm and Y. S. Yoon, "Analytical development in investigation of wrist kinematics." *J. Biomechanics*: 12:613 (1979).

and similarly,

$$\mathbf{q}_{C/A} = \mathbf{a}_{C/A} - \boldsymbol{\omega} \times (\boldsymbol{\omega} \times \mathbf{r}_{C/A}) \tag{5.119}$$

and

$$\mathbf{a}_{B/A} = \mathbf{a}_B - \mathbf{a}_A \qquad \mathbf{a}_{C/A} = \mathbf{a}_C - \mathbf{a}_A \tag{5.120}$$

Note that the $\mathbf{q}$ relative vectors are known, and the only unknowns in the aforementioned equations are the three components of $\boldsymbol{\alpha}$. The three scalar equations from either of the relative acceleration vector equations are linearly dependent, and both vector equations must be used to determine the angular velocity. The six scalar equations represent an overdetermined system of equations for the three components of the angular acceleration. As shown earlier, these equations can be solved using commercial software or by developing a computer program for a least-squares solution.

The angular acceleration may be determined by a direct vector solution without using the scalar equations in a manner similar to that used for the velocities. First, we cross the two acceleration equations, yielding

$$(\boldsymbol{\alpha} \times \mathbf{r}_{B/A}) \times (\boldsymbol{\alpha} \times \mathbf{r}_{C/A}) = \mathbf{q}_{B/A} \times \mathbf{q}_{C/A} \tag{5.121}$$

If the cross product of the right-hand side of this equation is not zero—that is, $\boldsymbol{\alpha}$ does not lie in the plane formed by $\mathbf{r}_{B/A}$ and $\mathbf{r}_{C/A}$—then

$$\boldsymbol{\alpha}[(\boldsymbol{\alpha} \times \mathbf{r}_{B/A}) \cdot \mathbf{r}_{C/A}] - \mathbf{r}_{C/A}[(\boldsymbol{\alpha} \times \mathbf{r}_{B/A}) \cdot \boldsymbol{\alpha}] = \mathbf{q}_{B/A} \times \mathbf{q}_{C/A} \tag{5.122}$$

$$\boldsymbol{\alpha}[\mathbf{q}_{B/A} \cdot \mathbf{r}_{C/A}] - 0 = \mathbf{q}_{B/A} \times \mathbf{q}_{C/A} \tag{5.123}$$

and

$$\boldsymbol{\alpha} = \frac{\mathbf{q}_{B/A} \times \mathbf{q}_{C/A}}{\mathbf{q}_{B/A} \cdot \mathbf{r}_{C/A}} \tag{5.124}$$

Once the angular velocity, angular acceleration, linear velocity, and acceleration of one point on a rigid body are known, the kinematic state of the rigid body is completely specified. If the two velocities and the two accelerations are given as a function of time or position, the motion of the body is completely described. As has been shown for plane motion, this type of complete description of motion is difficult, and many analyses are done using a position-by-position approach. If the angular velocity and acceleration and the linear velocity and acceleration of a point—call it point *A*—on a body are given for any instant of time, the acceleration of every other point on the body may be obtained as follows:

$$\begin{aligned} \mathbf{v}_P &= \mathbf{v}_A + \boldsymbol{\omega} \times \mathbf{r}_{P/A} \\ \mathbf{a}_P &= \mathbf{a}_A + \boldsymbol{\alpha} \times \mathbf{r}_{P/A} + \boldsymbol{\omega} \times (\boldsymbol{\omega} \times \mathbf{r}_{P/A}) \end{aligned} \tag{5.125}$$

In most cases, point *A* is chosen to coincide with the center of mass of the rigid body.

5.11.2 CONSTRAINTS TO THE GENERAL THREE-DIMENSIONAL MOTION OF A RIGID BODY

It has been shown earlier that the general three-dimensional motion of a rigid body may be fully determined under the following conditions:

A) Velocity

1. The linear velocities of at least three noncollinear points on the body are known. This specification allows us to express the motion in terms of two relative velocity equations that may be solved directly or by expansion into an overdetermined set of scalar equations for the angular velocity.

2. The linear velocity of a point on the rigid body and the angular velocity of the rigid body are known.

B) Acceleration (after determination of the angular velocity)

1. The linear accelerations of at least three noncollinear points on the body are known. Again, ths specification allows us to express the motion in terms of two relative acceleration equations. These equations may be solved directly or as an overdetermined set of scalar equations.

2. The linear acceleration of a point on the body and the angular velocity and acceleration are known.

Note that for both of the velocities to be fully determined, six kinematic variables must be known (one absolute velocity and two relative velocities or one absolute velocity and the angular velocity). As has been shown earlier, if one absolute velocity and two relative velocities are known, the problem is overdetermined. In a similar manner, if the acceleration is to be fully determined, six acceleration variables must be known in addition to the angular velocity. As for the case of plane motion, many of these variables are expressed in terms of constraints to the motion most of the time.

5.11.3 RIGID BODY WITH A FIXED POINT IN SPACE

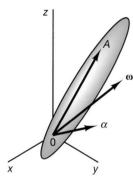

Figure 5.35

The simplest constraint to a motion occurs when a point on a rigid body is constrained such that it cannot move in space (i.e., its linear velocity and acceleration are zero). Such a constraint appears in ball-and-socket joints in which the socket is fixed in space. A body constrained to move in such a manner is shown in Figure 5.35. The origin of the coordinate system in Figure 5.35 has been selected at 0, the fixed point in space, to simplify the kinematic equations. It will be shown in Section 5.12 that the body always rotates about an axis that must pass through the point 0, but which will change in orientation with time. This axis is called an *instantaneous axis of rotation*. If $\boldsymbol{\omega}$ and $\boldsymbol{\alpha}$ are known, the linear velocity and acceleration may be determined for any point A on the rigid body by the relative velocity and acceleration equations, noting that the velocity and acceleration of point 0 are zero.

$$\mathbf{v}_A = \boldsymbol{\omega} \times \mathbf{r}_A$$
$$\mathbf{a}_A = \boldsymbol{\alpha} \times \mathbf{r}_A + \boldsymbol{\omega} \times (\boldsymbol{\omega} \times \mathbf{r}_A)$$
(5.126)

5.11.4 OTHER CONSTRAINTS

A more difficult constraint to express mathematically is shown in the clevis–collar connection shown in Figure 5.36. The collar B is free to rotate about the shaft C and may translate along it. The rod AB is attached by a ball-and-socket joint at point A and can

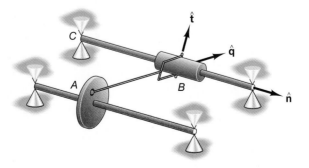

Figure 5.36

rotate about an axis through the clevis. A unit vector along the clevis axis $\hat{\mathbf{t}}$ is perpendicular to the plane formed by the shaft C and the rod AB. The unit vector along this axis and perpendicular to the plane of AB and C may be formed by the cross product between the two unit vectors, one along shaft C and one along rod AB, $\hat{\mathbf{n}} \times \hat{\mathbf{q}}$. Since $\hat{\mathbf{n}}$ and $\hat{\mathbf{q}}$ are, in general, not perpendicular to each other, a unit vector $\hat{\mathbf{t}}$ is obtained by dividing by the magnitude of $\hat{\mathbf{n}} \times \hat{\mathbf{q}}$:

$$\hat{\mathbf{t}} = \frac{\hat{\mathbf{n}} \times \hat{\mathbf{q}}}{|\hat{\mathbf{n}} \times \hat{\mathbf{q}}|}$$

It is apparent that rod AB is free to rotate only about the unit vectors $\hat{\mathbf{n}}$ and $\hat{\mathbf{t}}$. Since these two vectors are perpendicular, the rod may not rotate about an axis perpendicular to $\hat{\mathbf{n}}$ and $\hat{\mathbf{t}}$. Mathematically, this restriction can be written as a constraint on the angular velocity of the rod AB as follows:

$$\boldsymbol{\omega}_{AB} \cdot (\hat{\mathbf{n}} \times \hat{\mathbf{t}}) = 0$$

It may be noted that constraints in three dimensions require careful consideration in order to correctly express them in mathematical form.

In the previous example, if the rod AB had been connected by ball-and-socket joints at both A and B, the rod would have been free to rotate about its own axis $\hat{\mathbf{q}}$, and the angular velocity component along this axis would remain undetermined. A problem of this nature will be examined in Sample Problem 5.26.

Sample Problem 5.24

The position, linear velocities, and accelerations of three non-collinear points on a rigid body are given in the table below:

	r mm			v mm/s			a mm/s²		
	x	y	z	x	y	z	x	y	z
A	100	100	0	600	−400	100	850	1200	−240
B	300	300	0	200	0	0	200	200	0
C	220	180	0	440	−160	40	420	760	−140

Determine the angular velocity and the angular acceleration of the body.

Solution Consider only B relative to A and C relative to A. (An additional relation could be written for C relative to B, but we will not consider that relation here.) So, we can write

$$\mathbf{r}_{B/A} = \mathbf{r}_B - \mathbf{r}_A = 200\hat{\mathbf{i}} + 200\hat{\mathbf{j}} \text{ mm}$$

$$\mathbf{r}_{C/A} = 120\hat{\mathbf{i}} + 80\hat{\mathbf{j}} \text{ mm}$$

$$\mathbf{v}_{B/A} = \mathbf{v}_B - \mathbf{v}_A = -400\hat{\mathbf{i}} + 400\hat{\mathbf{j}} - 100\hat{\mathbf{k}} \text{ mm/s}$$

$$\mathbf{v}_{C/A} = -160\hat{\mathbf{i}} + 240\hat{\mathbf{j}} - 60\hat{\mathbf{k}} \text{ mm/s}$$

$$\boldsymbol{\omega} \times \mathbf{r}_{B/A} = \mathbf{v}_{B/A}$$

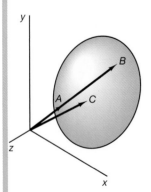

Expanding, we get

$$\omega_y 0 - \omega_z 200 = -400$$

$$\omega_z 200 - \omega_x 0 = 400$$

$$\omega_x 200 - \omega_y 200 = -100$$

Therefore,

$$\omega_z = 2 \qquad \omega_x - \omega_y = -0.5$$

$$\boldsymbol{\omega} \times \mathbf{r}_{C/A} = \mathbf{v}_{C/A}$$

Expanding, we get

$$\omega_y 0 - \omega_z 80 = -160$$

$$\omega_z 120 - \omega_x 0 = 240$$

$$\omega_x 80 - \omega_y 120 = -60$$

$$\omega_z = 2 \qquad \omega_x - 1.5\omega_y = -0.75$$

Solving the two equations for ω_x and ω_y, we obtain

$$\omega_x = 0 \qquad \omega_y = 0.5$$

Therefore,

$$\boldsymbol{\omega} = 0.5\hat{\mathbf{j}} + 2\hat{\mathbf{k}}$$

This equation may be solved directly from the vector equation:

$$\boldsymbol{\omega} = \frac{\mathbf{v}_{B/A} \times \mathbf{v}_{C/A}}{\mathbf{v}_{B/A} \cdot \mathbf{r}_{C/A}}$$

$$\boldsymbol{\omega} = \frac{(-400\hat{\mathbf{i}} + 400\hat{\mathbf{j}} - 100\hat{\mathbf{k}}) \times (-160\hat{\mathbf{i}} + 240\hat{\mathbf{j}} - 60\hat{\mathbf{k}})}{(-400\hat{\mathbf{i}} + 400\hat{\mathbf{j}} - 100\hat{\mathbf{k}}) \cdot (120\hat{\mathbf{i}} + 80\hat{\mathbf{j}})}$$

$$= \frac{(-8000\hat{\mathbf{j}} - 32000\hat{\mathbf{k}})}{(-16000)} = 0.5\hat{\mathbf{j}} + 2\hat{\mathbf{k}}$$

Now that the angular velocity is known, the angular acceleration may be determined:

$$\mathbf{a}_{B/A} = \mathbf{a}_B - \mathbf{a}_A = -650\hat{\mathbf{i}} - 1000\hat{\mathbf{j}} + 240\hat{\mathbf{k}} \text{ mm/s}^2$$

$$\mathbf{a}_{C/A} = -430\hat{\mathbf{i}} - 440\hat{\mathbf{j}} + 100\hat{\mathbf{k}} \text{ mm/s}^2$$

$$\mathbf{q}_{B/A} = \mathbf{a}_{B/A} - \boldsymbol{\omega} \times (\boldsymbol{\omega} \times \mathbf{r}_{B/A}) = \mathbf{a}_{B/A} - \boldsymbol{\omega} \times \mathbf{v}_{B/A}$$

$$\mathbf{q}_{B/A} = (-650\hat{\mathbf{i}} - 1000\hat{\mathbf{j}} + 240\hat{\mathbf{k}}) - (0.5\hat{\mathbf{j}} + 2\hat{\mathbf{k}})$$

$$\times (-400\hat{\mathbf{i}} + 400\hat{\mathbf{j}} - 100\hat{\mathbf{k}})$$

$$= 200\hat{\mathbf{i}} - 200\hat{\mathbf{j}} + 40\hat{\mathbf{k}}$$

$$\mathbf{q}_{C/A} = (-430\hat{\mathbf{i}} - 440\hat{\mathbf{j}} + 100\hat{\mathbf{k}}) - (0.5\hat{\mathbf{j}} + 2\hat{\mathbf{k}}) \times (-160\hat{\mathbf{i}} + 240\hat{\mathbf{j}} - 60\hat{\mathbf{k}})$$

$$= 80\hat{\mathbf{i}} - 120\hat{\mathbf{j}} + 20\hat{\mathbf{k}}$$

$$\boldsymbol{\alpha} \times \mathbf{r}_{B/A} = \mathbf{q}_{B/A}$$

Expanding, we get

$$-200\alpha_z = 200$$

$$200\alpha_z = -200$$

$$200\alpha_x - 200\alpha_y = 40$$

$$\alpha_z = -1 \text{ rad/s}^2 \qquad 200\alpha_x - 200\alpha_y = 40$$

$$\boldsymbol{\alpha} \times \mathbf{r}_{C/A} = \mathbf{q}_{C/A}$$

Expanding, we get

$$-80\alpha_z = 80$$

$$120\alpha_z = -120$$

$$80\alpha_x - 120\alpha_y = 20$$

$$\alpha_z = -1 \qquad 80\alpha_x - 120\alpha_y = 29$$

Solving the two equations, we obtain

$$\boldsymbol{\alpha} = 0.1\hat{\mathbf{i}} - 0.1\hat{\mathbf{j}} - \hat{\mathbf{k}} \text{ rad/s}^2$$

Again, this expression may be solved directly from the vector equation:

$$\boldsymbol{\alpha} = \frac{\mathbf{q}_{B/A} \times \mathbf{q}_{C/A}}{\mathbf{q}_{B/A} \cdot \mathbf{r}_{C/A}} = \frac{(200\hat{\mathbf{i}} - 200\hat{\mathbf{j}} + 40\hat{\mathbf{k}}) \times (80\hat{\mathbf{i}} - 120\hat{\mathbf{j}} + 20\hat{\mathbf{k}})}{(200\hat{\mathbf{i}} - 200\hat{\mathbf{j}} + 40\hat{\mathbf{k}}) \cdot (120\hat{\mathbf{i}} + 80\hat{\mathbf{j}})}$$

$$= \frac{(800\hat{\mathbf{i}} - 800\hat{\mathbf{j}} - 8000\hat{\mathbf{k}})}{8000} = 0.1\hat{\mathbf{i}} - 0.1\hat{\mathbf{j}} - \hat{\mathbf{k}} \text{ rad/s}^2$$

The angular velocity and acceleration can be calcualted by solving the overdetermined system of ordinary equations. This solution is obtained by forming an expanded coefficient matrix by premultiplying the matrix equation by the transpose of the coefficient matrix and solving the resulting system of equations:

$$[C][\omega] = [v_{\text{rel}}]$$

$$[C]^T[C][\omega] = [C]^T[v_{\text{rel}}]$$

$$[\omega] = [[C]^T[C]]^{-1}[C]^T[v_{\text{rel}}]$$

The solution is shown in the Computational Supplement.

Sample Problem 5.25

Consider the rigid body, shown in the figure on the next page, having the position and velocity of three points given as follows:

$$\mathbf{r}_A = 2\hat{\mathbf{i}} - \hat{\mathbf{j}} + 3\hat{\mathbf{k}} \text{ m} \qquad \mathbf{v}_A = 3\hat{\mathbf{i}} - 2\hat{\mathbf{j}} + \hat{\mathbf{k}} \text{ m/s}$$

$$\mathbf{r}_B = 3\hat{\mathbf{j}} - \hat{\mathbf{k}} \text{ m} \qquad \mathbf{v}_B = 19\hat{\mathbf{i}} + 10\hat{\mathbf{j}} + 5\hat{\mathbf{k}} \text{ m/s}$$

$$\mathbf{r}_C = \hat{\mathbf{i}} + 2\hat{\mathbf{j}} - 2\hat{\mathbf{k}} \text{ m} \qquad \mathbf{v}_C = 23\hat{\mathbf{i}} + 15\hat{\mathbf{j}} + 5\hat{\mathbf{k}} \text{ m/s}$$

Determine the angular velocity of the body.

Solution The relative position and velocity vectors are

$$\mathbf{r}_{B/A} = -2\hat{\mathbf{i}} + 4\hat{\mathbf{j}} - 4\hat{\mathbf{k}}$$

$$\mathbf{r}_{C/A} = -\hat{\mathbf{i}} + 3\hat{\mathbf{j}} - 5\hat{\mathbf{k}}$$

$$\mathbf{v}_{B/A} = 16\hat{\mathbf{i}} + 12\hat{\mathbf{j}} + 4\hat{\mathbf{k}}$$

$$\mathbf{v}_{C/A} = 20\hat{\mathbf{i}} + 15\hat{\mathbf{j}} + 5\hat{\mathbf{k}}$$

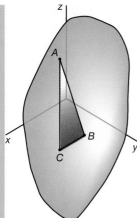

First, we will determine if a direct vector solution is possible; that is, we will check to see if the relative position vectors and the angular velocity vector are coplanar. First, we determine that

$$\mathbf{v}_{B/A} \times \mathbf{v}_{C/A} = 0$$

so $\boldsymbol{\omega}$ lies in the plane of $\mathbf{r}_{B/A}$ and $\mathbf{r}_{C/A}$, and a direct vector solution is not possible.

We then expand the relative velocity vector equations into scalar components, forming an overdetermined system of equations for the components of $\boldsymbol{\omega}$.

$$\boldsymbol{\omega} \times \mathbf{r}_{B/A} = \mathbf{v}_{B/A} \qquad -4\omega_y - 4\omega_z = 16$$

$$4\omega_x - 2\omega_z = 12$$

$$4\omega_x + 2\omega_y = 4$$

$$\boldsymbol{\omega} \times \mathbf{r}_{C/A} = \mathbf{v}_{C/A} \qquad -5\omega_y - 3\omega_z = 20$$

$$5\omega_x - \omega_z = 15$$

$$3\omega_x + \omega_y = 5$$

These equations can be solved by hand, as was shown in the previous example, or can be written in matrix notation and solved as an overdetermined systems of equations. First, we write

$$[C][\omega] = [v_{\text{rel}}]$$

where

$$[C] = \begin{bmatrix} 0 & -4 & -4 \\ 4 & 0 & -2 \\ 4 & 2 & 0 \\ 0 & -5 & -3 \\ 5 & 0 & -1 \\ 3 & 1 & 0 \end{bmatrix} \qquad [v_{\text{rel}}] = \begin{bmatrix} 16 \\ 12 \\ 4 \\ 20 \\ 15 \\ 5 \end{bmatrix}$$

The solution for $[\omega]$ in matrix notation is

$$[\omega] = ([C]^T[C])^{-1}[C]^T[v_{\text{rel}}]$$

This matrix equation can be solved using computational software (see the Computational Supplement), yielding

$$[\omega] = \begin{bmatrix} 3 \\ -4 \\ 0 \end{bmatrix} \text{rad/s}$$

Sample Problem 5.26

The mechanism shown in the acompanying figure converts rotational motion into transla-tional motion. Develop a relationship between the angular velocity of the wheel and the linear velocity of the collar on the shaft. The connections on bar AB at points A and B are both ball-and-socket connections.

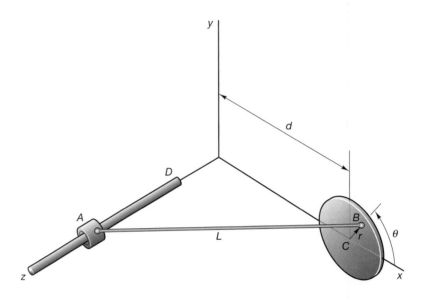

Let r be the distance from the disk center C to the socket B. Assume that the disk rotates with a constant angular velocity ω.

Solution   The geometry is shown in the figure. A general expression for A_z is needed, and this expression can be written in terms of θ as

$$A_z = \sqrt{L^2 - r^2 \sin^2\theta - (d + r\cos\theta)^2}$$

$$A_z = \sqrt{L^2 - r^2 - d^2 - 2dr\cos\theta}$$

The lengths L, r, and d are constant, so the velocity of the collar A, which is constrained to move only in the z-direction, may be obtained by differentiation of A_z with time:

$$v_A = \frac{dr \sin \theta \dot{\theta}}{\sqrt{L^2 - r^2 - d^2 - 2dr \cos \theta}}$$

Since $\dot{\theta}$ is equal to ω and is constant in time, $\theta = \omega t$.

The same solution may be obtained by use of the relative velocity relationships:

$$\mathbf{v}_B = \boldsymbol{\omega} \times \mathbf{r}_{B/C}$$
$$\mathbf{v}_B = \omega \hat{\mathbf{k}} \times r(\cos \theta \hat{\mathbf{i}} + \sin \theta \hat{\mathbf{j}})$$
$$\mathbf{v}_B = -r\omega \sin \theta \hat{\mathbf{i}} + r\omega \cos \theta \hat{\mathbf{j}}$$

The constraint for collar A is

$$v_A = v_A \hat{\mathbf{k}}$$

Using the relationship between A_z and θ, we can write that the position vector of B relative to A is

$$\mathbf{r}_{B/A} = (d + r \cos \theta)\hat{\mathbf{i}} + r \sin \theta \hat{\mathbf{j}} - \sqrt{L^2 - r^2 - d^2 - 2dr \cos \theta}\,\hat{\mathbf{k}}$$

The relative velocity $\mathbf{v}_{B/A}$ is

$$\mathbf{v}_{B/A} = \mathbf{v}_B - \mathbf{v}_A = -r\omega \sin \theta \hat{\mathbf{i}} + r\omega \cos \theta \hat{\mathbf{j}} - v_A \hat{\mathbf{k}}$$

The component of the angular velocity of AB that is parallel to the axis AB is indeterminate, as the link is able to freely rotate about that axis. It also has been noted that the scalar equations are not linearly independent. Therefore, the angular velocity of AB cannot be fully determined. However, the relative velocity of B to A must be perpendicular to the relative position of B to A. Therefore,

$$\mathbf{v}_{B/A} \cdot \mathbf{r}_{B/A} = 0$$

$$[-r\omega \sin \theta \hat{\mathbf{i}} + r\omega \cos \theta \hat{\mathbf{j}} - v_A \hat{\mathbf{k}}] \cdot [(d + r \cos \theta)\hat{\mathbf{i}} + r \sin \theta \hat{\mathbf{j}}$$
$$+ \sqrt{L^2 - r^2 - d^2 - 2rd \cos \theta}\,\hat{\mathbf{k}}] = 0$$

Performing the dot product and solving for v_A, we get

$$v_A = \frac{rd \sin \theta \omega}{\sqrt{L^2 - r^2 - d^2 - 2rd \cos \theta}}$$

which agrees with the previously obtained solution by differentiation of the parameter ω.

Again, this motion may be studied by plotting velocity and acceleration of the point A, using any software package. Consider the motion for a mechanism for the following values:

$$\omega = 10 \text{ rad/s} \quad L = 21 \text{ in.} \quad d = 9 \text{ in.} \quad r = 6 \text{ in.}$$

The results are shown in the Computational Supplement.

Problems

5.94 The positions and velocities of three points on a rigid body are

$$\mathbf{r}_A = \begin{pmatrix} 4 \\ -2 \\ 1 \end{pmatrix} m \qquad \mathbf{v}_A = \begin{pmatrix} -1 \\ 3 \\ 0 \end{pmatrix} m/s$$

$$\mathbf{r}_B = \begin{pmatrix} -2 \\ 0 \\ -3 \end{pmatrix} m \qquad \mathbf{v}_B = \begin{pmatrix} -1.2 \\ 0.8 \\ -0.8 \end{pmatrix} m/s$$

$$\mathbf{r}_C = \begin{pmatrix} 1 \\ 1 \\ 1 \end{pmatrix} m \qquad \mathbf{v}_C = \begin{pmatrix} -2.5 \\ 1.5 \\ 0 \end{pmatrix} m/s$$

Determine the angular velocity of the rigid body by solving an overdetermined system of linear equations (a) using the relative positions of B to A and of C to A and (b) using the relative position vectors of B to A, C to A, and C to B.

5.95 Determine if the angular velocity in Problem 5.94 can be obtained from a direct vector solution, and if possible, solve for the angular velocity of the rigid body in a direct vector solution.

5.96 If the linear accelerations of the three points on the rigid body in Problem 5.94 are

$$\mathbf{a}_A = \begin{pmatrix} 0.2 \\ 0.3 \\ 0.6 \end{pmatrix} m/s^2 \qquad \mathbf{a}_B = \begin{pmatrix} 2.46 \\ -1.44 \\ -2.28 \end{pmatrix} m/s^2$$

$$\mathbf{a}_C = \begin{pmatrix} 0.05 \\ -1.35 \\ -1.2 \end{pmatrix} m/s^2$$

Determine the angular acceleration of the rigid body by solving an overdetermined system of linear equations (a) using two relative acceleration relations and (b) using all three relative acceleration relations.

5.97 Determine if the angular acceleration in Problem 5.96 can be determined by a direct vector solution, and if so, determine the angular acceleration of the rigid body by a direct vector solution.

5.98 The position vectors and the linear velocities of the three points on a rigid body are

$$\mathbf{r}_A = \hat{\mathbf{i}} + 4\hat{\mathbf{j}} \text{ m} \qquad \mathbf{v}_A = 2\hat{\mathbf{j}} \text{ m/s}$$

$$\mathbf{r}_B = -2\hat{\mathbf{i}} + 2\hat{\mathbf{j}} + \hat{\mathbf{k}} \text{ m} \qquad \mathbf{v}_B = 5\hat{\mathbf{i}} - 6\hat{\mathbf{j}} - \hat{\mathbf{k}} \text{ m/s}$$

$$\mathbf{r}_C = -4\hat{\mathbf{i}} + \hat{\mathbf{j}} - \hat{\mathbf{k}} \text{ m} \qquad \mathbf{v}_C = 2.5\hat{\mathbf{i}} - 2\hat{\mathbf{j}} - 0.5\hat{\mathbf{k}} \text{ m/s}$$

Show that the angular velocity of the body cannot be obtained by a direct vector method, and determine the angular velocity by using an overdetermined linear system of equations.

5.99 A rigid body rotates about a fixed axis with a constant angular velocity $\boldsymbol{\omega} = 2\hat{\mathbf{i}} - \hat{\mathbf{j}} + 3\hat{\mathbf{k}}$ rad/s. If the acceleration of a point A on the rigid body is $\mathbf{a}_A = a_x\hat{\mathbf{i}} + 6\hat{\mathbf{j}} + 4\hat{\mathbf{k}}$ m/s², determine the x component of the acceleration of point A and the perpendicualr distance from the axis of rotation to point A.

5.100 A 30-inch rod AB is connected by two ball-and-socket joints to collars sliding along rods, as shown in Figure P5.100. If collar A moves along the hoizontal shaft at a constant velocity of 2 in/s, determine the velocity of collar B for any position d of collar A.

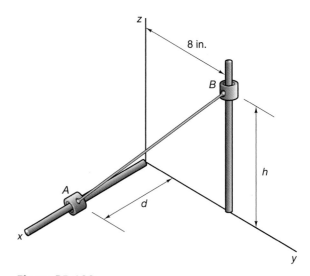

Figure P5.100

5.101 Plot the velocity of collar A in Problem 5.100 as a function of time for $d = 0$ to the point where $h = 0$.

5.102 Determine the angular velocity of the rod AB in Problem 5.101 for any position of the collar A.

5.103 The crank AB is used to convert rotational motion to translation of the collar C. The 150-mm rod BC is connected to the crank and the collar with ball-and-socket joints. If the crank has an angular acceleration of 2 rad/s² starting from rest at the position shown in Figure P5.103, determine the velocity and acceleration of the collar for any time.

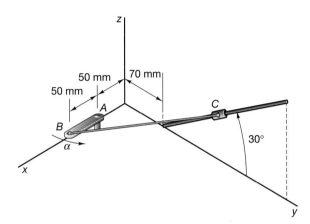

Figure P5.103

5.104 Determine an expression for the angular velocity and the angular acceleration of the rod BC in Problem 5.103.

5.105 A rod AB is connected to a collar by a ball-and-socket joint A which slides along a shaft C. The other end of the rod is connected to a clevis–collar at point B and slides along the shaft D. Determine the constraint to the angular velocity of the rod AB. See Figure P5.105.

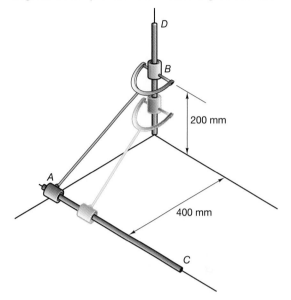

Figure P5.105

5.106 The collar at A in Problem 5.105 is moving with a constant linear velocity of 2 m/s to the right at the position shown in Figure P5.105. Determine the velocity of the collar at B and the angular velocity of the rod.

5.107 Determine the constraint to the angular velocity of rad AB in Problem 5.106 after the collar A has moved 20 mm to the right.

5.108 Determine the angular velocity of the rod and the linear velocity of the collar at B when the system is in the position described in Problem 5.107.

*5.12   INSTANTANEOUS HELICAL AXIS, OR SCREW AXIS

For plane motion, an instantaneous center of rotation was defined to be the intercept of the axis about which the body rotated at that instant of time with the plane of motion. This point was a point on the body or the body extended that had zero velocity at that instant. Since all points on a body move in the plane of motion during plane motion, all of the velocity vectors of all points on the body lie in that plane or in parallel planes, and therefore, the angular velocity vector is perpendicular to the plane of motion. In Section 5.6, a point C on a body or body extended that had zero velocity was determined by use of the following equation:

$$\mathbf{v}_C = 0 = \mathbf{v}_A + \boldsymbol{\omega} \times \mathbf{r}_{C/A}$$

where point C is the instantaneous center of rotation.

This equation allows us to determine the postion of the instantaneous center of rotation relative to a point A on the rigid body whose velocity is known. The angular velocity of the rigid body was determined by use of the methods in Sections 5.5 and 5.11.

In three-dimensional motion of a rigid body, $\boldsymbol{\omega}$ is not, in general, perpendicular to the velocity vector of any given point on that body. During this motion, there will not be a point on the rigid body that has zero velocity. Therefore, one can not define a point on the body as the instantaneous center of rotation. There will, however, be an axis in space, parallel to the $\boldsymbol{\omega}$ vector at any instant, along which the motion of the body is a *helical motion*. This axis is called the *instantaneous helical axis*, and in the case of plane motion, the intersection of this axis with the plane of motion is the instantaneous center of rotation of the body. In general three-dimensional motion, the body will, at any instant, appear to rotate about the helical axis and translate along it. In general, the helical axis will change at each instant, and both the angular velocity of the body and the linear velocity along the instantaneous axis will change. As in the case of the instantaneous center of rotation for plane motion, the notion of an instantaneous helical axis is of use conceptually and is not necessary in any analysis. The concept has, however, been found to be useful in understanding the relative motion of one rigid body to another. It is used in determining the axis of rotation of various joints in the human body in a manner similar to that shown in Section 5.7. Figure 5.37 shows the motion of a rigid body about an instantaneous helical axis at a given instant of time. As for the case of plane motion, the first step in the analysis of this problem is to determine the angular velocity of the rigid body. This expression may be obtained using the methods developed in Section 5.11. Guided by the methods of plane motion analysis, we separate the velocity of point A into a component parallel to the $\boldsymbol{\omega}$ vector and a component perpendicular to it.

$$\mathbf{v}_A = \mathbf{v}_\omega + \mathbf{v}_{\perp A} \tag{5.127}$$

Let $\hat{\mathbf{e}}_\omega$ be a unit vector in the direction of the $\boldsymbol{\omega}$ vector. If

$$\hat{\mathbf{e}}_\omega = \frac{\boldsymbol{\omega}}{|\boldsymbol{\omega}|} \tag{5.128}$$

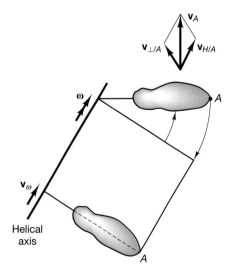

Figure 5.37

then

$$\mathbf{v}_\omega = (\mathbf{v}_A \cdot \hat{\mathbf{e}}_\omega)\hat{\mathbf{e}}_\omega \tag{5.129}$$

$$\mathbf{v}_{\perp A} = \mathbf{v}_A - \mathbf{v}_\omega \tag{5.130}$$

Now, the general relationship between the velocities of two points on the body or the body extended is

$$\mathbf{v}_B = \mathbf{v}_A + \boldsymbol{\omega} \times \mathbf{r}_{B/A} \tag{5.131}$$

The velocity of a point H on the helical axis is equal to the velocity of the body along the helical axis. Note that since the helical axis is a line in space, there are an infinite number of points along it. So, we can write

$$\mathbf{v}_H = \mathbf{v}_\omega = \mathbf{v}_A + \boldsymbol{\omega} \times \mathbf{r}_{H/A} \tag{5.132}$$

Separating $\mathbf{v}_A$ into its parallel and perpendicular parts, we obtain

$$\mathbf{v}_\omega = \mathbf{v}_\omega + \mathbf{v}_{\perp A} + \boldsymbol{\omega} \times \mathbf{r}_{H/A} \tag{5.133}$$

Therefore,

$$\boldsymbol{\omega} \times \mathbf{r}_{H/A} = -\mathbf{v}_{\perp A} \tag{5.134}$$

$\mathbf{r}_{H/A}$ is a vector from point A to some point on the helical axis. So, expanding into scalar components, we get

$$\omega_y z_{H/A} - \omega_z y_{H/A} = -v_{\perp Ax}$$
$$\omega_z x_{H/A} - \omega_x z_{H/A} = -v_{\perp Ay} \tag{5.135}$$
$$\omega_x y_{H/A} - \omega_y x_{H/A} = -v_{\perp Az}$$

These three equations are not linearly independent, but they do define a line in space that, at this instant, is the helical axis of rotation. The easiest way to see this line is to examine its intercepts with the coordinate planes. For example, the intercept with the xy-plane would be the point H having a zero coordinate in the z-direction, measured from point A:

$$z_{H/A} = 0$$

$$x_{H/A} = \frac{-v_{\perp Ay}}{\omega_z} \quad y_{H/A} = \frac{v_{\perp Ax}}{\omega_z}$$

This line may be seen in Figure 5.38. An easier method to locate the helical axis in space is to determine the vector perpendicular from point A to the helical axis. This perpendicular vector from A to a point H on the helical axis will be determined by a direct vector solution. We define this perpendicular vector as

$$\mathbf{r}_{H/A} = \mathbf{p}_{H/A} \tag{5.136}$$

where $\mathbf{p}_{H/A}$ is the vector perpendicular to the helical axis from point A. Therefore, $\mathbf{p}_{H/A}$ is perpendicular to both $\boldsymbol{\omega}$ and $\mathbf{v}_{\perp A}$. Eq. (5.134) may be written in terms of $\mathbf{p}_{H/A}$ as

$$\boldsymbol{\omega} \times \mathbf{p}_{H/A} = -\mathbf{v}_{\perp A} \tag{5.137}$$

If Eq. (5.137) is crossed with the angular velocity $\boldsymbol{\omega}$, the vector triple product is

$$\boldsymbol{\omega} \times (\boldsymbol{\omega} \times \mathbf{p}_{H/A}) = -\boldsymbol{\omega} \times \mathbf{v}_{\perp A} \tag{5.138}$$

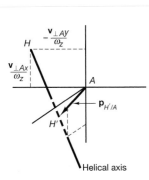

Figure 5.38

Using the vector triple product identity, we obtain

$$\boldsymbol{\omega}(\boldsymbol{\omega} \cdot \mathbf{p}_{H/A}) - \mathbf{p}_{H/A}(\boldsymbol{\omega} \cdot \boldsymbol{\omega}) = -\boldsymbol{\omega} \times \mathbf{v}_{\perp A} \qquad (5.139)$$

But, because $\boldsymbol{\omega}$ and $\mathbf{p}_{H/A}$ are perpendicular vectors, $(\boldsymbol{\omega} \cdot \mathbf{p}_{H/A}) = 0$; therefore,

$$\mathbf{p}_{H/A} = \frac{\boldsymbol{\omega} \times \mathbf{v}_{\perp A}}{\boldsymbol{\omega} \cdot \boldsymbol{\omega}} \qquad (5.140)$$

This vector locates the helical axis from point A in a unique manner that is very useful in many kinematic analyses.

5.12.1 MOTION OF A RIGID BODY HAVING A FIXED POINT IN SPACE

If one point on the helical axis is fixed in space, then the body cannot translate along the helical axis but only can rotate about that axis. This situation corresponds to the general motion of a body about a fixed point in space. The orientation of the helical axis—and therefore, the angular velocity vector—will change with time, but all helical axes will pass through this fixed point. This circumstance is the basis for Euler's theorem, which states that, *the most general displacement of a rigid body with a fixed point is equivalent to a rotation of the body about an axis through that fixed point.* If H were considered to be that fixed point on the body, then the velocity and acceleration of any point on the body could be written as

$$\mathbf{v}_P = \boldsymbol{\omega} \times \mathbf{r}_{P/H} \qquad (5.141)$$

$$\mathbf{a}_P = \boldsymbol{\alpha} \times \mathbf{r}_{P/H} + \boldsymbol{\omega} \times (\boldsymbol{\omega} \times \mathbf{r}_{P/H}) \qquad (5.142)$$

where

$$\boldsymbol{\alpha} = \frac{d\boldsymbol{\omega}}{dt} \qquad (5.143)$$

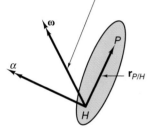

Instantaneous axis of rotation

Figure 5.39

The instantaneous axis of rotation is shown in Figure 5.39 for the case that point H is fixed in space. The helical axis is now called the instantaneous axis of rotation, is parallel to the angular velocity vector, and passes through the fixed point in space.

Sample Problem 5.27

The positions and linear velocities of a rigid body at a given time are

$$\mathbf{r}_A = 10\hat{\mathbf{i}} + 10\hat{\mathbf{k}} \qquad\qquad \mathbf{v}_A = 4\hat{\mathbf{i}} + 4\hat{\mathbf{j}} + 4\hat{\mathbf{k}}$$

$$\mathbf{r}_B = 12\hat{\mathbf{i}} + 3\hat{\mathbf{j}} + 8\hat{\mathbf{k}} \qquad \mathbf{v}_B = 3\hat{\mathbf{i}} + 10\hat{\mathbf{j}} + 12\hat{\mathbf{k}}$$

$$\mathbf{r}_C = 10\hat{\mathbf{i}} + 4\hat{\mathbf{j}} + 13\hat{\mathbf{k}} \qquad \mathbf{v}_C = -3\hat{\mathbf{i}} - 2\hat{\mathbf{j}} + 12\hat{\mathbf{k}}$$

All units are given in meters and meters/second. Determine the instantaneous helical axis.

Solution   The relative velocity vectors and the relative position vector for a direct vector solution are

$$\mathbf{r}_{C/A} = 4\hat{\mathbf{j}} + 3\hat{\mathbf{k}}$$

$$\mathbf{v}_{B/A} = -\hat{\mathbf{i}} + 6\hat{\mathbf{j}} + 8\hat{\mathbf{k}}$$

$$\mathbf{v}_{C/A} = -7\hat{\mathbf{i}} - 6\hat{\mathbf{j}} + 8\hat{\mathbf{k}}$$

$$\boldsymbol{\omega} = \frac{\mathbf{v}_{C/A} \times \mathbf{v}_{B/A}}{\mathbf{v}_{C/A} \cdot \mathbf{r}_{B/A}} = \frac{96\hat{\mathbf{i}} - 48\hat{\mathbf{j}} - 48\hat{\mathbf{k}}}{48}$$

$$\boldsymbol{\omega} = 2\hat{\mathbf{i}} - \hat{\mathbf{j}} + \hat{\mathbf{k}} \text{ rad/s}$$

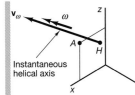

Instantaneous
helical axis

To determine the location of the instantaneous helical axis relative to point A, we construct a unit vector in the direction of the angular velocity:

$$\hat{\mathbf{e}}_\omega = \frac{\boldsymbol{\omega}}{|\boldsymbol{\omega}|} = 0.816\hat{\mathbf{i}} - 0.408\hat{\mathbf{j}} + 0.408\hat{\mathbf{k}}$$

The linear velocity of the body along the helical axis is

$$\mathbf{v}_\omega = (\mathbf{v}_A \cdot \hat{\mathbf{e}}_\omega)\hat{e}_\omega$$

$$\mathbf{v}_\omega = 3.264\hat{\mathbf{e}}_\omega = 2.663\hat{\mathbf{i}} - 1.332\hat{\mathbf{j}} + 1.332\hat{\mathbf{k}}$$

The component of the velocity of A perpendicular to the helical axis is

$$\mathbf{v}_{\perp A} = \mathbf{v}_A - \mathbf{v}_\omega = 1.337\hat{\mathbf{i}} + 5.332\hat{\mathbf{j}} + 2.668\hat{\mathbf{k}}$$

Therefore, the components of a vector from A to a point on the helical axis are

$$\boldsymbol{\omega} \times \mathbf{r}_{H/A} = -\mathbf{v}_{\perp A}$$

$$-z_{H/A} - y_{H/A} = -1.337$$

$$x_{H/A} - 2z_{H/A} = -5.332$$

$$2y_{H/A} + x_{H/A} = -2.668$$

An intercept of the helical axis with a plane parallel to the xy-plane and passing through A may be found by setting $z_{H/A} = 0$:

$$x_{H/A} = -5.332 \qquad y_{H/A} = 1.337 \qquad z_{H/A} = 0$$

The position vector to the point H in space is

$$\mathbf{r}_H = \mathbf{r}_A + \mathbf{r}_{H/A} = 4.668\hat{\mathbf{i}} + 1.337\hat{\mathbf{j}} + 10\hat{\mathbf{k}}$$

The perpendicular from point A to a point H' on the helical axis may be found from Eq. (5.140):

$$\mathbf{p}_{H'/A} = \frac{\boldsymbol{\omega} \times \bar{\mathbf{v}}_{\perp A}}{\boldsymbol{\omega} \cdot \boldsymbol{\omega}}$$

$$\mathbf{p}_{H'/A} = \frac{(2\hat{\mathbf{i}} - \hat{\mathbf{j}} + \hat{\mathbf{k}}) \times (1.337\hat{\mathbf{i}} + 5.332\hat{\mathbf{j}} + 2.668\hat{\mathbf{k}})}{(2\hat{\mathbf{i}} - \hat{\mathbf{j}} + \hat{\mathbf{k}}) \cdot (2\hat{\mathbf{i}} - \hat{\mathbf{j}} + \hat{\mathbf{k}})}$$

$$\mathbf{p}_{H'/A} = \frac{-8\hat{\mathbf{i}} - 4\hat{\mathbf{j}} + 12\hat{\mathbf{k}}}{6} = -1.333\hat{\mathbf{i}} - 0.667\hat{\mathbf{j}} + 2\hat{\mathbf{k}} \text{ m}$$

Now H' and H are both points on the helical axis. This fact may be verified, as a vector between them must be parallel to the helical axis:

$$\mathbf{r}_{H'/H} = \mathbf{p}_{H'/A} - \mathbf{r}_{H/A}$$

$$= 4\hat{\mathbf{i}} - 2\hat{\mathbf{j}} + 2\hat{\mathbf{k}} = 4.899\hat{\mathbf{e}}_\omega$$

*5.13 INSTANTANEOUS HELICAL AXIS OF ROTATION BETWEEN TWO RIGID BODIES

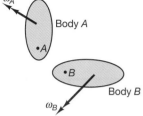

Figure 5.40

We have discussed the instantaneous helical axis of a body moving in general motion in space. In Section 5.7, we examined the instantaneous center of rotation between two bodies moving in plane motion, and we showed the application of this concept to an instantaneous joint center in biomechanics. A similar concept will be introduced here for the instantaneous helical axis of rotation between two bodies moving in general motion. Consider the two such bodies that are shown in Figure 5.40. Knowing the absolute linear velocities of three points on bodies A and B allows us to determine the absolute angular velocities of each body. The relative angular velocity of B relative to A at any instant is

$$\boldsymbol{\omega}_H = \boldsymbol{\omega}_B - \boldsymbol{\omega}_A \tag{5.144}$$

where $\boldsymbol{\omega}_H$ is the angular velocity of body B relative to body A about the helical axis. Now, we will seek a point H on body A that has a velocity of $\mathbf{v}_{HA}$ such that there is a coincident point H on body B, which has a velocity of

$$\mathbf{v}_{HB} = \mathbf{v}_{HA} + \mathbf{v}_\omega \tag{5.145}$$

where $\mathbf{v}_\omega$ is a velocity parallel to the relative angular velocity vector $\boldsymbol{\omega}_H$. The linear velocity of any point on body A or body A extended can be written as

$$\mathbf{v}_{HA} = \mathbf{v}_A + \boldsymbol{\omega}_A \times \mathbf{r}_{H/A} \tag{5.146}$$

In a similar manner, the linear velocity of any point on body B or body B extended can be written as

$$\mathbf{v}_{HB} = \mathbf{v}_B + \boldsymbol{\omega}_B \times \mathbf{r}_{H/B} \tag{5.147}$$

Considering the fact that point H is a coincident point on both body A and B at an instant of time, substitution of Eqs. (5.146) and (5.147) into Eq. (5.145) yields

$$\mathbf{v}_B + \boldsymbol{\omega}_B \times \mathbf{r}_{H/B} = \mathbf{v}_A + \boldsymbol{\omega}_A \times \mathbf{r}_{H/A} + \mathbf{v}_\omega \tag{5.148}$$

Now let us examine the vector relationship between point A on body A, point B on body B, and the coincident point H on the relative helical axis, as shown in Figure 5.41. The following vector relationship can be written from the vector diagram shown in Figure 5.41.

$$\mathbf{r}_{H/A} = \mathbf{r}_{B/A} + \mathbf{r}_{H/B} \tag{5.149}$$

Eq. (5.148) now can be written as

$$\boldsymbol{\omega}_H \times \mathbf{r}_{H/B} = \mathbf{v}_A - \mathbf{v}_B + \boldsymbol{\omega}_A \times \mathbf{r}_{B/A} + \mathbf{v}_\omega \tag{5.150}$$

All of the terms on the right side of Eq. (150) are known, except for the velocity along the helical axis. Let us introduce a notation for these known vectors as

$$\mathbf{v}_{\text{eff}} = \mathbf{v}_A - \mathbf{v}_B + \boldsymbol{\omega}_A \times \mathbf{r}_{B/A} \tag{5.151}$$

Eq.(5.150) now may be written as

$$\boldsymbol{\omega}_H \times \mathbf{r}_{H/B} = \mathbf{v}_{\text{eff}} + \mathbf{v}_\omega \tag{5.152}$$

Figure 5.41

The vector on the right side of Eq. (5.152) must be perpendicular to $\boldsymbol{\omega}_H$ in order to satisfy the definition of the cross product. We will now break $\mathbf{v}_{\text{eff}}$ into a component parallel to the angular velocity about the relative helical axis and a component perpendicular to the relative helical axis. The component parallel to the relative helical axis must be equal to the negative of $\mathbf{v}_\omega$ in order to satisfy the definition of the cross product. Therefore, the relative linear velocity along the relative helical axis is

$$\mathbf{v}_\omega = -(\mathbf{v}_{\text{eff}} \cdot \hat{\mathbf{e}}_{\omega H})\hat{\mathbf{e}}_{\omega H}$$

where

$$\hat{\mathbf{e}}_{\omega H} = \frac{\boldsymbol{\omega}_H}{|\boldsymbol{\omega}_H|} \tag{5.153}$$

If $\mathbf{v}_\omega$ is defined in this manner, the right side of Eq. (5.152) is perpendicular to $\boldsymbol{\omega}_H$ and may be written as

$$\mathbf{v}_P = \mathbf{v}_{\text{eff}} + \mathbf{v}_\omega \tag{5.154}$$

Eq. (5.150) can now be written in the form

$$\boldsymbol{\omega}_H \times \mathbf{r}_{H/B} = \mathbf{v}_P \tag{5.155}$$

Eq. (5.155) is now in the form of a cross product where the right hand side is known and one of the terms in the cross product is known. There are an infinite number of vectors from point B to a point H on the helical axis, and the general solution of Eq. (5.155 is a line in space. To obtain a direct vector solution of this equation, we will seek the vector from B that is perpendicular to the relative helical axis. If we define this vector as $\mathbf{p}_{H/B}$, Eq. (5.155) becomes

$$\boldsymbol{\omega}_H \times \mathbf{p}_{H/B} = \mathbf{v}_P \tag{5.156}$$

Taking the cross product of Eq. (5.156) with $\boldsymbol{\omega}_H$ yields

$$\boldsymbol{\omega}_H \times (\boldsymbol{\omega}_H \times \mathbf{p}_{H/B}) = \boldsymbol{\omega}_H \times \mathbf{v}_P \tag{5.157}$$

Expanding the triple vector product and realizing that $\mathbf{p}_{H/B}$ and $\boldsymbol{\omega}_H$ are perpendicular to each other gives the direct vector solution:

$$\mathbf{p}_{H/B} = -\frac{\boldsymbol{\omega}_H \times \mathbf{v}_P}{\boldsymbol{\omega}_H \cdot \boldsymbol{\omega}_H} \tag{5.158}$$

$\mathbf{v}_\omega$ was defined such that it is the velocity parallel to the angular velocity vector $\boldsymbol{\omega}$. Substituting Eq. (5.154) into Eq. (5.158) yields

$$\mathbf{p}_{H/B} = \frac{\boldsymbol{\omega}_H \times (\mathbf{v}_{\text{eff}} + \mathbf{v}_\omega)}{\boldsymbol{\omega}_H \cdot \boldsymbol{\omega}_H}$$

and $\boldsymbol{\omega}_H \times \mathbf{v}_\omega = 0$. Therefore an alternate form of equation Eq. (5.158) may be written

$$\mathbf{p}_{H/B} = \frac{\boldsymbol{\omega}_H \times \mathbf{v}_{\text{eff}}}{\boldsymbol{\omega}_H \cdot \boldsymbol{\omega}_H}$$

Sample Problem 5.28

Consider two rigid bodies, A and B, that are moving in space relative to each other. If the velocity of a point on each body and the angular velocity of each body are known, determine the angular velocity of B relative to A and the instantaneous relative helical axis, using the following information:

$$\mathbf{r}_A = \begin{pmatrix} 3 \\ 2 \\ -1 \end{pmatrix} \text{m} \quad \mathbf{v}_A = \begin{pmatrix} 0.1 \\ -0.4 \\ 0.3 \end{pmatrix} \text{m/s} \quad \boldsymbol{\omega}_A = \begin{pmatrix} 1 \\ 0 \\ -0.5 \end{pmatrix} \text{rad/s}$$

$$\mathbf{r}_B = \begin{pmatrix} 4 \\ 3 \\ 2 \end{pmatrix} \text{m} \quad \mathbf{v}_B = \begin{pmatrix} 0 \\ -0.2 \\ -0.1 \end{pmatrix} \text{m/s} \quad \boldsymbol{\omega}_B = \begin{pmatrix} 0.7 \\ -0.3 \\ 1.2 \end{pmatrix} \text{rad/s}$$

Solution First, we will determine the angular velocity of B relative to A:

$$\boldsymbol{\omega}_H = \boldsymbol{\omega}_B - \boldsymbol{\omega}_A = \begin{pmatrix} -0.3 \\ -0.3 \\ 1.7 \end{pmatrix} \text{rad/s}$$

Next, let us determine a relative position vector from A to B:

$$\mathbf{r}_{B/A} = \mathbf{r}_B - \mathbf{r}_A = \begin{pmatrix} 1 \\ 1 \\ 3 \end{pmatrix} \text{m}$$

The effective velocity defined by Eq. (5.151) is

$$\mathbf{v}_{\text{eff}} = \mathbf{v}_A - \mathbf{v}_B + \boldsymbol{\omega}_A \times \mathbf{r}_{B/A} = \begin{pmatrix} 0.6 \\ -3.7 \\ 1.4 \end{pmatrix} \text{m/s}$$

A unit vector along the helical axis is

$$\hat{\mathbf{e}}_{\omega H} = \frac{\boldsymbol{\omega}_H}{|\boldsymbol{\omega}_H|} = \begin{pmatrix} -0.171 \\ -0.171 \\ 0.97 \end{pmatrix} \text{rad/s}$$

The relative velocity component parallel to the helical axis and the position of the helical axis from point B are

$$\mathbf{v}_\omega = -(\mathbf{v}_{\text{eff}} \cdot \hat{\mathbf{e}}_{\omega H})\hat{\mathbf{e}}_{\omega H} = \begin{pmatrix} 0.323 \\ 0.323 \\ -1.833 \end{pmatrix} \text{m/s}$$

$$\mathbf{p}_{H/B} = -\frac{\boldsymbol{\omega}_H \times \mathbf{v}_{\text{eff}}}{\boldsymbol{\omega}_H \cdot \boldsymbol{\omega}_H} = \begin{pmatrix} -1.912 \\ -0.469 \\ -0.42 \end{pmatrix} \text{m}$$

Problems

5.109 The positions and velocities of three points on a rigid body are:

$$\mathbf{r}_A = \begin{pmatrix} 4 \\ -2 \\ 1 \end{pmatrix} m \quad \mathbf{r}_B = \begin{pmatrix} -2 \\ 0 \\ -3 \end{pmatrix} m \quad \mathbf{r}_C = \begin{pmatrix} 1 \\ 1 \\ 1 \end{pmatrix} m$$

$$\mathbf{v}_A = \begin{pmatrix} -1 \\ 3 \\ 0 \end{pmatrix} m/s \quad \mathbf{v}_B = \begin{pmatrix} -1.2 \\ 0.8 \\ -0.8 \end{pmatrix} m/s \quad \mathbf{v}_C = \begin{pmatrix} -2.5 \\ 1.5 \\ 0 \end{pmatrix} m/s$$

Determine the angular velocity of the rigid body and the instantaneous helical axis. Locate the helical axis in space by determining (a) it intercepts with the coordinate planes and (b) the perpendicular distance from A to the axis.

5.110 The position vectors and the linear velocities of three points on a rigid body are:

$$\mathbf{r}_A = \hat{\mathbf{i}} + 4\hat{\mathbf{j}} \text{ m} \qquad \mathbf{v}_A = 2\hat{\mathbf{j}} \text{ m/s}$$
$$\mathbf{r}_B = -2\hat{\mathbf{i}} + 2\hat{\mathbf{j}} + \hat{\mathbf{k}} \text{ m} \quad \mathbf{v}_B = 5\hat{\mathbf{i}} - 6\hat{\mathbf{j}} - \hat{\mathbf{k}} \text{ m/s}$$
$$\mathbf{r}_C = -4\hat{\mathbf{i}} + \hat{\mathbf{j}} - \hat{\mathbf{k}} \text{ m} \quad \mathbf{v}_C = 2.5\hat{\mathbf{i}} - 2\hat{\mathbf{j}} - 0.5\hat{\mathbf{k}} \text{ m/s}$$

Determine the angular velocity of the instantaneous helical axis and the perpendicular distance from point C to the helical axis.

5.111 The position vectors and the linear velocities of three points on a rigid body are:

$$\mathbf{r}_A = \begin{pmatrix} -2 \\ 4 \\ 9 \end{pmatrix} m \quad \mathbf{r}_B = \begin{pmatrix} 1 \\ -2 \\ -1 \end{pmatrix} m \quad \mathbf{r}_C = \begin{pmatrix} 3 \\ 0 \\ 2 \end{pmatrix} m$$

$$\mathbf{v}_A = \begin{pmatrix} 1 \\ 1 \\ 1 \end{pmatrix} m/s \quad \mathbf{v}_B = \begin{pmatrix} 2.8 \\ 2.4 \\ 0.7 \end{pmatrix} m/s \quad \mathbf{v}_C = \begin{pmatrix} 2.3 \\ 1.4 \\ 1.7 \end{pmatrix} m/s$$

Determine the angular velocity of the rigid body and the perpendicular distance from point C to the instantaneous helical axis.

5.112 Using the results of Problem 5.111, determine the linear velocity of a point on the rigid body located at $\mathbf{r}_D = -3\hat{\mathbf{i}} + 2\hat{\mathbf{j}} - 3\hat{\mathbf{k}}$ m using the instantaneous helical axis.

5.113 During a gait-analysis evaluation, three markers are placed on the thigh of the subject and three markers are placed on the lower leg (shank) of the subject. The data on the six markers are

$$\mathbf{r}_A = \begin{pmatrix} 2.30 \\ 0.80 \\ 0.60 \end{pmatrix} m \quad \mathbf{r}_B = \begin{pmatrix} 2.00 \\ 0.65 \\ 0.50 \end{pmatrix} m \quad \mathbf{r}_C = \begin{pmatrix} 2.50 \\ 0.50 \\ 0.55 \end{pmatrix} m/s$$

$$\mathbf{v}_A = \begin{pmatrix} 1.292 \\ -0.141 \\ 0.009 \end{pmatrix} m/s \quad \mathbf{v}_B = \begin{pmatrix} 1.394 \\ -0.349 \\ 0.017 \end{pmatrix} m/s \quad \mathbf{v}_C = \begin{pmatrix} 1.500 \\ 0 \\ 0 \end{pmatrix} m/s$$

$$\mathbf{r}_D = \begin{pmatrix} 2.50 \\ 0.45 \\ 0.50 \end{pmatrix} m \quad \mathbf{r}_E = \begin{pmatrix} 1.80 \\ 0.35 \\ 0.45 \end{pmatrix} m \quad \mathbf{r}_F = \begin{pmatrix} 1.75 \\ 0.25 \\ 0.50 \end{pmatrix} m$$

$$\mathbf{v}_D = \begin{pmatrix} 1.540 \\ 0 \\ 0 \end{pmatrix} m/s \quad \mathbf{v}_E = \begin{pmatrix} 1.638 \\ -0.699 \\ 0.022 \end{pmatrix} m/s \quad \mathbf{v}_F = \begin{pmatrix} 1.740 \\ -0.750 \\ 0.022 \end{pmatrix} m/s$$

Determine the angular velocity of the lower leg relative to the thigh and find the perpendicular distance from marker D to the instantaneous helical axis of the knee joint as seen in Figure P5.113. Make a sketch of the leg and show the orientation of the helical axis.

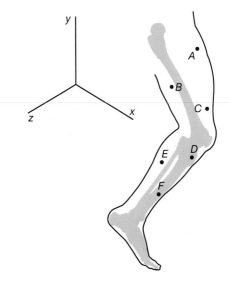

Figure P5.113

*5.14 MOTION WITH RESPECT TO A ROTATING REFERENCE FRAME OR COORDINATE SYSTEM

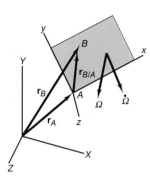

Figure 5.42

Previous sections have examined motion relative to fixed or translating coordinate systems. The unit base vectors of those coordinate systems were constant and did not depend upon position or time. The base vectors in normal, tangential, and curvilinear (polar, cylindrical, spherical, etc.) coordinates are, in general, functions of position and time. If rectilinear Cartesian coordinates are fixed to a rotating body, the absolute motion must be described relative to a frame of reference that is rotating as well as translating. As an example, consider the earth and sun. Since the earth rotates, all reference coordinate systems on the earth rotate with the earth about its own axis and about the sun. In most cases, the effects of these rotations may be ignored and the coordinate systems are assumed to be fixed. In many machines, where one part is rotating relative to another part, the ability to describe motion relative to rotating coordinate systems is very useful.

Consider two points A and B in space, as shown in Figure 5.42. Point A is also the origin of a coordinate system that is rotating and translating. The coordinate system X, Y, Z is fixed in space, and its unit base vectors, $\hat{\mathbf{I}}$, $\hat{\mathbf{J}}$, $\hat{\mathbf{K}}$, are constant. The coordinate system x, y, z is translating and rotating, and its unit base vectors, $\hat{\mathbf{i}}$, $\hat{\mathbf{j}}$, $\hat{\mathbf{k}}$ are changing as the coordinate system rotates with an angular velocity $\boldsymbol{\Omega}$ and an angular acceleration $\dot{\boldsymbol{\Omega}}$.

The positon vector $\mathbf{r}_B$ may be written as

$$\mathbf{r}_B = \mathbf{r}_A + \mathbf{r}_{B/A} \tag{5.159}$$

where $\mathbf{r}_{B/A}$ is the relative positon vector of B relative to A. This vector may be expressed in terms of the XYZ system or the xyz system. If the vector is expressed with components in the rotating coordinate system, the components change due to movement of point B in space or due to the rotation of the coordinate system. Point A, in general, will have an absolute (relative to the fixed or inertial coordinate system) linear velocity and acceleration.

The absolute velocity of point B may be written as

$$\mathbf{v}_B = \mathbf{v}_A + \frac{d\mathbf{r}_{B/A}}{dt} \tag{5.160}$$

If $\mathbf{r}_{B/A}$ is expressed in the rotating frame, then

$$\frac{d\mathbf{r}_{B/A}}{dt} = \dot{x}_{B/A}\hat{\mathbf{i}} + \dot{y}_{B/A}\hat{\mathbf{j}} + \dot{z}_{B/A}\hat{\mathbf{k}} + x_{B/A}\frac{d\hat{\mathbf{i}}}{dt} + y_{B/A}\frac{d\hat{\mathbf{j}}}{dt} + z_{B/A}\frac{d\hat{\mathbf{k}}}{dt} \tag{5.161}$$

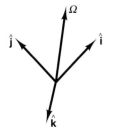

Figure 5.43

Consider the unit vectors to form a triad that is rotating at an angular velocity of $\boldsymbol{\Omega}$ about the $\boldsymbol{\Omega}$-axis, as illustrated in Figure 5.43. The length of the unit vectors is unity, and the vectors can only change direction. This system is similar to a rotating rigid body, and the derivatives of the unit vectors may be written as

$$\frac{d\hat{\mathbf{i}}}{dt} = \boldsymbol{\Omega} \times \hat{\mathbf{i}}, \quad \frac{d\hat{\mathbf{j}}}{dt} = \boldsymbol{\Omega} \times \hat{\mathbf{j}}, \text{ and } \frac{d\hat{\mathbf{k}}}{dt} = \boldsymbol{\Omega} \times \hat{\mathbf{k}} \tag{5.162}$$

The velocity of B relative to A in the rotating coordinate system may be written as

$$(\mathbf{v}_{B/A})_{xyz} = \dot{x}_{B/A}\hat{\mathbf{i}} + \dot{y}_{B/A}\hat{\mathbf{j}} + \dot{z}_{B/A}\hat{\mathbf{k}}$$

and the absolute relative velocity is

$$\frac{d\mathbf{r}_{B/A}}{dt} = (\mathbf{v}_{B/A})_{xyz} + x_{B/A}(\mathbf{\Omega} \times \hat{\mathbf{i}}) + y_{B/A}(\mathbf{\Omega} \times \hat{\mathbf{j}}) + z_{B/A}(\mathbf{\Omega} \times \hat{\mathbf{k}}) \quad (5.163)$$

or

$$\frac{d\mathbf{r}_{B/A}}{dt} = (\mathbf{v}_{B/A})_{xyz} + \mathbf{\Omega} \times \mathbf{r}_{B/A} \quad (5.164)$$

This expression can be generalized to the absolute time derivative of any vector in a rotating coordinate system as

$$\frac{d\mathbf{Q}}{dt} = (\dot{\mathbf{Q}}_{rel})_{xyz} + \mathbf{\Omega} \times \mathbf{Q} \quad (5.165)$$

The absolute velocity of B may now be written as

$$\mathbf{v}_B \quad = \quad \mathbf{v}_A \quad + \quad \mathbf{\Omega} \times \mathbf{r}_{B/A} \quad + \quad (\mathbf{v}_{B/A})_{xyz}$$

| absolute velocity of B | absolute velocity of A | relative velocity of B to A caused by rotation of xyz system | relative velocity of B to A in the rotating coordinate system | (5.166) |

The absolute velocity of B—that is, the velocity measured relative to the fixed coordinates XYZ—is equal to the absolute velocity of A plus the relative velocity due to the rotation of the coordinate system xyz plus the velocity of B relative to A in the rotating coordinate system. Note that if A and B had the same absolute velocity at a given instant, B would still have a relative velocity of B to A in the rotating coordinate system that is equal to the negative of the relative velocity of B to A caused by the rotation of the xyz system. That is, an observer at point A would see B rotating relative to him or herself in a direction opposite to his or her actual rotation in the XYZ system.

The absolute acceleration of point B may be obtained by careful differentiation of the absolute velocity of B:

$$\mathbf{a}_B = \mathbf{a}_A + \frac{d\mathbf{\Omega}}{dt} \times \mathbf{r}_{B/A} + \mathbf{\Omega} \times \frac{d\mathbf{r}_{B/A}}{dt} + \frac{d(\mathbf{v}_{B/A})_{xyz}}{dt} \quad (5.167)$$

$$\frac{d\mathbf{r}_{B/A}}{dt} = \mathbf{\Omega} \times \mathbf{r}_{B/A} + (\mathbf{v}_{B/A})_{xyz} \quad (5.168)$$

and

$$\mathbf{\Omega} \times \frac{d\mathbf{r}_{B/A}}{dt} = \mathbf{\Omega} \times (\mathbf{\Omega} \times \mathbf{r}_{B/A}) + \mathbf{\Omega} \times (\mathbf{v}_{B/A})_{xyz} \quad (5.169)$$

The term $\frac{d(\mathbf{v}_{B/A})_{xyz}}{dt}$ may be considered to be the absolute time derivative of a vector in a rotating coordinate system. So,

$$\frac{d(\mathbf{v}_{B/A})_{xyz}}{dt} = \mathbf{\Omega} \times (\mathbf{v}_{B/A})_{xyz} + (\mathbf{a}_{B/A})_{xyz}$$

where

$$(\mathbf{a}_{B/A})_{xyz} = \ddot{x}_{B/A}\hat{\mathbf{i}} + \ddot{y}_{B/A}\hat{\mathbf{j}} + \ddot{z}_{B/A}\hat{\mathbf{k}} \quad (5.170)$$

The absolute acceleration of B may be written

$$\mathbf{a}_B = \mathbf{a}_A + \dot{\boldsymbol{\Omega}} \times \mathbf{r}_{B/A} + \boldsymbol{\Omega} \times (\boldsymbol{\Omega} \times \mathbf{r}_{B/A}) + 2\boldsymbol{\Omega} \times (\mathbf{v}_{B/A})_{xyz} + (\mathbf{a}_{B/A})_{xyz} \quad (5.171)$$

Note that if point B is not moving relative to the rotating frame, the last two terms are zero and the acceleration of B has the same form as if A and B were points on a rigid body having angular velocity $\boldsymbol{\Omega}$ and an angular acceleration $\dot{\boldsymbol{\Omega}}$. The term $2\boldsymbol{\Omega} \times (\mathbf{v}_{B/A})_{xyz}$ is called the *Coriolis acceleration*, after the French military engineer, G. G. Coriolis (1792–1843), who was the first to examine this term. His observations were based upon the study of water wheels and were published in 1835.

When the observer is on the earth, he or she sees the acceleration relative to him or herself, and not the absolute acceleration. The term $(\mathbf{a}_{B/A})_{xyz}$ is the acceleration of B relative to an observer who is at A and is rotating with the frame of reference. If the origin of the rotating coordinate system is not moving, $\mathbf{a}_A = 0$, then the acceleration relative to an observer rotating with the coordinate system is

$$(\mathbf{a}_{B/A})_{xyz} = \mathbf{a}_B - \dot{\boldsymbol{\Omega}} \times \mathbf{r}_{B/A} - \boldsymbol{\Omega} \times (\boldsymbol{\Omega} \times \mathbf{r}_{B/A}) - 2\boldsymbol{\Omega} \times \mathbf{v}_{B/A} \quad (5.172)$$

The motion of the earth is dominated by its rotation about its own axis, the effects of its rotation about the sun, and motion of the solar system in the galaxy; other effects are small by comparison. The angular velocity of the earth about its polar axis is

$$\boldsymbol{\Omega} = \frac{2\pi \text{ rad/day}}{(60 \times 60 \times 24) \text{ s/day}} = 7.29 \times 10^{-5} \text{ rad/s}$$

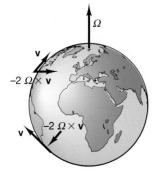

Figure 5.44

The angular acceleration of the earth is zero, and the effect of the Coriolis acceleration can be examined. In the northern hemisphere, an object moving in a horizontal plane relative to the earth's surface would be deflected to the *right* as observed from the earth, as illustrated in Figure 5.44. In the southern hemisphere, the object appears to be deflected to the *left*. Therefore, the flow of air masses from high-pressure regions to low-pressure regions in the northern hemisphere are deflected to the right and produce cyclonic motion, as shown in Figure 5.45. Therefore, in the northern hemisphere, the winds move counterclockwise around a low-pressure center and clockwise around a high-pressure center. In the southern hemisphere, the air masses flow in the opposite direction. In theory, a similar motion should be observable when it forms water in whirlpools or when it flows through sink drains, but both rotations are obscured, as other factors dominate the motion.

A story has been told that during World War I, in a naval battle near the Falkland islands, British gunners were firing shots that were missing by hundreds of yards to the left as they fired on German ships. The designers of the gun sights were aware of the Coriolis deflection and had designed these guns to correct the deflection of the shell trajectory to the right, as would be neccessary in the northern hemisphere. They had never thought that battles might occur at latitudes 50° south instead of 50° north. The British shells fell to the left of the targets at a distance equal to twice the Coriolis deflection.

In northern latitudes, an object dropped from rest will appear to deflect to the east. This eastward deflection was predicted by Newton in 1679 and confirmed by experiments by Robert Hooke. One of the most historically significant investigations of the Coriolis effect was done by the French physicist Jean Leon Foucault (1819–1868), who observed that a pendulum will appear to precess (or rotate) in its plane of observation due to the earth's rotation. Foucault's pendulum is displayed in many science museums throughout the world. If Foucault's pendulum were set in motion at the North Pole, the plane of oscillation would remain fixed in space, and the earth would rotate beneath it. To an observer on the earth, the plane of oscillation would appear to complete a full rotation each day.

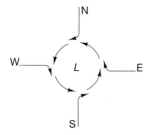

Figure 5.45

The two basic equations for rotating coordinates are shown again:

$$\mathbf{V}_P = \qquad \mathbf{V}_0 \qquad + \mathbf{\Omega} \times \mathbf{r}_{P/0} \qquad + \mathbf{v}_{P/0} \qquad (5.173)$$

| Absolute velocity of P | Absolute velocity of origin of rotating system | velocity due to rotation | relative velocity in rotating coordinate system |

$$\mathbf{A}_P = \qquad \mathbf{A}_0 \qquad + \dot{\mathbf{\Omega}} \times \mathbf{r}_{P/0} + \mathbf{\Omega} \times (\mathbf{\Omega} \times \mathbf{r}_{P/0}) + 2\mathbf{\Omega} \times \mathbf{v}_{P/0} + \mathbf{a}_{P/0}$$

| Absolute acceleration of P | Absolute acceleration of origin of rotating system | tangential acceleration due to rotation | normal acceleration due to rotating coordinate system | Coriolis acceleration | relative acceleration in rotating system |

$$(5.174)$$

In these equations, uppercase letters indicate absolute velocities and accelerations and lowercase letters indicate relativity to the rotating coordinate systems. In the same notation, a time derivative in the fixed coordinates may be related to that which is relative to a rotating system as

$$\frac{D}{Dt} = \qquad \frac{d}{dt} \qquad + \mathbf{\Omega} \times$$

| absolute time derivative | relative time derivative | angular velocity of rotating system | (5.175) |

The two general equations may be applied to the study of the motion of a particle in a rotating coordinate system and the motion of a rigid body in plane motion in a rotating coordinate system.

5.14.1 SLIDING CONTACT

Problems relating to a rigid body which slides with contact on another rigid body have to be considered carefully before a kinematic analysis is done. Two separate problems will be considered to examine the different analyses. Consider the problem of a bar sliding with one end in contact with a horizontal surface, while the other end is in contact with a vertical wall, as shown in Figure 5.46. If point A is moving to the right with a velocity of $\mathbf{V}$ and an acceleration of $\mathbf{A}$ to the right, determine the angular velocity and acceleration of the bar as a function of θ.

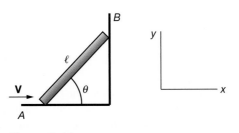

Figure 5.46

Using the methods from Section 5.8, we can write a general equation to relate the velocity of two points on a rigid body as

$$\mathbf{v}_B = \mathbf{v}_A + \boldsymbol{\omega} \times \mathbf{r}_{B/A} \tag{5.176}$$

The relative position vector is

$$\mathbf{r}_{B/A} = l\cos\theta\hat{\mathbf{i}} + l\sin\theta\hat{\mathbf{j}}$$

Therefore, Eq. (5.176) yields

$$v_B\hat{\mathbf{j}} = V\hat{\mathbf{i}} + \omega\hat{\mathbf{k}} \times (l\cos\theta\hat{\mathbf{i}} + l\sin\theta\hat{\mathbf{j}})$$

Expanding yields

$$\omega = \frac{V}{l\sin\theta}$$

$$v_B = \omega l\cos\theta = V\frac{\cos\theta}{\sin\theta}$$

In a similar manner, the acceleration of B may be written as

$$\mathbf{a}_B = \mathbf{a}_A + \boldsymbol{\alpha} \times \mathbf{r}_{B/A} - \omega^2\mathbf{r}_{B/A} \tag{5.177}$$

Expanding yields

$$\alpha = \frac{1}{l\sin\theta}(A - \omega^2 l\cos\theta)$$

$$a_B = \alpha l\cos\theta - \omega^2 l\sin\theta$$

This problem was solved using the basic relationships of two points on a rigid body.

This problem also could have been solved using the method presented in Section 5.10 by examining the trigonometric relationship from Figure 5.47

The distance *x* is positive to the left and

$$x = l\cos\theta$$

Differentiating this equation with respect to time yields

$$\dot{x} = -V = -l\sin\theta\omega$$

$$\omega = \frac{V}{l\sin\theta}$$

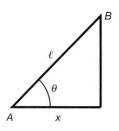

Figure 5.47

This agrees with the previous solution. The angular acceleration may be obtained by a second differentiation:

$$\ddot{x} = -A = -l\cos\theta\omega^2 - l\sin\theta\alpha$$

$$\alpha = \frac{1}{l\sin\theta}(A - l\cos\theta\omega^2)$$

Again, as expected, this solution agrees with the previous obtained functions.

Now let us consider a variation of this problem. Again, a bar is sliding on a surface as shown in Figure 5.48. The point *A* moves to the right with a velocity of *V* and an acceleration of *A*. Determine the angular velocity and the angular acceleration of the bar.

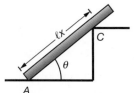

Figure 5.48

This problem, which appears similar to the one just analyzed by the method in Section 5.8, cannot be analyzed by the same method as it does **not** involve two points on the same body. The contact point C changes as the bar slides up and down and the relationships given in Eqs. (5.176) and (5.177) are **not** valid, as point C is not a fixed point on the rigid body and moves relative to the bar. This is an example of sliding contact between rigid bodies. We can analyze the angular velocity and acceleration using the methods from Section 5.10. The following trigonometric relationship holds:

$$\tan \theta = \frac{h}{x} \ \text{ or } \ x \tan \theta = h$$

Differentiating this relationship with respect to time and realizing that x is positive to the left yields

$$\dot{x} \tan \theta + x \sec^2 \theta \omega = 0$$

$$\dot{x} = -V \ \text{ and } \ x = \frac{h}{\tan \theta}$$

Solving for ω yields

$$\omega = \frac{V \tan \theta}{x \sec^2 \theta} = \frac{V}{h} \tan^2 \theta \cos^2 \theta = \frac{V}{h} \sin^2 \theta$$

The angular acceleration can be obtained by differentiation of the angular velocity:

$$\alpha = \frac{d\omega}{dt} = \frac{dV}{dt} \frac{\sin^2 \theta}{h} + \frac{2V}{h} \sin \theta \cos \theta \omega$$

$$\alpha = \frac{\sin \theta}{h} \left[A \sin \theta + \frac{2V^2 \sin^2 \theta \cos \theta}{h} \right]$$

A vector solution to this problem can be obtained using a rotating coordinate system as outlined in Section 5.14. We will select a coordinate system that is body fixed at A, a point on the rigid body, but we will not assume that point C is a point on the rigid body and thus the magnitude of $\mathbf{r}_{C/A}$ is not constant in magnitude. The coordinate system attached to the bar is shown in Figure 5.49.

The coordinate system is rotating with the angular velocity of the bar, $\boldsymbol{\omega}$, and an angular acceleration of $\boldsymbol{\alpha}$. Eq. (5.166) may be written as

$$\mathbf{v}_C = \mathbf{v}_A + \boldsymbol{\omega} \times \mathbf{r}_{C/A} + (\mathbf{v}_{C/A})_{\text{rel}}$$

Eq. (5.171) may be written as

$$\mathbf{a}_C = \mathbf{a}_A + \boldsymbol{\alpha} \times \mathbf{r}_{C/A} + \boldsymbol{\omega} \times (\boldsymbol{\omega} \times \mathbf{r}_{C/A}) + 2\boldsymbol{\omega} \times (\mathbf{v}_{C/A})_{\text{rel}} + (\mathbf{a}_{C/A})_{\text{rel}}$$

Now point C is a fixed point in space with zero absolute velocity and acceleration. The other known vectors can be written as

$$\mathbf{r}_{C/A} = l_x \hat{\mathbf{i}} \ \text{ where } l_x = \frac{h}{\sin \theta}$$

$$\mathbf{v}_A = V \cos \theta \hat{\mathbf{i}} - V \sin \theta \hat{\mathbf{j}}$$

$$\mathbf{v}_C = 0$$

$$\boldsymbol{\omega} = \omega \hat{\mathbf{k}}$$

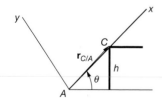

Figure 5.49

The velocity equation becomes

$$0 = V\cos\theta\hat{\mathbf{i}} - V\sin\theta\hat{\mathbf{j}} + \omega l_x\hat{\mathbf{j}} + (v_{C/A})_{\text{rel}}\hat{\mathbf{i}}$$

$$\boldsymbol{\omega} = \frac{V\sin\theta}{l_x} = \frac{V\sin^2\theta}{h}$$

$$(v_{C/A})_{\text{rel}} = -V\cos\theta$$

The known acceleration vectors can be written as

$$\mathbf{a}_A = A\cos\theta\hat{\mathbf{i}} - A\sin\theta\hat{\mathbf{j}}$$

$$\mathbf{a}_C = 0$$

$$\boldsymbol{\alpha} = \alpha\hat{\mathbf{k}}$$

The acceleration equation becomes

$$0 = A\cos\theta\hat{\mathbf{i}} - A\sin\theta\hat{\mathbf{j}} + \alpha l_x\hat{\mathbf{j}} - \omega^2 l_x\hat{\mathbf{i}} + 2\omega(v_{C/A})_{\text{rel}}\hat{\mathbf{j}} + (a_{B/A})_{\text{rel}}\hat{\mathbf{i}}$$

Equating vector components yields

$$\boldsymbol{\alpha} = \frac{1}{l_x}\left[A\sin\theta + 2\omega(v_{c/A})_{\text{rel}}\right]$$

$$\boldsymbol{\alpha} = \frac{\sin\theta}{h}\left[A\sin\theta + \frac{2V^2\sin^2\theta\cos\theta}{h}\right]$$

$$(a_{C/A})_{\text{rel}} = -A\cos\theta + \omega^2 l_x = -A\cos\theta + V^2\sin^3\theta$$

The two solutions agree, and either method could be used to solve this problem.

Sample Problem 5.29

A wheel of radius r rolls on a surface of radius R without slipping, as shown in the accompanying figure. Determine the velocity and acceleration of a point C if the wheel has an angular velocity $\boldsymbol{\omega}$ and an angular acceleration $\boldsymbol{\alpha}$.

Solution

Although there are many ways to solve this problem of plane motion, it is best treated as a rotating coordinate system attached to the center of the wheel. The wheel rotates at an angular velocity $\boldsymbol{\omega}$ and an angular acceleration $\boldsymbol{\alpha}$ about the origin of the rotating coordinate system. The motion of the wheel within the rotating coordinate system is the same as that of a wheel rotating about a fixed axis in space. The coordinate system, however, rotates around the center of a fixed coordinate system at the center of the larger circle. The choice of coordinates is shown in the next figure. The absolute velocity of point D, the point of contact, must be zero.

Examine the absolute velocity of the point D:

$$\mathbf{V}_D = \mathbf{V}_A + \boldsymbol{\Omega} \times \mathbf{r}_{D/A} + \mathbf{v}_{D/A} = 0$$

$\boldsymbol{\Omega}$ is the angular velocity of the center of the wheel about the center of the circle with radius R. Therefore,

$$\mathbf{V}_A = \Omega\hat{\mathbf{k}} \times (-R + r)\hat{\mathbf{j}}$$

The motion in the rotating coordinate system is described by

$$\mathbf{r}_{D/A} = -r\hat{\mathbf{j}}$$

$$\mathbf{v}_{D/A} = \omega\hat{\mathbf{k}} \times (-r\hat{\mathbf{j}})$$

$$\mathbf{V}_D = 0 = R\Omega\hat{\mathbf{i}} - r\Omega\hat{\mathbf{i}} + r\Omega\hat{\mathbf{i}} + r\omega\hat{\mathbf{i}}$$

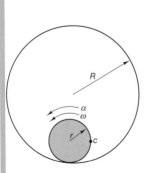

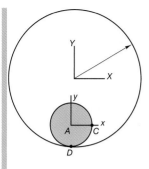

Therefore,

$$\Omega = -\frac{r\omega}{R}$$

The wheel is turning counterclockwise about its own central axis, and the center of the rotating coordinate system is rotating clockwise. This result conceptually makes sense if the motion is carefully examined. The absolute velocity of point C may now be obtained.

$$\mathbf{V}_C = \mathbf{V}_A + \mathbf{\Omega} \times \mathbf{r}_{C/A} + \mathbf{v}_{C/A}$$

$$\mathbf{V}_C = \left(-\frac{r\omega}{R}\hat{\mathbf{k}}\right) \times (-R + r)\hat{\mathbf{j}} + \left(-\frac{r\omega}{R}\hat{\mathbf{k}}\right) \times r\hat{\mathbf{i}} + \omega\hat{\mathbf{k}} \times r\hat{\mathbf{i}}$$

$$\mathbf{V}_C = \left(\frac{r^2\omega}{R} - r\omega\right)\hat{\mathbf{i}} - \left(\frac{r^2\omega}{R} - r\omega\right)\hat{\mathbf{j}}$$

The acceleration is handled in the same manner. Since the wheel rolls without slipping, the absolute acceleration of point D must be in the $\hat{\mathbf{j}}$ direction, so

$$\mathbf{A}_D = \mathbf{A}_A + \dot{\mathbf{\Omega}} \times \mathbf{r}_{D/A} + \mathbf{\Omega} \times (\mathbf{\Omega} \times \mathbf{r}_{D/A}) + 2\mathbf{\Omega} \times \mathbf{v}_{D/A} + \mathbf{a}_{D/A}$$

$$\mathbf{A}_D = A_D\hat{\mathbf{j}} \quad \mathbf{A}_A = \dot{\mathbf{\Omega}} \times (-R + r)\hat{\mathbf{j}} + \mathbf{\Omega} \times \mathbf{V}_A$$

$$= (R\dot{\Omega} - r\dot{\Omega})\hat{\mathbf{i}} + \left(\frac{r^2\omega\Omega}{R} - r\omega\Omega\right)\hat{\mathbf{j}}$$

$$\mathbf{r}_{D/A} = -r\hat{\mathbf{j}} \quad \mathbf{v}_{D/A} = r\omega\hat{\mathbf{i}} \quad \mathbf{a}_{D/A} = \boldsymbol{\alpha} \times \mathbf{r}_{D/A}$$

$$+ \boldsymbol{\omega} \times (\boldsymbol{\omega} \times \mathbf{r}_{D/A}) = r\alpha\hat{\mathbf{i}} + r\omega^2\hat{\mathbf{j}}$$

Now, the only unknowns are the acceleration of D in the $\hat{\mathbf{j}}$ direction and the angular acceleration of the rotating coordinate system. Performing the required vector operations yields

$$A_D\hat{\mathbf{j}} = (R\dot{\Omega} + r\alpha)\hat{\mathbf{i}} + \left(r\omega^2 - \frac{r^2\omega^2}{R}\right)\hat{\mathbf{j}}$$

Therefore,

$$\dot{\Omega} = -\frac{r\alpha}{R}$$

The acceleration of point C now may be obtained:

$$\mathbf{A}_C = \mathbf{A}_A + \dot{\mathbf{\Omega}} \times \mathbf{r}_{C/A} + \mathbf{\Omega} \times (\mathbf{\Omega} \times \mathbf{r}_{C/A}) + 2\mathbf{\Omega} \times \mathbf{v}_{C/A} + \mathbf{a}_{C/A}$$

$$\mathbf{r}_{C/A} = r\hat{\mathbf{i}}$$

$$\mathbf{v}_{C/A} = r\omega\hat{\mathbf{j}}$$

$$\mathbf{a}_{C/A} = r\alpha\hat{\mathbf{j}} - r\omega^2\hat{\mathbf{i}}$$

$$\mathbf{A}_C = -\left[r\alpha - \frac{r^2\alpha}{R} + r\omega^2 - \frac{2r^2\omega^2}{R} + \frac{r^3\omega^2}{R^2}\right]\hat{\mathbf{i}} + \left[r\alpha - \frac{r^2\alpha}{R} + \frac{r^2\omega^2}{R} - \frac{r^3\omega^2}{R^2}\right]\hat{\mathbf{j}}$$

Sample Problem 5.30	The disk shaft assembly in the accompanying figure rolls without slipping, such that at the instant shown, the angular velocity ω is 7 rad/s and the angular acceleration α is 10 rad/s². If the radius of the disk is 6 in. and the shaft is 30 in. long, determine the angular velocity and acceleration of the assembly around the center socket O, and the absolute velocity of point C on the disk.

Solution   A rotating coordinate system is attached to the disk, as shown in the figure, and this system rotates about the vertical axis Z at O. The motion within the rotating coordinate system is rotation about the x-axis; as if the x-axis were a fixed axis of rotation. x, y, z is a rotating coordinate system and X, Y, Z is the fixed coordinate system. The problem may be formulated in terms of the base vectors in either coordinate system, but all terms must be expressed in the same system. The angle that the shaft makes with the floor, β, is also the angle between the x-axis and the X-axis. This angle may be determined from the dimensions given in the statement of the problem. The angular velocity and acceleration are

$$\boldsymbol{\Omega} = \Omega\hat{\mathbf{K}} = \Omega(\sin\beta\hat{\mathbf{i}} + \cos\beta\hat{\mathbf{k}})$$

$$\dot{\boldsymbol{\Omega}} = \dot{\Omega}\hat{\mathbf{K}} = \dot{\Omega}(\sin\beta\hat{\mathbf{i}} + \cos\beta\hat{\mathbf{k}})$$

The absolute velocity of point D must be zero, as the disk is rolling without slipping, so

$$\mathbf{V}_D = \mathbf{V}_A + \boldsymbol{\Omega}\times\mathbf{r}_{D/A} + \mathbf{v}_{D/A}$$

$$\mathbf{V}_D = 0$$

$$\mathbf{V}_A = \boldsymbol{\Omega}\times L\hat{\mathbf{i}} = \Omega L\cos\beta\hat{\mathbf{j}}$$

$$\mathbf{r}_{D/A} = -r\hat{\mathbf{k}}$$

$$\mathbf{v}_{D/A} = \boldsymbol{\omega}\times\mathbf{r}_{D/A} = (-7\hat{\mathbf{i}})\times(-6\hat{\mathbf{k}}) = -42\hat{\mathbf{j}}$$

$$0 = \Omega L\cos\beta\hat{\mathbf{j}} + \Omega r\sin\beta\hat{\mathbf{j}} - 42\hat{\mathbf{j}}$$

$$\Omega = \frac{42}{(30)(0.981) + (6)(0.196)} = 1.372 \text{ rad/s}$$

$$\mathbf{V}_C = \mathbf{V}_A + \boldsymbol{\Omega}\times\mathbf{r}_{C/A} + \mathbf{v}_{C/A}$$

$$\mathbf{r}_{C/A} = r\hat{\mathbf{j}} \qquad \mathbf{v}_{C/A} = \boldsymbol{\omega}\times\mathbf{r}_{C/A} = -42\hat{\mathbf{k}}$$

$$\mathbf{V}_C = \Omega L\cos\beta\hat{\mathbf{j}} - \Omega r\cos\beta\hat{\mathbf{i}} + \Omega r\sin\beta\hat{\mathbf{k}} - 42\hat{\mathbf{k}}$$

$$\mathbf{V}_C = -8.1\hat{\mathbf{i}} + 40.4\hat{\mathbf{j}} - 40.4\hat{\mathbf{k}}$$

we proceed to analyze the acceleration in a similar manner, knowing that the absolute acceleration of the contact point D cannot have an acceleration in the $\hat{\mathbf{j}}$ direction for rolling without slipping. Our analysis is as follows:

$$\mathbf{A}_D = \mathbf{A}_A + \dot{\boldsymbol{\Omega}}\times\mathbf{r}_{D/A} + \boldsymbol{\Omega}\times(\boldsymbol{\Omega}\times\mathbf{r}_{D/A}) + 2\boldsymbol{\Omega}\times\mathbf{v}_{D/A} + \mathbf{a}_{D/A}$$

$$\mathbf{A}_A = \dot{\boldsymbol{\Omega}}\times L\hat{\mathbf{i}} + \boldsymbol{\Omega}\times(\boldsymbol{\Omega}\times L\hat{\mathbf{i}}) = \dot{\mathbf{V}}L\cos\beta\hat{\mathbf{j}}$$

$$+\Omega^2 L(-\cos^2\beta\hat{\mathbf{i}} + \sin\beta\cos\beta\hat{\mathbf{k}})$$

$$\dot{\boldsymbol{\Omega}}\times\mathbf{r}_{C/A} = r\dot{\Omega}\sin\beta\hat{\mathbf{j}}$$

$$\boldsymbol{\Omega}\times(\boldsymbol{\Omega}\times\mathbf{r}_{D/A}) = r\Omega^2(-\sin\beta\cos\beta\hat{\mathbf{i}} + \sin^2\beta\hat{\mathbf{k}})$$

$$2\boldsymbol{\Omega}\times\mathbf{v}_{D/A} = 2(42)\Omega(\cos\beta\hat{\mathbf{i}} - \sin\beta\hat{\mathbf{k}})$$

$$\mathbf{a}_{D/A} = \boldsymbol{\alpha}\times\mathbf{r}_{D/A} + \boldsymbol{\omega}\times(\boldsymbol{\omega}\times\mathbf{r}_{D/A}) = (-10\hat{\mathbf{i}})\times(-6\hat{\mathbf{k}}) - (7)^2(-6\hat{\mathbf{k}})$$

$$= -60\hat{\mathbf{j}} + 294\mathbf{k}$$

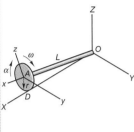

$$\tan(\beta) = r/L = 6/30$$
$$\beta = 11.31°$$

Setting the $\hat{\mathbf{j}}$-component equal to zero, we obtain

$$0 = L\dot{\Omega}\cos\beta + r\dot{\Omega}\sin\beta - 60$$

$$\Omega = 1.961 \text{ rad/s}^2$$

The acceleration of point C may now be obtained:

$$\mathbf{A}_C = \mathbf{A}_A + \dot{\mathbf{\Omega}} \times \mathbf{r}_{C/A} + \mathbf{\Omega} \times (\mathbf{\Omega} \times \mathbf{r}_{C/A}) + 2\mathbf{\Omega} \times \mathbf{v}_{C/A} + \mathbf{a}_{C/A}$$

$$\mathbf{r}_{C/A} = r\hat{\mathbf{j}} = 6\hat{\mathbf{j}}$$

$$\mathbf{v}_{C/A} = -42\hat{\mathbf{k}}$$

$$\mathbf{a}_{C/A} = \boldsymbol{\alpha} \times \mathbf{r}_{C/A} + \boldsymbol{\omega} \times (\boldsymbol{\omega} \times \mathbf{r}_{C/A}) = 60\hat{\mathbf{k}} - 294\hat{\mathbf{j}}$$

$$\mathbf{A}_A = \dot{\Omega}L\cos\beta\hat{\mathbf{j}} + \Omega^2L(-\cos^2\beta\hat{\mathbf{i}} + \sin\beta\cos\beta\hat{\mathbf{k}})$$

$$= -54.3\hat{\mathbf{i}} + 57.7\hat{\mathbf{j}} + 10.9\hat{\mathbf{k}}$$

$$\dot{\mathbf{\Omega}} \times \mathbf{r}_{C/A} = -11.5\hat{\mathbf{i}} + 2.3\hat{\mathbf{k}}$$

$$\mathbf{\Omega} \times (\mathbf{\Omega} \times \mathbf{r}_{C/A}) = -\Omega^2 r\hat{\mathbf{j}} = -11.3\hat{\mathbf{j}}$$

$$2\mathbf{\Omega} \times \mathbf{v}_{C/A} = 22.6\hat{\mathbf{j}}$$

$$\mathbf{A}_C = -65.6\hat{\mathbf{i}} - 225\hat{\mathbf{j}} + 73.2\hat{\mathbf{k}}$$

Sample Problem 5.31

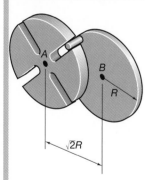

An intermittent motion device called the Geneva, or Malta, mechanism is shown in the accompanying figure. This mechanism allows intermittent rotation of disk A as disk B rotates at a constant rate. The Geneva wheel, disk A, is fitted with at least three equispaced, radial slots. Disk B has a pin which enters a radial slot and causes the Geneva wheel to turn through a portion of a revolution. When the pin leaves a slot, the Geneva wheel will remain stationary until the pin enters the next slot. A Geneva wheel needs a minimum of three slots to work, but the maximum number of slots is limited only by the size of the wheel. In this case, disk A will rotate $1/4$ turn for each full rotation of disk B. This information allows for a method of counting rotations and is useful for many machines. If disk B is rotating counterclockwise at a constant rate ω, determine the angular velocity and the angular acceleration of disk A when the pin P is engaged. Determine the velocity of the pin relative to the disk A during engagement.

Solution

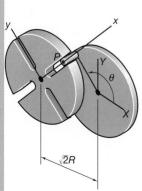

Angles θ and β designate the orientation of the two disks at any time during engagement. A relationship between these two angles can be established by examining the triangle formed from A, B, and P.

The distance between the centers $\sqrt{2}R$ is critical if the two disks are to properly engage and disengage. During the time of engagement, when θ is between $135°$ and $225°$, the angles β and θ may be related using trigonometry:

Law of sines:

$$x\sin\beta = R\sin(\pi - \theta) = R(\sin\pi\cos\theta - \cos\pi\sin\theta)$$

$$\frac{x}{\sin(\pi - \theta)} = \frac{R}{\sin\beta}$$

$$x\sin\beta = R\sin\theta$$

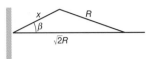

Law of cosines:

$$x^2 = R^2 + 2R^2 - 2\sqrt{2}R^2 \cos(\pi - \theta)$$

$$x^2 = R^2 + 2R^2 + 2\sqrt{2}R^2 \cos\theta$$

For any value of θ, the two values x and β may be determined:

$$x = R\sqrt{3 + 2\sqrt{2}\cos\theta}$$

$$\beta = \sin^{-1}\frac{\sin\theta}{\sqrt{3 + 2\sqrt{2}\cos\theta}}$$

Note that since θ may be given for any time during engagement as $\left(\frac{3\pi}{4} + \omega t\right)$ if t is measured such that at $t = 0$, the pin is just engaging disk A, the problem may be solved in parametric form by differentiating x and β with respect to time. The desired solution could then be obtained and would be valid for the values of θ between 135° and 225°. The parametric solution is shown below for comparison:

$$x = R\sqrt{3 + 2\sqrt{2}\cos\theta}$$

$$\dot{x} = -\frac{R\sqrt{2}\sin\theta\omega}{\sqrt{3 + 2\sqrt{2}\cos\theta}}$$

But,

$$\sin\beta = \frac{\sin\theta}{\sqrt{3 + 2\sqrt{2}\cos\theta}}$$

Therefore,

$$\dot{x} = -\sqrt{2}R\omega\sin\beta$$

if the relative velocity is express in terms of β. The angular velocity of the Geneva wheel may also be obtained:

$$\sin\beta = \frac{\sin\theta}{\sqrt{3 + 2\sqrt{2}\cos\theta}}$$

$$\cos\beta(\dot{\beta}) = \frac{\omega}{\sqrt{3 + 2\sqrt{2}\cos\theta}}\left[\frac{\cos\theta + \sqrt{2}\sin^2\theta}{3 + 2\sqrt{2}\cos\theta}\right]$$

$$\dot{\beta} = \frac{\omega(1 + \sqrt{2}\cos\theta)}{(3 + 2\sqrt{2}\cos\theta)}$$

An alternate approach uses a rotating coordinate system attached to A as shown in the figure. The problem can then be formulated in the rotating coordinate system. The angular velocity and acceleration of the rotating coordinate system is the angular velocity and acceleration of disk A, respectively:

$$\boldsymbol{\Omega} = \dot{\beta}\hat{\mathbf{k}} \text{ and } \boldsymbol{\Omega} = \ddot{\beta}\hat{\mathbf{k}}$$

The absolute velocity of A, the origin of the rotating coordinate system, is zero, and the absolute velocity of point P may be written first in the fixed coordinate system as

$$\mathbf{V}_A = 0$$

$$\mathbf{V}_P = \omega R(-\sin\theta\hat{\mathbf{I}} + \cos\theta\hat{\mathbf{J}})$$

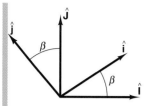

The base vectors $\hat{\mathbf{I}}$ and $\hat{\mathbf{J}}$ may be written in terms of the base vectors in the rotating coordinate system:

$$\hat{\mathbf{I}} = \cos \beta \hat{\mathbf{i}} - \sin \beta \hat{\mathbf{j}}$$
$$\hat{\mathbf{J}} = \sin \beta \hat{\mathbf{i}} + \cos \beta \hat{\mathbf{j}}$$

The velocity can be written in terms of the relative velocity as

$$\mathbf{V}_P = \mathbf{V}_A + \mathbf{\Omega} \times \mathbf{r}_{P/A} + \mathbf{v}_{P/A}$$

$$\mathbf{r}_{P/A} = x_{P/A}\hat{\mathbf{i}} \quad \text{and} \quad \mathbf{v}_{P/A} = \dot{x}_{P/A}\hat{\mathbf{i}}$$

Expressing $\mathbf{V}_P$ in the rotating coordinate system, we get

$$\mathbf{V}_P = \omega R([-\sin \theta \cos \beta + \cos \theta \sin \beta]\hat{\mathbf{i}} + [\sin \theta \sin \beta + \cos \theta \cos \beta]\hat{\mathbf{j}})$$

$$\mathbf{V}_P = 0 + \dot{\beta}\hat{\mathbf{k}} \times x_{P/A}\hat{\mathbf{i}} + \dot{x}_{P/A}\hat{\mathbf{i}} = \dot{x}_{P/A}\hat{\mathbf{i}} + \dot{\beta}x_{P/A}\hat{\mathbf{j}}$$

Equating components, we can write

$$\dot{x}_{P/A} = \omega R[-\sin \theta \cos \beta + \cos \theta \sin \beta]$$

If $\sin \beta$ and $\cos \beta$ are expressed as functions of θ, then

$$\dot{x}_{P/A} = -\frac{R\omega\sqrt{2}\cos \theta}{\sqrt{3 + 2\sqrt{2}\cos \theta}}$$

which agrees with the parametric solution. So,

$$\dot{\beta}x_{P/A} = \omega R[\sin \theta \sin \beta + \cos \theta \cos \beta]$$

But $x_{P/A}$ may be written as $R\sqrt{3 + 2\sqrt{2}\cos \theta}$:

$$\dot{\beta} = \frac{\omega R[\sin \theta \sin \beta + \cos \theta \cos \beta]}{R\sqrt{3 + 2\sqrt{2}\cos \theta}}$$

Using differences in angle relationships, these may be written as

$$\dot{\beta} = \frac{\omega \cos (\beta - \theta)}{\sqrt{3 + 2\sqrt{2}\cos \theta}} \quad \text{and} \quad \dot{x}_{P/A} = \omega R \sin (\beta - \theta)$$

The acceleration must be examined to determine the angular velocity of the disk A and the relative acceleration of the point P in the rotating system. The absolute acceleration of point P is

$$\mathbf{A}_P = -\omega^2 R(\cos \theta \hat{\mathbf{I}} + \sin \theta \hat{\mathbf{J}})$$
$$= -\omega^2 R([\cos \beta \cos \theta + \sin \beta \sin \theta]\hat{\mathbf{i}} + [-\sin \beta \cos \theta + \cos \beta \sin \theta]\hat{\mathbf{j}})$$
$$= -\omega^2 R[\cos (\beta - \theta)\hat{\mathbf{i}} - \sin (\beta - \theta)\hat{\mathbf{j}}]$$

$$\mathbf{a}_{P/A} = \ddot{x}_{P/A}\hat{\mathbf{i}}$$

$$\mathbf{A}_P = \mathbf{A}_A + \dot{\mathbf{\Omega}} \times \mathbf{r}_{P/A} + \mathbf{\Omega} \times (\mathbf{\Omega} \times \mathbf{r}_{P/A}) + 2\mathbf{\Omega} \times \mathbf{v}_{P/A} + \mathbf{a}_{B/A}$$

$$\mathbf{A}_P = 0 + \ddot{\beta}x_{P/A}\hat{\mathbf{j}} - (\dot{\beta})^2 x_{P/A}\hat{\mathbf{i}} + 2\dot{\beta}\dot{x}_{P/A}\hat{\mathbf{j}} + \ddot{x}_{P/A}\hat{\mathbf{i}}$$

Equating components, we write

$$-(\dot{\beta})^2 x_{P/A} + \ddot{x}_{P/A} = -\omega^2 R \cos(\beta - \theta)$$

$$\ddot{\beta} x_{P/A} + 2\dot{\beta}\dot{x}_{P/A} = \omega^2 R \sin(\beta - \theta)$$

Therefore,

$$\ddot{x}_{P/A} = -\omega^2 R \cos(\beta - \theta) + \frac{\omega^2 R \cos^2(\beta - \theta)}{\sqrt{3 + 2\sqrt{2}\cos\theta}}$$

$$\ddot{\beta} = \frac{\omega^2 \sin(\beta - \theta)}{\sqrt{3 + 2\sqrt{2}\cos\theta}} - \frac{2\omega^2 \sin(\beta - \theta)\cos(\beta - \theta)}{3 + 2\sqrt{2}\cos\theta}$$

This expression provides a general solution for the Geneva mechanism at any position or time during contact. If the mechanism was being used in a design, the general solution would be studied in detail using computational software. (See the Computational Supplement.)

For a full understanding of the use of rotating coordinate systems, the velocity portion of this solution will be formulated in terms of the base vectors in the fixed coordinate system. Remember, the problem may be written in terms of either rotating or fixed coordinates but all terms must be consistent. The relationship between the base vectors is

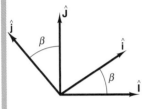

$$\hat{\mathbf{i}} = \cos\beta\hat{\mathbf{I}} + \sin\beta\hat{\mathbf{J}}$$

$$\hat{\mathbf{j}} = -\sin\beta\hat{\mathbf{I}} + \cos\beta\hat{\mathbf{J}}$$

The velocity equation can be written in terms of the base vectors of the fixed coordinate system:

$$\boldsymbol{\Omega} = \dot{\beta}\hat{\mathbf{K}}$$

$$\mathbf{V}_A = 0$$

$$\mathbf{V}_P = \omega R(-\sin\theta\hat{\mathbf{I}} + \cos\theta\hat{\mathbf{J}})$$

$$\mathbf{r}_{P/A} = x_{P/A}(\cos\beta\hat{\mathbf{I}} + \sin\beta\hat{\mathbf{J}})$$

$$\mathbf{v}_{P/A} = \dot{x}_{P/A}(\cos\beta\hat{\mathbf{I}} + \sin\beta\hat{\mathbf{J}})$$

$$\mathbf{V}_P = \mathbf{V}_A + \boldsymbol{\Omega} \times \mathbf{r}_{P/A} + \mathbf{v}_{P/A}$$

$$\omega R(-\sin\theta\hat{\mathbf{I}} + \cos\theta\hat{\mathbf{J}}) = \dot{\beta}\hat{\mathbf{K}} \times x_{P/A}(\cos\beta\hat{\mathbf{I}} + \sin\beta\hat{\mathbf{J}})$$
$$+ \dot{x}_{P/A}(\cos\beta\hat{\mathbf{I}} + \sin\beta\hat{\mathbf{J}})$$

Completing the vector product and equating components, we get

$$-\omega R \sin\theta = -\dot{\beta}x_{P/A}\sin\beta + \dot{x}_{P/A}\cos\beta$$

$$\omega R \cos\theta = \dot{\beta}x_{P/A}\cos\beta + \dot{x}_{P/A}\sin\beta$$

We may solve these two equations for $\dot{x}_{P/A}$ and $\dot{\beta}$, remembering that $x_{P/A} = R\sqrt{3 + 2\sqrt{2}\cos\theta}$:

$$\dot{\beta} = \frac{\omega\cos(\beta - \theta)}{\sqrt{3 + 2\sqrt{2}\cos\theta}}$$

$$\dot{x}_{P/A} = \omega R \sin(\beta - \theta)$$

These results agree with the solution formulated in terms of the base vectors in the rotating system, as required. The choice of which system to use is arbitrary, but the algebra is sometimes easier in one system than in the other, and conceptually, one system may be easier to work in than the other.

Problems

5.114 As shown in Figure P5.114, a disk rotates about the shaft BC with a constant angular velocity ω. The shaft BC is attached at an angle β to the shaft AB as shown and the shaft AB rotates at a constant angular velocity Ω. Determine the absolute linear velocity of point D on the disk.

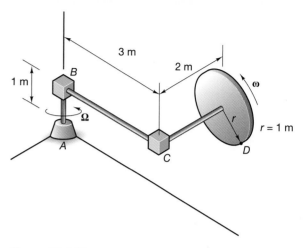

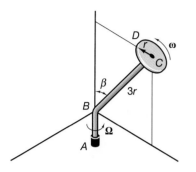

Figure P5.114

5.115 Determine the absolute acceleration of point D in Problem 5.114.

5.116 As shown in Figure P5.116, point C on a robotic arm is being extended at a rate of 30 in/s as the arm is being rotated at a constant rate of 3 rad/s as shown. Determine the absolute velocity and acceleration of point C at the position shown.

Figure P5.117

5.118 Determine the absolute velocity and acceleration of point D in Problem 5.117 if the rod ABC has an angular acceleration of $\dot{\Omega}$.

5.119 A disk of radius 200 mm rolls without slipping about a 400-mm shaft that rotates about a bearing attached to the flat bed of the truck, as seen in Figure P5.119. If the constant angular velocity of the disk is 3 rad/s and the truck is traveling at a velocity of 30 km/hr, determine the velocity and acceleration of point D on the disk.

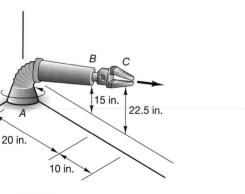

Figure P5.116

5.117 The rod ABC rotates at constant angular velocity of Ω while the disk-shaft CD rotates relative to ABC at a constant rate ω, as shown in Figure P5.117. Determine the absolute velocity and acceleration of the point D on the disk.

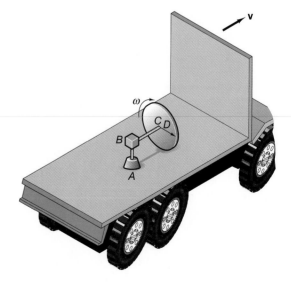

Figure P5.119

5.120 As seen in Figure P5.120, a collar C slides around a circular ring of radius 400 mm with a constant relative speed of 200 mm/s while the ring rotates about a shaft AB with a constant angular velocity of 3 rad/s. Determinet the absolute velocity and acceleration of the collar at any position θ and β.

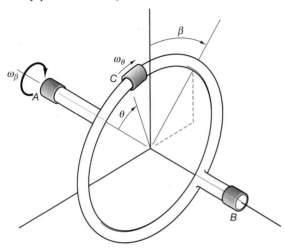

Figure P5.120

5.121 As depicted in Figure P5.121, a small bead slides in the slot cut at an angle α in the disk of radius R. The disk is accelerated from rest at a rate of $\dot{\Omega}$ rad/s² and the bead starts at the upper-left corner when the disk starts rotating. If the bead moves at a constant velocity of v in the slot, determine a general expression for the absolute velocity and acceleration of the bead at any time.

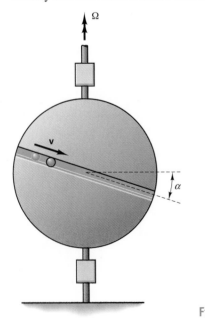

Figure P5.121

5.122 As shown in Figure P5.122, a robotic arm is used to control a sensor in space. At the position shown, the sensor E is moving at a speed of $v = 10$ mm/s relative to the component CD. If component CD is rotating relative to component ABC at a constant angular velocity of $\omega = 1.5$ rad/s and component ABC rotates at a constant angular velocity of $\Omega = 2$ rad/s, determine the absolute velocity and acceleration of the sensor at this position.

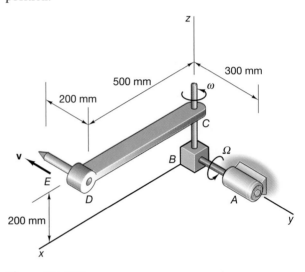

Figure P5.122

5.123 A satellite orbits the earth in a circular polar orbit at an altitude of h above the earth's surface, as seen in Figure P5.123. If the constant orbital angular velocity of the satellite is ω and the earth's radius is R and angular velocity is Ω, determine the relative velocity and acceleration of the satellite as it passes over an observer on North America at latitide θ.

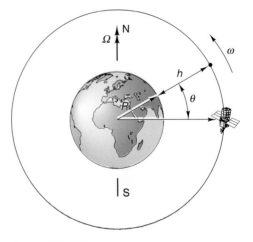

Figure P5.123

5.124 As shown in Figure P5.124, a disk of radius R is attached to a rotating shaft by a bar length L. If the disk rotates about the bar at a constant angular velocity $\dot{\theta}$ and the shaft rotates at a constant angular velocity $\dot{\beta}$, determine the absolute velocity and acceleration of point A on the rotating disk.

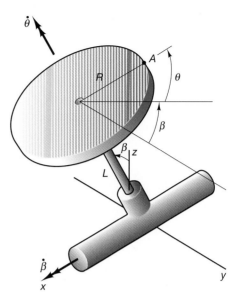

Figure P5.124

Chapter 6

DYNAMICS OF RIGID BODIES IN PLANE MOTION

High velocities are required for the amusement park loop roller to remain on the track. High forces are exerted on the passengers at the bottom of the loop.

6.1 INTRODUCTION

In Chapter 4, we analyzed the dynamics of a system of particles in terms of the linear momentum of the center of mass of the system and the angular momentum about the center of mass. In Chapter 5, we examined the kinematics of plane motion, noting that, for a rigid body in plane motion, the angular velocity and angular acceleration vectors must be perpendicular to the plane of motion. This condition, in fact, defines plane motion and greatly simplifies two-dimensional problems. A rigid body has only three degrees of freedom.

6.2 LINEAR AND ANGULAR MOMENTUM

In analyzing the dynamics of rigid bodies, we assume that a system of particles forms a rigid body and that motion of the system is restricted to a single plane. The linear momentum of a system of particles may be written as

$$\mathbf{L} = \sum_i m_i \mathbf{v}_i = M\mathbf{v}_{\text{cm}} \tag{6.1}$$

The angular momentum about a fixed point O or about the center of mass of the system may be expressed as

$$\mathbf{H}_O = \sum_i \mathbf{r}_i \times m_i \mathbf{v}_i$$

$$\mathbf{H}_{\text{cm}} = \sum_i \boldsymbol{\rho}_i \times m_i \dot{\boldsymbol{\rho}}_i \tag{6.2}$$

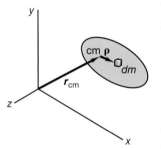

Figure 6.1

Consider the rigid body shown in Figure 6.1. For a rigid body, the particles form a fixed shape, and the summation in Eq. (6.1) is replaced with an integral over the distribution of mass. For rigid bodies of uniform distribution of mass, the integral can be considered to be an integration over the volume of the body. The linear momentum can now be written

$$\mathbf{L} = \int_M \mathbf{v}\,dm = M\mathbf{v}_{\text{cm}} \tag{6.3}$$

The velocity of an element of mass or of a point on the body relative to the center of mass is

$$\dot{\boldsymbol{\rho}} = \boldsymbol{\omega} \times \boldsymbol{\rho} \tag{6.4}$$

where ω is the angular velocity of the body.

The angular momentum about the center of mass is

$$\mathbf{H}_{\text{cm}} = \int_M [\boldsymbol{\rho} \times (\boldsymbol{\omega} \times \boldsymbol{\rho})]\,dm \tag{6.5}$$

Using the vector identity (triple vector product)

$$\mathbf{A} \times (\mathbf{B} \times \mathbf{C}) = \mathbf{B}(\mathbf{A}\cdot\mathbf{C}) - \mathbf{C}(\mathbf{A}\cdot\mathbf{B})$$

We may write the angular momentum about the center of mass as

$$\mathbf{H}_{\text{cm}} = \int_M \boldsymbol{\omega}(\boldsymbol{\rho}\cdot\boldsymbol{\rho})\,dm - \int_M \boldsymbol{\rho}(\boldsymbol{\omega}\cdot\boldsymbol{\rho})\,dm \tag{6.6}$$

Expanding the vector equation into component form yields

$$\mathbf{H}_{cm} = \int_M (x^2 + y^2 + z^2)dm[\omega_x \hat{\mathbf{i}} + \omega_y \hat{\mathbf{j}} + \omega_z \hat{\mathbf{k}}]$$
$$- \int_M (x\omega_x + y\omega_y + z\omega_z)[x\hat{\mathbf{i}} + y\hat{\mathbf{j}} + z\hat{\mathbf{k}}]dm \tag{6.7}$$

Collecting terms, we obtain

$$\mathbf{H}_{cm} = \left[\int_M (y^2 + z^2)dm\,\omega_x - \int_M xy\,dm\,\omega_y - \int_M zx\,dm\,\omega_z \right]\hat{\mathbf{i}}$$
$$+ \left[\int_M (x^2 + z^2)dm\,\omega_y - \int_M xy\,dm\,\omega_x - \int_M yz\,dm\,\omega_z \right]\hat{\mathbf{j}} \tag{6.8}$$
$$+ \left[\int_M (x^2 + y^2)dm\,\omega_z - \int_M xz\,dm\,\omega_x - \int_M yz\,dm\,\omega_y \right]\hat{\mathbf{k}}$$

The first integral in each of the terms of the right-hand side is called the *mass moment of inertia* and is a measure of the resistance to rotation about the individual axes. The mass moments of inertia are discussed in detail in Appendix A, and using consistent notation, we may write them as

$$I_x = \int_M (y^2 + z^2)dm$$
$$I_y = \int_M (x^2 + z^2)dm \tag{6.9}$$
$$I_z = \int_M (x^2 + y^2)dm$$

These equations describe the way the mass is distributed about the x-, y-, and z-axes. The other integrals are called the *product mass moments of inertia* and may be written as

$$I_{xy} = I_{yx} = -\int_M xy\,dm$$
$$I_{yz} = I_{zy} = -\int_M yz\,dm \tag{6.10}$$
$$I_{zx} = I_{xz} = -\int_M zx\,dm$$

Using the notation of the mass moments of inertia and the product mass moments of inertia, we see that the components of the angular momentum vector become

$$H_{cm_x} = I_x \omega_x + I_{xy} \omega_y + I_{xz} \omega_z$$
$$H_{cm_y} = I_{yx} \omega_x + I_y \omega_y + I_{yz} \omega_z \tag{6.11}$$
$$H_{cm_z} = I_{zx} \omega_x + I_{zy} \omega_y + I_z \omega_z$$

Note that each component of the angular momentum involves all the components of the angular velocity.

For a body to remain in plane motion, with the plane of motion parallel to the *xy*-plane, the angular velocity must be in the *z*-direction. In this case, the *x*- and *y*-components of the angular velocity are zero, and the angular momentum vector becomes

$$\mathbf{H}_{cm} = I_{zx}\omega_z\,\hat{\mathbf{i}} + I_{zy}\omega_z\,\hat{\mathbf{j}} + I_z\omega_z\,\hat{\mathbf{k}} \tag{6.12}$$

Therefore, in general, the angular momentum vector will not be in the *z*-direction, and although initially placed in a state of plane motion, the body will not remain in that state. Thus, the plane-motion problem will not reduce to the desired simple three-degree-of-freedom problem in two dimensions.

It can be shown (see Appendix A) that for every rigid body, there exists a set of orthogonal axes with origin at the center of mass, such that about these axes, the product mass moments of inertia will be zero. This special set of axes is called the *principal axes,* and the corresponding mass moments of inertia are called the *principal moments of inertia.* If a rigid body has an axis of symmetry of mass distribution, this axis will be a principal axis. Hence, for symmetrical bodies with a uniform distribution of mass, such as cylinders, spheres, rectangular plates, prisms, and bars, determining the principal axes presents no difficulty. Another interesting fact to note about the principal axes is that about one of them, the resistance to rotation is a maximum, and about another of them, the resistance to rotation is a minimum. The resistance to rotation about the third orthogonal axis is, in general, an intermediate value.

We will limit our study in this chapter to plane motion; that is, the body remains in a single plane—the *xy*-plane—and has only three degrees of freedom: translation in the *x*-direction, translation in the *y*-direction, and rotation about the *z*-axis. The *x*-, *y*-, and *z*-axes are centroidal axes, and the *z*-axis is a principal axis. This condition must hold, or else the body cannot remain in plane motion. The mass moment of inertia about the *z*-axis is

$$I_{zzcm} = \int_M (x^2 + y^2)dm$$

This integral can also be written in polar coordinates as

$$I_{zzcm} = \int_M r^2 dm$$

where $r^2 = x^2 + y^2$.

A convenient way to give the mass moment of inertia is to express the radius at which all the mass can be considered to be concentrated, called the *radius of gyration:*

$$k_z = \sqrt{\frac{I_{zzcm}}{M}}$$

6.2.1 ANGULAR MOMENTUM ABOUT A FIXED POINT ON A RIGID BODY

We have developed the equations of angular momentum about the center of mass of a rigid body. In general, dynamic analyses are referenced to the center of mass, but in the case where one point on the body is fixed in space, an alternative approach may be used. In

Chapter 4, the angular momentum of a system of particles about the origin of an inertial coordinate system was defined [Eq. (4.22)] as

$$\mathbf{H}_O = \mathbf{r}_{cm} \times M\mathbf{v}_{cm} + \mathbf{H}_{cm} \tag{6.13}$$

The equation of angular motion relative to the fixed origin is given [see Eq. (4.23)] as

$$\sum \mathbf{M}_O = \sum \mathbf{r} \times \mathbf{F} = \dot{\mathbf{H}}_O \tag{6.14}$$

If a body is rotating about a fixed point, the use of Eq. (6.14) eliminates the reactions at the point of rotation from the equation of angular momentum.

It is important to remember that for plane motion, the plane of motion must be perpendicular to a principal axis. That is to say, the body must be rotating about a principal axis. The mass moments of inertia about the principal axes of regular-shaped bodies are given in Appendix A.

6.3 EQUATIONS OF MOTION FOR RIGID BODIES IN PLANE MOTION

If all the forces acting on a rigid body are coplanar in a plane perpendicular to a principal axis, the body can be analyzed dynamically as a body in plane motion. The body has three degrees of freedom: two translations in the plane of motion and rotation about an axis perpendicular to the plane of motion. Hence, Eq. (4.16) can be written as

$$\sum \mathbf{F} = \dot{\mathbf{L}} = M\mathbf{a}_{cm} \tag{6.15}$$

If the z-axis of a coordinate system with origin at the center of mass is a principal axis, the angular momentum of a rigid body in plane motion is

$$\mathbf{H}_{cm} = I_{zz_{cm}}\omega_z \hat{\mathbf{k}} \tag{6.16}$$

It was shown in Chapter 4 that the moment about the center of mass is equal to the time rate of change of the angular momentum about the center of mass. Therefore,

$$\sum \mathbf{M}_{cm} = \dot{\mathbf{H}}_{cm}$$
$$\sum M_{cm}\hat{\mathbf{k}} = I_{zz_{cm}}\alpha_z\hat{\mathbf{k}} \tag{6.17}$$

Equations (6.15) and (6.17) are the equations of motion of a rigid body in plane motion. They may be written in scalar form as

$$\sum F_x = M\mathbf{a}_{cm_x}$$
$$\sum F_y = M\mathbf{a}_{cm_y} \tag{6.18}$$
$$\sum M_{cm} = I_{zz}\alpha_z$$

These three equations are used to determine the acceleration of the center of mass and the angular acceleration of the rigid body at any instant of time. In general, the forces acting on the body will be functions of time, position, or velocity, and Eqs. (6.18) can be viewed as three ordinary differential equations that may be coupled and linear or nonlinear. *Many dynamics problems require that the acceleration be determined only at a particular instant or only when the rigid body is at a particular position. Such an analysis, of course, does not trace the motion of the body during a period of time.* Note that if the body is at rest or in a state of constant motion, Eqs. (6.18) are equivalent to the two-dimensional equations of statics.

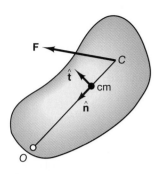

Figure 6.2

If the body is in plane motion rotating about a fixed point on itself, or the body extended, Eqs. (6.13) and (6.14) can be used. For plane motion, the angular momentum about the point O can be written as

$$\mathbf{H}_O = \mathbf{r}_{cm} \times M\mathbf{v}_{cm} + \mathbf{H}_{cm}$$

$$\mathbf{H}_O = \mathbf{r}_{cm} \times M(\omega \times \mathbf{r}_{cm}) + \mathbf{H}_{cm} \qquad (6.19)$$

$$\mathbf{H}_O = (r_{cm}^2 M + I_{zz_{cm}})\omega_z\hat{\mathbf{k}} = I_{zz_O}\omega_z\hat{\mathbf{k}}$$

The mass moment of inertia about the fixed point O is obtained by the use of the parallel-axes theorem. For plane motion about a fixed axis that does not pass through the center of mass, normal and tangential coordinates can be used to an advantage, as shown in Figure 6.2. If R_n and R_t are the reactions at point O, the equations of motion for the rigid body are

$$F_t + R_t = mr_{cm/O}\alpha$$

$$F_n + R_n = mr_{cm/O}\omega^2 \qquad (6.20)$$

$$r_{C/O}F_t = I_{zz_O}\alpha$$

Now suppose that the force $\mathbf{F}$ is applied to a point C such that the tangential component of the reaction R_t is zero. The vector from O to such a point C is $\mathbf{r}_{C/O}$.

$$F_t = mr_{cm/O}\alpha$$

$$r_{C/O}F_t = I_{zz_O}\alpha \qquad (6.21)$$

$$r_{C/O} = \frac{I_{zz_O}}{mr_{cm/O}} = \frac{(k_{zO})^2}{r_{cm/O}}$$

where k_{zO} is the radius of gyration about point O. The point C at which the force is applied is called the *center of percussion*. A force applied at the center of percussion will not cause a tangential component at the support pivot. The center of percussion is the "sweet spot" on a baseball bat or a golf club. The concept of the center of percussion is useful in the design of impact machines and connecting rods for engines.

In many problems, all the external forces acting on the body are known and the equations expressing those forces can be solved to determine the accelerations. However, if there are constraints on the motion of the body, there will be unknown constraint forces. The constraints will yield additional equations to solve for these additional unknowns. For a properly posed problem, there always will be sufficient equations to solve for the constraint forces and the accelerations. The application of dynamics to bodies moving in plane motion is straightforward and best mastered by examples.

The analysis of any problem should consist of the following steps:

1. Model the body as a rigid body, and construct a free-body diagram showing all the forces acting on the body.

2. List the unknown forces and accelerations.

3. Write the three equations of dynamics of plane motion.

4. Conceptualize the motions possible, and then determine whether there are constraints on the motions.

5. Solve the system of equations to determine the differential equations of motion.

Sample Problem 6.1

Consider an individual attempting to push a large box across the floor, as shown in the accompanying diagram. Determine the resulting movement of the box when its size and weight are known. Develop a general approach for any applied force, at any position, and for a particular coefficient of static and kinetic friction between the box and the floor.

Solution

The motion can be considered to be plane motion; a free-body diagram of the box, shown here.

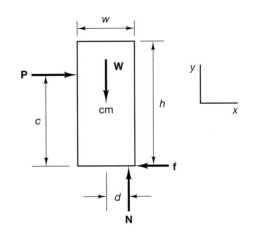

The free-body diagram included, the effects of static or kinetic friction between the box and the floor assumes that the individual pushes the box in the horizontal direction. The dimensions of the box (w and h), as well as the weight of the box, are known. In general, the force **P** and the point of application c also will be specified; however, in some problems, these must be determined in order to produce particular motions. The force between the box and the floor is distributed across the bottom of the box, but equivalent forces are shown acting at some point d from the centerline of the box. When the distance c is greater than $h/2$, the force **P** produces a moment about the center of mass that will cause the box to tip over in a clockwise manner. The only moment resisting this tipping is the moment of the normal force, and the moment arm of this force is limited to $w/2$.

The possible movements can now be conceptualized:

a) No movement could occur.

b) The box could slide.

c) The box could tip.

d) The box could tip and slide.

Assuming that the force **P** and its point of application are known, the unknowns in this problem are a_{cm_x}, a_{cm_y}, α, **f**, **N**, and d. There are six unknowns, but only the following three equations of plane dynamics:

$$P - f = \frac{W}{g} a_{cm_x}$$

$$N - W = \frac{W}{g} a_{cm_y}$$

$$Nd - f\frac{h}{2} - P\left(c - \frac{h}{2}\right) = I_{zz}\alpha,$$

where

$$I_{zz} = \frac{1}{12}\frac{W}{g}(w^2 + h^2)$$

(a) First consider the condition of no movement. In that case, the following three equations hold:

$$a_{cm_x} = 0$$

$$a_{cm_y} = 0$$

$$\alpha = 0$$

There are now six equations for the six unknowns, and the solution (for this statics case) is

$$f = P$$

$$N = W$$

$$d = Pc/W$$

To check whether this solution is correct, the required friction is compared to the maximum static friction force:

$$f \le \mu_s N$$

If this condition is satisfied, the box will not slide, but may tip. The condition required to prevent tipping is

$$d \le \frac{w}{2}$$

If both of the preceding conditions are satisfied, the statics solution is the correct solution.

(b) If the required friction is greater than the maximum friction, the box will slide. In this case, the following conditions would obtain:

$$a_{cm_y} = 0$$

$$\alpha = 0$$

$$f = \mu_k N$$

Again, there are six equations for the six unknowns, and the solution is

$$a_{cm_x} = \left(\frac{P}{W} - \mu_k\right)g$$

$$N = W$$

$$d = \left[\frac{\mu_k h}{2} + \frac{P}{W}\left(c - \frac{h}{2}\right)\right]$$

For this solution to satisfy the necessary conditions, the acceleration in the *x*-direction must be positive, as friction cannot drive motion. For tipping not to occur, we must have

$$d \le \frac{w}{2}$$

(c) Now suppose the box will tip, but not slide. In case (a), we found that the required friction force was less than the maximum static friction, and the required value of d was less than half the width of the box. In the current case, the normal force will act on the bottom-front edge of the box. If this point is designated by A, then the following equation holds:

$$d = \frac{w}{2}$$

$$\mathbf{a}_A = 0$$

$$\mathbf{a}_{cm} = \mathbf{a}_A + \boldsymbol{\alpha} \times \mathbf{r}_{cm/A} + \boldsymbol{\omega} \times (\boldsymbol{\omega} \times \mathbf{r}_{cm/A})$$

If the box starts from the rest, then $\omega = 0$, and it follows that

$$\mathbf{r}_{cm/A} = -\frac{w}{2}\hat{\mathbf{i}} + \frac{h}{2}\hat{\mathbf{j}} \quad \text{and} \quad \boldsymbol{\alpha} = \alpha\hat{\mathbf{k}}$$

$$\mathbf{a}_{cm} = -\alpha\frac{h}{2}\hat{\mathbf{i}} - \alpha\frac{w}{2}\hat{\mathbf{j}}$$

The three additional equations for this case are

$$a_{cm_x} = -\alpha\frac{h}{2}$$

$$a_{cm_y} = -\alpha\frac{w}{2}$$

$$d = \frac{w}{2}$$

The solution is obtained by analysis, wherein the coefficient matrix is inverted symbolically, as shown in the Computational Supplement.

The angular acceleration, normal force, and friction force for this case are

$$\alpha = -3g\frac{\left(Pc - W\frac{w}{2}\right)}{W(w^2 + h^2)}$$

$$N = W + \frac{3\left(Pc - W\frac{w}{2}\right)\frac{w}{2}}{(w^2 + h^2)}$$

$$f = P - \frac{3\left(Pc - W\frac{w}{2}\right)\frac{c}{2}}{(w^2 + h^2)}$$

Notice that the normal force increases and the required friction decreases. Notice also, that this solution is valid only for the start of tipping, as we have taken the angular velocity to be zero. To represent the full tipping of the box, the angular velocity would appear in the differential equation for the angular motion, and the resulting differential equation would be nonlinear.

(d) In this case, the friction force is known to be equal to the coefficient of kinetic friction times the normal force. The corner A will move in the x-direction, and it follows that

$$\mathbf{a}_A = a_A\hat{\mathbf{i}}$$

$$\mathbf{a}_{cm} = a_A\hat{\mathbf{i}} - \alpha\frac{h}{2}\hat{\mathbf{i}} - \alpha\frac{w}{2}\hat{\mathbf{j}}$$

Therefore,

$$a_{cm_x} = -\alpha \frac{w}{2}$$

The three additional equations are

$$a_{cm_y} = -\alpha \frac{w}{2}$$
$$f = \mu_k N$$
$$d = \frac{w}{2}$$

The solution is

$$\alpha = -\frac{g}{W} \frac{\left[P\left(c - \dfrac{h}{2}\right) - W\left(\dfrac{w}{2} - \mu_k \dfrac{h}{2}\right) \right]}{\left[\dfrac{w^2}{3} + \dfrac{h^2}{12} - \mu_k \dfrac{wh}{4} \right]}$$

$$N = W + \frac{w}{2} \frac{\left[P\left(c - \dfrac{h}{2}\right) - W\left(\dfrac{w}{2} - \mu_k \dfrac{h}{2}\right) \right]}{\left[\dfrac{w^2}{3} + \dfrac{h^2}{12} - \mu_k \dfrac{wh}{4} \right]}$$

$$a_{cm_x} = \frac{g}{W}(P - \mu_k N)$$
$$a_{cm_y} = -\alpha \frac{w}{2}$$
$$f = \mu_k N$$
$$d = \frac{w}{2}$$

We have now analyzed the problem in the most general manner, considering each possible movement separately for a given applied force, at a given position, and a given coefficient of friction. Next we consider an example in which the motion is restricted in a specific fashion.

Sample Problem 6.2

A uniform board of weight W is leaned against the cab of a truck with its base restricted by a block, as shown in the accompanying diagram. Determine the maximum constant acceleration that the truck can maintain without the board tipping.

Solution   We construct a free-body diagram of the board.

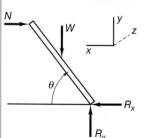

There are six unknowns in this problem: a_{cm_x}, a_{cm_y}, α, R_x, R_y, and N. The three equations of dynamics are

$$R_x - N = \frac{W}{g} a_{cm_x}$$
$$R_y - W = \frac{W}{g} a_{cm_y}$$
$$N \frac{l}{2} \sin\theta + R_x \frac{l}{2}\sin\theta - R_y \frac{l}{2}\cos\theta = I\alpha,$$

where

$$I = \frac{1}{12}\frac{W}{g}l^2$$

The statement of the problem indicates that the truck should accelerate in a manner such that there will be no motion of the board relative to the truck. Therefore, the remaining three equations are

$$a_{cm_x} = a_T \text{ (the acceleration of the truck)}$$

$$a_{cm_y} = 0$$

$$\alpha = 0$$

The acceleration of the truck is not known, so one additional condition must be specified. The maximum acceleration such that there will be no motion of the board relative to the truck will occur when the board just starts to rotate, or when $N = 0$. The system of equations to determine the unknowns is

$$R_x = \frac{W}{g}a_T$$

$$R_y - W = 0$$

$$R_x\frac{l}{2}\sin - R_y\frac{l}{2}\cos = 0$$

Solution of this system yields

$$R_y = W$$

$$R_x = W \cot \theta$$

$$a_T = g \cot \theta$$

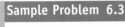

Sample Problem 6.3

Determine the maximum acceleration that a car can obtain when climbing a hill of slope θ if the coefficient of static friction between the tires and the road is μ_s. The center of mass of the car is toward the front of the car, due to the weight of the engine. (See the accompanying diagram.) Determine the maximum acceleration if the car is a (a) four-wheel-drive, (b) rear-wheel-drive, and (c) front-wheel-drive vehicle.

Solution

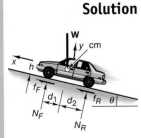

The free-body diagram is shown in the accompanying figure. The friction driving force is shown on all four wheels, as in the case of a four-wheel drive, and will be set to zero on the front wheels for a rear-wheel-drive car or set to zero on the rear wheels for a front-wheel-drive car.

The equations of motion are

$$f_F + f_R - W \sin \theta = \frac{W}{g}a_{cm_x}$$

$$N_F + N_R - W\cos \theta = \frac{W}{g}a_{cm_y}$$

$$-N_R d_2 + N_F d_1 + (f_F + f_R)h = I\alpha$$

The forces on the tires, f and N, represent the forces on both the driver and passenger sides. Assuming that the weight and position of the center of mass of the automobile are known, there are seven unknowns in this equation: the two normal forces, the two friction forces, and the three accelerations. Since the car is restricted to move up the slope,

$$a_{cm_y} = 0 \quad \text{and} \quad \alpha = 0$$

The two friction forces are related to the normal forces, or else will be zero on the front or the rear depending upon the kind of drive the automobile has.

(a) **Four-Wheel Drive.** Here, the forces of friction are

$$f_F = \mu_s N_F$$
$$f_R = \mu_s N_R$$

The three equations for the normal forces and the acceleration of the center of mass in the x-direction can be written as

$$\mu_s N_F + \mu_s N_R - \frac{W}{g} a_{cm_x} = W \sin \theta$$
$$N_F + N_R = W \cos \theta$$
$$(d_1 + \mu_s h)N_F + (-d_2 + \mu_s h)N_R = 0$$

These equations are solved analytically in the Computational Supplement. The maximum acceleration of the car on level ground is

$$a = \mu_s g$$

Note that the acceleration does not depend upon the mass of the car or the position of the center of mass.

(b) **Conventional Rear-Wheel Drive.** The friction forces in this case become

$$f_F = 0$$
$$f_R = \mu_k N_R$$

The equations for the normal forces and the acceleration are

$$0 + \mu_s N_R - \frac{W}{g} a_{cm_x} = W \sin \theta$$
$$N_F + N_R = W \cos \theta$$
$$(d_1 + 0)N_F + (-d_2 + \mu_s h)N_R = 0$$

Again, these equations are solved analytically in the Computational Supplement. The maximum acceleration of the car on level ground is

$$a = \frac{g \mu_s d_1}{(d_1 + d_2) - \mu_s h}$$

In this case, the acceleration depends on the location of the center of mass, but not upon the mass of the car.

(c) **Front-Wheel Drive.** For this kind of drive, the friction forces are

$$f_F = \mu_s N_F$$
$$f_R = 0$$

The equations for the normal forces and the acceleration are

$$\mu_s N_F + 0 - \frac{W}{g} a_{cm_x} = W \sin \theta$$

$$N_F + N_R = W \cos \theta$$

$$(d_1 + \mu_s h)N_F + (-d_2 + 0)N_R = 0$$

As before, these equations are solved analytically in the Computational Supplement. On level ground, the acceleration is

$$a = \frac{\mu_s g d_2}{d_1 + d_2 + \mu_s h}$$

As has been noted, the maximum acceleration of a four-wheel-drive vehicle on level ground is independent of the location of the center of mass of the vehicle. By contrast, the maximum acceleration of a rear-wheel-drive vehicle will increase as the center of mass moves back toward the rear wheels and as the height of the center of mass off the ground increases. This is the principle behind the design of most dragsters. For a front-wheel-drive vehicle, the maximum acceleration increases as the center of mass is moved toward the front wheels and is lowered closer to the ground.

Sample Problem 6.4

The uniform pendulum in the accompanying diagram is released from a horizontal position. Determine the motion of the pendulum. The motion is retarded by friction at the pin, which always opposes the motion.

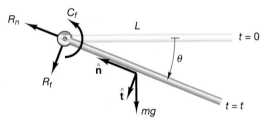

Solution The angular motion equation about the pivot point is

$$mg\frac{L}{2} \cos \theta - C_f \operatorname{sign}(\dot{\theta}) = I_{zz_0}\ddot{\theta} = \frac{1}{3}mL^2\ddot{\theta}$$

$$\ddot{\theta} = \frac{3}{2}\frac{g}{L} \cos \theta - M_f\frac{\dot{\theta}}{|\dot{\theta}|}$$

where

$$M_f = C_f\frac{3}{mL^2}$$

The friction moment always opposes the motion and therefore is in the direction opposite that of the angular velocity. The differential equation for the angular motion is nonlinear and can be solved numerically for a pendulum of a given length and for a specified friction moment, as is done in the Computational Supplement.

6.4 CONSTRAINTS ON THE MOTION

In most dynamic problems, the rigid body has *constraints* on its motion. For example, a wheel rolling along a flat plane has the constraint that its center will not move vertically. If the center of mass is at the center of the wheel, then the acceleration of the center of mass in the vertical direction is zero. An additional constraint may occur if the wheel *rolls without sliding*. This constraint is more difficult to express, as we need to capture the fact that the point of contact between the wheel and the ground will have no acceleration component tangent to the surface of the ground. The constraint, unfortunately, cannot be applied readily to the acceleration of the center of mass and the angular acceleration of the body. Instead, a kinematic analysis is required to develop the constraint equation.

Although we did not explicitly mention it, the motions illustrated in the three previous sample problems were constrained either by the physics of the situation or in the statement of the problem. We solved Sample Problem 6.1 by assuming four separate constraints. In part (a), we assumed that the box did not move. In part (b), we assumed that the angular acceleration of the box was zero. In part (c), we assumed that the lower-right corner of the box did not move—that is, that the box tipped about that point. We then expressed that constraint in a relation between the linear acceleration of the center of mass and the angular velocity and acceleration of the box. In part (d), which involved tipping and sliding, we assumed that the lower-right corner of the box translated along the surface of the floor and, therefore, that the acceleration of the point representing that corner was in the horizontal direction only.

In Sample Problem 6.2, the constraint on the motion was expressly given in the statement of the problem; that is, we wanted to find the maximum constant linear acceleration such that the board would not tip. We then sought a solution wherein the angular acceleration of the board was zero.

In Sample Problem 6.3, we assumed that the wheels would roll without slipping, because if the wheels did slip, the coefficient of kinetic friction, instead of the coefficient of static friction, would relate the friction force to the normal force. Since the coefficient of kinetic friction is approximately 20% less than that of static friction, the maximum acceleration is obtained at the point where slipping is impending. The car also was constrained to move up the incline. We incorporated that constraint by choosing the coordinates tangent and normal to the incline, thus constraining the motion such that the acceleration in the y-direction was zero.

The constraint on the motion in Sample Problem 6.4 was incorporated into the formulation of the equations of motion of the problem. We wrote the moment-angular acceleration equation related to point O, which was constrained to be fixed in space. All of the problems in this chapter are plane-motion problems; that is, if the xy-plane is the plane of motion, the linear acceleration in the z-direction is constrained to be zero, and the angular accelerations about the x- and y-axes are similarly constrained to be zero.

The motion of a ladder leaning against a building is constrained when the ladder slips. The bottom of the ladder will move along the ground while the top slides down the wall of the building. If a rigid body is pinned at a point so that that point cannot move, the body is constrained to rotate about the point. There is no set manner of approaching these constraints, but if the solution of a dynamics problem is approached in an orderly fashion, the number of constraints on the motion can be determined. In the dynamics of plane motion, only three equations of motion can be written for a single rigid body: one involving the acceleration of the center of mass in the x-direction, one involving the acceleration of the center of mass in the y-direction, and one involving the angular acceleration about the

z-axis. If all of the external forces are known, the two components of the linear acceleration and the single angular acceleration can be determined from these three equations. If,
however, there are reaction forces that are not known, there will be more unknowns than
equations to determine them. In statics, the constraint was that there be no accelerations,
so that the equations of motion became the equations of equilibrium, and we used these
equations to solve for the unknown reactions. In dynamics, we use the constraints on the
motion, together with the equations of motion, to determine the accelerations and the constraining forces.

Sample Problem 6.5	Consider a uniform ladder of weight W that slips when it is leaned against a building:
Solution	First, we model the problem by drawing a free-body diagram.

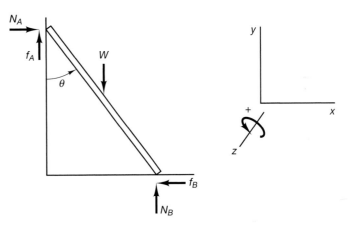

It is important to select the angle so that it increases in magnitude in a manner consistent
with the right-hand rule for angular velocity and acceleration, so that vector equations can
be formulated. Therefore, the angle θ between the ladder and the wall is chosen to increase
in value with a positive rotation about the *z*-axis. The equations of motion are then

$$N_A - f_B = \frac{W}{g} a_{cm_x}$$

$$N_B + f_A - W = \frac{W}{g} a_{cm_y}$$

$$-(N_A + f_B)\frac{l}{2}\cos\theta + (N_B - f_A)\frac{l}{2}\sin\theta = \frac{1}{12}\frac{Wl^2}{g}\alpha$$

Assuming that the weight and length of the ladder are known, when the ladder starts to slip
there are four unknown forces: the two normal forces and the two friction forces. However, the two friction forces will be known in terms of the normal forces if the ladder slips,
as the coefficient of kinetic friction will relate the friction to the normal force at both
points. Note that if the ladder were in equilibrium and slippage was not impending, it
would not be possible to determine the two normal and two friction forces from the three
static equilibrium equations. For the dynamics case being considered, there are five
unknowns: N_A, N_B, a_{cm_x}, a_{cm_y}, and α. Since there are only three equations of motion, there
must be two constraints on the motion. Examining the expected movement, we can see that

if point A stays in contact with the wall producing the normal force, then point A will move parallel to the wall, or in the vertical direction. In a similar manner, if point B stays in contact with the ground producing the normal force there, point B will move parallel to the ground, or in the horizontal direction.

The two constraints on the motion of the ladder are thus

$$\mathbf{a}_A = a_A \hat{\mathbf{j}}$$

$$\mathbf{a}_B = a_B \hat{\mathbf{i}}$$

These constraints are not in a form where they can be interpreted as constraints on either the acceleration of the center of mass or the angular acceleration. Therefore, the kinematics of the rigid body must be considered in order to obtain the proper constraints. Points A and B are points on the same rigid body, the ladder; therefore, for point A, the constraint can be written as

$$\mathbf{a}_{cm} = \mathbf{a}_A + \boldsymbol{\alpha} \times \mathbf{r}_{cm/A} + \boldsymbol{\omega} \times (\boldsymbol{\omega} \times \mathbf{r}_{cm/A})$$

Since the ladder starts to slip from rest, the angular velocity of the ladder is initially zero, and we have

$$\mathbf{r}_{cm/A} = \frac{l}{2}(\sin \hat{\mathbf{i}} - \cos \hat{\mathbf{j}})$$

$$\boldsymbol{\alpha} = \alpha \hat{\mathbf{k}}$$

$$\mathbf{a}_{cm} = a_A \hat{\mathbf{j}} + \alpha \hat{\mathbf{k}} \times \frac{l}{2}(\sin \hat{\mathbf{i}} - \cos \hat{\mathbf{j}})$$

$$= a_A \hat{\mathbf{j}} + \alpha \frac{l}{2} \sin \hat{\mathbf{j}} + \alpha \frac{l}{2} \cos \hat{\mathbf{i}}$$

$$= \alpha \frac{l}{2} \cos \hat{\mathbf{i}} + \left(a_A + \alpha \frac{l}{2} \sin\right)\hat{\mathbf{j}}$$

Now the constraint in terms of the acceleration of the center of mass is written as

$$a_{cm_x} = \alpha \frac{l}{2} \cos \theta$$

This constraint relates the acceleration of the center of mass in the x-directions to the angular acceleration of the body. In a similar manner, the constraint at point B can be written as

$$\mathbf{a}_{cm} = \mathbf{a}_B + \boldsymbol{\alpha} \times \mathbf{r}_{cm/B}$$

$$\mathbf{a}_{cm} = a_B \hat{\mathbf{i}} + \alpha \hat{\mathbf{k}} \times \frac{l}{2}(-\sin \theta \hat{\mathbf{i}} + \cos \theta \hat{\mathbf{j}})$$

$$= \left(a_B - \alpha \frac{l}{2} \cos \theta\right)\hat{\mathbf{i}} - \alpha \frac{l}{2} \sin \theta \hat{\mathbf{j}}$$

The other constraint now may be written as

$$a_{cm_y} = -\alpha \frac{l}{2} \sin \theta$$

Now there are five equations for the five unknowns, and the dynamics problem is solved easily at the instant slipping occurs. In the Computational Supplement, the seven equations (the three equations of motion, the two constraint equations, and the two relationships between the friction and the normal forces) are solved after they have been written in matrix form as:

$$
\begin{bmatrix}
-\mu_k & 0 & 1 & 0 & 0 & 0 & 0 \\
0 & -\mu_k & 0 & 1 & 0 & 0 & 0 \\
1 & 0 & 0 & -1 & -\dfrac{W}{g} & 0 & 0 \\
0 & 1 & 1 & 0 & 0 & -\dfrac{W}{g} & 0 \\
-\dfrac{l}{2}\cos\theta & \dfrac{l}{2}\sin\theta & -\dfrac{l}{2}\sin\theta & -\dfrac{l}{2}\cos\theta & 0 & 0 & -\dfrac{Wl^2}{12\,g} \\
0 & 0 & 0 & 0 & 1 & 0 & -\dfrac{l}{2}\cos\theta \\
0 & 0 & 0 & 0 & 0 & 1 & \dfrac{l}{2}\sin\theta
\end{bmatrix}
\begin{bmatrix}
N_A \\ N_B \\ f_A \\ f_B \\ a_{cm_x} \\ a_{cm_y} \\ \alpha
\end{bmatrix}
=
\begin{bmatrix}
0 \\ 0 \\ 0 \\ W \\ 0 \\ 0 \\ 0
\end{bmatrix}
$$

To examine the solution obtained in the Computational Supplement, numerical values of 50 lb for the weight of the ladder, 12 ft for the length of the ladder, and 35° for the initial angle are assumed. This problem was not initially examined in the static condition to see whether the ladder would actually slip; therefore, a low coefficient of kinetic friction of 0.2 is assumed, consistent with slipping. The values of the acceleration will change as the ladder slips, so the angular velocity term must be retained in the constraint equations. The differential equation will be nonlinear and will require a numerical solution.

6.4.1 ROLLING WITHOUT SLIDING

Consider a cylinder or wheel rolling along a surface, with sufficient friction to prevent sliding. In this interesting case, the direction in which the friction acts may not be easy to determine. However, the direction will be dictated by the constraint on the motion. Let us examine the general case of a uniform disk rolling without sliding under the action of the force **P**, applied at a point br above the center of the disk, as shown in Figure 6.3. Here, b, which can be negative or positive, represents the fraction of the radius where the horizontal force is placed. If b is negative, the force was placed below the center of the disk.

In the free-body diagram, friction is assumed to be acting to the left, resisting the tendency of the disk to slip to the right due to the pull of the force **P**. If, however, the moment arm from the center of mass, br, is large enough, the clockwise angular acceleration could be great enough for the point of contact of the disk with the floor to slip to the left, with friction acting in the opposite direction to oppose the sliding. We treat b as a variable that can have both positive and negative values. The equations of motion are

$$P - F = ma_{cm_x}$$

$$N - W = ma_{cm_y}$$

$$-P(br) - Fr = I\alpha$$

There are three equations for the five unknowns F, N, a_{cm_x}, a_{cm_y}, and α. The constraint on the motion is that the disk rolls without sliding or that, at the point of contact, denoted

D, the acceleration in the x-direction is zero. Writing the kinematic relationship between the point D and the center of mass yields

$$a_D\hat{\mathbf{j}} = a_{cm} + \boldsymbol{\alpha} \times r_{D/cm} + \boldsymbol{\omega} \times (\boldsymbol{\omega} \times \mathbf{r}_{D/cm})$$

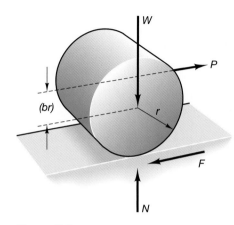

Figure 6.3

In this case, the center of mass is at the center of the disk and moves in rectilinear translation; therefore,

$$\mathbf{a}_{cm} = a_{cm_x}\hat{\mathbf{i}}$$

and $a_{cm_y} = 0$. This is another constraint on the motion. The relationship between the acceleration of point D and the acceleration of the center of mass can now be written as

$$a_D\hat{\mathbf{j}} = a_{cm_x}\hat{\mathbf{i}} + r\alpha\hat{\mathbf{i}} + r\omega^2\hat{\mathbf{j}}$$

Consequently,

$$a_{cm_x} = -r\alpha$$
$$a_{D_y} = r\omega^2$$

The equations of motion may now be written as

$$P - F = ma_{cm_x} = -mr\alpha$$
$$N = W$$
$$-P(br) - Fr = I\alpha$$

For a uniform disk, $I = mr^2/2$. To solve this system by hand, consider the first and third equations, and eliminate α. This yields the solution for the friction force $\mathbf{F}$ in terms of the applied force $\mathbf{P}$ and the variable b:

$$F = \frac{1}{3}P(1 - 2b)$$

The friction force will be positive (to the left) as long as $b < 1/2$—in other words, as long as the horizontal force acts below half the radius above the center of the wheel. Placing the force $\mathbf{P}$ below the center corresponds to b being negative, and the friction is in the correct direction.

For $b > 1/2$—for example, if $\mathbf{P}$ is placed at the top of the disk ($b = 1$)—friction would act in the opposite direction, to the right. In that case, the disk would tend to slip, due to the high clockwise moment of $\mathbf{P}$. This is, of course, the direction of friction when the axle drives the rear wheels of a car. Note that in the case where $b = 1/2$, the friction force between the disk and the ground is zero.

This analysis can be applied to the case of a sphere rolling without slipping under the influence of a horizontal force applied at some point br above the center of the sphere. The mass moment of inertia of a uniform sphere is $I = \frac{2}{5}mr^2$, and substituting this into the equations of motion yields

$$F = \frac{2}{7}P\left(1 - \frac{5}{2}b\right)$$

for the friction force. In this case, friction would be directed to the left as long as $b < 2/5$.

6.4.2 ROLLING AND SLIDING

Now consider an example in which it is not known whether sliding will occur. In this case, the coefficients of static and kinetic friction are known, and it is necessary to investigate whether sliding occurs. If the force $\mathbf{P}$ is applied as shown in Figure 6.3, the friction required to prevent slipping is $F = 1/3\ P(1-2b)$, where b is the fraction of the radius from the center of the wheel to the line of action of the horizontal force. If the horizontal force is applied at a point one-half the radius above the center of the cylinder, no friction will be required to prevent slipping. If the force is applied above this point, friction will act to the right, resisting the tendency of the cylinder to spin. The easiest way to approach problems in which it is not known whether slipping will occur is to assume that the cylinder will *not* slip and check to see whether there is a high enough coefficient of static friction to guarantee that slipping will not occur. If one examines the problem under this assumption, the test will be that

$$\frac{F}{N} \le \mu_s$$

If this condition is not satisfied, the direction and magnitude of the friction force are given by

$$F = \mu_k N$$

and there is no constraint to the motion other than $a_{cm_y} = 0$.

As an example, consider a disk rolling down an incline that makes an angle θ with the horizontal. The disk is under the influence only of the force of gravity. This example is illustrated in Figure 6.4. We assume that the disk rolls without sliding. The constraints on the motion are

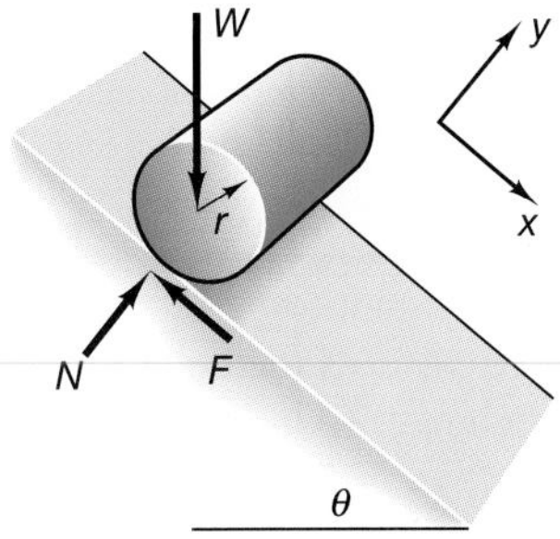

Figure 6.4

$$a_{cm_y} = 0 \quad \text{and} \quad a_{cm_x} = -r\alpha$$

The equations of motion are

$$W \sin\theta - F = ma_{cm_x}$$

$$N - W\cos\theta = 0$$

$$-Fr = I\alpha = \frac{1}{2}mr^2\alpha$$

and, using the constraint $a_{cm_x} = -r\alpha$, we obtain

$$F = \frac{1}{2}ma_{cm_x}$$

This relationship can be substituted into the first equation of motion, upon which the friction force determined to be

$$F = \frac{1}{3}W\sin\theta$$

The normal force is $N = W\cos\theta$, and the ratio of friction to the normal force is

$$\frac{F}{N} = \frac{1}{3}\tan\theta$$

For example, if the coefficient of static friction is 0.5, the cylinder will roll without sliding. As long as the angle between the slope and the horizontal is less than 56.3°, we have

$$\frac{F}{N} \leq \mu_s$$

$$\frac{1}{3}\tan\theta \leq 0.5$$

$$\tan\theta \leq 1.5$$

$$\theta \leq 56.3°$$

In this example, we will take the incline to be 60°, so that the disk will slip, and we will assume that the coefficient of kinetic friction is 0.45. Then the friction is

$$F = 0.45N = 0.45W\cos 60$$

and the dynamic equations become

$$W\sin 60 - 0.45\,W\cos 60 = \frac{W}{g}a_{cm_x}$$

$$-0.45W\cos 60 = \frac{1}{2}\frac{W}{g}r\alpha$$

Solving these gives $a_{cm_x} = 0.641\,g$ and $\alpha = -0.45g/r$.

Sample Problem 6.6

A yo-yo is rewound by pulling it across the floor by applying a constant force **P**, as illustrated in the accompanying diagram. If the inner radius of the yo-yo is r_1, the outer radius is r_2, and the radius gyration is k, determine the minimum coefficient of friction for the yo-yo to roll up the string without slipping.

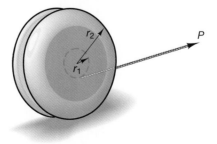

Solution The free-body diagram for the yo-yo is shown here.

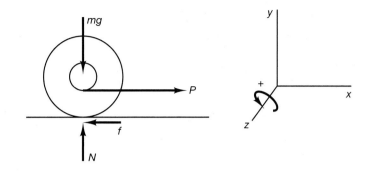

The equations of motion are

$$N - mg = 0$$

$$P - f = ma_{cm_x}$$

$$Pr_1 - fr_2 = mk^2\alpha$$

If the yo-yo rolls without slipping, the constraint on the motion is

$$a_{cm_x} = -r_2\alpha$$

The system of two equations for the friction force and the angular velocity are solved analytically in the Computational Supplement.

Sample Problem 6.7

A ball is thrown with an initial velocity of V_o parallel to the rough plane, as shown in the accompanying diagram. The initial angular velocity is zero. Determine when the sphere will roll without sliding, and find the linear velocity of the ball at that time.

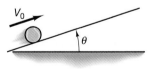

Solution Let the mass of the ball be m, the radius be r, and the coefficient of kinetic friction between the ball and the plane be μ_k. Shown here is a free-body diagram of the ball.
The equations of motion are

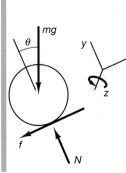

$$-mg \sin \theta - f = ma_{cm_x}$$

$$N - mg \cos \theta = ma_{cm_y}$$

$$-fr = \frac{2}{5} mr^2\alpha$$

The constraint on the motion is $a_{cm_y} = 0$. Also, while the ball rolls and slides, $f = \mu_k N = \mu_k mg \cos \theta$.

Therefore, the linear and angular accelerations are constant until the ball reaches a point where it rolls without sliding. The linear and angular accelerations when sliding are

$$a_{cm_x} = -g(\sin \theta + \mu_k \cos \theta)$$

$$\alpha = -\frac{5g}{2r}\mu_k \cos \theta$$

The linear velocity of the center of mass and the angular velocity while the wheel slides are

$$v_{cm_x} = -g(\sin \theta + \mu_k \cos \theta)t + V_0$$

$$\omega = -\frac{5g}{2r}\mu_k \cos \theta t$$

The wheel will stop slipping when $v_{cm_x} = -r\omega$; that is,

$$v_{cm_x} = -g(\sin \theta + \mu_k \cos \theta)t + V_0 = \frac{5g}{2}\mu_k \cos \theta t$$

$$-g\left(\sin \theta + \frac{7}{2}\mu_k \cos \theta\right)t + V_0 = 0$$

$$t = \frac{V_0}{g(\sin \theta + \frac{7}{2}\mu_k \cos \theta)}$$

$$v_{cm_x} = V_0\left[1 - \frac{(\sin \theta + \mu_k \cos \theta)}{(\sin \theta + \frac{7}{2}\mu_k \cos \theta)}\right]$$

If the slope of the incline were zero (a level surface), the ball would roll without sliding when

$$t_1 = \frac{2}{7}\frac{V_0}{\mu_k g}$$

$$v_{cm_x}(t_1) = \frac{5}{7}V_0$$

Sample Problem 6.8

In Sample Problem 6.7, a ball was rolled up an inclined plane. If instead, the ball is rolled up a curved surface (see the accompanying diagram), the differential equations will be non-linear. In this case, normal and tangential coordinates are used to formulate the problem.

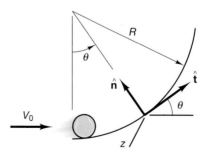

Solution The equations of motion are

$$-f - mg \sin \theta = ma_{cm_t} = mR\ddot{\theta}$$

$$N - mg \cos \theta = ma_{cm_n} = mR\dot{\theta}^2$$

$$-fr = \frac{2}{5} mr^2 \alpha$$

The nonlinear differential equations for the tangential acceleration $R\ddot{\theta}$ and the angular acceleration α of the ball are

$$\ddot{\theta} = -\frac{g}{R} (\sin \theta + \mu_k \cos \theta) - \mu_k \dot{\theta}^2$$

$$\alpha = -\frac{5\mu_k}{2r} [g \cos \theta + R\dot{\theta}^2]$$

The nonlinear differential equation for the angular acceleration of the center of mass moving as a particle is solved in the Computational Supplement for specified numerical values.

6.5 COMPUTATIONAL METHODS FOR PLANE DYNAMIC SYSTEMS

In most problems in plane dynamics, the accelerations will change during the motion, and the solution will involve linear or nonlinear coupled differential equations. Solving these equations is best accomplished using modern computational software. We present examples in this section.

Sample Problem 6.9

Consider the ladder sliding down a wall in Sample Problem 6.5. The acceleration of the ladder was determined for the instant that the ladder started to slip. The equations of motion were given as

$$N_A - f_B = \frac{W}{g} a_{cm_x}$$

$$N_B + f_A - W = \frac{W}{g} a_{cm_y} \qquad (6.22)$$

$$-(N_A + f_B)\frac{l}{2} \cos \theta + (N_B - f_A)\frac{l}{2} \sin \theta = \frac{1}{12} \frac{Wl^2}{g} \alpha$$

These three equations are valid for any position of the ladder, from the initial position until the ladder strikes the ground. However, it is clear that the angular acceleration is dependent upon the angle θ and that the differential equations will be nonlinear. Less apparent is the fact that the normal forces and, therefore, the friction forces are also dependent upon this angle. The friction forces are related to the normal forces by the coefficient of kinetic friction. Hence, the unknowns in the equations are the two normal forces, the two components of the acceleration of the center of mass of the ladder, and the angular acceleration of the ladder. Consequently, as previously noted, there are five unknowns and only three equations of motion. The two constraints on the motion are

$$\mathbf{a}_A = a_A \hat{\mathbf{j}}$$

$$\mathbf{a}_B = a_B \hat{\mathbf{i}} \qquad (6.23)$$

These constraints state that, as long as the top of the ladder (point A) is in contact with the wall, it will move only in the y-direction, and in a similar manner, the base of the ladder (point B) will move only in the x-direction. These constraints can be written in terms of the acceleration of the center of mass as

$$\mathbf{a}_{cm} = \mathbf{a}_{A \text{ or } B} + \boldsymbol{\alpha} \times \mathbf{r}_{cm/A \text{ or } B} + \boldsymbol{\omega} \times (\boldsymbol{\omega} \times \mathbf{r}_{cm/A \text{ or } B})$$

$$\mathbf{a}_{cm} = \mathbf{a}_{A \text{ or } B} + \boldsymbol{\alpha} \times \mathbf{r}_{cm/A \text{ or } B} - \omega^2 \mathbf{r}_{cm/A \text{ or } B}$$

(6.24)

The constraints on the acceleration of the center of mass are

$$a_{cm_x} = \alpha \frac{l}{2} \cos\theta - \omega^2 \frac{l}{2} \sin\theta$$

$$a_{cm_y} = -\alpha \frac{l}{2} \sin\theta - \omega^2 \frac{l}{2} \cos\theta$$

(6.25)

The friction forces are related to the normal forces by

$$f_A = \mu_k N_A$$

$$f_B = \mu_k N_B$$

(6.26)

The equations of motion, the constraint equations, and the friction relationships constitute a system of seven coupled algebraic and nonlinear differential equations that describe the motion of the system. Equation (6.26) and the first two formulas of Eq. (6.22) will be used to form an equation relating the angular acceleration to the two components of linear acceleration. The resulting equation is

$$[-(1 - \mu_k^2)a_{cm_x} - 2\mu_k a_{cm_y} - 2\mu_k g] \cos\theta$$

$$+ [(1 - \mu_k^2)a_{cm_y} - 2\mu_k a_{cm_x} + (1 - \mu_k^2)g] \sin\theta = \frac{1}{6}(1 + \mu_k^2)\alpha l$$

(6.27)

Substitution of Eq. (6.25) into Eq. (6.27) gives a single nonlinear differential equation for the angle θ, namely

$$\alpha = \frac{3}{(2 - \mu_k^2)l} \left[(1 - \mu_k^2)g \sin\theta - 2\mu_k g \cos\theta + 2\mu_k \frac{l}{2} \omega^2\right]$$

where

$$\alpha = \frac{d^2\theta}{dt^2}$$

$$\omega = \frac{d\theta}{dt}$$

(6.28)

If the surfaces are smooth, Eq. (6.28) simplifies to

$$\alpha = \frac{3}{2} \frac{g}{l} \sin\theta$$

(6.29)

Therefore, even in the absence of friction, the differential equation of motion is nonlinear and must be solved by some numerical integration method.

A numerical example of the solution of Eq. (6.28) is given in the Computational Supplement. Consistent with the example shown in Sample Problem 6.5, the weight of the ladder is 50 lb, the length is 12 ft, and the initial angle is 35°. The coefficient of kinetic friction between the wall and the ladder and between the floor and the ladder is 0.2. The values of the angular velocity and the angle θ should be checked to ensure that the angular velocity is positive and the angle is increasing in value. If the angular velocity is negative as the motion progresses, then friction drives the motion instead of resisting slip, and the ladder will not slip. The method of numerical integration is Euler's method using a seeded iteration in the Computational Software.

Note that the dynamics of the ladder do not depend upon the weight of the ladder, but do depend upon the length and upon the assumption that the center of mass of the ladder is at midlength. If an engineer were asked to design the end friction pads on the ladder and write the safety instructions for the ladder's use, consideration would have to be given to the position and mass of the individual climbing the ladder. In the absence of friction, the ladder would slip regardless of the initial angle. This case can be examined easily once the problem is entered into a computational software program by changing the coefficient of kinetic friction. The case is shown in the Computational Supplement with an initial angle of 1°. The time taken for the ladder to hit the ground can be found from the graphs by determining when the angle reaches 1.57 rad.

Problems

6.1 An 8-ft board leans against the back of a truck at an angle $\theta = 60°$, as shown in Figure P6.1. If the board weighs 20 lb and the coefficient of static friction between the board and the bed of the truck is 0.6, determine the maximum deceleration of the truck if the board does not slip. Neglect friction between the board and the back of the truck.

Figure P6.1

6.2 A uniform 200-kg door hangs from rollers, as shown in Figure P6.2. Determine the minimum value of h so that a 1500-N force can be placed such that the door will move on the rollers without tipping. Ignore friction at the rollers.

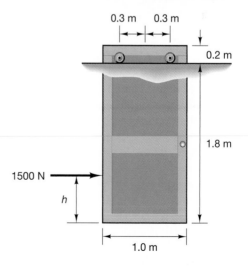

Figure P6.2

6.3 If $h = 1.2$ m, determine the acceleration of the door in Problem 6.2.

6.4 A homogeneous cylinder of diameter d and height h slides down an inclined plane, as shown in Figure P6.4. If the coefficient of kinetic friction is μ_k, determine the minimum ratio d/h such that the cylinder does not tip while sliding down the plane.

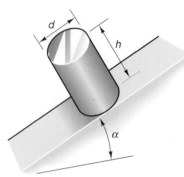

Figure P6.4

6.5 A uniform rod of mass m is supported by two cords, as shown in Figure P6.5. If the rod is released from rest at an angle θ, determine the tension in the cords and the acceleration of the mass at the instant of release.

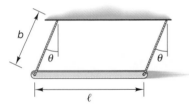

Figure P6.5

6.6 Two 2×4 in. boards are used to prevent a 3-ft-diameter, 200-lb cylinder from rolling when it is on the bed of a truck. (See Figure P6.6.) Determine the maximum deceleration that the truck can have before the cylinder begins to roll over the board.

Figure P6.6

6.7 Determine the maximum acceleration that the truck in Problem 6.6 can have while climbing up a hill with a 10% grade before the cylinder begins to roll over the board.

6.8 A 0.2-kg yo-yo with a radius of gyration of 25 mm is released from rest, as shown in Figure P6.8. If the inner radius of the yo-yo is 10 mm and the outer radius is 40 mm, determine the tension in the cord as a function of time.

Figure P6.8

6.9 A ball of mass m is released at an initial angle β inside a bowl of radius R. (See Figure P6.9.) If the radius of the ball, modeled as a sphere, is r, determine an expression for the velocity of the center of mass of the ball as a function of the position of the center of mass. Assume that the ball rolls without slipping.

Figure P6.9

6.10 A ball of radius r and mass m rolls without slipping on a flat surface that ends in a curved surface of radius R. (See Figure P6.10.) If the initial velocity of the ball is v_0, determine the angle β at which the ball leaves the surface.

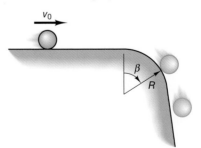

Figure P6.10

6.11 A uniform rod of mass m and length l is released from rest from the position shown in Figure P6.11. The spring has a spring constant k and an unstretched length l. Determine the equation of motion of the rod.

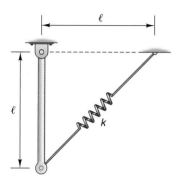

Figure P6.11

6.12 Solve the equation of motion of the rod in Problem 6.11 for the numerical values $m = 2$ kg, $l = 0.2$ m, and $k = 50$ N/m. Show by numerical experimentation that the period of oscillation increases as the stiffness of the spring decreases and that the amplitude of the oscillation decreases as the stiffness of the spring decreases.

6.13 A semicircular disk of mass m and radius r rotates about point O. If the disk is released from rest at an angle θ_0 (See Figure P6.13.), determine the reaction forces at point O as a function of the angle θ.

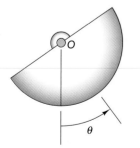

Figure P6.13

6.14 An inverted pendulum of length l and mass m (see Figure P6.14) is released from an initial angular position θ_0. (a) Determine the angular velocity at any angle θ, and (b) determine the reactions at the pinned point.

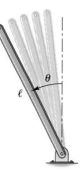

Figure P6.14

6.15 A detail of the pinned connection of an inverted pendulum is shown in Figure P6.15. If the frictional moment at the pin is $M_f = \mu_k r F$, where μ_k is the coefficient of kinetic friction between the pin and the pendulum bar, r is the radius of the pin, and F is the reaction force acting on the pin at any time, develop the equation of motion of the inverted pendulum. (*Hint: Remember that the frictional moment will always oppose motion.*)

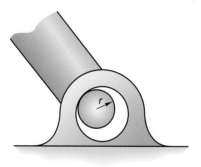

Figure P6.15

6.16 A hand brake is used to stop a 50-kg disk with a 300 mm-radius. (See Figure P6.16.) If the initial angular velocity of the disk is 4 rad/s and the force **F** increases linearly at a rate of 5 N/s, determine the time required to stop the disk. The coefficient of kinetic friction between the disk and the brake is 0.3.

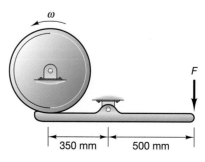

| 350 mm | 500 mm |

Figure P6.16

6.17 A uniform ring of small cross-sectional area A and mean radius r is rotated in a horizontal plane at a constant angular velocity ω. (See Figure P6.17.) If the mass per unit volume is ρ, determine the tension per unit cross-sectional area (hoop stress) in the ring. (*Hint: Construct a free-body diagram of a small element of the ring.*)

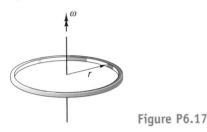

Figure P6.17

6.18 A disk of mass m and radius r has an offset center of mass and is placed on an incline plane. (See Figure 6.18.) If the disk is released from rest and has a radius of gyration of k_{cm}, develop the equations of motion of the disk. Assume that the disk rolls without slipping.

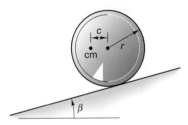

Figure P6.18

6.19 If for the disk in Problem 6.18, $m = 4$ kg, $r = 200$ mm, $c = 50$ mm, $k_{cm} = 150$ mm, and $\beta = 20°$, determine the velocity of the center of the disk after one complete revolution.

6.20 A uniform rod of mass m is released from rest at the position shown in Figure P6.20. Determine the equations of motion of the rod. The coefficient of kinetic friction between the rod and the inclined plane is μ_k, and friction can be neglected between the rod and the vertical wall.

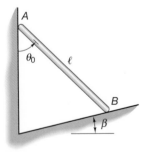

Figure P6.20

6.21 Develop the equations of motion of the uniform rod in Problem 6.20 if friction can be neglected between the rod and the inclined plane and the coefficient of friction between the rod and the wall is μ_k.

6.22 Determine the motion of the rod in Problem 6.20 from the initial position θ_0 to a final position $(\pi/2) + \beta$ using the numerical values $m = 3$ kg, $l = 800$ mm, $\theta_0 = 30°$, $\beta = 10°$, and $\mu_k = 0.1$.

6.23 Determine the motion of the rod in Problem 6.21 from the initial position θ_0 to a final position $(\pi/2) + \beta$ using the numerical values $m = 3$ kg, $l = 800$ mm, $\theta_0 = 30°$, $\beta = 10°$, and $\mu_k = 0.1$.

6.24 Develop the equations of motion for the uniform rod in Problem 6.20 if friction can be neglected between the rod and all surfaces.

6.25 Determine the motion of the rod in Problem 6.24 from the initial position θ_0 to a final position using the numerical values $m = 5$ kg, $l = 1.20$ m, $\theta_0 = 30°$, $\beta = 30°$, and $\mu_k = 0.4$. (*Note: Rod rotates to upright position.*)

6.26 A uniform rod of mass m is mounted on a vertical shaft and the motor applies a constant moment M to the rod. (See Figure P6.26.) If the rod starts from rest, determine the equations of motion of the rod, and determine the reactions at point A between the rod and the shaft.

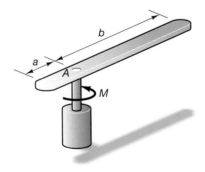

Figure P6.26

6.27 For the numerical values $m = 5$ kg, $a = 100$ mm, $b = 300$ mm, and $M = 2$ Nm, determine (a) the time required for the rod in Problem 6.26 to reach an angular velocity of 6 rad/s and (b) the reactions between the rod and the shaft at that time.

6.6 SYSTEMS OF RIGID BODIES OR PARTICLES

A system composed of rigid bodies or particles is analyzed by creating free-body diagrams of each body or particle with the interacting and external forces shown and then writing the equations of motion for each body or particle. For each rigid body in plane motion, three scalar equations of motion may be written: the two linear momentum equations in the plane of motion and the angular momentum equation about the axis perpendicular to the plane of motion. These three equations, originally set forth in Eq. (6.18), are

$$\sum F_x = Ma_{cm_x}$$

$$\sum F_y = Ma_{cm_y} \qquad (6.30)$$

$$\sum M_{cm_z} = I_{zz}\alpha_z$$

If the system contains particles, two equations of motion can be written for each particle, namely, the linear momentum equations for the particle in the plane of motion:

$$\sum F_x = Ma_{cm_x}$$

$$\qquad\qquad (6.31)$$

$$\sum F_y = Ma_{cm_y}$$

The sum of forces shown in Eqs. (6.30) and (6.31) include all the forces acting on the rigid body or particle. Some of these forces may be internal to the system, but external to the particular rigid body or particle. In general, the internal forces will be unknown quantities and are determined by examining the constraints of motion of the system—that is, constraints that relate the motion of one rigid body or particle to the motion of another rigid body or particle. As more rigid bodies and particles are added to the system, the number of simultaneous equations that must be solved to determine the dynamic parameters at a given instant or in a specific position increases. The use of matrices to solve this system of equations then becomes necessary. If the motion of the system over time is to be determined, the system will be treated as a coupled system of linear or nonlinear differential equations.

Sample Problem 6.10

Consider the system of a pulley and two masses shown in the accompanying diagram. Designating the inner and outer radii of the pulley as r_i and r_o, respectively, and the masses and moments of inertia as m_A, m_B, and I_{zz}, write the equations of motion of the system at any instant.

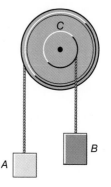

Solution   A right-hand rectangular coordinate system is established as in the diagram.

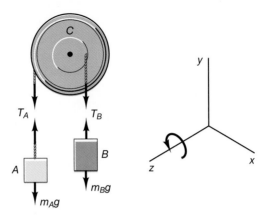

The two masses, A and B, will be modeled as particles moving in rectilinear motion in the y-direction. The pulley will be modeled as a rigid body rotating about a fixed point in space (its geometric center and center of gravity). The tension in the cables acting on A and B are designated by T_A and T_B, respectively. The three equations of motion are as follows:

Particle A: $T_A - m_A g = m_A a_A$

Particle B: $T_B - m_B g = m_B a_B$

Pulley C: $T_A r_O - T_B r_i = I_{zz}\alpha$

This system of equations has five unknowns: T_A, T_B, a_A, a_B and α. To determine the values of these unknowns, two additional equations must be developed from the system constraints. The accelerations of the masses are related to the angular acceleration of the pulley by

$$a_A = -r_O\alpha$$

$$a_B = r_i\alpha$$

Solving the linear system of five equations and five unknowns yields the following for the angular acceleration of the pulley:

$$\alpha = \frac{r_o m_A g - r_i m_B g}{I_{zz} + r_o^2 m_A + r_i^2 m_B}$$

Examining this result, we see that the numerator is the torque applied to the pulley and the denominator is an equivalent mass moment of inertia of the system. If the torque is positive— that is, if the moment caused by mass A is greater than the moment applied by mass B—then the angular acceleration will be positive, and the pulley will rotate in a counter-clockwise manner. The other unknowns are obtained by back substitution into the system of equations, yielding

$$a_A = -r_o\alpha = -r_o\frac{r_o m_A g - r_i m_B g}{I_{zz} + r_o^2 m_A + r_i^2 m_B}$$

$$a_B = r_i\alpha = r_i\frac{r_o m_A g - r_i m_B g}{I_{zz} + r_o^2 m_A + r_i^2 m_B}$$

$$T_A = m_A\left[g - r_o\frac{r_o m_A g - r_i m_B g}{I_{zz} + r_o^2 m_A + r_i^2 m_B}\right]$$

$$T_B = m_B\left[g + r_i\frac{r_o m_A g - r_i m_B g}{I_{zz} + r_o^2 m_A + r_i^2 m_B}\right]$$

If the system is rotating counterclockwise, the tension in cable B will be greater than that in cable A. In that case, the acceleration is constant, and the linear and angular velocities are obtained by simple integration of the acceleration, as shown in Chapters 1 and 5.

Sample Problem 6.11

A wheel rolls down an incline without slipping. A cord is wound around the outside of the wheel and attached to a mass, as shown in the accompanying diagram. As the wheel rolls down the incline, the mass is pulled into the wheel. The system will be in a steady state of

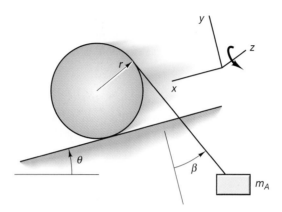

acceleration until the mass is completely drawn in. Determine the angular velocity of the wheel.

Solution

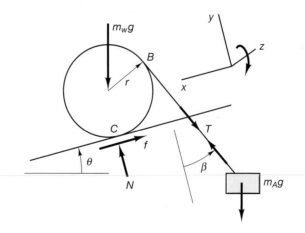

The free-body diagram of the wheel and the mass are shown in the preceding diagram. Identify a point on the cord at point B on the wheel. The tangential acceleration of this point is equal to the acceleration of the mass A. The equations of motion for the wheel are

$$m_w g \sin \theta - f - T \sin \beta = m_w a_{\text{cm}_x}$$

$$N - m_w g \cos \theta - T \cos \beta = m_w a_{\text{cm}_y}$$

$$Tr - fr = I_{zz}\alpha$$

The equations for the mass A are

$$T \sin \beta + m_A g \sin \theta = m_A a_{A_x}$$

$$T \cos \beta - m_A g \cos \theta = m_A a_{A_y}$$

There are five equations of motion and nine unknown quantities: the two components of the acceleration of the center of mass of the wheel and the acceleration of the mass A, the angular acceleration of the wheel, the tension in the cord, and the friction and normal forces between the wheel and the ground and the angle β that the cord makes with the normal to the plane. Therefore, there must be four scalar constraints. One of these comes from the observation that acceleration of the center of mass (the center of the wheel) is parallel to the inclined plane and the acceleration of the center of mass in the y-direction is zero. Thus, we have

$$\mathbf{a}_{cm} = a_{cm_x}\hat{\mathbf{i}}$$

A second constraint is that the wheel rolls without slipping. Therefore, the acceleration of the point of contact between the wheel and the plane is normal to the plane and is related to the acceleration of the center of mass by the equations

$$\mathbf{a}_{cm} = \mathbf{a}_C + \alpha\hat{\mathbf{k}} \times r\hat{\mathbf{j}} + \boldsymbol{\omega} \times (\boldsymbol{\omega} \times r\hat{\mathbf{j}})$$

$$a_{cm_x} = -r\alpha$$

The last constraint on the motion is more difficult to recognize. The acceleration of point B on the cord is equal to the acceleration of the center of the wheel plus the tangential acceleration of point B relative to the center of the wheel. This condition can be written as

$$\mathbf{a}_A = \mathbf{a}_B = \mathbf{a}_{cm} + \boldsymbol{\alpha} \times \mathbf{r}_{B/cm}$$

$$\mathbf{r}_{B/cm} = -r \cos \beta\hat{\mathbf{i}} + r \sin \beta\hat{\mathbf{j}} \qquad \boldsymbol{\alpha} = \alpha\hat{\mathbf{k}}$$

$$\mathbf{a}_A = (a_{cm_x} - r\alpha \sin \beta)\hat{\mathbf{i}} - r\alpha \cos \beta\hat{\mathbf{j}}$$

Accordingly, the four scalar constraint equations are

$$a_{cm_x} = r\alpha$$

$$a_{cm_y} = 0$$

$$a_{Ax} = (a_{cm_x} - r\alpha \sin \beta)$$

$$a_{Ay} = -r\alpha \cos \beta$$

Now we have a system of nine equations and nine unknowns. Since the system is nonlinear in β, a solution is shown in the Computational Supplement using the following numerical values:

$$m_w = 40 \text{ kg} \qquad m_B = 10 \text{ kg} \qquad r = 0.40 \text{ m} \qquad \theta = 30°$$

The results of solving this system of nonlinear algebraic equations are

$$T = 109.96 \text{ N} \qquad \alpha = -5.064 \text{ rad/s}^2 \quad f = 150.47 \text{ N} \qquad N = 443.97 \text{ N}$$

$$\beta = -0.327 \text{ rad} \quad a_{cm} = 2.026 \text{ m/s}^2 \qquad a_{Ax} = 1.375 \text{ m/s}^2 \quad a_{Ay} = 1.918 \text{ m/s}^2$$

Notice that the angle β is -0.327 radian and the mass A swings forward of the normal to the plane. This is $18.7°$ in front of the normal to the plane, but still $11.3°$ behind the vertical in space. Notice, in addition, that the angular acceleration of the wheel is negative, indicating that the wheel rolls down the incline.

Sample Problem 6.12

Consider the belt drive system shown in the accompanying figure. If a constant moment M_a is applied to the drive wheel A, determine the angular acceleration of wheel B. Assume that there is no belt slippage, and express the solution generally in terms of the radii and mass moments of inertia of the two drums.

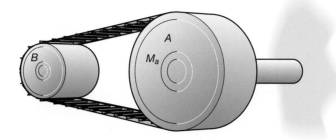

Solution

Let r_a and I_a be the radius and moment of inertia, respectively, of drive wheel A and r_b and I_b be the radius and moment of inertia, respectively, of wheel B. Assume that both wheels are of uniform mass distribution; that is, the center of mass is located at the center of the wheel. The acceleration of the center of mass is zero, and the free-body diagram of the system is shown here. Note that the tension in the upper part of the belt is not the same as the tension in the lower part of the belt. A moment is required to accelerate wheel B, and this moment is generated by the difference in the belt tensions. The angular momentum equations for the two wheels are

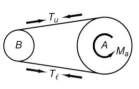

$$M_a - (T_l - T_u)r_a = I_a \alpha_a$$

$$(T_l - T_u)r_b = I_b \alpha_b$$

$$\Delta T = (T_l - T_u)$$

Note that only the difference in the tensions of the lower and upper belts appears in the aforementioned equations. The actual tensions would include the pretension in the belts, which cannot be determined, as the dynamics of the system is independent of this pretension. There are three unknowns—ΔT, α_a and α_b—in the two equations. Therefore, we conclude that there must be an equation of constraint. The constraint to the motion is that the linear acceleration of the belt at the two contact points must be equal:

$$a_{A_c} = a_{B_c}$$

$$r_a \alpha_a = r_b \alpha_b$$

Solution of the three equations yields

$$\alpha_a = \frac{M_a}{\left(I_a + \frac{r_a^2}{r_b^2} I_b \right)}$$

$$\alpha_b = \frac{r_a}{r_b} \frac{M_a}{\left(I_a + \frac{r_a^2}{r_b^2} I_b \right)}$$

$$\Delta T = \frac{M_a}{r_a} \left[1 - \frac{I_a}{\left(I_a + \frac{r_a^2}{r_b^2} I_b \right)} \right]$$

The accelerations in the three previous sample problems shown are constant throughout the motion described in each problem, and the velocities and displacements of the rigid bodies and particles comprising the system are obtained by simple integration of the accelerations subject to initial conditions. This concept was shown in Chapters 1 and 5 and will not be repeated here. Although the free-body diagram in this sample problem has been constructed for a particular position or an instant of time, it is also applicable to the complete motion, as the accelerations are not dependent upon position or velocity. The forces and moments causing the accelerations may be functions of time, and the differential equation for the motion is still solved by direct integration and does not require special methods.

Problems

6.28 Determine the angular acceleration of the pulley in Figure P6.28 if the pulley has a radius of 300 mm, a mass of 2 kg, and a radius of gyration of 250 mm.

Figure P6.28

6.29 Determine the angular acceleration for each of the 100-lb disks of equal radius of 0.4 ft in Figure P6.29.

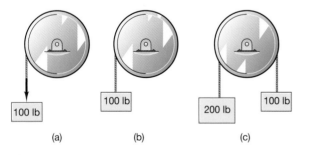

(a) (b) (c)

Figure P6.29

6.30 A rectangular crate, shown in Figure P6.30, has a mass of 100 kg and is attached to a 70-kg block by a cable system. If the coefficient of static friction between the crate and the floor is 0.4 and the coefficient of kinetic friction is 0.32, determine the acceleration of the crate. For what range of values of the distance between the cable and the floor h will the crate slide without tipping? Neglect the mass of the pulley and cable.

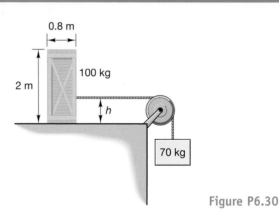

Figure P6.30

6.31 To determine the coefficient of kinetic friction between a shaft on which a 50-kg, 1-m diameter disk rotates and its bearing, shown in Figure P6.31, a 40-kg mass is attached to the end of the rope hung off of the disk. If the frictional moment at the bearing is equal to $M_f = \mu_k R_a r$, where r is the radius of the shaft, R_a is the reaction at the bearing of the shaft, and the mass B falls 4 m in 1.3 s, determine the coefficient of kinetic friction.

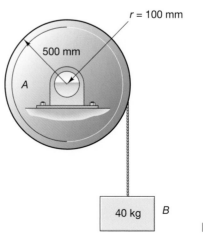

Figure P6.31

6.32 In Figure P6.32, a 2-kg collar A slides on a smooth rod, and a 4-kg bar BC is attached to the collar by a cord. Determine the minimum force **P** applied to the collar such that the cord and the bar will lie on a straight line—that is, the force necessary to begin to lift the bar from the surface.

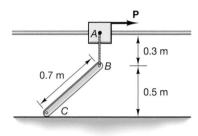

Figure P6.32

6.33 The 5-kg solid disk A in Figure P6.33 rotates on a smooth pin O. Block B in the figure has a mass of 2 kg, and the mass of the pulley attached to block B is negligible. Determine the angular acceleration of the disk and the tension in the cord.

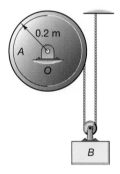

Figure P6.33

6.34 In Figure P6.34, gears A and B have masses of 2 kg and 5 kg and radii of gyration of 100 mm and 400 mm, respectively. If gear A is driven by a moment of 3 N · m, determine the angular acceleration of each gear and the contacting force between the gears.

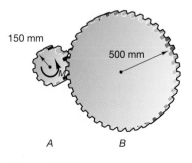

Figure P6.34

6.35 A small gear A drives two identical larger gears B and C, as shown in Figure P6.35. Develop a relationship between the moment M applied to gear A and the angular acceleration of B and C in terms of the gear radii and their mass moments of inertia.

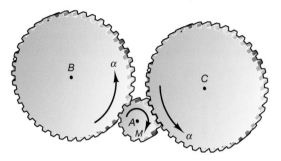

Figure P6.35

6.36 A small gear A drives two larger gears B and C, as shown in Figure P6.36. Develop a relationship between the moment M applied to gear A and the angular acceleration of C in terms of the gear radii and their mass moments of inertia.

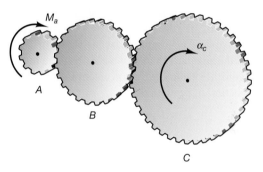

Figure P6.36

6.37 A bar of length l and mass m is attached to a block of mass M that slides on a smooth horizontal surface, as shown in Figure P6.37. Develop the differential equations of motion for the acceleration **a** of the block and the angular acceleration **α** of the bar, neglecting friction on the pin between the bar and the block, if the bar is released in the position shown.

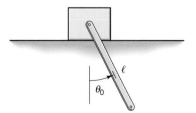

Figure P6.37

6.38 Solve the differential equations of motion for the system shown in Figure P6.37, using the following parameters and initial values: $l = 0.4$ m, $m = 4$ kg, $M = 3$ kg, $\theta_0 = -30°$, and $\dot{\theta}_0 = 0$.

6.39 Determine the moment as a function of position that must be applied to a 10-kg disk A, shown in Figure P6.39, in order to maintain a constant angular velocity of the disk while it drives the 8-kg oscillating bar CD.

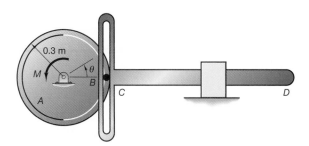

Figure P6.39

6.40 A mechanism consisting of a 500-mm, 5-kg bar AB and a 300-mm, 3-kg BC, is driven by a varying force acting on a piston of negligible mass, as shown in Figure P6.40. If the angular velocity of bar BC is to be maintained at a constant value of 2 rad/s, determine the value of **P** as a function of position θ.

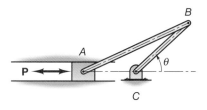

Figure P6.40

6.41 Determine the magnitude and sense of the force **P** in Problem 6.40 when the mechanism is in the position (a) $\theta = 90°$ and (b) $\theta = 270°$.

6.42 In Figure P6.42, a belt of negligible mass passes between two 0.4-m-radius identical cylinders, and the belt is pulled to the right by a 200-N force. If the mass of the cylinders is 50 kg and the cylinders are initially at rest, determine whether the belt slips on either of the cylinders, and find the angular acceleration of the cylinders. The coefficients of friction between the belt and the

cylinders are $\mu_s = 0.6$ and $\mu_k = 0.48$, and cylinder A slides freely in the slot.

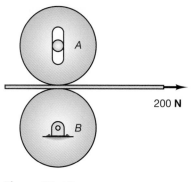

Figure P6.42

6.43 Determine the angular accelerations of the two disks in Problem 6.42 if cylinder A has a radius 0.3 m and a mass of 20 kg. The mass and radius of cylinder B are as specified in Problem 6.42.

6.44 The 150-mm-radius unbalanced disk A, shown in Figure P6.44, has a mass of 30 kg, and the center of mass is 50 mm from the axis of rotation. The radius of gyration of the disk about the axis of rotation is 120 mm, and the mass of block B is 25 kg. If the system is released from rest at the position shown in the figure, determine the equation of motion of the disk. Neglect the mass of the pulley in the system.

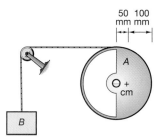

Figure P6.44

6.45 Determine the angular displacement as a function of time of the disk in Problem 6.44 during one full revolution of the disk.

6.46 A cable of negligible mass is wrapped around a 20-kg, 300 mm-diameter disk A, passes without slipping over a 3-kg, 50-mm-diameter pulley B, and is connected to a 15-kg block C, as shown in Figure P6.46. The coefficient of static friction between the disk and the 30° inclined plane is 0.6. Determine the angular acceleration of the disk if the system is released from rest at the position shown in the figure.

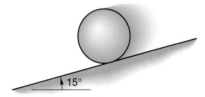

Figure P6.46

6.47 Solve Problem 6.46 if the coefficient of static friction between the disk and the plane is 0.2 and the coefficient of kinetic friction is 0.15.

6.48 In Figure P6.48, a 30-kg sphere of radius 100 mm rolls down an inclined plane without slipping. Determine the angular acceleration of the sphere and the minimum coefficient of static friction to prevent slipping.

Figure P6.48

6.49 Determine the time required for the sphere in Problem 6.48 to roll 2 m.

6.50 The 12-in-diameter, 20-lb disk in Figure P6.50 has a radius of gyration of 5 in. and rolls without slipping on its 6-in.-diameter inner shaft on inclined guides on both sides of the disk. A cord is wrapped around the disk and is attached to a 5-lb block. Determine the angular acceleration of the disk and the tension in the cord.

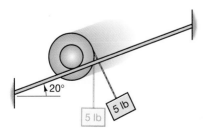

Figure P6.50

6.51 The mechanical system in Figure P6.51 consists of a 6-kg uniform bar A, a 5-kg uniform bar B, and a 2-kg collar C that slides on a smooth rod due to a spring force. If the unstretched length of the spring is 600 mm and the spring constant is 500 N/m, determine the equation of motion for the system if it is released from rest at the position shown in the figure.

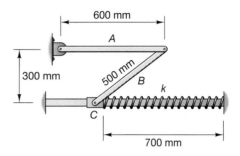

Figure P6.51

6.52 A 50-kg disk and shaft, shown in Figure P6.52, has a radius of gyration of 200 mm. The disk radius is 300 mm, and the radius of the shaft is 50 mm. A spring with a spring constant of 2 kN/m is attached to the shaft by cords wrapped around the axis of the shaft. Another cord is wrapped around the disk in a shallow groove, passes over a pulley of negligible mass, and is attached to a 30-kg block. If the spring is unstretched when the system is released from rest, determine the angular acceleration of the disk as a function of the displacement of the block. Assume that the disk rolls without slipping.

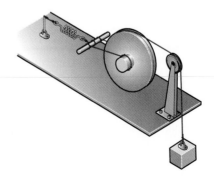

Figure P6.52

6.53 A 30-kg homogeneous 300-mm diameter disk is lowered onto a conveyor belt that is moving at a constant velocity of 3 m/s, as shown in Figure P6.53. If the disk is initially at rest and the coefficient of kinetic friction between the belt and the disk is 0.3, determine the angular acceleration of the disk during the time that it is slipping on the belt and number of revolutions before the disk reaches its final velocity.

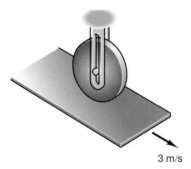

3 m/s

Figure P6.53

6.54 A homogeneous bar of length L and mass m is released from a horizontal position while one end of the bar is in contact with a smooth inclined plane, as depicted in Figure P6.54. Assuming that the bar remains in contact with the plane, determine the differential equation for the angular acceleration of the bar.

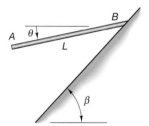

Figure P6.54

6.55 Numerically solve the differential equation in Problem 6.54 to obtain the angle θ as a function of time from release until the bar hits the plane, using the following values: $L = 0.6$ m, $m = 4$ kg, and $\beta = 60°$.

6.56 Determine the differential equation for the angular acceleration of the bar in Problem 6.54 if the coefficient of kinetic friction between the bar and the inclined plane is μ_k.

6.57 Numerically solve the differential equation in Problem 6.56 to obtain the angle θ as a function of time from release until the bar hits the plane, using the following values: $L = 0.6$ m, $m = 4$ kg, $\beta = 60°$, and $\mu_k = 0.3$.

6.58 The 10-kg, 100-mm-radius disk shown in Figure P6.58 rotates in a counterclockwise manner in a vertical plane at a constant rate of 10 rad/s; the disk is driven by a motor. A 6-kg, 600-mm homogeneous bar is connected to the disk with a pin, as shown in the figure. Determine the differential equation for the angular acceleration of the bar and the equation for the motor torque to drive the disk. Neglect friction at the pin and between the rod and the horizontal plane.

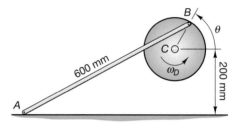

Figure P6.58

6.59 Numerically solve the differential equation of motion for Problem 6.58 and graph the horizontal displacement of point A for one complete revolution of the disk.

6.60 Two identical homogeneous 4-kg, 2-m bars are pinned at point B, as shown in Figure P6.60. If bar AB is released from rest in a horizontal position, develop the equations of motion and constraint equations. Point C slides on a smooth horizontal surface.

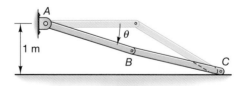

Figure P6.60

6.61 Develop the differential equation of motion for bar AB in Problem 6.60.

6.62 In Figure P6.62, a 150-mm-radius, 4-kg pulley A has a radius of gyration of 120 mm, and a 3-kg, 100-mm-radius pulley B has a radius of gyration of 80 mm. If a 2-kg mass is hung on pulley A, determine the time that it takes for the 10-kg mass to move 800 mm.

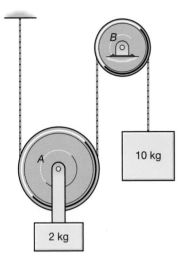

Figure P6.62

6.63 Determine the time that it takes for the 10-kg mass in Problem 6.62 to move 800 mm if the 2-kg mass is removed from pulley A.

6.64 A bar of length $3R$ is attached to a homogeneous disk of radius R, as illustrated in Figure P6.64. The mass of the bar is m, and the mass of the disk is $3m$. If the disk rolls without slipping and the friction is negligible between the bar and the horizontal surface, determine the angular acceleration of the disk at the instant that the bar is released from the position shown in the figure.

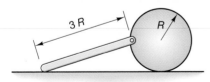

Figure P6.64

6.65 Develop the differential equation for the angular acceleration of the disk in Problem 6.64.

6.66 Numerically solve the differential equation in Problem 6.65, and graph the angular position of the disk for the first 5 s of movement for the values $R = 100$ mm and $m = 4$ kg.

6.67 The 4-kg bar in Figure P6.67 has a torsional spring at its pivot point, with a spring constant of 10 Nm/rad. If the spring is untwisted when the bar is released from the horizontal position, determine the differential equation for the angular acceleration of the bar.

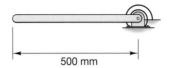

500 mm

Figure P6.67

6.68 Determine the angular position of the bar in Problem 6.67 for the first 2 s of movement. (Use $\Delta t = 0.0001$.)

6.69 Select the linear damping coefficient c if a damper is to be placed at the pivot point of the bar in Problem 6.67 such that the bar oscillates 2 times before coming to rest in an equilibrium position. $M_d = -c\omega$, where M_d is the damping moment in Nm, c is the damping coefficient in Nms/rad, and ω is the angular velocity in rad/s.

6.70 A uniform bar of mass m and length l is pivoted at the left end and supported by a spring with an unstretched length of l and a spring constant k at the right end, as shown in Figure P6.70. If the spring is unstretched when the bar is released from rest in a horizontal position, determine the differential equation for the angular acceleration of the bar.

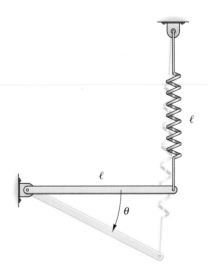

Figure P6.70

6.71 Numerically solve the differential equation in Problem 6.70 and graph the angular position of the bar with time for 2 oscillations of the bar. Use the following numerical data: $m = 5$ kg, $l = 0.6$ m, and $k = 300$ N/m.

6.72 Increase the spring constant in Problem 6.71 by a factor of 10 and examine the change in the period of oscillation. Decrease the spring constant by a factor of 10 and examine the change in the period of oscillation.

6.73 The pulleys in Figure P6.73 turn freely on their axes and are initially at rest when a 5-Nm torque is applied to pulley A. The masses and radii of gyration of the pulleys are $m_A = 3$ kg, $k_{A_z} = 180$ mm, $m_B = 8$ kg, $k_{B_z} = 370$ mm, $m_C = 6$ kg, and $k_{C_z} = 350$ mm. If the belts do not slip on the pulleys, determine the angular velocity of pulley C 2 s after the torque is applied.

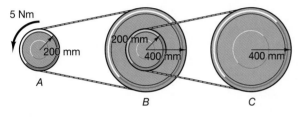

Figure P6.73

6.74 The mechanism in Figure P6.74 consists of a rod AB, a connecting rod BC, and a disk that rolls without slipping on a horizontal surface. The rod AB is 200 mm long, and the connecting rod BC is 500 mm long and connects to a 4-kg, 120-mm-radius disk. If the rod AB is driven at a constant angular velocity of 4 rad/s, determine the torque required to drive the mechanism if the masses of the two rods can be neglected.

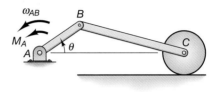

Figure P6.74

6.75 Determine the position of the disk in Problem 6.74 during a complete revolution of the rod AB.

6.7 D'ALEMBERT'S PRINCIPLE

In his *Traité de Dynamique,* written in 1743, the French mathematician Jean le Rond d'Alembert rewrote Newton's equation of linear momentum, Eq. (6.15), as

$$\sum \mathbf{F} = m\mathbf{a}$$

$$\sum \mathbf{F} + (-m\mathbf{a}) = 0 \tag{6.32}$$

The inertia vector ($m\mathbf{a}$) has units of force, and when its direction is reversed by the minus sign, it is referred to as the *inertial force vector,* or the inertia force. This inertial force vector acts through the center of mass in the direction opposite to the acceleration vector. In a similar manner, the angular momentum Eq. (6.17) can be written as

$$\sum \mathbf{M}_{cm} + (-\dot{\mathbf{H}}_{cm}) = 0 \tag{6.33}$$

The treatment of the inertia vectors ($m\mathbf{a}$) and ($I\alpha$) as a reversed inertial force and a reversed inertial moment, respectively, is referred to as *d'Alembert's principle.* For a particle, d'Alembert's principle states that *the applied forces together with the forces of inertia form a system in equilibrium.* Treating the derivative of the linear momentum vectors in this manner establishes an analogy to static equilibrium and has led to the appearance of an "equilibrium" state that is termed "dynamic equilibrium." This appellation is a contradiction of terms, and while d'Alembert's principle is used in many introductory texts, it is of only limited importance at this level. Other acceleration terms viewed as reversed effective forces give rise to the *centrifugal force* ($-m\boldsymbol{\omega} \times (\boldsymbol{\omega} \times \mathbf{r})$) and the *Coriolis force* ($-2m(\boldsymbol{\omega} \times \mathbf{v}_r)$).

D'Alembert's principle does have applications in advanced dynamics, particularly in the development of the Lagrange equations of motion through the methods of virtual work. However, the principle offers little advantage for the analysis of the dynamics of plane motion. A variation of this principle considers the right-hand side of the Newton equations of motion as an *equivalent force system.* Again, the system is reduced to a static equivalence. Consider the rigid body shown in Figure 6.5(a) and the inertia vectors shown in Figure 6.5(b).

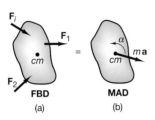

Figure 6.5

Using the methods of equivalent force systems shown in Chapter 3 of Statics, we can say that the system shown in the free-body diagram (FBD) is equivalent to that shown in the mass–acceleration diagram (MAD). Note that this approach does not change the inertia vectors to reversed inertia forces, but treats them as an equivalent force and moment system. In Statics, two force systems were said to be equivalent if they produced the same resultant force and the same moment about any point on the body. If a point A is selected, the force systems are equivalent if

$$\sum_i \mathbf{F}_i = m\mathbf{a}_{cm}$$

(6.34)

$$\sum_i \mathbf{r}_{i/A} \times \mathbf{F}_i = \mathbf{r}_{cm/A} \times m\mathbf{a}_{cm} + I_{cm_z}\alpha_z\hat{\mathbf{k}}$$

If d'Alembert's principle is used—that is, bringing the inertia terms to the left side of the equation and treating them as inertial forces and moments—the equivalent free-body diagram with the inertial force and inertial moment (or couple) shown in Figure 6.5 would appear as shown in Figure 6.6.

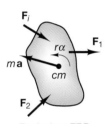

Equivalent **FBD**

Figure 6.6

The summation of moments about any point would yield Eq. (6.34). It is apparant that treating plane dynamics in this manner offers no conceptual advantages and may cause confusion. Inertial forces and moments are *not* forces and moments, and there is no such thing as "dynamic equilibrium." The continued use of d'Alembert's principle in introductory courses in dynamics seems to be based on reducing computational difficulties by allowing moments to be taken about any point. Since it has been shown that computational difficulties associated with systems of equations are better addressed with modern computational methods, dynamic problems now will be solved by direct application of the principles of dynamics and not by a reduction to static equivalence. One advantage to the concept of inertial forces and moments, other than its use in virtual work does exist, however, and this advantage is that these terms may be compared to external forces, and some problems can then be modeled in a quasi-static manner.

Problems

These problems are additional problems in plane dynamics that can be formulated using either the traditional Newton's equations or d'Alembert's principle.

6.76 In Figure P6.76, a bar is lifted in a vertical plane by a constant moment of 60 Nm applied at point A. If the mass of the bar is 10 kg, the mass of the two links is negligible and the system starts from rest when the links are horizontal, determine the angular acceleration of the links as a function of the angle θ.

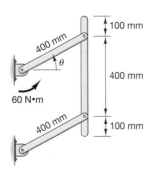

Figure P6.76

6.77 The coefficient of static friction between the tires of a car and the snow-covered road in Figure P6.77 is 0.3. Determine the maximum acceleration of the car when it goes up an incline of 10° for (a) a rear-wheel-drive car, (b) a front-wheel-drive car, and (c) a four-wheel-drive car.

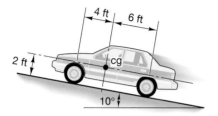

Figure P6.77

6.78 In Figure P6.78, a 100-lb crate rests on a 20-lb cart, and the crate and cart are pulled forward by 40-lb block *C*. If the coefficient of static friction between the cart and the crate is 0.6, determine whether the crate will tip or slip on the cart. Ignore friction between the cart and the floor.

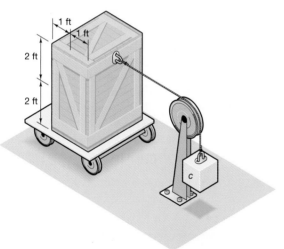

Figure P6.78

6.79 Determine the acceleration of *C* in Problem 6.78 if the effective coefficient of kinetic friction between the cart and the floor is 0.1.

6.80 Determine the acceleration of *C* in Problem 6.78 if the pulley can be modeled as a homogeneous disk of weight 10 lb and having a radius of 6 in.

6.81 A board is leaned against the side of a box, as shown in Figure P6.81. Determine the acceleration of the box necessary for the board to maintain this position if friction is neglected.

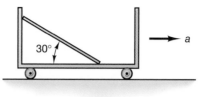

Figure P6.81

6.82 A 3-kg disk is placed on a conveyor belt that is moving 500 mm/s to the left, as depicted in Figure P6.82. The mass of the 400-mm connecting arm is 2 kg, and the disk is not rotating when it is placed on the conveyor belt. If the coefficient of kinetic friction between the disk and the belt is 0.3, determine the angular acceleration of the disk while slipping occurs and the time required for the disk to reach a constant angular velocity.

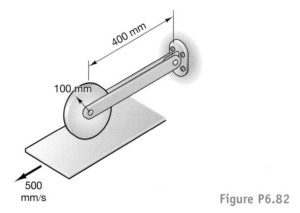

Figure P6.82

6.83 The 10-kg pulley *C* in Figure P6.83 is modeled as a homogeneous disk of radius 200 mm. Determine the angular acceleration of the pulley if the system is released from rest in the position shown in the figure. Ignore friction at the shaft of the pulley.

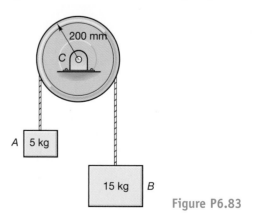

Figure P6.83

6.84 Determine the tension in the cable to block B in Problem 6.83 if there is a frictional moment on the pulley resisting rotation of 4 N · m.

6.85 A drum of mass m and radius r is set on two shafts A and B, as shown in Figure P6.85. Neglecting the mass of the shafts, determine an expression for the acceleration of the drum in terms of the coefficient of static friction μ_k between the drum and shaft A if the shaft is rotated at a high angular velocity.

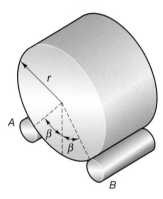

Figure P6.85

6.86 A homogeneous bar of length 2 m and of mass 10 kg is resting on a smooth horizontal surface, as depicted in Figure P6.86. If a 500-N force is applied perpendicularly to the rod at one end, determine the angular acceleration of the rod, the linear acceleration of the center of the rod, and the point on the rod that has no acceleration.

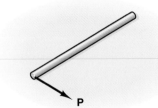

Figure P6.86

6.87 The 5-kg pulley in Figure P6.87 has an outer radius of 100 mm, an inner radius of 50 mm, and a radius of gyration of 75 mm. If a 50-N force is applied in a vertical direction to the cord wrapped around the inner radius, determine the angular acceleration of the pulley and the acceleration at the center of mass.

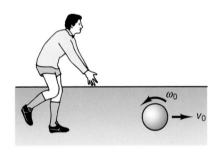

Figure P6.87

6.88 A spherical ball of radius r is thrown onto a horizontal surface with an initial linear and angular velocity, as shown in Figure P6.88. Determine the time when the ball will start rolling without sliding, and determine the linear velocity of the ball at that time. Denote the coefficient of kinetic friction as μ_k.

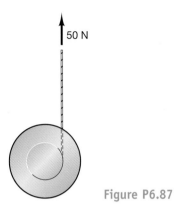

Figure P6.88

6.89 A spherical ball of radius r and mass m is released from rest on an inclined plane, as illustrated in Figure P6.89. Determine the minimum coefficient of static friction for the ball to roll without slipping on the plane.

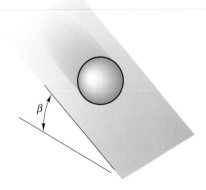

Figure P6.89

6.90 In Figure P6.90, gear A is stationary, and gear B rotates around it. The mass of gear B is 4 kg, and its centroidal radius of gyration is 60 mm. The connecting homogeneous bar has a mass of 2 kg, and friction can be neglected at the pins. If the system is released from rest at the position shown in the figure, determine the angular acceleration of gear B at the time of release.

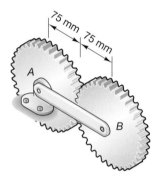

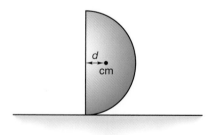

Figure P6.90

6.91 Write the differential equation for the angular acceleration of the connecting bar AB in Problem 6.90 for any position.

6.92 A hemisphere of mass m and radius r is released from the position shown in Figure P6.92. Determine the minimum coefficient of static friction for the hemisphere to roll without sliding. The distance to the center of mass of the hemisphere d is $\left(\frac{3}{8}\right)r$, and the mass moment of inertia about the center of mass is $I_{z_{cm}} = \frac{83}{320}mr^2$.

Figure P6.92

6.93 A cylinder of mass m and radius r is on a plate that is given a constant acceleration to the right, as illustrated in Figure P6.93. Determine the angular acceleration of the disk and the minimum coefficient of static friction for the cylinder to roll without sliding.

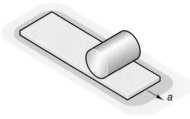

Figure P6.93

6.94 A 1-m bar has a mass of 20 kg and is released from rest in the position shown in Figure P6.94. Determine the differential equation for the angular acceleration of the bar if the horizontal surface is smooth.

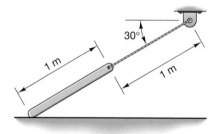

Figure P6.94

6.95 Write the differential equation of motion for the bar in Problem 6.94 if the coefficient of kinetic friction between the surface and the bar is 0.1.

6.96 A uniform rod of length $1.5r$ and mass m slides inside a smooth fixed cylinder of radius r, as shown in Figure P6.96. If the rod is released from rest at an angle θ_0, write the differential equation for the angular acceleration of the rod.

Figure P6.96

6.97 Numerically solve the differential equation for Problem 6.96 for the following values: $r = 0.4$ m, $m = 3$ kg, and $\theta_0 = 45°$.

Chapter 7

POWER, WORK, ENERGY, IMPULSE, AND MOMENTUM OF A RIGID BODY

A diver uses conservation of angular momentum to alter his angular velocity by changing the mass moment of inertia of his body.

7.1 POWER, WORK, AND ENERGY OF A RIGID BODY

The concepts of work and energy are very powerful tools for determining the velocity and position of a particle in motion. For a single particle, we defined the power in Chapter 3 as

$$\mathbf{F} \cdot \mathbf{v} = \frac{d}{dt}(1/2\, m\mathbf{v} \cdot \mathbf{v}) \tag{7.1}$$

The left side of the equation is the power, or the rate of doing work, and the right side is the rate of change of the kinetic energy of the particle with respect to time. For a system of particles, the sum of the power acting on all the particles is equal to the sum of the time rate of change of the kinetic energies of all the particles:

$$\sum_i \left[\left(\mathbf{F}_i + \sum_j \mathbf{f}_{ij} \right) \cdot \mathbf{v}_i \right] = \sum_i \frac{d}{dt}(1/2\, m_i \mathbf{v}_i \cdot \mathbf{v}_i) \tag{7.2}$$

We examined the time rate of change of the kinetic energy of the particles relative to the center of mass of the system by introducing the velocity of each particle relative to the center of mass.

$$\mathbf{v}_i = \mathbf{v}_{cm} + \dot{\boldsymbol{\rho}}_i$$

where $\dot{\boldsymbol{\rho}}_i$ is the velocity of the ith particle relative to the center of mass. Therefore, the resulting kinetic energy is

$$\sum_i \frac{d}{dt}(1/2\, m_i \mathbf{v}_i \cdot \mathbf{v}_i) = \frac{d}{dt}\left(1/2\, M\mathbf{v}_{cm} \cdot \mathbf{v}_{cm} + \sum_i 1/2\, m_i\, \dot{\boldsymbol{\rho}}_i \cdot \dot{\boldsymbol{\rho}}_i \right) \tag{7.3}$$

Two factors prevented further reduction of Eq. (7.3) for the power of a system of particles. First, as was pointed out earlier, the power of the internal force of one particle is not the negative of the power of that internal force acting on another particle, because the velocities of the two particles are not equal and opposite. Second, the second term in the expression for kinetic energy required that the motion of each particle relative to the center of mass or in absolute space be known.

For a system of particles, work and energy concepts were used only in examples in which it could be observed by other means that the kinetic energy of the system was conserved and in which the velocities of each of the particles could be determined. The most practical applications were elastic collisions within a system of particles. Therefore, questions may arise as to the usefulness of the concept of work and energy when working with a system of particles.

However, for a system of particles that form a rigid body, a work–energy concept can be developed that is as powerful as that used for a single particle. The development of this concept is based upon the equations derived in Chapter 4 for a system of particles. Although derivations of equations of this type usually are of little interest to the student, many useful concepts can be fully understood only by a careful review of the derivation of the final equations. For a rigid body, the constraint on the system is that any point on the body can rotate only relative to any other point on the rigid body. That is, the magnitude of the relative position vector between any two points on the same rigid body must be constant. Therefore, for any point i on a rigid body, the velocity of that point can be related to the velocity of the center of mass and the angular velocity of the body by the equation

$$\mathbf{v_i} = \mathbf{v}_{cm} + \dot{\boldsymbol{\rho}}_i = \mathbf{v}_{cm} + (\boldsymbol{\omega} \times \boldsymbol{\rho}_i) \tag{7.4}$$

where $\boldsymbol{\omega}$ is the angular velocity of the body.

The subscript notation to designate a point on the body will be used initially to make the development clearer, and later, integration, rather than a summation, over the total mass of the rigid body will be used. The power term will keep the summation notation, as there are not, in general, external forces acting on each point of the rigid body, with the exception of gravitational attraction, which can be treated as a single force—that is, the weight—acting at the center of mass of the body.

Substituting the constraint equation Eq. (7.4) into Eq. (7.2) yields

$$\sum_i \left(\mathbf{F}_i + \sum_j \mathbf{f}_{ij} \right) \cdot (\mathbf{v}_{cm} + \boldsymbol{\omega} \times \boldsymbol{\rho}_i)$$

$$= \frac{d}{dt} \left[1/2 \, M \mathbf{v}_{cm} \cdot \mathbf{v}_{cm} + \sum_i 1/2 \, m_i (\boldsymbol{\omega} \times \boldsymbol{\rho}_i) \cdot (\boldsymbol{\omega} \times \boldsymbol{\rho}_i) \right] \qquad (7.5)$$

Consider the power term first (the left-hand side of the equation), which can be written as

$$P = \sum_i \mathbf{F}_i \cdot (\mathbf{v}_{cm} + \boldsymbol{\omega} \times \boldsymbol{\rho}_i) + \sum_i \sum_j \mathbf{f}_{ij} \cdot \mathbf{v}_{cm} + \sum_i \sum_j \mathbf{f}_{ji} \cdot (\boldsymbol{\omega} \times \boldsymbol{\rho}_i) \qquad (7.6)$$

For any system of particles, Newton's third law shows that

$$\sum_i \sum_j \mathbf{f}_{ij} = 0 \qquad (7.7)$$

From the properties of the triple scalar product, the last term in Eq. (7.6) can be written as

$$\boldsymbol{\omega} \cdot \sum_i \sum_j \boldsymbol{\rho}_i \times \mathbf{f}_{ij} \qquad (7.8)$$

It was shown in Chapter 4 that this term is zero. Therefore, for a rigid body, the power associated with the internal forces cancels, and only the external forces do work. The power due to the external forces may be treated in two different manners, both of which are useful. Consider the power of an external force to be the force times the velocity of the point on which it acts. This leads to the traditional concept that the work done by a force is the integral over the path of the scalar product of the force and the displacement vector of the point of application:

$$P = \sum_i \mathbf{F}_i \cdot \mathbf{v}_i \qquad (7.9)$$

This equation will be called *definition A of the power acting on a rigid body*. However, the concept of the center of mass can be used, and definitions of power and work can be related to the displacement of the center of mass

$$P = \sum_i \mathbf{F}_i \cdot (\mathbf{v}_{cm} + \boldsymbol{\omega} \times \boldsymbol{\rho}_i) = \left(\sum_i \mathbf{F}_i \right) \cdot \mathbf{v}_{cm} + \sum_i \mathbf{F}_i \cdot (\boldsymbol{\omega} \times \boldsymbol{\rho}_i) \qquad (7.10)$$

The second term on the right-hand side can be rewritten, using the triple scalar product vector identity, as

$$\sum_i \mathbf{F}_i \cdot (\boldsymbol{\omega} \times \boldsymbol{\rho}_i) = \boldsymbol{\omega} \cdot \sum_i (\boldsymbol{\rho}_i \times \mathbf{F}_i)$$

But

$$\sum_i (\boldsymbol{\rho}_i \times \mathbf{F}_i) = \sum \mathbf{M}_{cm} \qquad (7.11)$$

Therefore,

$$P = \left(\sum \mathbf{F} \right) \cdot \mathbf{v}_{cm} + \left(\sum \mathbf{M}_{cm} \right) \cdot \boldsymbol{\omega}$$

This equation will be called *definition B of the power acting on a rigid body.* It allows the power and the work of all the external forces to be treated as if the forces are acting on the center of mass plus the power associated with the moment of those forces about the center of mass. Note that this way of treating power and work is more consistent with the manner of solution of the dynamic problems of rigid bodies.

The kinetic energy of a system of particles constituting a rigid body can be written as

$$T = 1/2\, M\mathbf{v}_{cm} \cdot \mathbf{v}_{cm} + 1/2 \int_M (\boldsymbol{\omega} \times \boldsymbol{\rho}) \cdot (\boldsymbol{\omega} \times \boldsymbol{\rho})\, dm \tag{7.12}$$

The summation over all the mass particles making up the rigid body is replaced by an integration over the entire mass distribution. If the mass density of the body is a constant, the integration may be considered as integration over the volume. The power is equal to the rate of change of the kinetic energy with respect to time. Examining the second term in the kinetic energy equation and using the triple scalar product identity, yields

$$\mathbf{A} \cdot (\mathbf{B} \times \mathbf{C}) = \mathbf{C} \cdot (\mathbf{A} \times \mathbf{B})$$

Letting $\mathbf{A} = (\boldsymbol{\omega} \times \boldsymbol{\rho})$, $\mathbf{B} = \boldsymbol{\omega}$, and $\mathbf{C} = \boldsymbol{\rho}$, we obtain

$$(\boldsymbol{\omega} \times \boldsymbol{\rho}) \cdot (\boldsymbol{\omega} \times \boldsymbol{\rho}) = \boldsymbol{\rho} \cdot [(\boldsymbol{\omega} \times \boldsymbol{\rho}) \times \boldsymbol{\omega}] \tag{7.13}$$

Now, using the triple vector product vector identity, we get

$$(\boldsymbol{\omega} \times \boldsymbol{\rho})\boldsymbol{\omega} = -\boldsymbol{\omega} \times (\boldsymbol{\omega} \times \boldsymbol{\rho}) = \boldsymbol{\rho}(\boldsymbol{\omega} \cdot \boldsymbol{\omega}) - \boldsymbol{\omega}(\boldsymbol{\omega} \cdot \boldsymbol{\rho})$$

From this relationship, we may write the kinetic energy of the rigid body as

$$T = 1/2\, M\mathbf{v}_{cm} \cdot \mathbf{v}_{cm} + 1/2 \int_M \boldsymbol{\rho} \cdot [\boldsymbol{\rho}(\boldsymbol{\omega} \cdot \boldsymbol{\omega}) - \boldsymbol{\omega}(\boldsymbol{\omega} \cdot \boldsymbol{\rho})]\, dm$$

The last term may be written as

$$1/2 \int_M [(\boldsymbol{\rho} \cdot \boldsymbol{\rho})(\boldsymbol{\omega} \cdot \boldsymbol{\omega}) - (\boldsymbol{\rho} \cdot \boldsymbol{\omega})^2]\, dm \tag{7.14}$$

Expanding this equation results in

$$1/2 \int_M [(x^2 + y^2 + z^2)(\omega_x^2 + \omega_y^2 + \omega_z^2) - (\omega_x x + \omega_y y + \omega_z z)^2]\, dm$$

The mass moments of inertia and the product moments of inertia were defined previously, and the kinetic energy now may be written as

$$T = 1/2\, M\mathbf{v}_{cm} \cdot \mathbf{v}_{cm}$$
$$+ 1/2 \left[I_{xx}\omega_x^2 + I_{yy}\omega_y^2 + I_{zz}\omega_z^2 + 2I_{xy}\omega_x\omega_y + 2I_{yz}\omega_y\omega_z + 2I_{zx}\omega_z\omega_x \right] \tag{7.15}$$

If the kinetic energy is expressed about the principal axes, the product-mass-moment-of-inertia terms are all zero. Again, remember that kinetic energy, power, and work are all scalar quantities, and one cannot independently divide any of these into quantities in coordinate directions, as can be done with vector quantities.

The power equation for a rigid body now can be written in terms of the time rate of change of the kinetic energy as

$$P = \frac{dT}{dt}$$

so that

$$\left(\sum \mathbf{F}\right) \cdot \mathbf{v}_{cm} + \left(\sum \mathbf{M}_{cm}\right) \cdot \boldsymbol{\omega} = \frac{dT}{dt} \tag{7.16}$$

or

$$\sum_i \mathbf{F}_i \cdot \mathbf{v}_i = \frac{dT}{dt}$$

For plane motion about a principal axis perpendicular to the plane of motion (the z-axis if the plane of motion is the xy-plane), the power equation becomes

$$\left(\sum \mathbf{F}\right) \cdot \mathbf{v}_{cm} + \left(\sum M_{cm_z}\right)\omega_z = \frac{d}{dt}\left[1/2\, Mv_{cm}^2 + 1/2\, I_{zz}\omega_z^2\right] \tag{7.17}$$

Treating each force at its point of application, we may write this equation as

$$\sum_i \mathbf{F}_i \cdot \mathbf{v}_i = \frac{d}{dr}\left[1/2\, Mv_{cm}^2 + 1/2\, I_{zz}\omega_z^2\right] \tag{7.18}$$

For plane motion, the concept of work–energy becomes

$$\int_{\bar{\mathbf{r}}_1}^{\bar{\mathbf{r}}_2} \mathbf{F} \cdot d\mathbf{r}_{cm} + \int_{\theta_1}^{\theta_2} M_{cm_z}\, d\theta = \left[1/2\, Mv_{cm}^2 + 1/2\, I_{zz}\omega_z^2\right]_1^2 = T_2 - T_1 \tag{7.19}$$

This equation relates the total work done on the rigid body to the change in the total kinetic energy of the body. Previously, it was shown that the equations of motion for plane motion can be written as

$$\sum \mathbf{F} = M\frac{d\mathbf{v}_{cm}}{dt}$$

$$\sum M_{cm_z} = I_{zz}\frac{d\omega_z}{dt} \tag{7.20}$$

The power of all the forces considered as acting at the center of mass and the power of the moment about the center of mass can be examined by forming the dot product of these two equations with the linear velocity of the center of mass or the angular velocity of the body, yielding

$$\sum \mathbf{F} \cdot \mathbf{v}_{cm} = M\mathbf{v}_{cm} \cdot \frac{d\mathbf{v}_{cm}}{dt} = \frac{d}{dt}\left[1/2\, M\mathbf{v}_{cm} \cdot \mathbf{v}_{cm}\right]$$

$$\sum M_{cm_z}\omega_z = I_{zz}\omega_z\frac{d\omega_z}{dt} = \frac{d}{dt}\left[1/2\, I_{zz}\omega_z^2\right] \tag{7.21}$$

This construction allows the power and work–energy of linear translation to be decoupled from that of rotation when these terms are related to the center of mass. In many problems, working with these equations separately is a great advantage. The pair of work–energy equations becomes

$$\int_{\bar{\mathbf{r}}_{cm_1}}^{\bar{\mathbf{r}}_{cm_2}} \sum \mathbf{F} \cdot d\mathbf{r}_{cm} = \left[1/2\, M\mathbf{v}_{cm} \cdot \mathbf{v}_{cm}\right]_1^2$$

$$\int_{\theta_1}^{\theta_2} M_{cm_z} d\theta = \left[1/2\, I_{zz}\omega_z^2\right]_1^2 \tag{7.22}$$

Note that the previous form of the work–energy relative to the center of mass is the sum of these two equations.

An alternative way to write the work–energy equation is

$$\int_{\bar{r}_1}^{\bar{r}_2} \mathbf{F}_i \cdot d\mathbf{r}_i = \left[1/2\, Mv_{cm}^2 + 1/2\, I_{zz}\omega_z^2\right]_1^2 = T_2 - T_1 \tag{7.23}$$

where each force is handled separately with respect to the work it does. This is the more convenient form of work–energy when there are forces that do no work, such as friction, for example, when a wheel is rolling without sliding. Note that in this form the kinetic energy associated with translation of the center of mass cannot be separated from the kinetic energy associated with rotation of the body. Since the work–energy will yield only one equation for the two unknowns—the linear velocity of the center of mass and the angular velocity of the body—another relationship between these two unknowns is needed. This relationship usually takes the form of a constraint on the motion, such as rolling without sliding or rotation about a fixed axis in space.

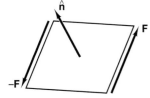

Figure 7.1

7.1.1 WORK DONE BY A COUPLE

The moment of a couple is a free vector, and the work that the couple does can be examined by considering the work done by the two forces forming the couple. A couple is shown in Figure 7.1. If the plane of the couple is translated through a distance $d\mathbf{r}$, no net work is done by the forces making up the couple. If the plane of the couple is rotated by an amount $d\theta$ about an axis perpendicular to the plane of the couple, the work done by the couple is

$$dU = M\, d\theta \tag{7.24}$$

Sample Problem 7.1

A ball rolls down an incline without slipping, as shown in the acccompanying diagram. If the ball is released from rest, determine its velocity after it has rolled a distance s.

Solution   A free-body diagram of the ball is shown here.
The mass moment of inertia is

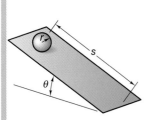

$$I = \frac{2}{5}mr^2$$

The normal and friction forces do no work; only the weight does work, equal to

$$U_{1\rightarrow2} = mgs \sin\theta$$

When the ball is released, its kinetic energy is zero, so the change in kinetic energy equals the final kinetic energy,

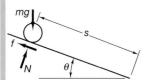

$$T_2 = \frac{1}{2}\left[mv_{cm}^2 + \frac{2}{5}mr^2\omega^2\right]$$

Since the ball is rolling without sliding, the velocity of the contact point with the ground is zero, and the linear and angular velocities are related by

$$v_{cm} = -r\omega$$

Equating the work to the change in kinetic energy yields

$$mgs \sin\theta = \frac{1}{2}mv_{cm}^2\left[1 + \frac{2}{5}\right]$$

$$v_{cm} = \sqrt{\frac{10}{5}gs \sin\theta}$$

7.2   SYSTEMS OF RIGID BODIES AND PARTICLES

When a system of rigid bodies or a system of rigid bodies and particles is analyzed, the rigid bodies may be analyzed separately or the system as a whole analyzed. The principle of work and energy can be applied to each body separately or to the system. When the system is analyzed as a whole, the kinetic energy is the sum of the kinetic energy of each body. The work of all external forces must be included, as must the work due to internal forces, such as spring connections, friction, etc. Other internal forces, such as pin connection forces, forces transmitted through inextensible cords, or meshed-gear forces, occur in pairs equal, opposite, and collinear. If the displacements of the two forces are equal, the work done by these forces is zero, and the work done on the system is due only to external forces.

Sample Problem 7.2

The pulley system shown in the accompanying diagram is released from rest. If the frictional moment on the pulley is $\mathbf{M}_f = 5 \ \text{N} \cdot \text{m} \ \hat{\mathbf{k}}$, determine the velocity of the mass A after it has dropped 0.5 m. The inertial properties of the system are $m_A = 20$ kg, $m_B = 10$ kg, and $I_p = 5 \ \text{kg} \cdot \text{m}^2$, where the radius of the pulley is 400 mm.

Solution

The pulley will be modeled as a rigid body and the two blocks treated as particles in rectilinear motion. The initial kinetic energy of the system is zero, and, treating the system as a whole, we find that the work–energy principle yields

$$M_f\theta \ + \ m_A g x_A \ + \ m_B g x_B \ = \ \frac{1}{2}[I_p\omega^2 \ + \ m_A v_A^2 \ + \ m_B v_B^2]$$

The constraints on the motion are as follows:

$$\theta \ = \ -\frac{x_A}{r}$$

$$\omega \ = \ -\frac{v_A}{r}$$

$$v_B \ = \ -v_A$$

Substituting the given properties of the system yields

$$-5\left(\frac{0.5}{0.4}\right) \ + \ 20(9.81)0.5 \ - \ 10(9.81)0.5 \ = \ \frac{1}{2}\left[5\,\frac{v_A^2}{0.4^2} \ + \ 20v_A^2 \ + \ 10v_A^2\right]$$

$$v_A \ = \ 1.18 \ \text{m/s}$$

Note that the tension in the inextensible cord does no work on the system.

7.3   CONSERVATION OF ENERGY

In Section 3.3, it was shown that if all the forces acting on a particle are conservative forces, the work done during any movement of the particle can be equated to the change in the potential energy of the system. The total energy of the system at any time is

$$E \ = \ V \ + \ T \tag{7.25}$$

where V is the potential energy of the system and T is the kinetic energy of the system. The principle of conservation of the total energy applies to a system of rigid bodies and particles under the influence of only conservative forces and is expressed as

$$\Delta E \ = \ 0 \tag{7.26}$$

Sample Problem 7.3

Solution

A yo-yo is released as shown in the accompanying diagram. Determine its velocity after it has dropped a distance d. The yo-yo has a radius r to the connection at the string, a radius of gyration k_g, and a mass m.

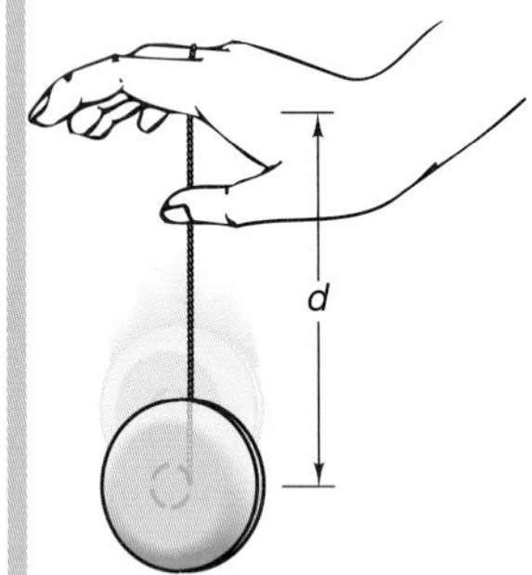

The potential energy of the yo-yo is $V_1 = mgd$ if the reference plane of the gravitational potential is taken to be a distance d below the position at which the yo-yo is released. The initial kinetic energy is zero. The potential energy at the position shown is zero, and the kinetic energy is

$$T_2 = \frac{1}{2}(mk_g^2\omega^2 + mv^2)$$

The constraint on the motion is

$$\omega = -\frac{v}{r}$$

$$T_2 = \frac{1}{2}mv^2\left(\frac{k_g^2 + r^2}{r^2}\right)$$

Since the energy of the system is conserved, it follows that

$$E_1 = V_1 = mgd$$

$$E_2 = T_2 = \frac{1}{2}mv^2\left(\frac{k_g^2 + r^2}{r^2}\right)$$

$$v = \sqrt{2gd\left(\frac{r^2}{k_g^2 + r^2}\right)}$$

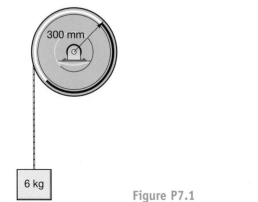

Problems

7.1 A cord is wrapped around a 5-kg, 300-mm-radius pulley and attached to a 6-kg mass as illustrated in Figure P7.1. If the system is released from rest in the position shown, determine the velocity of the mass after it has dropped 1 m.

7.2 A 3-ft-long, 6-lb pendulum is pinned at one end and released from rest in a horizontal position. (See Figure P7.2.) Determine the angular velocity of the pendulum and the reaction on the pin when when the pendulum passes through the vertical position. Neglect friction at the pin connection.

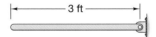

Figure P7.2

7.3 Determine the angular velocity of the pendulum in Problem 7.2 if the friction at the pin causes a constant pin frictional couple of 0.5 lb · ft.

7.4 A 5-kg, 400-mm disk is at rest when it is brought into contact with a belt moving at a constant velocity of 2 m/s,

Figure P7.1

as shown in Figure P7.4. If the coefficient of kinetic friction between the belt and the disk is 0.3, determine the number of revolutions the disk will make before it reaches a constant angular velocity.

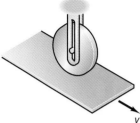

Figure P7.4

7.5 A double pulley has a radius of gyration of 220 mm and a mass of 8 kg. The cable wrapped around the inner radius of 200 mm is attached to a spring with spring constant of 200 N/m. A 6-kg mass is attached to the cable wrapped around the outer radius of 300 mm, and the spring is unstretched when the system is released from rest at the position shown in Figure P7.5. Determine the maximum velocity of the 6-kg mass.

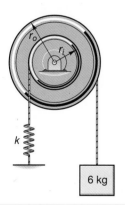

Figure P7.5

7.6 Determine the maximum displacement of the 6-kg mass in Problem 7.5.

7.7 If there is a constant bearing frictional moment of 5 N m acting on the double pulley in Problem 7.5, determine the maximum displacement of the 6-kg mass.

7.8 A uniform bar of mass m is supported by two cords as shown in Figure P7.8. If the rod is released from rest at an angle θ, determine the tension in the cables when the bar is at its lowest position.

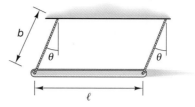

Figure P7.8

7.9 A 0.2-kg yo-yo has a centroidal radius of gyration of 30 mm and is released from rest as shown in Figure P7.9. If the inner radius of the yo-yo is 10 mm and the outer radius is 40 mm, determine the angular velocity of the yo-yo after it has moved down the string a distance of 100 mm.

Figure P7.9

7.10 A ball of mass m is released at an initial angle β inside a bowl of radius R. (See Figure P7.10.) If the radius of the ball, modeled as a sphere, is r, determine an expression for the angular velocity and position of the ball as a function of time. Assume that the ball rolls without slipping.

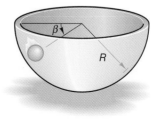

Figure P7.10

7.11 A uniform rod of mass m and length l is released from rest from the position shown in Figure P7.11. The spring has a spring constant k and an unstretched length l. Determine a general equation relating the angular velocity of the rod to its angular position.

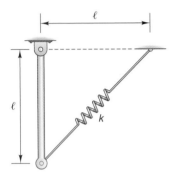

Figure P7.11

7.12 A homogeneous disk of radius r and mass m is released from rest in the position shown in Figure P7.12. Determine the angular velocity of the disk when the line AB of length d is vertical.

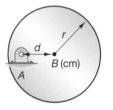

Figure P7.12

7.13 An inverted pendulum of length l and mass m is released from rest at an initial angular position θ_0. (See Figure P7.13.) Determine the angular velocity at any angle θ.

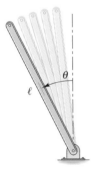

Figure P7.13

7.14 A detailed diagram of the pinned connection of the inverted pendulum in Problem 7.13 is shown in Figure P7.14. If a constant friction force f acts on the pin, determine the angular velocity of the pendulum when it passes the horizontal position.

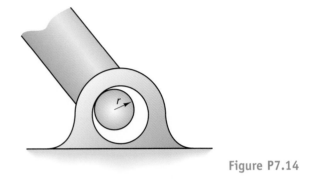

Figure P7.14

7.15 Determine the angular velocity of the inverted pendulum in Problem 7.14 when $\theta = 180°$.

7.16 A hand brake is used to stop a 50-kg disk with a 300-m radius. (See Figure P7.16.) If the initial angular velocity of the disk is 10 rad/s and the force $\mathbf{F} = 200$ N is constant, determine the number of revolutions the disk will turn after the force is applied before the disk comes to rest. The coefficient of kinetic friction between the brake and the disk is 0.3.

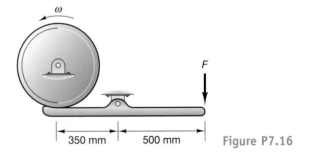

350 mm   500 mm   Figure P7.16

7.17 A bar of length $2R$ and mass m is welded to a disk of radius R and mass m. If the system is released from rest in the position shown in Figure P7.17, determine the angular velocity of the system when the bar passes through the vertical position, and determine the reactions at point A.

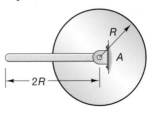

Figure P7.17

7.18 If a constant frictional moment M_f acts at point A on the system in Problem 7.17, determine the angular velocity of the system when the bar passes through the vertical position.

7.19 Determine the maximum angle through which the disk in Problem 7.18 will rotate.

7.20 In Figure P7.20, determine the angular velocity of each of the 100-lb disks of equal radius of 0.4 ft after each disk has rotated 90° if each is released from rest.

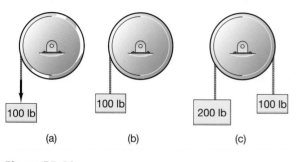

(a)   (b)   (c)

Figure P7.20

7.21 A uniform rod of mass m is mounted on a vertical shaft, and a motor applies a constant moment M to the rod. (See Figure P7.21.) If the rod starts from rest, determine the angular velocity of the rod after three revolutions, and determine the reactions at point A between the rod and the shaft.

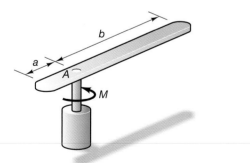

Figure P7.21

7.22 To determine the coefficient of kinetic friction between a shaft on which a 50-kg, 1-m-diameter disk rotates and a bearing, a 40-kg mass is attached to the end of a rope. (See Figure P7.22.) If the frictional moment at the bearing is equal to $M_f = \mu_k R_A r$, where r is the radius of the shaft and R_A is the reaction at the bearing of the shaft, determine the coefficient of kinetic friction if the mass B has a velocity of 5.5 m/s after it falls 1 m.

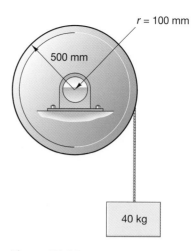

Figure P7.22

7.23 A 5-kg solid disk A rotates on a smooth pin O, as shown in Figure P7.23. The block B has a mass of 2 kg, and the mass of the pulley attached to B is negligible. Determine the angular velocity of the disk if the block is released from rest and moves down 0.5 m.

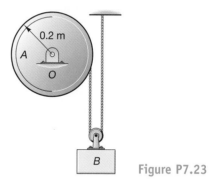

Figure P7.23

7.24 In Figure P7.24, gears A and B have masses of 2 kg and 5 kg and radii of gyration of 100 mm and 400 mm, respectively. If gear A is driven by a moment of 3 N m, determine the number of revolutions that A will turn before it reaches an angular velocity of 10 rad/s.

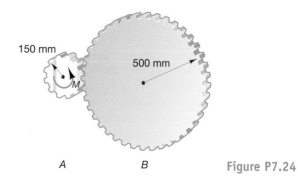

Figure P7.24

7.25 The belt system shown in Figure P7.25 consists of two disks A and B and a belt of negligible mass. Disk A, of mass 2 kg and radius 200 mm, is driven by a moment $M_A = 5$ N·m. Disk B has a mass of 4 kg and a radius of 300 mm, and the constant frictional moment at the bearing is $M_f = 0.5$ N·m. Determine the number of revolutions that A turns to reach an angular velocity of 6 rad/s if the system starts from rest.

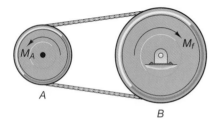

Figure P7.25

7.26 A semicircular disk is released from rest at the position shown in Figure P7.26 and rotates in a vertical plane. The center of mass of the disk is located a distance $d = 4R/(3\pi)$ from the axis of rotation, and the mass moment of inertia about the center of mass is

$$I_{cm_z} = \left(\frac{1}{2} - \frac{16}{9\pi^2}\right)mR^2$$

Determine the angular velocity of the disk as the center of mass reaches its lowest point.

Figure P7.26

7.27 A slender, homogeneous bar with a mass m and a length $2l$ is released from rest when θ is equal to θ_0. The spring is unstretched when the bar is vertical. (See Figure P7.27.) The point A moves in a smooth horizontal guide, and the point B moves in a smooth vertical guide. The spring constant k is such that the bar is in equilibrium when $\theta = 45°$. Determine the spring constant and develop a relationship for the angular velocity of the bar at any angle θ.

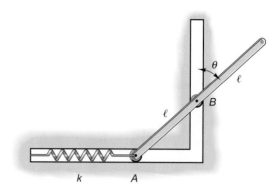

Figure P7.27

7.28 Develop the differential equation for the angular acceleration for the bar in Problem 7.27. Solve the nonlinear differential equation for θ as a function of time for $m = 10$ Kg and $l = 2$ m. Plot the angular velocity versus the angle using both the energy solution from Problem 7.27 and the results of the solution of the differential equation.

7.29 Two identical bars of mass m and length l are released from rest at the position shown in Figure P7.29. Determine the velocity of the center of mass of the bars as they pass through the horizontal position. Neglect friction in the horizontal guide.

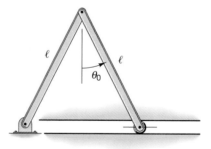

Figure P7.29

7.30 Determine a general expression for the angular velocity of the bar as a function of θ for the system shown in Problem 7.29.

7.31 A bar AB has a length l and a mass m, and a bar BC has a length $2l$ and a mass $2m$. A disk C has a radius R and a mass M and rolls without slipping on a horizontal surface. If the system is released from rest when bar AB is in the horizontal position, as shown in Figure P7.31, determine the angular velocity of the disk when AB passes throught the vertical position.

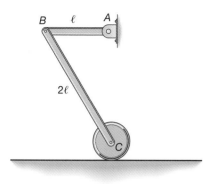

Figure P7.31

🖳 **7.32** Using computational software, create a graph for the linear velocity and angular velocity of link AB in Problem 7.31 for $\theta = 0$ to $180°$ and for system properties of $m = 3$ kg, $l = 400$ mm, $M = 2$ kg, and $R = 100$ mm.

7.33 A crank AB is driven by a constant moment of $M_{AB} = 3$ N·m. If the system starts from rest at the position shown in Figure P7.33, determine the angular velocity of member CD after the crank has completed one revolution. The mass per unit length of each of the bars is 20 kg/m. (*Hint: The kinetic energy of bars AB and CD is $T = \frac{1}{2}I_0\omega^2$, where the mass moment of inertia is related to the fixed point on the bar, but the kinetic energy of BC is $T = \frac{1}{2}(I_{cm}\omega^2 + mv_{cm}^2)$.*)

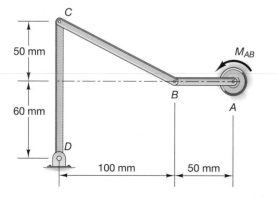

Figure P7.33

🖳 **7.34** Determine the angular velocity of CD in Problem 7.33 after the crank AB has revolved through an angle θ.

7.35 A 20-kg, 300-mm-radius disk rolls without slipping down an incline onto a circular path as shown in Figure P7.35. Determine the normal force under the disk as it passes point B if it is released from rest at point A.

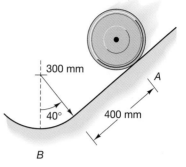

Figure P7.35

7.36 A 10-kg, 100-mm-radius disk is suspended by a cord and a spring, with a spring constant of 60 N/m. (See Figure P7.36.) If the spring is unstretched when the system is released from rest, determine the linear velocity of the disk after it has dropped 500 mm.

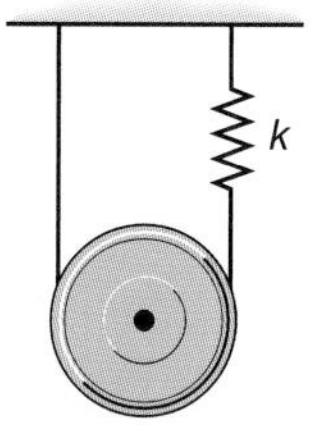

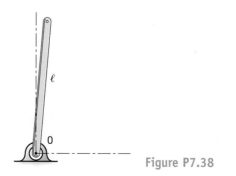

Figure P7.36

7.37 If the spring in Problem 7.36 is stretched 100 mm when the system is released, determine the linear velocity of the disk after it has dropped 400 mm.

7.38 A bar of mass m and length l is released from an almost vertical position, as shown in Figure P7.38. Determine the spring constant of a torsional spring at the pin O which will ensure that the angular velocity of the bar is zero when the bar reaches a horizontal position.

Figure P7.38

7.39 A 4-kg disk O is driven by a constant moment $M = 15$ Nm and is connected to a 3-kg piston by a 2-kg connecting rod AB. If the spring constant is $k = 500$ N/m, determine the angular velocity of the connecting rod after the disk rotates one-fourth of a revolution if it starts from rest at the postion shown in Figure P7.39.

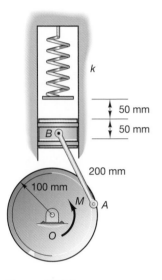

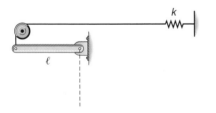

Figure P7.39

7.40 Determine the velocity of the piston in Problem 7.39 after the disk has rotated through one complete revolution.

7.41 Determine the angular velocity of the homogeneous bar shown in Figure P7.41 when the bar passes through the vertical position. The bar is released from rest in the horizontal position when the spring is unstretched.

Figure P7.41

7.42 Two bars are homogeneous and have a mass of ρ kg/unit length. Determine the angular velocity of the bars when they become colinear if the system is released from rest in the position shown in Figure P7.42.

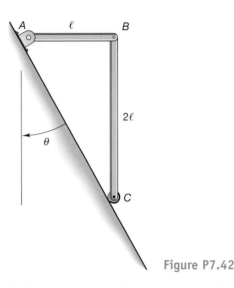

Figure P7.42

7.43 A homogeneous bar of mass m is released from rest when the bar is horizontal and the spring is unstretched. (See Figure P7.43.) Determine the velocity of the bar at any angle θ.

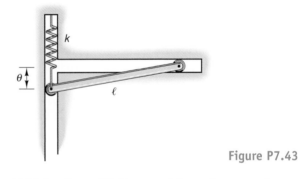

Figure P7.43

7.44 In Figure P7.44, gear A is stationary and gear B rotates around it in a vertical plane. The gears, connected by a 5-kg bar of length 200 mm, have a mass of 10 kg, a radius of 100 mm, and a radius of gyration about their center of mass of 70 mm. If the system is released from rest when the bar is horizontal, determine the angular velocity of gear B as a function of the angle θ.

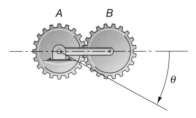

Figure P7.44

7.45 A small gear A drives two identical larger gears B and C as shown in Figure P7.45. A constant torque M is applied to the small gear as the system starts from rest. Determine an expression for the angular velocity of the larger gears in terms of the constant moment, the mass moment of inertia of the gears and the radii of the gears.

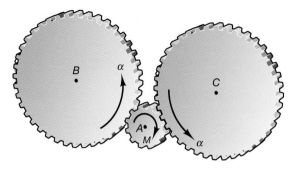

Figure P7.45

7.4 IMPULSE AND MOMENTUM

In Section 3.5, we saw that the first integral of the vector equation of linear momentum with respect to time yielded the principle of impulse and momentum for a particle. In Chapter 4, we extended these ideas to a system of particles and defined the angular momentum relative to the center of mass of the system and about a fixed point is space. The concept of impulse is of most value when the force is known as a function of time or when the force acts over a short period of time, as in an impact between two particles or rigid bodies. The linear and angular momentum for a rigid body in plane motion are, respectively,

$$\mathbf{L} = m\mathbf{v}_{cm} = m\dot{x}_{cm}\hat{\mathbf{i}} + m\dot{y}_{cm}\hat{\mathbf{j}}$$

$$\mathbf{H}_{cm} = I_{cm_z}\omega_z\hat{\mathbf{k}} \tag{7.27}$$

The angular momentum about a fixed point in space is defined as

$$\mathbf{H}_0 = I_{cm_z}\omega_z\hat{\mathbf{k}} + \mathbf{r}_{cm/0} \times m\mathbf{v}_{cm} \tag{7.28}$$

Equation (7.28) can be written as

$$\mathbf{H}_0 = I_{cm_z}\omega_z\hat{\mathbf{k}} + \mathbf{r}_{cm/0} \times m(\boldsymbol{\omega} \times \mathbf{r}_{cm/0}) \tag{7.29}$$

We will use the triple vector product identity to write the second term on the right side of Eq. (7.29) as

$$\mathbf{r}_{cm/0} \times (m\omega_z\hat{\mathbf{k}} \times \mathbf{r}_{cm/0}) = m\omega_z\left[\hat{\mathbf{k}}(\mathbf{r}_{cm/0} \cdot \mathbf{r}_{cm/0}) - \mathbf{r}_{cm/0}(\mathbf{r}_{cm/0} \cdot \hat{\mathbf{k}})\right] \tag{7.30}$$

The second term on the right side of Eq. (7.30) is zero for plane motion, and Eq. (7.29) can be written as

$$\mathbf{H}_0 = \left[I_{cm_z} + m(\mathbf{r}_{cm/0} \cdot \mathbf{r}_{cm/0})\right]\omega_z\hat{\mathbf{k}} = I_0\omega\hat{\mathbf{k}} \tag{7.31}$$

The mass moment of inertia about an axis in the z-direction passing through the point 0 is defined by the parallel-axis theorem (see Appendix A) as

$$I_{0z} = I_{cm_z} + mr_{cm/0}^2 \tag{7.32}$$

The equations of motion are

$$\sum \mathbf{F} = \frac{d}{dt}(m\mathbf{v}_{cm})$$

$$\sum M_{cm} = \frac{d\mathbf{H}_{cm}}{dt}$$

(7.33)

If there is a point on the body or body extended that is fixed in space, the angular momentum equation can be written relative to this point:

$$\sum \mathbf{M}_0 = \frac{d\mathbf{H}_0}{dt}$$

(7.34)

The first integral of Eq. (7.33) yields the following two vector equations for the linear and angular impulse and momentum:

$$\int_{t_1}^{t_2} \sum \mathbf{F} dt = m\mathbf{v}_{cm_2} - m\mathbf{v}_{cm_1}$$

$$\int_{t_1}^{t_2} \mathbf{M}_{cm} dt = \mathbf{H}_{cm_2} - \mathbf{H}_{cm_1}$$

(7.35)

We can express the latter equation as

$$\int_{t_1}^{t_2} \mathbf{M}_0 dt = \mathbf{H}_{0_2} - \mathbf{H}_{0_1}$$

Equations (7.35) are called the ***principle of linear and angular impulse and momentum for a rigid body.*** If no external force or moment is applied to the rigid body, linear and angular momentum are conserved.

Sample Problem 7.4

A ball is thrown with an initial velocity v_0 parallel to a rough plane, as shown in the accompanying figure. (See also Sample Problem 6.7.) The initial angular velocity is zero. Using the principle of impulse and momentum, determine the time at which the sphere will roll without sliding.

Solution

Initially, the ball will slide, and it will not begin to roll until its angular velocity is $\omega = -v_{cm_x}/r$. The free-body diagram of the ball is reproduced from Sample Problem 6.7. The linear and angular impulse equations are

$$-mg \sin \theta \Delta t - f\Delta t = m(v - V_0)$$

$$N\Delta t - mg \cos \theta \Delta t = 0$$

$$-fr\Delta t = \frac{2}{5} mr^2(\omega - 0)$$

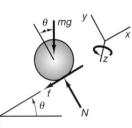

While the ball rolls and slides, $f = \mu_k N = \mu_k mg \cos \theta$.
Solving for Δt when $\omega = -v_{cm_x}/r$ yields

$$\Delta t = \frac{V_0}{g\left(\sin \theta + \frac{7}{2}\mu_k \cos \theta\right)}$$

Sample Problem 7.5

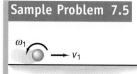

A hard rubber ball is modeled with high skin tension such that its elastic properties and the surface friction make impact elastic and slip negligible. The ball, marketed as a "super ball," displays some unusual motions when thrown with an initial velocity and spin. Referring to the accompanying diagram, determine the general expressions for the spin and velocity of the ball after each bounce.

Solution

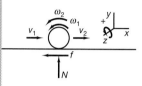

Let v_1 be the initial velocity in the x-direction and ω_1 be the inital angular momentum. A free-body diagram of the ball is as shown. During impact, the linear impulse and momentum in the x-direction are given by

$$-\int f\,dt = mv_2 - mv_1$$

The angular impulse and momentum are given by

$$-r\int f\,dt = I(\omega_2 - \omega_1) = \frac{2}{5}mr^2(\omega_2 - \omega_1)$$

Eliminating the impulse from the two equations yields

$$v_2 = \frac{2}{5}r(\omega_2 - \omega_1) + v_1$$

If the energy is conserved, we have

$$\frac{2}{5}mr^2\omega_1^2 + mv_1^2 = \frac{2}{5}mr^2\omega_2^2 + mv_2^2$$

Solving the two equations for the linear and angular velocities after impact yields

$$v_2 = \frac{1}{7}\left[3v_1 - 4\omega_1 r\right]$$

$$\omega_2 = -\frac{1}{7}\left[3\omega_1 + \frac{10v_1}{r}\right]$$

Now consider a special repeatable motion with the following initial conditions:

$$v_1 = v_1$$

$$\omega_1 = \frac{3}{4}\frac{v_1}{r}$$

The velocities after the next two bounces are

$$v_2 = 0$$

$$\omega_2 = -\frac{7}{4}\frac{v_1}{r}$$

$$v_3 = v_1$$

$$\omega_3 = \frac{3}{4}\frac{v_1}{r}$$

and the motion repeats itself. The motion of the "super ball" is shown in the diagram. There are many other interesting motions of such a ball.

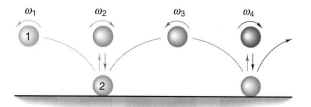

| **Sample Problem 7.6** | Two disks are brought into contact with one another when disk A is rotating with an initial angular velocity of ω_{A_0} and disk B is at rest, as shown in the accompanying diagram. The friction between the disks will give an impulse to both disks until a steady state is reached. Determine the angular velocities of the disks at steady state. |

Solution   Free-body diagrams of the disks are as shown. The angular impulse–momentum equations for the two disks are

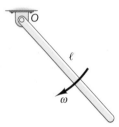

$$-r_A \int f\, dt = I_A(\omega_A - \omega_{A_0}) = \frac{1}{2} m_A r_A^2 (\omega_A - \omega_{A_0})$$

$$-r_B \int f\, dt = I_B(\omega_B - 0) = \frac{1}{2} m_B r_B^2 (\omega_B - 0)$$

The steady state will be reached when the linear velocities of the point of contact on the two disks are equal. These velocities are

$$\mathbf{v}_{PA} = \omega_A \hat{\mathbf{k}} \times (-r_A \hat{\mathbf{j}}) = r_A \omega_A \hat{\mathbf{i}}$$

$$\mathbf{v}_{PB} = \omega_B \hat{\mathbf{k}} \times (r_B \hat{\mathbf{j}}) = -r_B \omega_B \hat{\mathbf{i}}$$

Setting the two equations equal to each other gives

$$r_A \omega_A = -r_B \omega_B$$

Eliminating the impulse and solving for the steady–state angular velocities yields

$$\omega_A = \left(\frac{m_A}{m_A + m_B}\right)\omega_{A_0}$$

$$\omega_B = -\frac{r_A}{r_B}\left(\frac{m_A}{m_A + m_B}\right)\omega_{A_0}$$

Problems

7.46 A bar of length l and mass m is pinned at one end, as shown in Figure P7.46. If the angular velocity of the bar is ω, determine the angular momentum of the bar about the center of mass and about point O.

7.47 A disk of mass m and radius r is pulled by a yoke of negligible mass, as shown in Figure P7.47. If the force $\mathbf{P}$ is constant, determine the linear velocity of the disk as a function of time if the disk starts from rest and rolls without slipping on the horizontal surface.

Figure P7.46

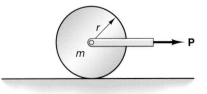

Figure P7.47

7.48 If the mass of the disk shown in Figure P7.47 is 10 kg and the radius is 400 mm, determine the angular velocity of the disk after 2 s if $\mathbf{P} = 15 \sin(\pi t/2)$ N if the disk is at rest when the force is applied.

7.49 The identical uniform disks in the two systems shown in Figure P7.49 have masses of 20 kg and radii of 800 mm. If the systems are released from rest and friction is ignored, determine the angular velocities of the disks after 5 s.

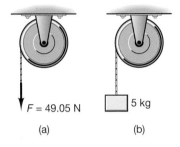

$F = 49.05$ N

(a)

5 kg

(b)

Figure P7.49

7.50 A belt drive system consists of a 25-kg disk A and a 10-kg disk B, as shown in Figure P7.50. If the disk B is driven by a constant moment $M = 50$ N·m when it starts from rest, determine its angular velocity after 4 s. Neglect the mass of the belt.

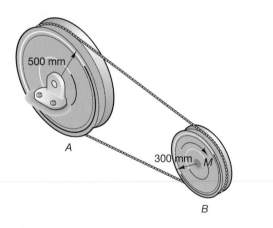
500 mm

A

300 mm M

B

Figure P7.50

7.51 A 50-kg diver leaves a platform in a laid-out position with an initial angular velocity of $\omega_0 = 2\pi/3$ rad/s, as illustrated in Figure P7.51. Determine her angular velocity in a pike position and in a tuck position. Approximate the moment of inertia in the laid-out position as a uniform cylinder of length 2.2 m, in the pike position as a uniform cylinder of length 1.1 m, and in the tuck position as as sphere of radius 0.5 m.

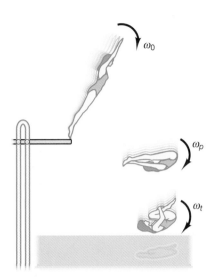
ω_0

ω_p

ω_t

Figure P7.51

7.52 A 30-kg mass slides out on a 2-m rod with a mass of 5 kg. (See Figure P7.52.) The rod rotates in a horizontal plane, and the mass is released from the inner position when the rod is rotating about a vertical shaft with an initial angular velocity of 10 rad/s. Determine the angular velocity of the system as a function of the position x of the sliding mass.

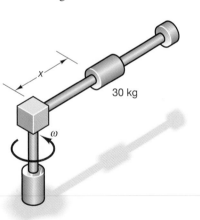
x

30 kg

ω

Figure P7.52

7.53 Determine the initial conditions for the "super ball" in Sample Problem 7.5 such that the horizontal distance traveled in the second bounce will be one-half the distance traveled before the initial bounce.

7.54 A uniform disk of mass m and radius r is released from rest at time $t = 0$ on an inclined plane and rolls without slipping. (See Figure P7.54.) Determine an expression for the velocity of the center of the disk with time t.

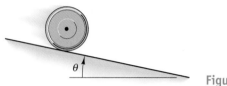

Figure P7.54

7.55 A cord of negligible mass is wrapped around a cylinder of radius r and mass m, as shown in Figure P7.55. If the system is released from rest at time $t = 0$, determine an expression for the velocity of the center of the cylinder as a function of time.

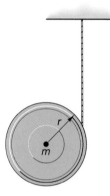

Figure P7.55

7.56 A 5-ft 2-in., 110-lb figure skater starts her spin with her arms extended to the sides. (See Figure P7.56.) She increases her angular velocity by bringing her arms in close. Suppose that each arm weighs 8 lb and is 2 ft long. The radius of her body when her arms are in close is 8 in. and does not vary that much when the arms are

extended. Model the skater's body as an 8-in.-radius, 94-lb cylinder with two 2-ft cylinders when her arms are extended and an 8-in., 110-lb cylinder when her arms are in close. If her initial angular velocity was 2.5 rad/s, determine her final angular velocity if friction with the ice is ignored.

7.57 A homogeneous sphere of radius r and mass m is thrown along an inclined plane as shown in Figure P7.57. The sphere is given an initial linear velocity of v_0 and no angular velocity. If the coefficient of kinetic friction between the sphere and the surface is μ_k, determine the time that elapses before the sphere stops slipping. What is the linear velocity of the center of the sphere and the angular velocity at the time it stops slipping?

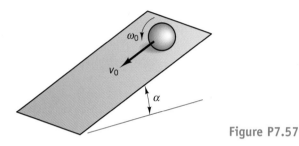

Figure P7.57

7.58 Determine the time that elapses for the sphere in Figure P7.57 before it stops slipping if its initial linear velocity is zero and its initial angular velocity is ω_0. Also, determine the linear velocity of the center of the sphere and its angular velocity at the time it stops slipping.

7.59 A belt passes over two homogeneous pulleys without slipping, as shown in Figure P7.59. Pulley A has a mass of 20 kg and a radius of 300 mm, pulley B has a mass of 10 kg and a radius of 200 mm, and the constant tension in the belt as it goes into pulley A is 100 N. If the system starts from rest and after 0.5 s, the velocity of the belt is 3 m/s, determine the constant tension on the belt as it leaves pulley B. Also, determine the tension in the portion of the belt between the two pulleys.

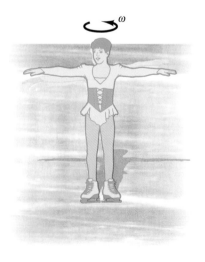

Figure P7.56

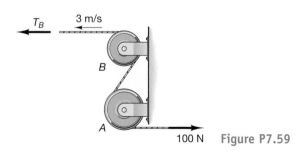

Figure P7.59

7.60 For the system shown in Figure P7.59, determine the velocity after 2 s if the system starts from rest and the tension T_B is the constant value of 200 N.

7.61 A 150-kg, 200-mm-radius homogeneous disk is rotating at a constant angular velocity of 10 rad/s when the brake force $P(t) = 20(1 + \sin \pi t)$ N is applied as shown in Figure P7.61. Determine the angular velocity of the disk after 2 s if the coefficient of kinetic friction between the brake and the disk is 0.2. Neglect the mass of the brake bar.

7.62 Determine the angular velocity of the disk in Problem 7.61 after 3 s if the braking force is given by $P(t) = 10t + t^3$ N.

7.63 A system consists of two identical 150-mm, 6-kg homogeneous disks. (See Figure P7.63.) Determine the angular velocity of disk A 3 s after the system is released from rest.

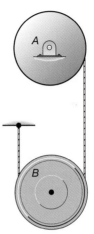

Figure P7.63

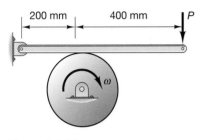

Figure P7.61

7.5 ECCENTRIC IMPACT ON A SINGLE RIGID BODY

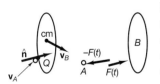

Figure 7.2

In Section 3.6, we examined the concept of a central impact between two particles and, using the principle of linear momentum and an empirical constant, called the coefficient of restitution, analyzed the postimpact dynamics of the particles. We then applied the analysis to billiard balls, racket balls, and the reconstruction of an accident. In the current section, we will consider first the case when a single rigid body is given an impact in such a manner that the impulsive force of the impact does not pass through the center of mass of the body. This is called an *eccentric impact*. In Section 7.6, we will broaden our study of eccentric impacts to the case of an impact between two bodies.

Accordingly, consider the impact of a particle of mass m_A with a rigid body of mass m_B, as shown in Figure 7.2. We will let $\mathbf{v}_A$ be the initial velocity of the particle and $\mathbf{v}_B$, be the initial velocity of the center of mass of the rigid body. The velocity of the point of contact Q on the rigid body is

$$\mathbf{v}_{QB} = \mathbf{v}_B + \boldsymbol{\omega}_B \times \mathbf{r}_{Q/\text{cm}B} \tag{7.36}$$

where $\boldsymbol{\omega}_B$ is the initial angular velocity of the rigid body and $\mathbf{r}_{Q/\text{cm}B}$ is the relative position vector from the center of mass of the rigid body to the point of contact, Q. The velocity of the particle at the point of contact is its initial velocity. A normal vector $\hat{\mathbf{n}}$ is taken perpendicular to the surface of the rigid body at the point of contact. We will break the initial velocity of the particle and the velocity of point Q on the rigid body into components in the normal and tangential directions of the point of contact. The components of the velocities in the tangential direction will be unaffected by the impact, and we will relate the relative approach velocity in the normal direction to the relative separation velocity in the normal direction by the coefficient of restitution e according to the equation

$$(\mathbf{v}'_{QB})_n - (\mathbf{v}'_A)_n = e[(\mathbf{v}_A)_n - (\mathbf{v}_{QB})_n] \tag{7.37}$$

where the primed velocities are the postimpact velocities and the unprimed velocities are the preimpact velocities.

Now, let us consider the impulse–momentum equations for the particle and the rigid body. The linear impulse–momentum equations are

$$\int_{t_1}^{t_2} (-F\hat{\mathbf{n}})dt = m_A(\mathbf{v}'_A - \mathbf{v}_A)$$

$$\int_{t_1}^{t_2} (F\hat{\mathbf{n}})dt = m_B(\mathbf{v}'_B - \mathbf{v}_B) \tag{7.38}$$

The angular impulse–momentum equation for the rigid body B is

$$\mathbf{r}_{Q/cmB} \times \int_{t_1}^{t_2} (F\hat{\mathbf{n}})dt = I_{Bcm}(\omega'_B - \omega_B)\hat{\mathbf{k}} \tag{7.39}$$

The impulse–momentum equation in the n-direction can be found from Eq. (7.38) by taking the dot product of both sides with the unit vector $\hat{\mathbf{n}}$. We obtain

$$\int_{t_1}^{t_2} -Fdt = m_A(v'_{An} - v_{An})$$

$$\int_{t_1}^{t_2} Fdt = m_B(v'_{Bn} - v_{Bn}) \tag{7.40}$$

The equivalent scalar angular impulse–momentum equation can be obtained from Eq. (7.39) by taking the dot product of both sides with the unit vector $\hat{\mathbf{k}}$, yielding

$$X_B \int_{t_1}^{t_2} Fdt = I_{Bcm}(\omega'_B - \omega_B)$$

where

$$X_B = \hat{\mathbf{k}} \cdot (\mathbf{r}_{Q/cmB} \times \hat{\mathbf{n}}) \tag{7.41}$$

If we eliminate the impulsive force from Eqs. (7.40) and (7.41), we obtain the conservation-of-momentum equations for the system:

$$m_B v'_{Bn} + m_A v'_{An} - m_B v_{Bn} - m_A v_{An} = 0$$

$$X_B[m_B(v'_{Bn} - v_{Bn})] = I_{Bcm}(\omega'_B - \omega_B) \tag{7.42}$$

Eq. (7.36) can be used in Eq.(7.37) to express the normal component of the velocity of point Q on the rigid body in terms of the velocity of the center of mass and the angular velocity of the body:

$$(v_{QB})_n = v_{Bn} + \hat{\mathbf{n}} \cdot (\omega_B \hat{\mathbf{k}} \times \mathbf{r}_{QcmB}) \tag{7.43}$$

The scalar triple product satisfies the identity

$$X_B = \hat{\mathbf{k}} \cdot (\mathbf{r}_{Q/cmB} \times \hat{\mathbf{n}}) = \hat{\mathbf{n}} \cdot (\hat{\mathbf{k}} \times \mathbf{r}_{Q/cmB}) \tag{7.44}$$

Therefore, Eq. (7.43) may be written as

$$(v_{QB})_n = v_{Bn} + X_B \omega_B \tag{7.45}$$

Now Eq. (7.37) can be written as

$$v'_{Bn} + X_B\omega'_B - v'_{An} = e(v_{An} - v_{Bn} - X_B\omega_B) \tag{7.46}$$

If the initial velocity of the particle is known and the initial linear and angular velocities of the rigid body are known, Eqs. (7.42) and (7.46) form a system of three equations for the three unknown normal components of the postimpact velocities. The tangential components of the linear velocities of the particle and the rigid body are unchanged during the impact.

If there is a fixed point on the rigid body in space—that is, if the rigid body is pinned—the impact equations can be simplified by expressing the conservation of angular momentum about the fixed point. Since the velocity of that point is zero, the velocity of the impact point in the normal direction can be written as

$$(v_{QB})_n = X_{B0}\omega_B \tag{7.47}$$

where

$$X_{B0} = \hat{\mathbf{k}} \cdot (\mathbf{r}_{Q/0} \times \hat{\mathbf{n}})$$

Equation (7.46) can be written as

$$X_{B0}\omega'_B - v'_{An} = e(v_{An} - X_{B0}\omega_B) \tag{7.48}$$

The angular momentum of the system is conserved about the point 0, and the conservation equation is

$$I_{B0}\omega'_B + X_{B0}m_A v'_{An} = I_{B0}\omega_B + X_{B0}m_A v_{An} \tag{7.49}$$

where I_{B0} is the mass moment of inertia about the fixed point 0. For this case, Eqs. (7.48) and (7.49) form a system of two equations for the postimpact normal component of the velocity of the particle and the postimpact angular velocity of the rigid body, respectively.

Sample Problem 7.7

A homogeneous rod of mass 5 kg and length 2 m rests on a horizontal smooth table when it is hit at its end by a particle of mass 1 kg moving at a velocity of 3 m/s perpendicular to the rod. Determine the final linear velocities of the center of mass of the rod and the particle and the angular velocity of the rod if the collision is elastic.

Solution

The problem can be modeled as shown in the figure. The initial velocities of the mass and the bar are

$$\mathbf{v}_A = 3\hat{\mathbf{i}} \qquad \mathbf{v}_B = 0 \qquad \omega_B = 0$$

Since the normal direction of the contact is the *x*-direction, conservation of momentum for the system in that direction yields

$$m_A\mathbf{v}_A + m_B\mathbf{v}_B = m_A\mathbf{v}'_A + m_B\mathbf{v}'_B$$

$$1(3) = v'_{Ax} + 5v'_{Bx}$$

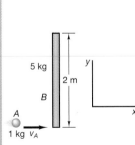

Note that only the scalar momentum equation in the *x*-direction has been written, as the velocities in the tangential direction are unchanged by the collision and therefore, are zero. Conservation of angular momentum of the system about the center of mass of the rod yields

$$\frac{l}{2}m_A v_{Ax} + I_B\omega_B = \frac{l}{2}m_A v'_{Ax} + I_B\omega'_B$$

$$(1)(1)(3) = (1)(1)v'_A + \frac{1}{12}(5)(2)^2\omega'_B$$

The final equation relates the relative approach velocity at the point of impact to the relative separation velocity by the coefficient of restitution:

$$v'_{Bx} + \frac{l}{2}\omega'_B - v'_A = e\left(v_A - v_{Bx} - \frac{l}{2}\omega_B\right)$$

$$v'_{Bx} + \omega'_B - v'_A = 1(3)$$

The system of three equations for the three unknown postimpact velocities may be written in matrix notion as

$$\begin{bmatrix} 1 & 5 & 0 \\ 1 & 0 & \frac{5}{3} \\ -1 & 1 & 1 \end{bmatrix}\begin{bmatrix} v'_{Ax} \\ v'_{Bx} \\ \omega'_B \end{bmatrix} = \begin{bmatrix} 3 \\ 3 \\ 3 \end{bmatrix}$$

The equations are solved easily by matrix inversion, yielding

$$\begin{bmatrix} v'_{Ax} \\ v'_{Bx} \\ \omega'_B \end{bmatrix} = \begin{bmatrix} -0.333 \text{ m/s} \\ 0.667 \text{ m/s} \\ 2 \text{ rad/s} \end{bmatrix}$$

A general analytic solution of the problem could be formed symbolically for any masses, length of the rod, coefficient of restitution, and point of impact on the rod.

Sample Problem 7.8

A homogeneous rod of length l and mass m_A is pinned at the top and hangs at rest in a vertical position when it is struck by a mass m_B moving in a horizontal direction with a velocity of v_B, as shown in the accompanying diagram. Determine the postimpact linear velocity of the mass and the linear velocity and angular velocity of the rod if the mass strikes the bar at midlength and the coefficient of restitution is e.

Solution   Conservation of angular momentum about the fixed pivot yields

$$\frac{l}{2}m_B v_B = \frac{l}{2}m_B v'_B + \frac{1}{3}m_A l^2 \omega'_A$$

The separation velocities can be related to the approach velocities by

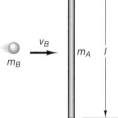

$$\frac{l}{2}\omega'_A - v'_B = e v_B$$

The postimpact velocities can be determined symbolically in matrix notation as

$$\begin{bmatrix} v'_B \\ \omega'_A \end{bmatrix} = \begin{bmatrix} \frac{l}{2}m_B & \frac{1}{3}m_A l^2 \\ -1 & \frac{l}{2} \end{bmatrix}^{-1}\begin{bmatrix} \frac{l}{2}m_B v_B \\ e v_B \end{bmatrix}$$

Solution of the matrix equation yields

$$\begin{bmatrix} v'_B \\ \omega'_A \end{bmatrix} = \begin{bmatrix} -\dfrac{(4m_A e - 3m_B)v_B}{(4m_A + 3m_B)} \text{ m/s} \\ \\ \dfrac{(1 + e)6m_B v_B}{l(4m_A + 3m_B)} \text{ rad/s} \end{bmatrix}$$

The impulsive force at the pin can be determine from the impulse–momentum equation in the *x*-direction:

$$\int F_{0x}\, dt - \frac{l}{2} m_A \omega_A' + m_B v_B' - m_B v_B = -\frac{m_A m_B (1 + e)}{(4m_A + 3m_B)}$$

Note that the impulsive force is negative, indicating that it acts in the negative *x*-direction. The impulsive force for this case is twice as large for an elastic impact compared to a plastic impact.

Problems

7.64 A long, slender, homogeneous bar of mass m_A hangs at rest hinged at the top when it is struck at a distance c below the hinge point by a particle of mass m_B moving in a horizontal direction with a velocity of v_B. (See Figure P7.64.) Determine the postimpact linear velocity of the particle and the postimpact angular velocity of the bar in terms of the mass of the particle and the bar, respectively, and the length of the bar, the coefficient of restitution, and the distance c.

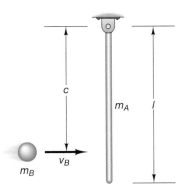

Figure P7.64

7.65 For the bar shown in Figure P7.64, determine the distance c such that the impulsive force at the hinge will be zero. (This point is called the *center of percussion*.)

7.66 A 20-kg, 200-mm-radius homogeneous disk hangs at rest from a pinned connection as shown in Figure P7.66. If the disk is struck by a 2-kg object moving 2 m/s upward at an angle of 30° to the horizontal and the coefficient of restitution for the impact is 0.3, determine the postimpact angular velocity of the disk.

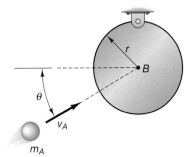

Figure P7.66

7.67 In Problem 7.66, determine the impulsive reactions at the pinned connection.

7.68 A homogeneous disk of mass m_B and radius r rests on a horizontal surface when it is struck by a mass m_A moving in the horizontal direction at a velocity v_A along a trajectory at a distance $3r/2$ above the surface, as shown in Figure P7.68. If there is sufficient friction such that the disk does not slip during impact, determine the postimpact angular velocity of the disk for a coefficient of restitution of e.

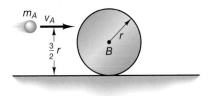

Figure P7.68

7.69 Determine the frictional impulsive force between the disk and the horizontal surface in Problem 7.68.

7.70 A homogeneous bar of mass m_A and length l lies at rest on a smooth horizontal surface when it is struck by a mass m_B moving at a velocity v_B along a line that makes an angle θ to the axis of the bar. If the impact occurs at a distance c from the end of the bar, as shown in Figure P7.70, and the coefficient of restitution is e, determine a general expression for the postimpact linear velocity of the mass B and the linear and angular velocities of the bar.

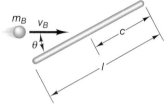

Figure P7.70

7.71 Determine the postimpact velocities of the bar and mass in Problem 7.70 for the following conditions: $m_A = 10$ kg, $l_A = 1.5$ m, $m_B = 2$ kg, $c = 1$ m, $\theta = 65°$, $e = 0.3$, and $v_B = 3$ m/s.

7.72 For the conditions given in Problem 7.71, determine a position on the bar or along the extended axis of the bar that has zero velocity at the instant after impact.

7.73 A homogeneous bar of mass m and length l is pinned at the top and released from a horizontal position. (See Figure P7.73.) Determine the postimpact angular velocity of the bar if it hits a stop at the midlength when in the vertical position. Take the coefficient of restitution to be e.

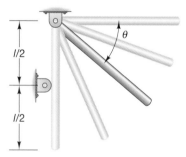

Figure P7.73

7.74 Determine the general solution for Problem 7.73 if the stop is placed at a distance c from the hinge point.

7.75 Determine the distance c in Problem 7.74 such that there will be no impulse at the hinge.

7.76 A uniform rod of length l and mass m is released from rest at a height h above the corner of the table, as shown in Figure P7.76. If the end of the rod strikes the corner of the table and the coefficient of restitution of the impact is e, determine an expression for the postimpact velocity of the center of mass of the rod, the angular velocity of the rod, and the impulse between the rod and the table.

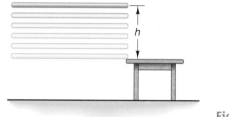

Figure P7.76

7.77 Determine numerical values for the postimpact velocities of the rod in Problem 7.76 if $l = 1.5$ m, $m = 4$ kg, and $h = 2$ m. Investigate the dependency of the postimpact velocities on the coefficient of restitution by varying e from 0 to 1.

7.78 The velocity of the center of mass of the uniform rod shown in Figure P7.78 when it impacts the inclined plane is v. If the length of the bar is l and the mass of the bar is m, determine the postimpact linear and angular velocities of the rod in terms of the coefficient of restitution e.

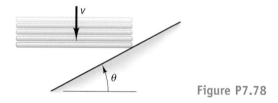

Figure P7.78

7.79 In Problem 7.78, if $v = 5$ m/s, $l = 0.6$ m, $m = 2$ kg, and $e = 0.8$, determine the postimpact velocities for a range of inclined-plane angles from 5° to 60°. Plot the results as a function of the angle.

7.80 A uniform disk of radius r and mass m impacts the corner of a table with a vertical velocity v, as shown in Figure P7.80. If the coefficient of restitution is e, determine the postimpact velocity of the disk.

Figure P7.80

7.81. In Problem 7.80, investigate the dependency of the impact impulse on the impact postion given by the angle β, which varies from $0°$ to $80°$, for the numerical values $r = 300$ mm, $m = 3$ kg, $v = 2$ m/s, and $e = 0.3$.

7.82 The "super ball" examined in Sample Problem 7.5 is thrown against an inclined plane as shown in Figure P7.82. Determine the postimpact linear and angular velocities of the ball.

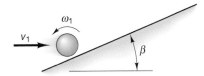

Figure P7.82

7.6   ECCENTRIC IMPACT

When two bodies that cannot be modeled as particles undergo an eccentric impact (i.e., the impulsive forces do not pass through the center of mass of the bodies), the situation presents special problems. Although the equations of linear impulse and momentum and angular impulse and momentum are applicable, they are insufficient to determine the final linear and angular velocities of the contacting bodies. The introduction of the coefficient of restitution was an empirical method to estimate the energy loss during the impact. If the coefficient of restitution is unity, the impact is elastic and no energy is lost. If the coefficient of restitution is zero, the impact is plastic and the two bodies move as one after the impact. These two extremes are applicable in only a limited number of cases, and therefore, many texts do not present a development of eccentric impact. In Section 7.5, we examined the eccentric impact between a particle and a single rigid body.

The single biggest application of eccentric impact between two bodies is the reconstruction of traffic accidents. The number of software programs currently being used by police organizations, engineers, and others who reconstruct accidents in order to analyze the large number of motor vehicle accidents that occur in the United States makes a discussion of the topic of eccentric impact necessary in any text on dynamics. The application of impulse and momentum principles to motor vehicle accidents is more difficult than that of eccentric impact on rigid bodies, as the bodies are not rigid and deform during the collision. The large amount of damage that the vehicles incur absorbs most of the energy of the collision. The difficulty that arises is estimating the coefficient of restitution. In the case of minimum damage to the vehicles, this coefficient is close to unity; however, in most serious accidents the damage is extensive, and the coefficient of restitution approaches zero.

Therefore, as we develop the equations for eccentric impact, the analyst must take extra care to understand the assumptions that are made and the sensitivity of the answers to these assumptions. For classroom exercises, the assumptions may seem unimportant, but when they are applied to actual cases, the analyses frequently have large legal implications and may result in unwarranted blame placed on people involved in accidents. It is the responsibility of the engineer to make sure that the people who purchase commercial accident reconstruction programs do not misuse them. The National Highway Traffic Safety Administration has encouraged the development of programs, starting with the work by Calspan in the early 1970s, and many commercial programs exist, such as the Simulated Linear Accident Momentum (SLAM); Crash 1,2,3; and Edcrash programs. These programs and other similar programs determine the energy lost during a collision by performing an analysis of the crush damage to the vehicles instead of assuming a coefficient of restitution. Crush damage estimates the energy loss based on experimentally measured stiffnesses of test vehicles.

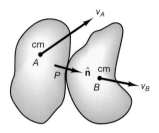

Figure 7.3

Consider the impact of the two bodies shown in Figure 7.3. The normal and tangential directions are designated at the point of contact between the two bodies. In many analyses, the normal direction will be chosen to be normal to the undeformed surfaces, but other analyses leave it as a direction to be specified after an examination of the bodies following impact. This is the direction that the impulsive impact force is assumed to act along, and in addition, it is assumed that the direction and point of application of the impulsive force remain constant throughout the impact. If the point of impact is designated by P and the direction of the impulsive force on the body B is given by the unit vector $\hat{\mathbf{n}}$, the linear and angular impulse–momentum equations yield

$$\int_{t_1}^{t_2} (-F\hat{\mathbf{n}})dt = m_A(\mathbf{v}'_A - \mathbf{v}_A)$$

$$\int_{t_1}^{t_2} (F\hat{\mathbf{n}})dt = m_B(\mathbf{v}'_B - \mathbf{v}_B)$$

$$\int_{t_1}^{t_2} [\mathbf{r}_{P/Acm} \times (-F\hat{\mathbf{n}})]dt = I_{Acm}(\omega'_A - \omega_A)\hat{\mathbf{k}}$$

$$\int_{t_1}^{t_2} [\mathbf{r}_{P/Bcm} \times (F\hat{\mathbf{n}})]dt = I_{Bcm}(\omega'_B - \omega_B)\hat{\mathbf{k}}$$

(7.50)

where, again, the primed values refer to the postimpact linear velocity of the center of mass and the angular velocity of the body.

The impulsive force is an unknown function of time during the impact, and therefore, Eqs. (7.50) cannot be solved to determine the final linear and angular velocities of the bodies. If the two bodies adhere to one another and move as a single body after impact, the final linear velocity of the combined bodies can be obtained easily from the two linear momentum equations by eliminating the impulse.

The linear impulse–momentum equation in Eq. (7.50) may be obtained by using the scalar product with $\hat{\mathbf{n}}$:

$$-\int F\,dt = m_A(v'_{An} - v_{An})$$

$$\int F\,dt = m_B(v'_{Bn} - v_{Bn})$$

(7.51)

The equivalent scalar angular impulse–momentum equations can be obtained from Eq. (7.50) by taking the scalar product of the angular impulse–momentum with the unit vector $\hat{\mathbf{k}}$. When we do so, we obtain

$$-X_A \int F\,dt = I_{Acm}(\omega'_A - \omega_A)$$

$$X_B \int F\,dt = I_{Bcm}(\omega'_B - \omega_B)$$

(7.52)

where

$$X_A = \hat{\mathbf{k}} \cdot (\mathbf{r}_{P/Acm} \times \hat{\mathbf{n}}) \text{ and } X_B = \hat{\mathbf{k}} \cdot (\mathbf{r}_{P/Bcm} \times \hat{\mathbf{n}})$$

For impact between particles, the coefficient of restitution was defined as the ratio of the restitution impulse to the deformation impulse. The period of deformation was defined to be from the initial impact until the velocities of the two particles were equal, and the coefficient of restitution was shown to be equal to the ratio of the relative separation velocity to the relative approach velocity. If P is the point of impact, then the components of the velocities in the $\hat{\mathbf{n}}$ direction are related by

$$e = \frac{(v'_{BPn} - v'_{APn})}{(v_{APn} - v_{BPn})} \tag{7.53}$$

where the primed values are the velocity components in the normal direction after impact.

The velocity of point P on two rigid bodies is related to the velocities of the center of masses by

$$\mathbf{v}_{AP} = \mathbf{v}_A + \boldsymbol{\omega}_A \times \mathbf{r}_{P/Acm}$$
$$\mathbf{v}_{BP} = \mathbf{v}_B + \boldsymbol{\omega}_B \times \mathbf{r}_{P/Bcm} \tag{7.54}$$

The tangential component of the center of mass of both bodies is unchanged during the impact. Given the initial dynamic states of the two bodies in plane motion, the six post-impact linear and angular velocities can be determined from the six scalar equations obtained by eliminating the impulsive force from Eqs. (7.50) and (7.51) and substituting Eq. (7.54) into Eq. (7.53). The six equations are

$$v'_{At} = v_{At}$$
$$v'_{Bt} = v_{Bt}$$
$$m_A v'_{An} + m_B v'_{Bn} - m_A v_{An} - m_B v_{Bn} = 0$$
$$X_A[m_A(v'_{An} - v_{An})] = I_{Acm}(\omega'_A - \omega_A) \tag{7.55}$$
$$X_B[m_B(v'_{Bn} - v_{Bn})] = I_{Bcm}(\omega'_B - \omega_B)$$
$$v'_{Bn} + X_B\omega'_B - v'_{An} - X_A\omega'_A = e(v_{An} + X_A\omega_A - v_{Bn} - X_B\omega_B)$$

where

$$X_A = \hat{\mathbf{k}} \cdot (\mathbf{r}_{P/Acm} \times \hat{\mathbf{n}})$$

and

$$X_B = \hat{\mathbf{k}} \cdot (\mathbf{r}_{P/Bcm} \times \hat{\mathbf{n}})$$

are the scalar products. The value of the coefficient of restitution is estimated, as is the unit normal vector $\hat{\mathbf{n}}$. The unit tangential vector is determined by

$$\hat{\mathbf{t}} = \hat{\mathbf{k}} \times \hat{\mathbf{n}} \tag{7.56}$$

The preimpact components of the velocities of the center of masses are obtained by the appropriate dot product with the normal and unit vectors. As was shown in Section 3.6, the loss of energy during the impact can be related to the coefficient of restitution.

In Sample Problem 3.11, the vehicles in an accident were modeled as particles, and an impulse–momentum method was used to analyze the impact of the particles. This analysis was the middle of the accident reconstruction process, and the method of work–energy was used to analyze postimpact and preimpact speeds relating the kinetic energy to the negative work during the skids. To form a complete reconstruction of the accident vehicles by means of eccentric-impact analysis, some method of dealing with the preimpact and postimpact vehicular motion must be developed. We will not try to use work–energy methods for this purpose but will develop a method to determine the equations of rate of change with respect to time of the angular and linear momentum of the vehicles. Such an analysis is included in this chapter to complete the reconstruction of the accident begun in Chapter 3.

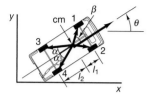

Figure 7.4

Consider a vehicle oriented at an angle θ to the x-axis in an inertial coordinate system, as shown in Figure 7.4. The center of mass of the vehicle is shown, and relative vectors from the center of mass to each of the tires will be designated by $\mathbf{r}_{i/cm}$. The position vectors to the four tires of the vehicle are

$$
\begin{aligned}
\mathbf{r}_{1/cm} &= l_1\big[\cos(\theta + \beta)\hat{\mathbf{i}} + \sin(\theta + \beta)\hat{\mathbf{j}}\big] \\
\mathbf{r}_{2/cm} &= l_1[\cos(\theta - \beta)\hat{\mathbf{i}} - \sin(\theta - \beta)\hat{\mathbf{j}}] \\
\mathbf{r}_{3/cm} &= l_2\big[-\cos(\alpha - \theta)\hat{\mathbf{i}} + \sin(\alpha - \theta)\hat{\mathbf{j}}\big] \\
\mathbf{r}_{4/cm} &= l_2\big[-\cos(\alpha + \theta)\hat{\mathbf{i}} - \sin(\alpha + \theta)\hat{\mathbf{j}}\big]
\end{aligned}
\tag{7.57}
$$

We will assume that the combined weight of the vehicle and its occupants is divided such that the weight distribution on the tires will depend on the position of the center of mass of the vehicle. For example, 60% of the weight of a conventional vehicle will be on the front tires, and 40% of the weight will be on the rear tires. Friction always opposes motion, and the friction force on each tire will oppose the velocity of the tire and will depend upon the coefficient of kinetic friction and the percent of the vehicle's weight on that tire. The friction force on the ith tire may be written as

$$
\mathbf{F}_i = -\mu_k(W_i)\,\frac{\mathbf{v}_i}{|\mathbf{v}_i|}
\tag{7.58}
$$

The moment about the center of mass of the friction force for any tire is

$$
\mathbf{M}_i = \mathbf{r}_{i/cm} \times \mathbf{F}_i
\tag{7.59}
$$

The velocity of any tire is given by

$$
\mathbf{v}_i = \mathbf{v}_{cm} + \boldsymbol{\omega} \times \mathbf{r}_{i/cm}
\tag{7.60}
$$

The moment of the friction force involves the cross product of the position vector of the tire and the velocity of the tire:

$$
\mathbf{r}_{i/cm} \times \mathbf{v}_i = \mathbf{r}_{i/cm} \times \mathbf{v}_{cm} + \mathbf{r}_{i/cm} \times (\boldsymbol{\omega} \times \mathbf{r}_{i/cm})
\tag{7.61}
$$

The triple vector product may be written as

$$
\mathbf{r}_{i/cm} \times (\boldsymbol{\omega} \times \mathbf{r}_{i/cm}) = \boldsymbol{\omega}(\mathbf{r}_{i/cm} \cdot \mathbf{r}_{i/cm}) - \mathbf{r}_{i/cm}(\mathbf{r}_{i/cm} \cdot \boldsymbol{\omega})
\tag{7.62}
$$

The second term on the right side of Eq. (7.62) is zero for plane motion. Therefore, Eq. (7.59) can be written as

$$
\mathbf{M}_i = -\mu_k(W_i)\,\frac{\big[l_i^2\boldsymbol{\omega} + \mathbf{r}_{i/cm} \times \mathbf{v}_{cm}\big]}{|\mathbf{v}_i|}
\tag{7.63}
$$

The equations of motion are

$$
\frac{W}{g}\frac{d\mathbf{v}_{cm}}{dt} = \sum_{i=1}^{4}\mathbf{F}_i
$$

$$
I_{cmz}\frac{d\boldsymbol{\omega}}{dt} = \sum_{i=1}^{4}\mathbf{M}_i
\tag{7.64}
$$

Equation (7.64) forms two coupled, nonlinear differential equations for the linear velocity of the center of mass and the angular velocity of the vehicle. These equations must be solved numerically for each case.

Sample Problem 7.9

In the accompanying diagram, a Pontiac (vehicle B), while waiting to make a left turn is rear-ended by a Toyota, (vehicle A). Determine the postimpact dynamics if, while reconstructing the accident, you base your analysis on the following data: Witnesses report that the Pontiac was stopped waiting to make a left turn and the speed of the Toyota was 50 mph at the time of impact—that is, the Toyota did not brake before impact. This could be verified by the absence of skid marks. The direction of the velocity vector of A is assumed to be $-10°$ off the horizontal. The direction of the normal vector $\hat{\mathbf{n}}$ is assumed to be 5° from the horizontal. This direction can be obtained by examining the damage to the two vehicles. The coefficient of restitution is obtained by examining the damage to the vehicles and from the position of the vehicles after the accident is over. The values found for the various parameters are

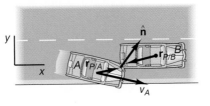

$$m_A = 82.3 \text{ lb-s}^2/\text{ft} \qquad\qquad m_B = 87.6 \text{ lb-s}^2/\text{ft}$$

$$I_A = 1{,}686 \text{ lb-ft-s}^2 \qquad\qquad I_B = 1{,}795 \text{ lb-ft-s}^2$$

$$\mathbf{r}_{P/A} = 6\hat{\mathbf{i}} + 2\hat{\mathbf{j}} \text{ ft} \qquad\qquad \mathbf{r}_{P/B} = -7\hat{\mathbf{i}} - 2.2\hat{\mathbf{j}} \text{ ft}$$

$$\mathbf{v}_A = 73.3(\cos10°\hat{\mathbf{i}} - \sin10°\hat{\mathbf{j}}) \text{ ft/s} \qquad \mathbf{v}_B = 0$$

$$\omega_A = 0 \qquad\qquad \omega_B = 0$$

$$\hat{\mathbf{n}} = \cos5°\hat{\mathbf{i}} + \sin5°\hat{\mathbf{j}}$$

$$e = 0.8$$

In an actual reconstruction, the engineer should check for sensitivity of the solution to any variation in the direction of the unit vector $\hat{\mathbf{n}}$, the velocities, and the coefficient of restitution. The four equations for the postimpact normal component of the velocity and the angular velocity are solved using matrix notation.

Solution   The unit vector in the tangential direction is

$$\hat{\mathbf{t}} = \hat{\mathbf{k}} \times \hat{\mathbf{n}} = -\sin5°\hat{\mathbf{i}} + \cos5°\hat{\mathbf{j}}$$

The initial velocities are expressed, in normal and tangential components to the impact plane, in a general form as

$$v_{At} = \mathbf{v}_A \cdot \hat{\mathbf{t}} \qquad v_{An} = \mathbf{v}_A \cdot \hat{\mathbf{n}} \qquad v_{Bt} = \mathbf{v}_B \cdot \hat{\mathbf{t}} \qquad v_{Bn} = \mathbf{v}_B \cdot \hat{\mathbf{n}}$$

The triple scalar products are

$$X_A = \hat{\mathbf{k}} \cdot (\mathbf{r}_{P/A} \times \hat{\mathbf{n}}) \qquad X_B = \hat{\mathbf{k}} \cdot (\mathbf{r}_{P/B} \times \hat{\mathbf{n}})$$

The postimpact linear and angular velocities are related to the preimpact velocities by the matrix equation

$$
\begin{bmatrix}
m_A & m_B & 0 & 0 \\
X_A m_A & 0 & -I_A & 0 \\
0 & X_B m_B & 0 & -I_B \\
-1 & 1 & -X_A & X_B
\end{bmatrix}
\begin{bmatrix}
v'_{An} \\
v'_{Bn} \\
\omega'_A \\
\omega'_B
\end{bmatrix}
$$

$$
=
\begin{bmatrix}
m_A v_{An} + m_B v_{Bn} \\
X_A m_A v_{An} - I_A \omega_A \\
X_B m_B v_{Bn} - I_B \omega_B \\
(v_{An} + X_A \omega_A - v_{Bn} - X_B \omega_B)
\end{bmatrix}
$$

Using the initial conditions, we can obtain the postimpact angular velocities and the component of the linear velocities in the normal direction by solving this system of linear equations. The postimpact linear velocities are

$$
\mathbf{v}'_A = v'_{An}\hat{\mathbf{n}} + v_{At}\hat{\mathbf{t}} \qquad \mathbf{v}'_B = v'_{Bn}\hat{\mathbf{n}} + v_{Bt}\hat{\mathbf{t}}
$$

For the given initial conditions, the postimpact velocities are

$$
\mathbf{v}'_A = 13.4\hat{\mathbf{i}} - 17.9\hat{\mathbf{j}} \text{ ft/s} \qquad \boldsymbol{\omega}'_A = 4.2\hat{\mathbf{k}} \text{ rad/s}
$$
$$
\mathbf{v}'_B = 55.2\hat{\mathbf{i}} + 4.8\hat{\mathbf{j}} \text{ ft/s} \qquad \boldsymbol{\omega}'_B = 4.3\hat{\mathbf{k}} \text{ rad/s}
$$

The Pontiac is pushed into the lane of oncoming traffic and rotates such that the passenger's side is exposed to the oncoming traffic. The accident analyzed in this sample problem actually happened and the passenger in the front seat of the Pontiac died. One of the people reconstructing the accident, who was not trained in any engineering discipline, erroneously testified in his deposition that the Pontiac could not have gone into oncoming traffic unless the driver had turned his wheels to the left while stopped. If this were the case, the driver of the Pontiac would have been, in part, to blame for the death. Fortunately, the analysis was corrected in court by an engineer, and no blame was attributed to the driver of the Pontiac.

Problems

7.83 A homogeneous bar of mass m and length L is falling with a velocity of v when its end strikes an identical bar, as shown in Figure P7.83. Determine the postimpact angular velocities of both bars if the impact is elastic.

7.84 Two identical bars of length L and mass m are pinned at midlength. Bar A has an angular velocity ω_A, as shown in Figure P7.84, when it strikes bar B. Determine the postimpact angular velocities of the bars in terms of these parameters and the coefficient of restitution, e.

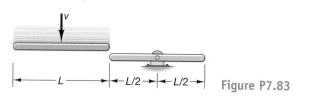

Figure P7.83

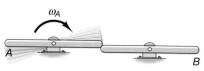

Figure P7.84

7.85 Determine the impulse between the two bars in Problem 7.83.

7.86 Determine the impulse between the two bars in Problem 7.84.

7.87 A 300-mm, 2-kg bar swings from a horizontal position and strikes a 3-kg disk of radius 50 mm that rests on a horizontal plane. (See Figure P7.87.) If the disk rolls without sliding after impact and the coefficient of restitution is 0.7, determine the angular velocities of the bar and the disk after impact.

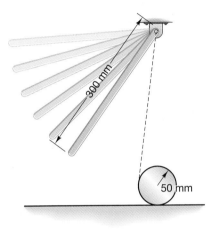

Figure P7.87

7.88 Determine the postimpact velocities for an accident between the two vehicles in Sample Problem 7.9 if the initial velocity of vehicle A was as shown, but the velocity of vehicle B was

$$\mathbf{v}_B = -44\hat{\mathbf{i}} \text{ ft/s}$$

7.89 An 85-slug car with a yaw mass moment of inertia of 1,700 lb-ft-s² is traveling in a straight line at a velocity of 25 mph when it hits a batch of ice and an angular velocity of 0.2 rad/s is initiated clockwise, as viewed from the top. (See Figure P7.89.) If the coefficient of kinetic friction between the tires and the ice is 0.3, determine the motion of the car as it skids to a stop.

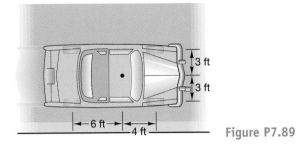

Figure P7.89

7.90 Vehicle A is traveling at 40 mph when vehicle B runs a stop sign while going 30 mph. (See Figure P7.90.) Vehicle B strikes vehicle A at right angles at the passenger tire. If the coefficient of restitution is estimated to be 0.2, determine the postimpact velocities of the two vehicles.

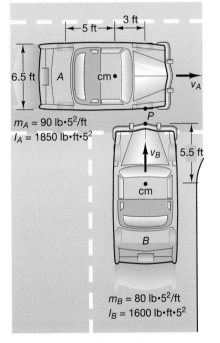

Figure P7.90

7.91 If the coefficient of kinetic friction between the tires and the road is 0.7, determine the final position of vehicle A in Problem 7.90 after the accident.

Chapter 8 THREE-DIMENSIONAL DYNAMICS OF RIGID BODIES

The spinning top is one of the most studied problems in dynamics, and only with new computational software can the nonlinear differential equations be solved.

8.1 INTRODUCTION

Many of the concepts of three-dimensional kinematics of rigid bodies were introduced in Chapter 5, but only solutions of dynamics problems for plane motion—that is, motion with just three degrees of freedom—were presented in Chapters 6 and 7. Although many motions can be modeled two dimensionally, complex machines, airplanes, automobiles, and the human body, in fact, most things move in a three-dimensional manner. The major difficulty in extending our study to three dimensions is that we will now have six degrees of freedom. The three translational degrees of freedom were considered in Chapters 1 and 2 when we studied particle dynamics; now we need to examine the three rotational degrees of freedom.

We begin our study of three-dimensional dynamics by first considering descriptions of finite rotations of coordinate systems. We then consider the mass-moment-of-inertia tensor and the principal axes and principal mass moments of inertia. Finally, we develop the Euler equations of motion and particular solutions of these equations.

8.2 ROTATIONAL TRANSFORMATION BETWEEN COORDINATE SYSTEMS

In Section 2.2 of Statics, we showed that finite rotations are not vectors and cannot be expressed in components about coordinate axes. The problem is that three-dimensional finite rotations depend on the order of rotation and therefore do not satisfy the laws of vector addition. However, angular velocity and angular acceleration are related to *infinitesimal* rotations and *are* vectors. Because it is difficult to express angular position in three-dimensional space, we will seek a manner of expressing the position and orientation of a rigid body in space by examining coordinate transformations.

8.2.1 COORDINATE TRANSFORMATIONS

Consider a (three-dimensional) translation from x–y–z coordinates to x'–y'–z' coordinates, as shown in Figure 8.1. In terms of the old coordinates, the new coordinates are given by

$$\mathbf{r}' = \mathbf{r} - \mathbf{r}_0 \tag{8.1}$$

This vector equation can be written in scalar form as

$$x' = x - x_0$$
$$y' = y - y_0 \tag{8.2}$$
$$z' = z - z_0$$

or in matrix notation as

$$[x'] = [x] - [x_0] \tag{8.3}$$

We now can write the translational transformation from the primed to the unprimed coordinates:

$$[x] = [x'] + [x_0] \tag{8.4}$$

This matrix equation is called the *inverse transformation;* note the change in sign from the previous transformation.

Now let us consider a rotational transformation of coordinates. First, we examine rotation about a single axis. The primed coordinates are rotated relative to the unprimed coordinates. Consider a rotation about the z-axis as shown in Figure 8.2. The components of

Figure 8.1

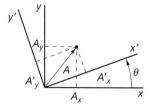

Figure 8.2

the vector **A** are designated by A_x and A_y in the x–y coordinate system and by A'_x and A'_y in the x'–y' coordinate system. We now seek to express the components in the primed coordinate system in terms of those in the unprimed system. From Figure 8.2, we see that

$$A'_x = A_x \cos\theta + A_y \sin\theta$$
$$A'_y = -A_x \sin\theta + A_y \cos\theta \qquad (8.5)$$
$$A'_z = A_z$$

We can extend this transformation to a three-dimensional one using the direction cosines between the coordinate axes:

$$A'_x = \cos(x', x)A_x + \cos(x', y)A_y + \cos(x', z)A_z$$
$$A'_y = \cos(y', x)A_x + \cos(y', y)A_y + \cos(y', z)A_z \qquad (8.6)$$
$$A'_z = \cos(z', x)A_x + \cos(z', y)A_y + \cos(z', z)A_z$$

The nine direction cosines are designated by

$$\begin{aligned}
\lambda_{xx} &= \cos(x', x) & \lambda_{xy} &= \cos(x', y) & \lambda_{xz} &= \cos(x', z) \\
\lambda_{yx} &= \cos(y', x) & \lambda_{yy} &= \cos(y', y) & \lambda_{yz} &= \cos(y', z) \\
\lambda_{zx} &= \cos(z', x) & \lambda_{zy} &= \cos(z', y) & \lambda_{zz} &= \cos(z', z)
\end{aligned} \qquad (8.7)$$

where $\cos(y', x)$ is the cosine of the angle between the y' and x axes.

We next introduce an *orthogonal transformation matrix* composed of these nine direction cosines:

$$[R] = \begin{bmatrix} \lambda_{xx} & \lambda_{xy} & \lambda_{xz} \\ \lambda_{yx} & \lambda_{yy} & \lambda_{yz} \\ \lambda_{zx} & \lambda_{zy} & \lambda_{zz} \end{bmatrix} \qquad (8.8)$$

The components of the vector **A** in the primed coordinate system are related to the components in the unprimed coordinate system by the matrix relationship

$$[\mathbf{A}'] = [R][\mathbf{A}] \qquad (8.9)$$

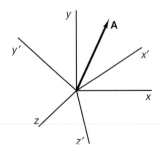

Figure 8.3

To describe and understand finite rotations in space, it is best to begin with the relationship between two rotated coordinate systems. (See Figure 8.3.) We shall examine the *transformation* of one right-handed coordinate system to another right-handed system by rotation, where the relationship between the two systems is described by the direction cosines of each of the primed coordinate axes to the three unprimed axes. The base-unit vectors in the primed coordinate system can be expressed in the unprimed coordinate system as

$$\hat{\mathbf{i}}' = \lambda_{xx}\hat{\mathbf{i}} + \lambda_{xy}\hat{\mathbf{j}} + \lambda_{xz}\hat{\mathbf{k}}$$
$$\hat{\mathbf{j}}' = \lambda_{yx}\hat{\mathbf{i}} + \lambda_{yy}\hat{\mathbf{j}} + \lambda_{yz}\hat{\mathbf{k}} \qquad (8.10)$$
$$\hat{\mathbf{k}}' = \lambda_{zx}\hat{\mathbf{i}} + \lambda_{zy}\hat{\mathbf{j}} + \lambda_{zz}\hat{\mathbf{k}}$$

Note in Eq. (8.8) that the first row of the $[R]$ matrix consists of the components of $\hat{\mathbf{i}}'$ in the unprimed system, the second row consists of the components of $\hat{\mathbf{j}}'$, and the third row is made up of the components of $\hat{\mathbf{k}}'$. The unit vectors in the primed coordinate system can be related to those in the unprimed coordinate system by the matrix equation

$$\begin{bmatrix} \hat{\mathbf{i}}' \\ \hat{\mathbf{j}}' \\ \hat{\mathbf{k}}' \end{bmatrix} = [R] \begin{bmatrix} \hat{\mathbf{i}} \\ \hat{\mathbf{j}} \\ \hat{\mathbf{k}} \end{bmatrix} \qquad (8.11)$$

The columns of the matrix $[R]$ are the components of the unprimed base-unit vectors in the primed coordinates. We can relate the unprimed base-unit vectors to the primed base-unit vectors by the set of equations

$$\hat{\mathbf{i}} = \lambda_{xx}\hat{\mathbf{i}}' + \lambda_{yx}\hat{\mathbf{j}}' + \lambda_{zx}\hat{\mathbf{k}}'$$

$$\hat{\mathbf{j}} = \lambda_{xy}\hat{\mathbf{i}}' + \lambda_{yy}\hat{\mathbf{j}}' + \lambda_{zy}\hat{\mathbf{k}}' \tag{8.12}$$

$$\hat{\mathbf{k}} = \lambda_{xz}\hat{\mathbf{i}}' + \lambda_{yz}\hat{\mathbf{j}}' + \lambda_{zz}\hat{\mathbf{k}}'$$

The primed base-unit vectors are related to the unprimed base-unit vectors by the orthogonal rotation matrix $[R]$, and the unprimed base-unit vectors are related to the primed base-unit vectors by the orthogonal rotation matrix $[R]^T$—the transpose of $[R]$.

Clearly, the sum of the squares of the elements of any row or column of $[R]$ equals unity. Since the base vectors are orthogonal, the sum of the products of corresponding elements of any two rows or columns is equal to zero. Another important characteristic of this matrix is that the determinant is equal to $+1$. If we transformed from a right-handed coordinate system to a left-handed coordinate system, the determinant of the transformation matrix would be equal to -1.

The components of the vector $\mathbf{A}$ in the unprimed coordinate system are related to those in the primed coordinate system by multiplying both sides of Eq. (8.9) by the inverse of the transformation matrix, yielding

$$[\mathbf{A}] = [R]^{-1}[\mathbf{A}'] \tag{8.13}$$

The inverse transformation from the primed to the unprimed system can also be written in matrix notation, as

$$[A] = [R]^T[A'] \tag{8.14}$$

where $[R]^T$ is the transpose of the rotational transformation matrix $[R]$.

Among the interesting characteristics of the transformation matrix is that the inverse of the orthogonal rotational transformation matrix is its transpose; that is,

$$[R]^{-1} = [R]^T \qquad [R]^T[R] = [R][R]^T = [I] \tag{8.15}$$

where $[I]$ is the unity matrix. The transformation and inverse transformations are usually written as

$$[x'] = [R][x]$$
$$[x] = [R]^T[x'] \tag{8.16}$$

Next, we consider a transformation of coordinates that involves a translation and a rotation, as shown in Figure 8.4. We can see that we can go from the unprimed coordinates to the double-primed coordinates by two different routes: (a) We could translate the origin and then rotate the coordinates, or (b) we could rotate first and then translate the origin. These two transformations are respectively given by the following equations:

(a)

$$[x'] = [x] - [x_0] \rightarrow [x''] = [R][x'] = [R][x] - [R][x_0] \tag{8.17}$$

(b)

$$[x'] = [R][x] \rightarrow [x''] = [x'] - [x_0'] = [R][x] - [x_0'] \tag{8.18}$$

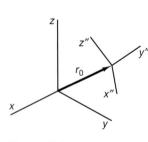

Figure 8.4

Since the relationship between the unprimed coordinates and the double-primed coordinates is unique, the inverse of Eqs. (8.17) and (8.18) are, respectively,

$$[x] = [R]^T [x''] + [x_0]$$
$$[x] = [R]^T [x'' + x_0']$$

(8.19)

where $[x_0'] = [R][x_0]$ is the vector from the origin expressed in the rotated coordinates. To invert a double transformation of this type, we also have to invert the order of the transformation.

Now let us consider two rotational transformations $[A]$ and $[B]$ such that

$$[x'] = [A][x] \quad \rightarrow \quad [x''] = [B][x']$$

(8.20)

Equation 8.20 is equivalent to

$$[x''] = [B][A][x]$$

(8.21)

Note that if we performed the transformation in the reverse order, we would obtain a different final coordinate orientation,

$$[x'''] = [A][B][x]$$

(8.22)

There are two different equivalent rotational transformations:

$$[C] = [B][A]$$

(8.23)

$$[D] = [A][B]$$

(8.24)

Thus, matrix multiplication is not commutative; that is,

$$[C] \neq [D]$$

(8.25)

The inverse transformations can be written as

$$[x] = [C]^T [x'']$$
$$[x] = [D]^T [x''']$$

(8.27)

The transpose of the product of two matrices is the transposed product of the transposed matrices:

$$[C]^T = [[B][A]]^T = [A]^T [B]^T$$

(8.28)

Finally, in a similar manner, we obtain

$$[D]^T = [[A][B]]^T = [B]^T [A]^T$$

(8.29)

Sample Problem 8.1

Develop the transformation for a right-handed rotation about an unprimed z-coordinate axis of θ degrees.

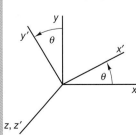

Solution   The transformation is illustrated in the accompanying figure.

Examining the direction cosines for this rotation, we see that the transformation matrix is

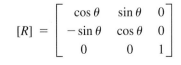

$$[R] = \begin{bmatrix} \cos\theta & \sin\theta & 0 \\ -\sin\theta & \cos\theta & 0 \\ 0 & 0 & 1 \end{bmatrix}$$

When the coordinate system is rotated about an axis, the diagonal element in the rotation matrix corresponding to that axis is equal to unity, and the other elements in the corresponding row and column are zero.

Sample Problem 8.2

Develop the transformation matrix corresponding to a rotation α about the x-axis followed by a rotation β about the rotated y'-axis.

Solution Let the initial rotation about the x-axis be

$$[A] = \begin{bmatrix} 1 & 0 & 0 \\ 0 & \cos\alpha & \sin\alpha \\ 0 & -\sin\alpha & \cos\alpha \end{bmatrix}$$

The second rotation about the y'-axis will be designated by

$$[B] = \begin{bmatrix} \cos\beta & 0 & -\sin\beta \\ 0 & 1 & 0 \\ \sin\beta & 0 & \cos\beta \end{bmatrix}$$

Therefore, the full transformation matrix is

$$[C] = [B][A]$$

Multiplying the two matrices yields

$$[C] = \begin{bmatrix} \cos\beta & \sin\alpha\sin\beta & -\cos\alpha\sin\beta \\ 0 & \cos\alpha & \sin\alpha \\ \sin\beta & -\sin\alpha\cos\beta & \cos\alpha\cos\beta \end{bmatrix}$$

Problems

8.1 Given

$$\mathbf{r}_0 = \begin{bmatrix} 2 \\ -1 \\ 4 \end{bmatrix} \text{ m} \qquad [R] = \begin{bmatrix} 0.707 & 0.612 & 0.354 \\ -0.707 & 0.612 & 0.354 \\ 0 & -0.5 & 0.866 \end{bmatrix}$$

develop the transformation from the x-coordinates to the x''-coordinates as (a) first a translation $\mathbf{r}_0$ and then a rotation $[R]$ and (b) first the rotation $[R]$ and then the translation $\mathbf{r}_0'$. (c) Develop the corresponding inverse transformations.

8.2 Develop the transformation matrix to rotate a coordinate system 30° about the x-axis, followed by a 60° rotation about the z'-axis. Both of these rotations are positive as defined by the right-hand rule. Determine an equivalent transformation matrix to go from the unprimed to the double-primed coordinate system, and show that the rows of this matrix are the base-unit vectors expressed in the original coordinate system.

8.3 Develop the transformation matrix in Problem 8.2 for the same rotations in the inverse order; that is, rotate 60° about the z-axis, followed by a 30° rotation about the x'-axis.

8.4 Develop the general orthogonal rotation transformation for a positive rotation θ about the y-axis.

8.5 Develop the general orthogonal rotation transformation for a positive rotation θ about the x-axis, followed by a positive rotation β about the y'-axis.

8.6 Develop the orthogonal rotation transformation for the rotations in Problem 8.5, taken first about the y-axis and then about the x'-axis.

8.7 Show that the orthogonal rotation transformations in Problems 8.5 and 8.6 are equal for infinitesimal rotations $d\theta$ and $d\beta$. (*Therefore, infinitesimal rotations are not dependent on the order of rotation and may be treated as vectors with unique components.*)

🖥 **8.8** A rotational transformation matrix is obtained by a rotation α about the x-axis, a rotation β about the y'-axis, and, finally, a rotation γ about the z''-axis. Determine the angles α, β, and γ if the matrix is

$$[R] = \begin{bmatrix} 0.353 & 0.918 & -0.177 \\ -0.353 & 0.306 & 0.884 \\ 0.866 & -0.25 & 0.433 \end{bmatrix}$$

8.9 Show that the product of the orthogonal rotational transformation matrix given in Problem 8.8 and its transpose is equal to the unity matrix. That is, show that

$$[R] [R]^T = [I]$$
$$[R]^T [R] = [I]$$

8.10 If the orthogonal rotational transformation matrix

$$[R] = \begin{bmatrix} 0.707 & 0.612 & 0.354 \\ -0.707 & 0.612 & 0.354 \\ 0 & -0.5 & 0.866 \end{bmatrix}$$

is obtained by a rotation θ about the x-axis followed by a rotation β about the z'-axis, determine the two angles of rotation.

8.3 EULERIAN ANGLES

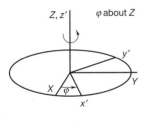

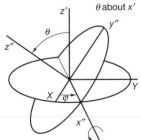

We have seen that finite rotations are dependent on the order of rotation and therefore cannot be treated as components of a rotation vector. In Section 8.2, we also saw that the orientation of a rigid body could be given by an orthogonal rotational transformation generated by three rotations. In 1776, the mathematician Leonhard Euler (1707–1783) proposed that a specific sequence of rotations be used to specify the orientation of a body rotating about a fixed point. The three rotation angles are called the ***Eulerian angles***, and although there are many variations of the rotation sequence, we will examine an initial rotation about the Z-axis, followed by a rotation about the x'-axis and, finally, a rotation about the z''-axis. We will designate this sequence as a (z'', x', Z) sequence, given in the order of the matrix multiplication. Eulerian angles of this type were studied in detail when the motion of a heavy top was analyzed by F. Klein and A. Sommerfeld in the late 1800s. The (z'', x', Z) sequence is shown in Figure 8.5

In the sequence, we first rotate an angle ϕ about the Z-axis, using the right-hand rule to designate a positive rotation. This is followed by a rotation θ about the new x'-axis, again using the right-hand rule. The final rotation is about the new z''-axis, using the right-hand rule once more. We will show how the orientation of a body in space can be defined in terms of these three successive orthogonal rotational transformations.

We define the first transformation as

$$[Z] = \begin{bmatrix} \cos\phi & \sin\phi & 0 \\ -\sin\phi & \cos\phi & 0 \\ 0 & 0 & 1 \end{bmatrix} \tag{8.30}$$

Note that the 1 on the diagonal corresponding to the cosine of the angle between the z- and z'-axes clearly indicates that this is a rotation about the z-axis. The other elements in the same row or column are zero, since the sum of the squares of the elements of any row or column of an orthogonal rotation transformation matrix must be equal to unity.

The second transformation, from the (x', y', z') coordinates to the (x'', y'', z'') coordinates is a rotation about the x'-axis and is given by

$$[x'] = \begin{bmatrix} 1 & 0 & 0 \\ 0 & \cos\theta & \sin\theta \\ 0 & -\sin\theta & \cos\theta \end{bmatrix} \tag{8.31}$$

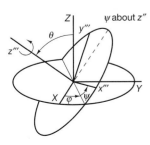

Figure 8.5

Again, the element in the first row and first column designates the matrix to be a rotation about the x'-axis. The final rotation is about the z''-axis, and the transformation matrix will have the same appearance as Eq. (8.30):

$$[z''] = \begin{bmatrix} \cos\psi & \sin\psi & 0 \\ -\sin\psi & \cos\psi & 0 \\ 0 & 0 & 1 \end{bmatrix} \tag{8.32}$$

The full transformation matrix is the ordered product of the foregoing three transformations:

$$[T] = [z''][x'][Z] \tag{8.33}$$

We will expand this matrix in full, but will use the notation c for the cosine and s for the sine, to improve the appearance of the final matrix, which now becomes

$$[T] = \begin{bmatrix} c\psi c\phi - c\theta s\phi s\psi & c\psi s\phi + c\theta c\phi s\psi & s\psi s\theta \\ -s\psi c\phi - c\theta s\phi c\psi & -s\psi s\phi + c\theta c\phi c\psi & c\psi s\theta \\ s\theta s\phi & -s\theta c\phi & c\theta \end{bmatrix} \tag{8.34}$$

Note in Figure (8.5) that the x', x''-axis lies in the first plane of rotation and in the final plane of rotation and is the intersection of these two planes. This line of intersection is called the **line of nodes**. The three axes of rotation consist of an original axis of the body, an axis of the body in its final orientation, and the line of nodes. Now, given a particular transformation matrix, we wish to tease out the three angles of rotation, assuming a (z'', x'', Z) sequence. The only element in the transformation matrix $[T]$ that involves exactly one angle is the T_{33} element, which is equal to the cosine of θ. (Note that the cosine of an angle and of the negative of the angle are equal, so that two possible sets of Eulerian angles are found for each particular transformation.)

As mentioned earlier, this sequence of rotations has been used to study the motion of a heavy top. The first rotation, about the Z-axis, corresponds to **precession** of the top—that is, the rotation of the top's z'''-axis about the Z-axis in space. The second rotation corresponds to **nutation** of the top's z'''-axis to the horizontal axis, or the wobble of the top. The final rotation is the **spin** of the top about its own z'''-axis.

One can obtain a transformation from one Cartesian coordinate system to another by means of three successive rotations in a specific sequence; therefore, there are many Eulerian angles. If the sequence is (z'', y', X) or some other sequence involving each of the coordinate axes, the angles are sometimes called Cardan angles, named after the mechanist Cardano. Consider, for example, the transformation sequence (z'', x', Y) yielding the transformation matrix. If the sequential angles of rotation are designated as (γ, α, β), corresponding to the rotations (z'', x', Y), respectively, the three transformation matrices are

$$[Y] = \begin{bmatrix} \cos\beta & 0 & -\sin\beta \\ 0 & 1 & 0 \\ \sin\beta & 0 & \cos\beta \end{bmatrix}$$

$$[x'] = \begin{bmatrix} 1 & 0 & 0 \\ 0 & \cos\alpha & \sin\alpha \\ 0 & -\sin\alpha & \cos\alpha \end{bmatrix} \tag{8.35}$$

$$[z''] = \begin{bmatrix} \cos\gamma & \sin\gamma & 0 \\ -\sin\gamma & \cos\gamma & 0 \\ 0 & 0 & 1 \end{bmatrix}$$

The complete Cardan transformation corresponding to this sequence is

$$[C] = \begin{bmatrix} c\gamma c\beta + s\gamma s\alpha s\beta & s\gamma c\alpha & -c\gamma s\beta + s\gamma s\alpha c\beta \\ -s\gamma c\beta + c\gamma s\alpha s\beta & c\gamma c\alpha & s\gamma s\beta + c\gamma s\alpha c\beta \\ c\alpha s\beta & -s\alpha & c\alpha c\beta \end{bmatrix} \qquad (8.36)$$

If the angles are limited to

$$0 \leq \beta < 2\pi, \quad -\frac{\pi}{2} \leq \alpha \leq \frac{\pi}{2}, \quad 0 \leq \gamma < 2\pi$$

we can see that we can describe the orientation of the body with the preceding sequence of rotations. If, however, we allow $\alpha = \pm\pi/2$ the z-axis aligns with the original Y-axis, and the angles β and γ are undefined. It can be shown that at the angle $(\beta - \gamma)$ is defined when $\alpha = \pi/2$ and the angle $(\beta + \gamma)$ is defined when $\alpha = -\pi/2$. The condition $\alpha = \pm\pi/2$ is termed **gimbal lock** in discussions of gyroscope suspensions.

This sequence of angles is used to describe motion at the knee joint in biomechanical analyses of human movement. The first rotation is about the Y-axis of the femur and corresponds to flexion/extension of the knee. The rotation about the line of nodes, the x'-axis, corresponds to abduction/adduction of the knee, and the final rotation, about the z''-axis of the tibia, corresponds to internal/external rotation of the knee.

The sequence of angles is chosen to describe the motion of the body in relevant physical terms. Aircraft use a Cardan sequence corresponding to the pitch, roll, and yaw of the plane. Although the resulting angles will differ with the choice of sequence, each sequence is mathematically correct, and the only difficulty arises if two of the axes of rotation become parallel during the motion and the rotations are not unique.

8.3.1 PRECESSION OF THE EQUINOXES

The earth's polar axis is tilted to its axis of rotation around the sun, *the normal to the ecliptic*. This is called the *obliquity of the ecliptic*, and if the earth is considered a heavy top, this would be the nutation angle. The angle of tilt is approximately 23.5° to the vertical. This points the North Pole away from the sun for six months of the year while the southern hemisphere gets the solar rays of summer. This tilt or obliquity accounts for the seasons and why the seasons are reversed in the northern and southern hemispheres. The seasons are marked by four crucial moments: the winter and summer solstice and the spring and fall equinoxes. In the northern hemisphere, the shortest day of the year is the winter solstice, which falls on December 21st of each year, marking the first day of winter. The same day marks the summer solstice for the southern hemisphere. The summer solstice in the northern hemisphere occurs on June 21st and is the first day of summer. The equinoxes are the two times of the year on which day and night are of equal lengths (March 21 and September 23). See Figure 8.6.

Ancient cultures considered the equinoxes to be important days of the year and had methods to measure them carefully and observe the constellation in which the sun rose in the morning of the spring equinox. For the past 2000 years, the sun has risen in the constellation Pisces, but soon the sun will pass out of Pisces into the constellation Aquarius. This change is due to the precession of the earth's north–south polar axis about an axis parallel to the normal to the ecliptic. The approximate period of this precession is 25,776 years. This precession is called the precession of the equinoxes (the equinoxes and winter and summer solstice can be considered to be attached to the earth) and is resulting in a change in the dates corresponding to the astrological signs. (Astrology is the divination of the supposed influences of the stars upon human affairs and terrestrial events.) The dates currently given in horoscopes are approximately 2000 years old:

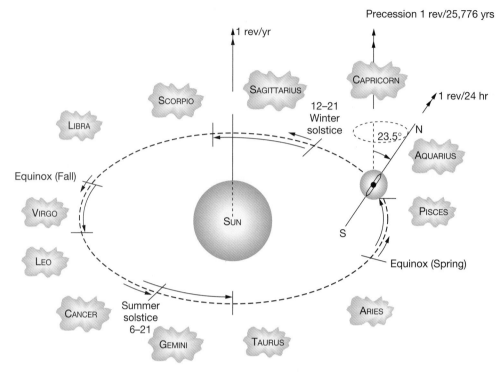

Precession of the equinoxes

Figure 8.6

Aries (March 21–April 19)   *Starting on the spring equinox.*
Taurus (April 20–May 20)
Gemini (May 21–June 21)
Cancer (June 22–July 22)   *Starting on the summer solstice in the*
                                         northern hemisphere.

Leo (July 23–Aug. 22)
Virgo (Aug. 23–Sept. 22)
Libra (Sept. 23–Oct. 23)   *Starting on the fall equinox.*
Scorpio (Oct. 24–Nov. 21)
Sagittarius (Nov. 22–Dec. 21)
Capricorn (Dec. 22–Jan. 19)   *Starting on the winter solstice in the*
                                         northern hemisphere.

Aquarius (Jan. 20–Feb. 18)
Pisces (Feb. 19–March 20)

The difficulty with the interpretation of your horoscope is that the spring equinox is occurring currently (the sun rising on March 21st) between the constellations Pisces and Aquarius instead of between Pisces and Aries, as shown in the dates above. During February 19th to March 20th, the sun rises in the constellation of Aquarius—not Pisces, as is assumed in the current horoscope. Therefore, the astrological signs are approximately one month out of phase due to the precession of the equinoxes—that is, if you were born in late February or early March, you were born under the sign of Aquarius and not under the sign of Pisces.

A full understanding of the precession of the equinoxes requires a detailed study of the forces and moments acting on the earth and includes the study of the "wobble" of the earth, known as the Chandler wobble. This wobble is thought to be due to the fact that the earth is not a rigid body but is an elastic body. In theory, this wobble should damp out over a period of 10 to 20 years, but there seems to be random excitation keeping the wobble going. Present speculation points to deep earthquakes, or the mantle phenomena underlying them, as possibly producing discontinuous changes in the earth's moment of inertial large enough to keep exciting the free-body motion.

| Sample Problem 8.3 |

Consider the transformation matrix given in Problem 8.8:

$$[R] = \begin{bmatrix} 0.353 & 0.918 & -0.177 \\ -0.353 & 0.306 & 0.884 \\ 0.866 & -0.25 & 0.433 \end{bmatrix}$$

Determine the Eulerian angles using Eq. (8.34).

Solution   We will examine the Eulerian angles from Eq. (8.34). The T_{33} element gives

$$\theta = \cos^{-1}(0.433)$$
$$= \pm 64.34°$$

We will consider the angles corresponding to the positive value of θ. From the T_{31} element, we obtain

$$\sin \theta \sin \phi = 0.866$$

$$\sin \phi = \frac{0.866}{0.901} = 0.961$$

$$\phi = 73.88° \quad \text{or} \quad 106.12°$$

Examining the T_{32} element $= -0.25$, we can conclude that $\cos \phi$ must be positive and the correct choice of ϕ is 73.88°. From the T_{13} element, we obtain

$$\sin \psi \sin \theta = -0.177$$

$$\sin \psi = \frac{-0.177}{0.901} = -0.196$$

$$\psi = -11.33° \quad \text{or} \quad 191.33°$$

Since the T_{23} element is positive, the angle ψ is $-11.33°$. Therefore, one sequence of Eulerian angles is

$$\phi = 73.88°, \qquad \theta = 64.34°, \qquad \psi = -11.33°$$

To obtain the second set of Eulerian angles, we select the negative value of $\theta = -64.34°$. Examining the T_{31} element, we obtain

$$\sin \phi = -0.961$$
$$\phi = -73.88° \quad \text{or} \quad 253.88°$$

Since the element T_{32} is negative, $\cos \phi$ must be negative, and $\phi = 253.88°$. Examining the T_{13} element, we get

$$\sin \psi = 0.196$$
$$\psi = 11.33° \quad \text{or} \quad 168.67°$$

Finally, examining T_{23}, we conclude that $\psi = 168.67°$. Therefore, the second possible set of Eulerian angles is

$$\phi = 253.88°, \quad \theta = -64.34°, \quad \psi = 168.67°$$

Examination of the physical aspects of the problem will indicate the right choice of Eulerian angles, which may be obtained using an overdetermined system of nonlinear algebraic equations. (See the Computational Supplement.)

8.4 ANGULAR MOTION

The angular velocity of a body can be written in terms of the angular velocities about each of the body's axes of rotation. Examining the Eulerian angles shown in Figure 8.5, we see that the nonorthogonal components of the angular velocity vector are given by

$$\boldsymbol{\omega} = \dot{\boldsymbol{\phi}} + \dot{\boldsymbol{\theta}} + \dot{\boldsymbol{\psi}} \tag{8.37}$$

where $\dot{\phi}$ is the rate of precession, $\dot{\theta}$ is the rate of nutation, and $\dot{\psi}$ is the spin rate. This vector can be expressed in terms of the x''', y''', z''' coordinate system—that is, the final coordinate system. Its components are

$$\omega_x''' = \dot{\phi} \sin \theta \sin \psi + \dot{\theta} \cos \psi$$
$$\omega_y''' = \dot{\phi} \sin \theta \cos \psi - \dot{\theta} \sin \psi \tag{8.38}$$
$$\omega_z''' = \dot{\phi} \cos \theta + \dot{\psi}$$

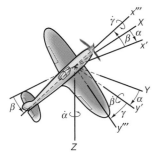

Figure 8.7

Equations (8.38) are obtained by treating the angular velocities about the three axes of rotation as vectors and expressing these vectors in the final orthogonal coordinates. Note that the angular velocities are vectors even though the finite angles are not vectors.

Let us now examine the Cardan sequence (x'', y', Z), as shown in Figure 8.7, and designate the first rotation about the Z-axis as α, the second rotation about the y'-axis (the line of nodes) as β, and the rotation about the x''-axis as γ. The x'''-axis is directed along the long axis of the plane, the y'''-axis along the right wing, and the z'''-axis toward the bottom of the plane. The first rotation α is the heading angle, the second rotation β is the attitude angle (or climb angle), and the third rotation γ is the bank angle. Again, we wish to express the angular velocity of the plane in terms of these three rates of rotation as

$$\boldsymbol{\omega} = \dot{\boldsymbol{\alpha}} + \dot{\boldsymbol{\beta}} + \dot{\boldsymbol{\gamma}} \tag{8.39}$$

The components of the angular velocity can be expressed in the set of axes attached to the plane as

$$\omega_x''' = \dot{\gamma} - \dot{\alpha} \sin \beta$$
$$\omega_y''' = \dot{\beta} \cos \gamma + \dot{\alpha} \cos \beta \sin \gamma \tag{8.40}$$
$$\omega_z''' = \dot{\alpha} \cos \beta \cos \gamma - \dot{\beta} \sin \gamma$$

Problems

8.11 Determine the precession, nutation, and spin for a (z'', x', Z) Euler transformation expressed as

$$[T] = \begin{bmatrix} 0.280 & 0.945 & 0.171 \\ -0.929 & 0.222 & 0.296 \\ 0.242 & -0.242 & 0.940 \end{bmatrix}$$

8.12 Using Eq. (8.36), determine the Cardan angles for the (z'', x', Y) transformation given in Problem 8.11.

8.13 Develop a general Euler transformation for a (y'', x', Z) Cardan transformation.

8.14 Develop the equations for the angular velocity in the body coordinate system for a (z'', x', Y) transformation.

8.15 Differentiate Eq. (8.38) to determine the angular acceleration in the body coordinate system.

8.16 Differentiate Eq. (8.40) to determine the angular acceleration in the body coordinate system.

8.5 JOINT COORDINATE SYSTEM

In 1983, Grood and Suntay (*J. Biomechanical Engineering*) introduced the concept of a **joint coordinate system** to describe the Euler or Cardan rotations for a joint in the human body. Presenting Euler angles in a form that is much easier for a clinical researcher in biomechanics to understand, they determined the angles of rotation using a vector analysis instead of the matrix analysis discussed in Section 8.3. Note that the Euler or Cardan angles involve a rotation about one of the base coordinate axes, followed by a rotation about an intermediate axis (the line of nodes) that Grood and Suntay called a **floating axis**, and, finally, a rotation about one of the body or segment coordinate axes. If we concentrate on where these axes are located, we can begin to give clinical definitions of the rotations. Grood and Suntay used the knee joint to introduce the joint coordinate system. We will designate the coordinates in the thigh or femur with capital letters and the subscript f and the coordinates in the lower leg or tibia with the subscript t. The two coordinate systems are shown in Figure 8.8 with the tibia in a general rotation with respect to the femur—that is, the distal body segment relative to the proximal body segment.

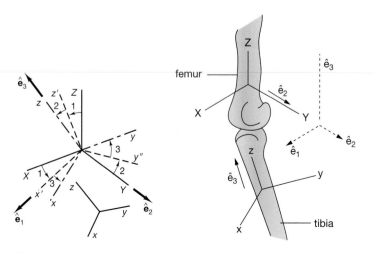

Figure 8.8

Grood and Suntay decided to use a (z_t, x', Y_f) Cardan transformation—that is, a rotation first about the medial–lateral axis of the femur (knee flexion/extension), followed by an intermediate rotation about the x' floating axis (knee abduction/adduction), and, finally, a rotation about the inferior–superior axis of the tibia (knee internal/external rotation). Note that the floating axis (the line of nodes) x' is perpendicular to the Y_f-axis and the z_t-axis. We will choose a coordinate system (x', Y, z) as a nonorthogonal *joint coordinate system*. We will denote this system as a $(1, 2, 3)$ system, so as not to confuse the notation for the axes with that for the order of rotations. The base-unit vectors in the joint coordinate system are defined as

$$\hat{\mathbf{e}}_2 = \hat{\mathbf{J}}_f$$
$$\hat{\mathbf{e}}_3 = \hat{\mathbf{k}}_t$$
$$\hat{\mathbf{e}}_1 = \frac{\hat{\mathbf{e}}_2 \times \hat{\mathbf{e}}_3}{|\hat{\mathbf{e}}_2 \times \hat{\mathbf{e}}_3|}$$

(8.41)

The last unit vector is obtained by dividing the cross product by the magnitude of the cross product of two unit vectors that are not perpendicular to each other.

We have numbered the unit vectors such that they resemble a right-hand x', Y, z coordinate system, and this does not represent the order of the Euler rotations. Note the three vectors form a nonorthogonal coordinate system. Now we will use vector algebra to determine the clinical angles of flexion and extension about the femoral medial–lateral axis, abduction and adduction about the floating axis, and internal and external tibia rotation about the distal–proximal tibial axis. To determine these angles, we need to realize that the dot product of two unit vectors can be used to determine the cosine of the angle between the vectors. The cross product can be used to determine the sine of the angle between the two vectors.

The balance of the analysis is dependent upon the fact that the floating axis $\hat{\mathbf{e}}_1$ lies in the XZ-plane of the femur and in the xy-plane of the tibia. Therefore, if the X-axis of the femur and the x-axis of the tibia are in the anterior direction, knee flexion will be positive. The angle will be negative if there is hyperextension, or back bending, of the knee. Hyperextension is also called *genu recurvatum*. The angle is defined as

Knee Flexion(+)/Extension(−)

$$\theta_{F/E} = \left(\frac{\hat{\mathbf{I}}_f \times \mathbf{e}_1}{|\hat{\mathbf{I}}_f \times \mathbf{e}_1|} \cdot \mathbf{J}_f \right) a \sin \left(|\hat{\mathbf{I}}_f \times \mathbf{e}_1| \right)$$

or

(8.42)

$$\theta_{F/E} = -a \sin \left(\hat{\mathbf{K}}_f \cdot \hat{\mathbf{e}}_1 \right)$$

The relationship between the two approaches can be seen to be $\sin \alpha = -\cos (\alpha + 90°)$.

Tibial internal–external rotation is computed in a similar manner; however, a positive rotation is internal rotation of the right knee and external rotation of the left knee. We have

Internal Rotation (+) Right External Rotation (+) Left

$$\theta_{I/E} = \left(\frac{\hat{\mathbf{e}}_1 \times \hat{\mathbf{i}}_t}{|\hat{\mathbf{e}}_1 \times \hat{\mathbf{i}}_t|} \cdot \hat{\mathbf{k}}_t \right) a \sin \left(|\hat{\mathbf{e}}_1 \times \hat{\mathbf{i}}_t| \right)$$

or

(8.43)

$$\theta_{I/E} = -a \sin \left(\hat{\mathbf{e}}_1 \cdot \hat{\mathbf{j}}_t \right)$$

Tibial abduction (valgus, or the tibia angling laterally relative to the femur) and adduction (varus, or the tibia angling medially relative to the femur) are computed using the second and third unit vectors of the joint coordinate system. These two vectors are not orthogonal if the knee is abducted or adducted. Excessive abduction of the knees is called knock-knees, and excessive adduction of the knees is called bowlegs. The angle will be positive for adduction of the right leg and positive for abduction of the left knee:

Adduction (+) Right   Abduction (+) Left

$$\theta_{AB/AD} = -a\sin(\hat{\mathbf{e}}_2 \cdot \hat{\mathbf{e}}_3)$$

(8.44)

We now have defined the joint angles and also have established a nonorthogonal coordinate system for the joint that is clinically relevant. The axes of rotation will be different for different joints and should be chosen to maximize communication with the clinical community.

Problems

8.17 Develop the equations for head movement relative to the thoracic cage (cervical spine movement) using a joint coordinate system for the following sequence of rotations: First rotate about the vertical Z-axis of the thoracic cage, then rotate about a floating x'-axis, and, finally, rotate about the y-axis of the head. Thus, the sequence of rotations is a (y, x', Z) sequence.

8.18 For head movement relative to the thoracic cage, develop a joint coordinate system for the following sequence of rotations: First rotate about the medial–lateral Y-axis of the thoracic cage, then rotate about the anterior floating x'-axis, and, finally, rotate about the z-axis of the head. In other words, the sequence is a (z, x', Y) sequence.

8.19 Using the pelvis as the base reference, develop a joint coordinate system for the hip joint on the basis of a (z, x', Y) rotation sequence.

8.20 Using the thoracic cage as a base reference, develop a joint coordinate system for the shoulder using the Eulerian sequence (z, x', Z).

8.6 EQUATIONS OF MOTION

In Section 6.2, we developed the following equations of motion:

$$\sum \mathbf{F} = \frac{d\mathbf{L}}{dt} = \frac{d}{dt}(M\mathbf{v}_{\text{cm}})$$

$$\sum \mathbf{M}_{\text{cm}} = \frac{d\mathbf{H}_{\text{cm}}}{dt}$$

(8.44)

$$\sum \mathbf{M}_0 = \frac{d\mathbf{H}_0}{dt} = \frac{d}{dt}(\mathbf{r}_{\text{cm}/0} \times M\mathbf{v}_{\text{cm}} + \mathbf{H}_{\text{cm}})$$

The third equation relates the moment about a fixed point in space to the change in the angular momentum about that point. This is applicable when the rigid body rotates about a fixed point on the body or the body extended.

The angular-momentum vector about the center of mass is defined as

$$H_{\text{cm}} = H_{\text{cm}_x}\hat{\mathbf{i}} + H_{\text{cm}_y}\hat{\mathbf{j}} + H_{\text{cm}_z}\hat{\mathbf{k}}$$

(8.45)

where the scalar components are

$$H_{cm_x} = I_x\omega_x + I_{xy}\omega_y + I_{xz}\omega_z$$
$$H_{cm_y} = I_{yx}\omega_x + I_y\omega_y + I_{yz}\omega_z \quad\quad (8.46)$$
$$H_{cm_z} = I_{zx}\omega_x + I_{zy}\omega_y + I_z\omega_z$$

The major difficulty in three-dimensional dynamics is that if the centroidal coordinate axes are not fixed to the body—that is, if the coordinate system is not rotating—the mass moments of inertia and the product mass moments of inertia are functions of time. For example, the derivative, with respect to time, of the x-component of the angular momentum is

$$\frac{dH_{cm_x}}{dt} = \dot{I}_x\omega_x + I_x\dot{\omega} + \dot{I}_{xy}\omega_y + I_{xy}\dot{\omega}y + \dot{I}_{xz}\omega_z + I_{xz}\dot{\omega}_z \quad\quad (8.47)$$

Clearly, then, dealing with the 18 derivatives involved in the rate-of-change of angular momentum would render the problem intractable and the resulting differential equations unsolvable.

In Section 5.14, rotating coordinate systems were introduced, and the time derivative in the fixed coordinate system was related to that relative to the rotating coordinate system by Eq. (5.175), or

$$\frac{D}{dt} = \frac{d}{dt} + \quad\quad \Omega x$$

absolute   relative time   angular velocity $\quad\quad (8.48)$
time   derivative   of rotating system
derivative

If the body rotates about a fixed point in space and the body-fixed coordinates are fixed to that point in space, the third equation of Eqs. (8.44) can be written as

$$\sum \mathbf{M}_0 = (\dot{\mathbf{H}}_0)_{rel} + \Omega \times \mathbf{H}_0 \quad\quad (8.49)$$

Expanding Eq. (8.49) into scalar components yields

$$\sum M_x = I_x\frac{d\omega_x}{dt} + I_{xy}\frac{d\omega_y}{dt} + I_{xz}\frac{d\omega_z}{dt}$$
$$\quad + \Omega_y\,(I_{zx}\omega_x + I_{zy}\omega_y + I_z\omega_z)$$
$$\quad - \Omega_z\,(I_{yz}\omega_x + I_y\omega_y + I_{yz}\omega_z)$$

$$\sum M_y = I_{yx}\frac{d\omega_x}{dt} + I_y\frac{d\omega_y}{dt} + I_{yz}\frac{d\omega_z}{dt}$$
$$\quad + \Omega_z(I_x\omega_x + I_{xy}\omega_y + I_{xz}\omega_z) \quad\quad (8.50)$$
$$\quad - \Omega_x(I_{zx}\omega_x + I_{zy}\omega_y + I_z\omega_z)$$

$$\sum M_z = I_{zx}\frac{d\omega_x}{dt} + I_{zy}\frac{d\omega_y}{dt} + I_z\frac{d\omega_z}{dt}$$
$$\quad + \Omega_x(I_{yx}\omega_x + I_y\omega_y + I_{yz}\omega_z)$$
$$\quad - \Omega_y(I_{xy}\omega_y + I_x\omega_x + I_{xz}\omega_z)$$

If the rotating coordinate system is fixed to the rigid body, the angular velocity of the rotating coordinate system is equal to the angular velocity of the body. That is,

$$\boldsymbol{\Omega} = \boldsymbol{\omega} \tag{8.51}$$

The rate-of-change of the linear momentum with respect to time is

$$\sum \mathbf{F} = (\dot{\mathbf{L}})_{\text{rel}} + \boldsymbol{\omega} \times \mathbf{L} \tag{8.52}$$

and the rate-of-change of the angular momentum is

$$\dot{\mathbf{H}}_{\text{cm}} = (\dot{\mathbf{H}}_{\text{cm}})_{\text{rel}} + \boldsymbol{\omega} \times \mathbf{H}_{\text{cm}}$$

or $\tag{8.53}$

$$\dot{\mathbf{H}}_0 = (\dot{\mathbf{H}}_0)_{\text{rel}} + \boldsymbol{\omega} \times \mathbf{H}_0$$

The three scalar equations for the moment and rate-of-change of the angular momentum about the center of mass or a fixed point on the body in space are

$$
\begin{aligned}
M_x &= I_x \dot{\omega}_x + I_{xy}(\dot{\omega}_y - \omega_x \omega_z) + I_{xz}(\dot{\omega}_z + \omega_x \omega_y) \\
&\quad + (I_z - I_y)\omega_y \omega_z + I_{yz}(\omega_y^2 - \omega_z^2) \\
M_y &= I_{xy}(\dot{\omega}_x + \omega_y \omega_z) + I_y \dot{\omega}_y + I_{yz}(\dot{\omega}_z - \omega_x \omega_y) \\
&\quad + (I_x - I_z)\omega_x \omega_z + I_{xz}(\omega_z^2 - \omega_x^2) \\
M_z &= I_{xz}(\dot{\omega}_x - \omega_y \omega_z) + I_{zy}(\dot{\omega}_y + \omega_x \omega_z) + I_z \dot{\omega}_z \\
&\quad + (I_y - I_x)\omega_y \omega_x + I_{xy}(\omega_x^2 - \omega_y^2)
\end{aligned}
\tag{8.54}
$$

If the moments are specified as a function of time, the direct dynamics problem to determine the angular velocity and the angular position involves solving coupled simultaneous nonlinear differential equations. This will require numerical solutions and the use of computational software. However, if the inverse dynamics problem is considered—that is, if the angular position and therefore the angular velocity and the angular acceleration of the body are given as a function of time—the problem is easily solved.

Sample Problem 8.4

A homogeneous ring of mass m and radius R is supported by a smooth collar as shown in the Figure (b). If the collar is attached to a vertical shaft that is rotated at constant angular velocity, determine the angle β that the ring will reach. Determine the minimum angular velocity for the ring to leave the vertical position.

Solution   We will attach the coordinate system to the ring with x directed outward to the plane of the ring and y upward at the collar, as shown in the figure (b). The principal mass moments of inertia about point A are

$$I_{xxA} = mR^2 + mR^2 = 2mR^2$$

$$I_{yyA} = \frac{1}{2}mR^2$$

$$I_{zzA} = \frac{1}{2}mR^2 + mR^2 = \frac{3}{2}mR^2$$

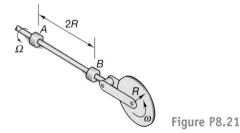

(a)

(b)

The angular velocity of the ring in the ring coordinate system is

$$\omega_x = \omega \sin \beta$$
$$\omega_y = \omega \cos \beta$$
$$\omega_z = \dot{\beta}$$

The only moment is about the z axis and therefore, from Eq. (8.54) we obtain

$$M_z = I_z \dot{\omega}_z + (I_y - I_x)\omega_y \omega_x$$

$$-mgR \sin \beta = \frac{3}{2} mR^2 \ddot{\beta} - \frac{3}{2} mR^2 \omega^2 \sin \beta \cos \beta$$

The differential equation of motion becomes

$$\ddot{\beta} + \left(\frac{2g}{3R} - \omega^2 \cos \beta\right) \sin \beta = 0$$

The stable position that the ring will reach can be obtained examining when $\ddot{\beta} = 0$:

$$\beta = \cos^{-1}\left(\frac{2g}{3R\omega^2}\right)$$

If the ring is to remain in the vertical position, then

$$\beta = 0$$

$$\omega = \sqrt{\frac{2g}{3R}}$$

Problems

8.21 A disk of mass m and radius R is held by the forked rod, as shown in Figure P8.21. The disk and rod are subjected to constant angular velocities ω and Ω, respectively. Determine the reactions at the bearings A and B.

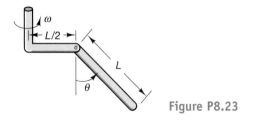

Figure P8.21

8.22 Determine the reactions at the bearings in Problem 8.21 if the disk is replaced by a sphere of mass M and radius R.

8.23 A bar of mass m, radius r, and length L is attached by a clevis to a shaft with a 90° bend, as shown in Figure P8.23. If the shaft is rotated at a constant rate ω about the vertical axis, determine a relationship between the angular velocity and the angle the bar makes with the vertical. Plot this relationship so that the angle the bar makes with the vertical can be determined for any angular velocity. (Assume the values $L = 1$ m and $r = 0.1$ m.)

Figure P8.23

8.24 A disk of mass m rotates with a constant angular velocity ω about a bent shaft that rotates about the vertical axis with a constant angular velocity Ω as shown in Figure P8.24. Neglecting the weight of the shaft, determine the moment and the force that act at point A.

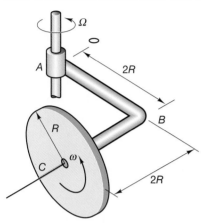

Figure P8.24

8.25 The disk in the system shown in Figure P8.21 rotates at a constant angular velocity ω. The shaft AB starts rotating from rest at a constant angular acceleration Ω. Determine an expression for the moment on the shaft AB required to produce this angular acceleration.

8.26 The disk in the system shown in Figure P8.24 rotates at a constant angular velocity ω. The shaft AB starts rotating from rest at a constant angular acceleration Ω. Neglecting the weight of the shaft, determine the moment and the force that act at point A.

8.27 A thin disk of mass m and radius R is hinged at the end of a shaft that is rotating at a constant angular velocity of ω about its vertical axis. (See Figure P8.27.) Determine the steady-state angle β of the plane of the disk with the vertical axis. Also determine the maximum constant angular velocity required for the disk to remain vertical.

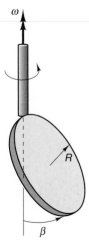

Figure P8.27

8.28 Develop a differential equation for $\beta(t)$ if the disk in Problem 8.27 rotates about the shaft, which is being accelerated at a constant rate starting from rest.

8.29 Numerically solve the differential equation for the angular position of $\beta(t)$, and plot the angular position for the first 5.5 s for Problem 8.28 for a 300-mm-radius disk that is accelerated at a constant rate of 2 rad/s² for the first 3 s and then has a constant angular velocity of 6 rad/s.

8.30 A uniform square plate of mass m is attached to a hinge that rotates at a constant angular velocity ω, as shown in Figure P8.30. Determine the angle β as a function of the angular velocity. Also, determine the range of angular velocity required for the plate to hang in a vertical position.

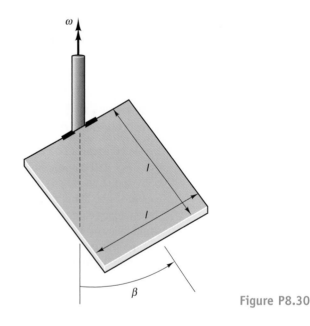

Figure P8.30

8.31 Develop a differential equation for $\beta(t)$ if the disk in Problem 8.30 rotates about the shaft, which is being accelerated at a constant rate starting from rest.

8.7 EULER'S EQUATIONS OF MOTION

One system of equations used to determine the motion of a rigid body moving in three dimensions is known as *Euler's equations of motion.* These equations were developed by the earlier mentioned brilliant Swiss mathematician Leonhard Euler, called by some the most prolific mathematician in history, even though he was totally blind the last 17 years of his life. Euler's equations of motion consisted of Newton's second law [Eq. (8.52)] and the equations of angular momentum [Eqs. (8.54)]. Euler greatly simplified the equations for rotational motion by choosing the *xyz* coordinate system to coincide with the principal axes centered at the center of mass or a set of parallel axes centered at a point fixed in space. With that choice, product mass moments of inertia vanish, and Eqs. (8.54) reduce to

$$
\begin{aligned}
M_x &= I_x\dot{\omega}_x + (I_z - I_y)\omega_y\omega_z \\
M_y &= I_y\dot{\omega}_y + (I_x - I_z)\omega_z\omega_x \\
M_z &= I_z\dot{\omega}_z + (I_y - I_x)\omega_x\omega_y
\end{aligned}
\tag{8.55}
$$

Equations (8.55) generally are known as *Euler's equations of motion.* They are used commonly to solve the rotational motion of a rigid body. Note, however, that they are still nonlinear and therefore are difficult to solve without use of numerical integration techniques.

8.7.1 STABILITY OF ROTATION ABOUT A PRINCIPAL AXIS

Let us consider a body, free of any applied moments, rotating about one of its principal axes. We will assume an *xyz* coordinate system with origin at the center of mass of the body and aligned with the principal axes of the body. We also will assume the body's free rotation to be about the *x*-axis; that is, the angular velocity about this axis is constant, and the angular velocities about the *y*- and *z*-axes are zero. Now, suppose the body is given a slight disturbance and then allowed to rotate freely, but no longer about the *x*-axis. Let us assume the original angular velocity about the *x*-axis to be Ω. The angular velocities reflecting the slight perturbation are

$$
\begin{aligned}
\omega_x &= \Omega + \delta\omega_x \\
\omega_y &= \delta\omega_y \\
\omega_z &= \delta\omega_z
\end{aligned}
\tag{8.56}
$$

If Ω is much larger than the perturbation angular velocities, the first of Euler's equations of motion becomes

$$
\dot{\omega}_x = 0
\tag{8.57}
$$

where higher-order terms are neglected. Since the angular acceleration about the *x*-axis is zero, the angular velocity is equal to its value before the perturbation. The other Euler's equations then become

$$
\begin{aligned}
I_{yy}\dot{\omega}_y + (I_{xx} - I_{zz})\,\Omega\omega_z &= 0 \\
I_{zz}\dot{\omega}_z + (I_{yy} - I_{xx})\,\Omega\omega_y &= 0
\end{aligned}
\tag{8.58}
$$

Differentiating the first of these two equations and substituting the second equation into the differentiated first equation yields

$$\ddot{\omega}_y + \frac{(I_{yy} - I_{xx})(I_{zz} - I_{xx})}{I_{yy}I_{zz}} \Omega^2 \omega_y = 0 \tag{8.59}$$

Equation (8.59) is the differential equation defining the perturbation about the y-axis. This differential equation has the form

$$\frac{d^2y}{dt^2} + \lambda y = 0 \tag{8.60}$$

and its solution, obtained using an integrating factor, is

$$y = Ae^{\sqrt{-\lambda}t} + Be^{-\sqrt{-\lambda}t} \tag{8.61}$$

If λ is positive, the solution is sinusoidal, in which case the solution of Eq. (8.59) is **stable** and the body will experience small oscillations about the x-axis of rotation. However, if λ is negative, the solution is exponential in nature, and the rotational motion about the x-axis is said to be **unstable**. Examining Eq. (8.59) now shows that

$$\lambda = \frac{(I_{yy} - I_{xx})(I_{zz} - I_{xx})}{I_{yy}I_{zz}} \Omega^2 \tag{8.62}$$

The stability of the free rotation depends upon the relative values of the principal mass moments of inertia. If I_{xx} is the largest or smallest principal mass moment of inertia, λ is positive and the rotation is stable. However, λ is negative if $I_{yy} > I_{xx} > I_{zz}$ or $I_{yy} < I_{xx} < I_{zz}$; that is, the mass moment of inertia about the x-axis is an intermediate value. The disturbance will grow exponentially if the body is rotated about an axis with an intermediate principal mass moment of inertia. If the mass moment of inertia about the x-axis is equal to either the mass moment of inertia about the y- or z-axis, then $\lambda = 0$, and the perturbed components of the angular velocity are constant.

8.7.2 MOTION OF AN AXISYMMETRIC OBJECT

An object is said to be *axisymmetric* if it has an axis of rotational symmetry. Cylinders, disks, and cones are examples of axisymmetric objects, and their axis of rotational symmetry is a principal axis of the mass-moment-of-inertia tensor. Let us consider two coordinate systems: an *xyz* coordinate system with the z-axis coincident with the axis of rotational symmetry of the object and an *XYZ* inertial-reference coordinate system. We will relate the two coordinate systems using a rotational transformation ϕ about the Z-axis (a precession angle), followed by a rotational transformation θ about the x-axis (a nutation angle). The body will spin through an angle ψ relative to the *xyz* system about the z-axis. It can be seen that this pair of coordinate systems describes the position of the body with Eulerian angles. The transformation is illustrated in Figure 8.9. The angular velocity of the rotating coordinate system, expressed in the *xyz* coordinate system, is

$$\begin{aligned}
\Omega_x &= \dot{\theta} \\
\Omega_y &= \dot{\phi} \sin \theta \\
\Omega_z &= \dot{\phi} \cos \theta
\end{aligned} \tag{8.63}$$

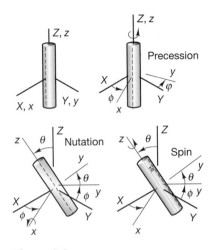

Figure 8.9

The angular velocity vector of the body can be written as

$$\boldsymbol{\omega} = \dot{\boldsymbol{\phi}} + \dot{\boldsymbol{\theta}} + \dot{\boldsymbol{\psi}} \qquad (8.64)$$

The angular velocity vector of the body can be expressed in the *xyz* coordinate system as

$$
\begin{aligned}
\omega_x &= \dot{\theta} \\
\omega_y &= \dot{\phi} \sin \theta \\
\omega_z &= \dot{\phi} \cos \theta + \dot{\psi}
\end{aligned}
\qquad (8.65)
$$

Note that, since the *z*-axis is an axis of rotational symmetry, we do not have to spin the *xyz* coordinate system with the body. All *x*- and *y*-axes on the body through its center of mass are principal axes, $I_{xx} = I_{yy}$. Accordingly, the equations of motion [see Eqs. (8.50)] for an axisymmetric body are

$$\sum M_x = I_{xx}\ddot{\theta} - I_{xx}(\dot{\phi}\sin\theta)(\dot{\phi}\cos\theta) + I_{zz}(\dot{\phi}\sin\theta)(\dot{\psi} + \dot{\phi}\cos\theta)$$

$$\sum M_y = I_{xx}\frac{d}{dt}(\dot{\phi}\sin\theta) - I_{zz}\dot{\theta}(\dot{\phi}\cos\theta + \dot{\psi}) + I_{xx}(\dot{\phi}\cos\theta)\dot{\theta} \qquad (8.66)$$

$$\sum M_z = I_{zz}\frac{d}{dt}(\dot{\phi}\cos\theta + \dot{\psi})$$

A wide range of applications arises for an axisymmetric body in a uniform gravitational field when one point on the axis of symmetry is fixed in space.

The **gyroscopic motion** described by Eq. (8.66) is one of the most important and interesting problems of dynamics. Systems governed by this equation range from a child's top to gyroscopic navigational systems. Gyroscopic effects also are an important design consideration in examining bearing forces for shafts of rotors that are subjected to forced precessions. The gyroscope, which is discussed in more detail later, has important engineering applications, including inertial guidance systems, gyro compasses, and large gyros to stabilize the motion of ships.

Sample Problem 8.5

Consider a thin disk of mass m and radius r mounted on a rigid shaft of negligible mass, as shown in the figure (a). The plane of the disk is not normal to the shaft, so the principal z-axis of the disk does not coincide with the axis of the shaft. If the shaft is rotating at a constant angular velocity ω, determine the reactions at the bearings.

Solution   The principal moments of inertia of the disk are

$$I_z = \frac{1}{2} mr^2 \qquad I_x = I_y = \frac{1}{4} mr^2$$

The angular velocity in the body coordinate system is

$$\boldsymbol{\omega} = \omega \sin \theta \hat{\mathbf{j}} + \omega \cos \theta \hat{\mathbf{k}}$$

Using Euler's equations of motion [Eqs. (8.55)], we obtain, for the moments,

$$M_x = \frac{1}{4} mr^2 \omega^2 \sin \theta \cos \theta$$

$$M_y = M_z = 0$$

A free-body diagram of the shaft and disk is shown in the figure (b), with the positive x-axis into the paper.

The reactions at the bearings are

$$R_l = R_r = \frac{mr^2 \omega^2 \sin \theta \cos \theta}{4l}$$

If the angle θ is small, the reactions are

$$R_l = R_r = \frac{mr^2 \omega^2 \theta}{4l}$$

The rate-of-change of the angular momentum vector with respect to time is sometimes called the *gyroscopic moment,* in a manner similar to the way inertial forces are treated as actual forces. Conceptually, this is useful if the shaft is considered a deformable body. The deformed shaft would appear as shown in the figure (c). Using beam theory, we can treat the shaft as a beam with zero slope at both ends due to moments at those points. The loading is a concentrated moment (the gyroscopic moment) at the center of the shaft.

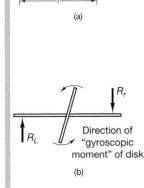

(a)

(b)

Direction of "gyroscopic moment" of disk

(c)

8.7.3 HEAVY AXISYMMETRIC TOP

The motion of a top has been studied extensively and has become a classical problem in rigid-body dynamics. Consider the top shown in Figure 8.10, where the origin of the xyz coordinate system is taken at the base of the top. The applied moment to the top, expressed in the xyz coordinate system, is

$$\mathbf{M} = mgl \sin \theta \hat{\mathbf{i}} \tag{8.67}$$

Euler's equations (8.66) for this case become

$$mgl \sin \theta = I_{xx}\ddot{\theta} - I_{xx}(\dot{\phi} \sin \theta)(\dot{\phi} \cos \theta) + I_{zz}(\dot{\phi} \sin \theta)(\dot{\psi} + \dot{\phi} \cos \theta)$$

$$0 = I_{xx}\frac{d}{dt}(\dot{\phi} \sin \theta) - I_{zz}\dot{\theta}(\dot{\phi} \cos \theta + \dot{\psi}) + I_{xx}(\dot{\phi} \cos \theta)\dot{\theta} \tag{8.68}$$

$$0 = I_{zz}\frac{d}{dt}(\dot{\phi} \cos \theta + \dot{\psi})$$

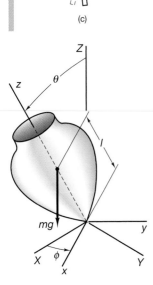

Figure 8.10

The third equation of Eqs. (8.68) can be integrated directly, yielding

$$(\dot{\phi}\cos\theta + \dot{\psi}) = p_\psi \qquad \text{(constant)} \qquad (8.69)$$

The quantity $I_{zz}p_\psi$ called the *generalized momentum* and is constant in this case. The second equation of Eqs. (8.68) can be written as

$$\sin\theta\left[I_{xx}\frac{d}{dt}(\dot{\phi}\sin\theta) - I_{zz}p_\psi\dot{\theta} + I_{xx}(\dot{\phi}\cos\theta)\dot{\theta}\right] = \frac{d}{dt}[I_{xx}\dot{\phi}\sin^2\theta + I_{zz}p_\psi\cos\theta] \qquad (8.70)$$

Another first integral can be taken leading to the relation

$$[I_{xx}\dot{\phi}\sin^2\theta + I_{zz}p_\psi\cos\theta] = p_\phi \qquad \text{(constant)} \qquad (8.71)$$

The quantity p_ϕ also is called the generalized momentum. We can use the two constant generalized momenta to derive a single differential equation for the nutation angle θ if we will change the constants as follows to simplify the algebra:

$$
\begin{aligned}
p_\psi &= \frac{I_{xx}}{I_{zz}}a \\
p_\phi &= I_{xx}b
\end{aligned}
\qquad (8.72)
$$

Equation (8.71) can then be written as

$$\dot{\phi} = \frac{b - a\cos\theta}{\sin^2\theta} \qquad (8.73)$$

Using Eqs. (8.72) and (8.73), we can write Eq. (8.69) as

$$\dot{\psi} = \frac{I_{xx}}{I_{zz}}a - \frac{b\cos\theta - a\cos^2\theta}{\sin^2\theta} \qquad (8.74)$$

The first equation in Eqs. (8.65) can be written as the single second-order nonlinear differential equation

$$\frac{d^2\theta}{dt^2} = \frac{1}{I_{xx}}[mgl\sin\theta + I_{xx}(\dot{\phi}^2\sin\theta\cos\theta - \dot{\phi}\sin\theta a)] \qquad (8.75)$$

Equations (8.73), (8.74), and (8.75) form three coupled differential equations for the Eulerian angles and can be solved numerically. The constants a and b are determined from the initial conditions.

We have presented a method for completely solving the motion of the top numerically, but a large amount of progress had been made previously to gain insight into the motion of the top by considering its energy. If the kinetic energy connected with the spin axis is large compared to the gravitational potential energy, the top is referred to as a *fast top*. The effects of the gravitational torques—namely, the precession and accompanying nutation, will be small perturbations on the spin rotation of the top. Given an initial nutation angle of 10°, initial precession rate of 3 rad/s, and initial spin of 100 rad/s, the nutation and precession for a top are shown in Figures 8.11(a) and (b), respectively. The ratio b/a for this example was 0.989.

Consider the case of steady precession of the top at a constant nutation angle. We can imagine that we could obtain this state by first giving the top a specific spin rate while holding it a particular nutation angle and then giving it the proper precession rate. If the

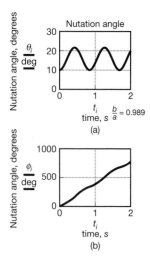

Figure 8.11

precession rate is not proper, the motion will not be stable, and the nutations will be superimposed on the constant rate. The first equation of Eqs. (8.68) can be written as

$$mgl = -I_{xx}\dot{\phi}^2 \cos\theta + I_{zz}\dot{\phi}(\dot{\psi} + \dot{\phi}\cos\theta) \tag{8.76}$$

Using Eqs. (8.65), we can write Eq. (8.76) in terms of the angular velocity about the z-axis, which is a constant:

$$(I_{xx}\cos\theta)\dot{\phi}^2 - I_{zz}\omega_z\dot{\phi} + mgl = 0 \tag{8.77}$$

We can view Eq. (8.77) as a quadratic equation to determine the proper precession rate. Solving this equation yields

$$
\begin{aligned}
\dot{\phi}_1 &= \frac{I_{zz}\omega_z}{2I_{xx}\cos\theta}\left(1 - \sqrt{1 - \frac{4I_{xx}\cos\theta mgl}{I_{zz}^2\,\omega_z^2}}\right)\\[2mm]
\dot{\phi}_2 &= \frac{I_{zz}\omega_z}{2I_{xx}\cos\theta}\left(1 + \sqrt{1 - \frac{4I_{xx}\cos\theta mgl}{I_{zz}^2\,\omega_z^2}}\right)
\end{aligned}
\tag{8.78}
$$

The first solution corresponds to a slow precession, and the second corresponds to a fast precession. For a large spin rate, we can approximate the square root as

$$\sqrt{1-x} \approx 1 - \frac{x}{2} \quad \text{for} \quad x \ll 1$$

For slow precession with a high spin rate, we obtain

$$\dot{\phi}_1 \approx \frac{mgl}{I_{zz}\,\omega_z} \tag{8.79}$$

where $\omega_z \approx \dot{\psi}$ since $\dot{\phi}$ is very small. By contrast, fast precession with a high spin rate is not seen experimentally, because it is almost impossible to create the proper initial precession rate.

Fast or slow precession is possible only if the radicand (the quantity under the radical) is positive. Therefore, stable motion is possible only if

$$\omega_z^2 > \frac{4I_{xx}mgl}{I_{zz}^2}\cos\theta \tag{8.80}$$

For a "standing-up" top ($\cos\theta > 0$), the motion becomes unstable for a spin rate below that required by the inequality in Eq. (8.80). However, for a gyro pendulum ($\cos\theta < 0$), the motion is stable for all spin rates.

8.7.4 GYROSCOPIC MOTION WITH STEADY PRECESSION

Let us consider the special case of gyroscopic motion in which the nutation angle is a constant and there is steady precession; that is,

$$
\begin{aligned}
\theta &= \theta_0\\
\dot{\phi} &= p \text{ (constant)}
\end{aligned}
\tag{8.81}
$$

In this case, Eqs. (8.66) reduce to

$$\sum M_x = -I_{xx}(\dot{\phi}^2 \sin\theta \cos\theta) + I_{zz}(\dot{\phi}\sin\theta)(\dot{\psi} + \dot{\phi}\cos\theta)$$
$$\sum M_y = 0 \tag{8.82}$$
$$\sum M_z = 0$$

We can gain a greater insight into Eq. (8.82) by assuming a constant nutation angle of 90°, in which case

$$\sum M_x = I_{zz}\dot{\phi}\dot{\psi} \tag{8.83}$$

If we designate the x-axis as the moment axis, the Z- or y-axis as the precession axis, and the z-axis as the spin axis, we see that these axes form an orthogonal coordinate system, as shown in Figure 8.12. Note that the gyroscopic moment about the x-axis is

$$mgc = I_{zz}\dot{\phi}\dot{\psi} \tag{8.84}$$

We can write the precession rate in terms of the angular momentum about the z-axis as

$$\dot{\phi} = \frac{mgc}{H_{0z}} \tag{8.85}$$

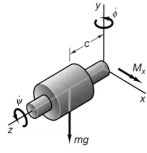

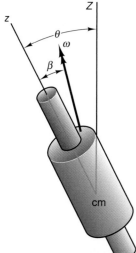

Figure 8.12

8.7.5 MOTION OF AN AXISYMMETRIC BODY SUBJECTED TO NO EXTERNAL FORCES

In this section, we consider an axisymmetric body subjected only to gravitational attraction— that is, its weight—and no external forces. This situation arised in the study of projectiles when air resistance can be neglected and in the study of space vehicles or satellites after burnout of their rockets. In these cases, the angular momentum about the center of mass is a constant, so that its derivative

$$\dot{\mathbf{H}}_{cm} = 0 \tag{8.86}$$

The angular momentum vector is at an angle with the axisymmetric axis z, as shown in Figure 8.13. The axis in the direction of the angular momentum vector will be designated as the Z-axis, or the axis of precession. This axis is fixed in space. The components of the angular momentum vector in the x- and y-axes are

$$H_x = I_{xx}\omega_x = -H_{cm}\sin\theta$$
$$H_z = I_{zz}\omega_z = H_{cm}\cos\theta \tag{8.87}$$

The components of the angular velocity vector of the body are

$$\omega_x = \frac{-H_{cm}\sin\theta}{I_{xx}}$$
$$\omega_z = \frac{H_{cm}\cos\theta}{I_{zz}} \tag{8.88}$$

Figure 8.13

The angular velocity vector makes an angle β with the z-axis given by

$$\tan\beta = \frac{-\omega_x}{\omega_z} = \frac{I_{zz}}{I_{xx}}\tan\theta \tag{8.89}$$

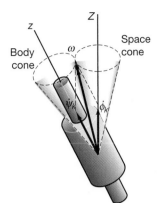

Figure 8.14

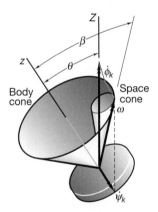

Figure 8.15

There are two special cases of motion of an axisymmetric body that do not involve precession. The first occurs when the angular momentum vector is aligned with the z-axis of the body—that is, when $\theta = 0$. In this case, the body keeps spinning about its axis of symmetry, and by Eq. (8.62), the motion is stable. The second case occurs when the angular momentum vector is aligned with a transverse body axis—that is, when $\theta = 90°$. In this case, the body keeps spinning about this transverse axis. The two cases are respectively seen when, in a football game, the quarterback throws a perfect spiral and the placekicker kicks the football end over end for a field goal.

We can visualize the general motion of an axisymmetric body by setting up two imaginary cones, one with an axis coincident with the z-axis of the body and the other with an axis coincident with the Z-axis, the precession axis. These cones are shown in Figure 8.14. The **body cone** is the cone created by the angular velocity vector's motion about the z-axis of the body and has a half-cone angle of β. The **space cone** is created by precession of the angular velocity vector around the Z-axis and has a half-cone angle of $|\theta - \beta|$. The body cone rolls on the space cone.

The general motion of the space cone can be characterized by the relative value of the mass moment of inertia about the z-axis with respect to that about the x-axis:

1. $I_{zz} < I_{xx}$. This occurs when the axisymmetric body is an elongated body such as a rocket, a football, or the body shown in Figure 8.14. From Eq. (8.89), we see that in this case $\theta > \beta$, and the space cone and the body cone are tangent externally. The spin and precession are both positive relative to the z- and Z-axis, respectively (i.e., positive defined by the right-hand rule). The precession is said to be **direct**.

2. $I_{zz} > I_{xx}$. This occurs in the case of a flattened body as shown in Figure 8.15. From Eq. (8.89), we see that in this case $\beta > \theta$, and the space cone is interior to the body cone. The precession is positive relative to the Z-axis, but the spin is negative relative to the z-axis. The precession and the spin thus have opposite senses, and the precession is said to be **retrograde**.

8.7.6 THE GYROSCOPE

The motion of a gyroscope is governed by Eqs. (8.66), which are repeated here as follows:

$$\sum M_x = I_{xx}\ddot{\theta} - I_{xx}(\dot{\phi}\sin\theta)(\dot{\phi}\cos\theta) + I_{zz}(\dot{\phi}\sin\theta)(\dot{\psi} + \dot{\phi}\cos\theta)$$

$$\sum M_y = I_{xx}\frac{d}{dt}(\dot{\phi}\sin\theta) - I_{zz}\dot{\theta}(\dot{\phi}\cos\theta + \dot{\psi}) + I_{xx}(\dot{\phi}\cos\theta)\dot{\theta} \qquad (8.91)$$

$$\sum M_z = I_{zz}\frac{d}{dt}(\dot{\phi}\cos\theta + \dot{\psi})$$

Figure 8.16

A gyroscope consists of a rotor mounted on a set of gimbal rings or gimbals, as shown in Figure 8.16. The gyroscope shown in the figure is free of moments that might be applied to its base due to the bearings on the gimbals if friction is neglected. If the gyro is given a high angular velocity, its angular momentum will be conserved, and the gimbal rings will assume positions such that the gyro remains vertical. This is called a **free gyroscope** and is used in a gyrocompass. If the bearings on the x-axis are fixed, the gyro will resist any moment about that axis. In this way, a gyroscope can be used as a stabilizer.

Problems

8.32 A slender bar of mass m and length $2l$ is attached to a shaft of negligible weight, as shown in Figure P8.32. If the shaft rotates at a constant angular velocity of ω, determine the reactions at the bearings, and show a loading diagram of the shaft, modeled as a flexible beam.

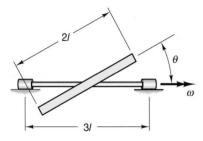

Figure P8.32

8.33 Assume that the shaft in the system shown in Figure P8.32 starts from rest and is accelerated by a constant angular acceleration of α rad/s². Determine the reactions at the bearings.

8.34 Plot the reactions at the bearings for the shaft in Problem 8.33 for the first two seconds, for the following parameters: $l = 1$ m, $m = 2$ kg, $\theta = 30°$, and $\alpha = 3$ rad/s².

8.35 A 3000-kg, 3-m-diameter, 5-m-long steam turbine assembly is misaligned on its shaft by a small angle of 5°, as shown in Figure P8.35. If the assembly is accelerated at a constant rate of $\alpha = \pi$ rad/s² until it reaches an operating speed of 3600 rpm, determine the bearing forces during start-up and at the operating speed. Model the turbine assembly as a cylinder.

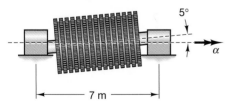

Figure P8.35

8.36 A group of engineers wants to play a practical joke on a bellhop at a hotel. The engineers construct a battery-operated rotor system such that the assembly has an angular momentum vector of 50 kg m²/s directed toward the front of the case, as shown in Figure P8.36.

Describe the moment the bellhop must apply to the case when he reaches the end of the corridor and wishes to turn left. If he applies a moment about the vertical axis of the case, as would be natural, describe the movement of the case.

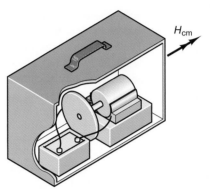

Figure P8.36

8.37 A 2000-kg turbine on a ship is mounted along the long axis of the ship and rotates at a constant rate of 3600 rpm. (See Figure P8.37.) The turbine has a mass radius of gyration along its long axis of 0.4 m and about its transverse axes of 0.6 m. The bearings are separated by a distance of 3 m. Determine the reactions at the bearings when (a) the ship makes a turn to the left of 0.2 rad/s, (b) rolling along its long axis 0.4 rad/s, and (c) pitching forward at a rate of 0.3 rad/s due to wave swells.

Figure P8.37

8.38 A thin rectangular plate of mass m, as shown in Figure P8.38, rotates about a horizontal axis through its diagonal with a constant angular velocity ω. Determine the reactions at the bearings.

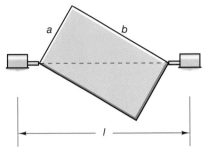

Figure P8.38

8.39 Show that the condition for steady precession for the top shown in Figure P8.39 is $(I_{zz}\omega_z - I_{xx}\dot\phi \cos\theta)\dot\phi = mgl$, and show that if $\dot\psi$ is very large compared to the rate of precession, $\dot\phi$, the condition for steady precession is $I_{zz}\dot\psi\dot\phi \approx mgl$.

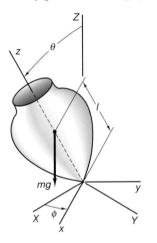

Figure P8.39

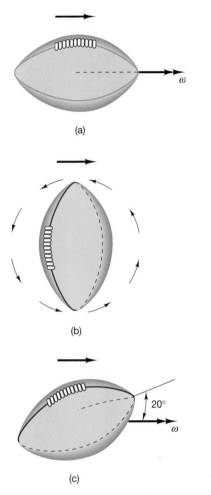

Figure P8.41

💻 **8.40** Consider the top in Problem 8.39 with a mass of 0.085 kg and radii of gyration of 20 mm about the z-axis and 45 mm about the x-axis. The distance from the tip of the top to its center of mass is 35 mm. If the spin rate of the top is 200 rad/s at a constant nutation angle of 30°, determine the two possible rates of precession. Determine the error if the approximation developed in Problem 8.39 is used.

8.41 Figure P8.41 shows three common flight configurations of a football. Case (a) is a perfect spiral pass with a spin rate of 150 rev/min. Case (b) is an end-over-end place kick rotating at a rate of 150 rev/min. Case (c) is a kicked football wobbling with an angular velocity of 150 rev/min about a horizontal axis, but with its axisymmetric axis at an angle of 20°. The mass moment of inertia about the long axis is one-third that about the transverse axis. Determine the orientation of the axis of precession and the rates of spin and precession in each case.

8.42 A satellite can be modeled as a 600-kg, 8-m-long and 3-m-diameter cylinder. (See Figure P8.42.) If the nutation angle is 30° and the spin rate is 7 rad/s, determine the rate of precession of the satellite.

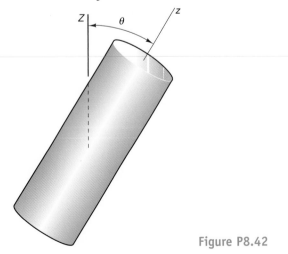

Figure P8.42

8.43 If the satellite in Figure P8.42 precesses at a rate of 1 rev/s, determine the spin rate of the satellite.

8.44 The rocket shown in Figure P8.44 is in moment-free steady precession with a nutation angle of 25° and a spin rate of 5 rev/sec. Determine the precession rate if the mass moments of inertia are

$$I_{xx} = 12,000 \text{ kg} \cdot \text{m}^2$$
$$I_{zz} = 3000 \text{ kg} \cdot \text{m}^2$$

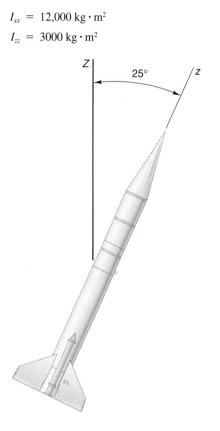

Figure P8.44

8.45 A coin tossed into the air undergoes moment-free steady precession. (See Figure P8.45.) Show that the rate of precession is related to the rate of spin by

$$\dot{\phi} = -\frac{2\dot{\psi}}{\cos \theta}.$$

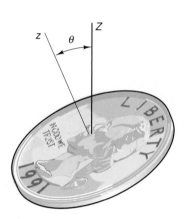

Figure P8.45

Chapter 9 VIBRATION

Rotor blade vibrations must be considered in the design of helicopters, such as the one shown assisting in the rescue of an alpine skier.

9.1 INTRODUCTION

Vibration is very common in engineering systems and forms an important subset of dynamics. We already have solved many vibration problems in the previous chapters, without calling specific attention to them. The kinematic description of Sample Problem 1.2 and the swinging of a pendulum in Sample Problem 2.13 are examples of vibration. Vibration is the study of the repetitive motion of objects relative to some stationary frame of reference or nominal position (usually, the equilibrium position). Vibration is evident everywhere and in many cases greatly affects the nature of engineering design. The vibrational properties of engineering devices are often limiting factors in their performance. Vibration can be harmful and should then be avoided, or it can be extremely useful and desired. For instance, vibration induced in a building or car causes fatigue and failure. Yet vibration purposely is introduced by designers into certain machines used to sort parts and by musicians to create music. Vibration is often taught as an advanced undergraduate course in mechanical, aerospace, civil, and engineering mechanics curricula. The purpose of this chapter is to provide a brief introduction to the field of vibration analysis and thus form a bridge to later courses specializing in vibration. We present an introductory theory of vibration, starting with a free-body diagram, progressing to the notion of summation of forces to produce equations of motion, and then solving and analyzing the equations of motion.

Typical examples of vibration familiar to most are the motion of a guitar string, the quality of ride of an automobile or motorcycle, the motion of an airplane's wings, and the swaying of a large building due to a wind. Vibrational analysis is extremely important in the design of machines and structures, as the effective loads imposed through vibration can easily exceed static loads by orders of magnitude. Hence, engineering designs that do not account for vibration can often lead to catastrophic failure. The concept of vibration leads to a unique phenomenon, called resonance, which can be extremely useful, but is also the source of many failures when it is not properly accounted for. Failures due to resonance that may be familiar to the reader include

- The Tacoma Narrows Bridge collapse
- The collapse of some buildings and bridges in mild earthquakes
- The collapse of a hotel skywalk as the result of dancing

Another familiar example of resonance is the shaking of an automobile steering wheel because of an out-of-balance wheel and tire.

The physical explanation of the phenomenon of vibration concerns the interplay between potential energy and kinetic energy. A vibrating system must have a component that stores potential energy and releases it as kinetic energy in the form of motion (vibration) of a mass. The motion of the mass then gives up kinetic energy to the potential-energy storing device. Vibration can occur in many directions and can be the result of the interaction of many objects. Here, however, we will focus on simple, basic models.

9.2 UNDAMPED SINGLE-DEGREE-OF-FREEDOM SYSTEMS

From Chapter 1, the fundamental kinematic quantities used to describe the motion of a particle are displacement, velocity, and acceleration vectors. In addition, Newton's law, discussed in Chapter 2, states that the motion of a mass is determined by the net force acting on the mass. An easy device to use in thinking about vibration is a spring (such as one used to pull a storm door shut or an automobile suspension spring) with one end attached to a

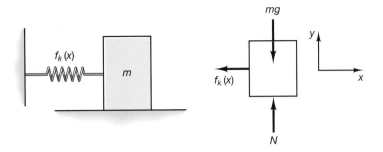

Figure 9.1

fixed object and a mass attached to the other end. A schematic of this arrangement is given in Figure 9.1 and is referred to as a spring-and-mass system. Such a system was first introduced in Problem 1.9. Let the spring rest on a frictionless surface, and consider a free-body diagram of the mass displaced slightly from equilibrium. Ignoring the mass of the spring itself, we find that the forces acting on the mass consist of the force of gravity pulling downward (mg), the normal force N pushing upwards, and the force of the spring pulling back to the left, f_k. The summation of forces in the y-direction yields, simply, $N = mg$, as the mass is constrained in static equilibrium in this direction. Writing Newton's law in the x-direction gives

$$m\ddot{x}(t) = -f_k(x(t)) \quad \text{or} \quad m\ddot{x} + f_k(x) = 0 \tag{9.1}$$

which is a second-order differential equation with respect to time, subject to two initial conditions (the initial velocity and position), as discussed in Section 1.5. Eq. (9.1) is the equation of motion for a spring–mass system.

The nature of the spring force $f_k(x)$ can be deduced by performing a simple static experiment, as illustrated in Figure 9.2. With no mass attached, the spring hangs in static equilibrium under the force of gravity in a position labeled $x_0 = 0$. As successively more mass is attached to the spring, the force of gravity causes the spring to stretch further. If the value of the mass is recorded, along with the value of the displacement of the end of the spring each time more mass is added, a plot of the force (mass, denoted by m, times the acceleration due to gravity, denoted by g) versus this displacement, denoted by x, yields a force-deflection curve similar to that illustrated in Figure 9.3.

Note that in the region of values for x between 0 and about 20 mm the curve is a straight line. This indicates that, for deflections less than 20 mm and forces less than 1000 N, the force that is applied by the spring to the mass is proportional to the stretch of the spring. The constant of proportionality is the slope of the straight line between 0 and 20 mm.

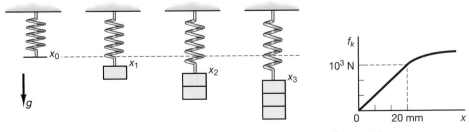

Figure 9.2

Figure 9.3

For the particular spring of Figure 9.3, the constant is 50 N/mm, or 5×10^4 N/m. Thus, the equation that describes the force applied by the spring to the mass is

$$f_k(x) = kx \tag{9.2}$$

The value of the slope, denoted by k, is called the *stiffness* of the spring and is a property that characterizes the spring in all situations in which the displacement is less than 20 mm. The relationship between f_k and x in Eq. (9.2) is *linear*; that is, the curve is a straight line, and f_k depends linearly on x. If the displacement of the spring is larger then 20 mm, the relationship between f_k and x becomes *nonlinear*, as indicated in Figure 9.3. Nonlinear systems are much more difficult to analyze and usually must be solved numerically, as discussed in Section 1.5.

9.2.1 LINEAR VIBRATION

First, we will assume that displacements (and forces) are limited to the linear range. In that case, the resulting equation of motion forms the *linear* vibration problem obtained by substitution of Eq. (9.2) into Eq. (9.1) resulting in

$$m\ddot{x}(t) + kx(t) = 0 \tag{9.3}$$

which is a second-order linear differential equation subject to two initial conditions, as discussed and solved in Section 1.5. It is usual to write Eq. (9.3) in monic (leading coefficient of one) form by dividing by the mass to get

$$\ddot{x}(t) + \omega_n^2 x(t) = 0 \tag{9.4}$$

where $\omega_n = \sqrt{k/m}$ is the *angular natural frequency* in units of radians per second (rad/s). In this form, the natural frequency of the system is obvious. As discussed in Section 1.5, Eqs. (9.3) and (9.4) have a solution of the form

$$x(t) = A \sin(\omega_n t + \phi) \tag{9.5}$$

which is the sine function in its most general form and which describes periodic phenomena. Here, the constant A is the *amplitude* or maximum value of the periodic function, ω_n determines the interval in time during which the function repeats itself, and ϕ (called the *phase*), in radians, determines the initial value of the sine function. It is standard to measure the time t in seconds (s). The values of the amplitude A and the phase ϕ are determined by the initial conditions specified for the motion.

The equation of motion (9.4) is of second order, so that solving it involves integrating twice. Thus, there are two constants of integration to evaluate. These are the constants A and ϕ. The physical significance, or interpretation, of these constants is that they are determined by the initial state of motion of the spring–mass system. Recall from Newton's laws that if no motion is imparted to the mass, it will stay at rest. If, however, the mass is displaced to a position x_0 at time $t = 0$, the potential energy in the spring will result in motion. Also, if the mass is given an initial velocity v_0 at time $t = 0$, motion will result. The last two statements specify the *initial conditions*, which, when substituted into the solution, Eq. (9.5), yield

$$x_0 = x(0) = A \sin(\omega_n 0 + \phi) = A \sin(\phi) \tag{9.6}$$

and

$$v_0 = \dot{x}(0) = \omega A \cos(\omega_n 0 + \phi) = \omega A \cos(\phi) \tag{9.7}$$

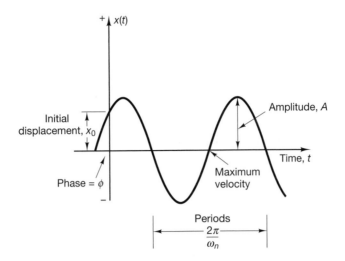

Figure 9.4

Solving these two simultaneous equations for the two unknowns A and ϕ gives

$$A = \frac{\sqrt{\omega_n^2 x_0^2 + v_0^2}}{\omega_n} \qquad \text{and} \qquad \phi = \tan^{-1} \frac{\omega_n x_0}{v_0} \tag{9.8}$$

Hence, the solution of the equation of motion for the spring–mass system is

$$x(t) = \frac{\sqrt{\omega_n^2 x_0^2 + v_0^2}}{\omega_n} \sin\left[\omega_n t + \tan^{-1} \frac{\omega_n x_0}{v_0}\right] \tag{9.9}$$

and is plotted in Figure 9.4. This solution is called the *free response* of the system, because no force external to the system is applied after $t = 0$. The motion of the spring–mass system is called *simple harmonic motion* or *oscillatory motion* and is discussed in detail in the next section. The spring–mass system also is referred to as a *simple harmonic oscillator*, as well as an *undamped single-degree-of-freedom system*.

Many systems can be modeled by a simple spring and mass. Once a system is modeled as a linear spring and mass, its solution is given by Eq. (9.9), where the values of m and k determine the frequency of vibration and the initial conditions determine the phase and amplitude.

The angular natural frequency ω_n is measured in radians per second, denoted rad/s, and describes the repetitiveness of the oscillation. As indicated in Figure 9.4, the time the cycle takes to repeat itself is the *period T*, which is related to the natural frequency by

$$T = \frac{2\pi \text{ rad}}{\omega_n \text{ rad/s}} = \frac{2\pi}{\omega_n} \text{ s}$$

This equation results from the elementary definition of the period of a sine function. Quite often, the frequency is measured and discussed in terms of hertz (Hz), which is the number of cycles per second. The frequency in hertz, denoted f, is related to the frequency in radians per second, denoted ω_n, by

$$f = \frac{\omega_n}{2\pi} = \frac{\omega_n \text{ rad/s}}{2\pi \text{ rad/cycle}} = \frac{\omega_n \text{ cycles}}{2\pi \text{ s}} = \frac{\omega_n}{2\pi} \text{ Hz}$$

Sample Problem 9.1

Consider a small spring about 30 mm ($1\frac{1}{6}$ in.) long, welded to a stationary table (ground) so that it is fixed at the point of contact, with a 12-mm ($\frac{1}{2}$-in.) bolt welded to the other end, which is free to move. The mass of this system is about 49.2×10^{-3} kg (equivalent to about 1.73 oz). The stiffness of the spring can be measured using the method suggested in Figure 9.2. Such a method yields a spring constant of $k = 857.8$ N/m. Calculate the natural frequency and period of the system. Also, determine the maximum amplitude of the response if the spring is initially deflected 10 mm.

Solution   The natural frequency is

$$\omega_n = \sqrt{k/m} = \sqrt{\frac{857.8\ \text{N/m}}{49.2 \times 10^{-3}\ \text{kg}}} = 132\ \text{rad/s}$$

In hertz, this becomes

$$f = \frac{\omega_n}{2\pi} = 21\ \text{Hz}$$

The period is

$$T = \frac{2\pi}{\omega_n} = \frac{1}{f} = 0.048\ \text{s}$$

To determine the maximum value of the displacement response, note from Figure 9.4 that this corresponds to the value of the constant A. Assuming that no initial velocity is given to the spring ($v_0 = 0$), Eq. (9.8) yields

$$A = \frac{\sqrt{\omega_n^2 x_0^2 + v_0^2}}{\omega_n} = x_0 = 10\ \text{mm}$$

Note that the maximum value of the velocity response is $\omega_n A$ or $\omega x_0 = 1{,}320$ mm/s, and the acceleration response has a maximum value $\omega_n^2 A = \omega_n^2 x_0 = 174.24 \times 10^3$ mm/s². Since $v_0 = 0$, the phase is $\phi = \tan^{-1}\frac{\omega_n x_0}{0} = \pi/2$, or 90°. Hence, in this case, the response is

$$x(t) = 10\sin(132t + \pi/2) = 10\cos(132t)\ \text{mm}$$

Sample Problem 9.2

Plot the position, velocity, and acceleration of a linear spring–mass system, and note the relationship between the amplitude and phase of each.

Solution   Straightforward differentiation of the position given by Eq. (9.5), to obtain the velocity and then the acceleration, yields the given plots.

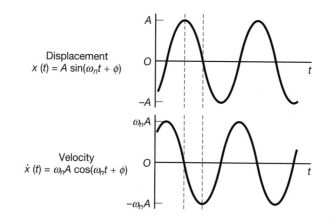

Displacement
$x(t) = A\sin(\omega_n t + \phi)$

Velocity
$\dot{x}(t) = \omega_n A\cos(\omega_n t + \phi)$

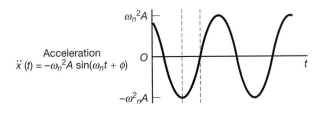

Acceleration
$\ddot{x}(t) = -\omega_n^2 A \sin(\omega_n t + \phi)$

Note that the amplitude of the velocity is scaled from that of the position by a factor of ω_n and that of the acceleration by a factor of ω_n^2. Also, note that the velocity is 90° out of phase with the position and the acceleration is 180° out of phase with the position.

The fundamental kinematic properties of a particle moving in one dimension are displacement, velocity, and acceleration. For the harmonic motion of a simple spring–mass system, these are given by the expressions next to the plots in Sample Problem 9.2. For systems with a natural frequency larger than 1 rad/s, the relative amplitude of the velocity response is larger than that of the displacement response by a factor of ω_n, and the acceleration response is larger by a factor of ω_n^2. For systems with a frequency less than 1, the velocity and acceleration have smaller relative amplitudes than the displacement does.

Sample Problem 9.3

Consider the pendulum illustrated in Sample Problem 2.13. What determines the natural frequency of the linearized pendulum equation?

Solution   From the free-body diagram (following Sample Problem 2.13), the linearized equation of motion (approximating $\sin\theta$ with θ) is

$$\ddot{\theta}(t) + \frac{g}{L}\theta(t) = 0$$

A comparison with Eq. (9.4) yields the result that the natural frequency of the pendulum is

$$\omega = \sqrt{g/L}$$

Thus, the natural frequency is determined by the acceleration due to gravity and the length of the pendulum. Early clock makers used this relationship to set the "speed" of a clock to fit the actual time by adjusting the length L.

9.2.2 NONLINEAR VIBRATION

If the spring represented in Figure 9.3 is forced or stretched beyond the 20-mm mark, the spring force $f_k(x)$ becomes a nonlinear function of x, and the corresponding equation of motion becomes nonlinear, requiring numerical methods for its solution and introducing new physical behavior. In this case, the equation of motion is of the form of Eq. (9.1), where $f_k(x)$ can take on a variety of functional forms, depending on the nature of the spring. A common nonlinearity is cubic of the form

$$\frac{F_k(x)}{m} = f_k(x) = \omega_n^2 x(t) \pm \beta^2 x^3(t) \qquad (9.10)$$

where the plus sign is used in the case of a hardening spring (if the curve shown in Figure 9.3 were to curve up after 20 mm) and the minus sign is used in the case of a softening spring (if the curve in Figure 9.3 curves down after 20 mm, as is in fact shown). In either case, two new features arise: First, Eq. (9.1) no longer has a closed-form solution but rather must be solved numerically; and second, the equilibrium (or rest) position of the system is no longer just at $x(t) = 0$. In what follows, we examine both the nature of the multiple equilibrium positions and the numerical solution of the nonlinear equation of motion.

Consider the definition of the *equilibrium position* of a system. The general state–space model of Eq. (9.1) is written following the procedure in the previous sections by defining the two state variables $x_1 = x(t)$ and $x_2 = \dot{x}(t)$. Then Eq. (9.1) can be written as the first-order pair

$$\dot{x}_1(t) = x_2(t)$$
$$\dot{x}_2(t) = -f(x_1, x_2) \tag{9.11}$$

This state–space or first-order form of the equation is used both for numerical integration (as in Section 1.5) and for formally defining an equilibrium position. By defining the state vector $\mathbf{x}$ used in Eqs. (9.11) and a nonlinear vector function $\mathbf{F}$ as

$$\mathbf{x} = \begin{bmatrix} x_1 \\ x_2 \end{bmatrix}, \quad \mathbf{F} = \begin{bmatrix} x_2(t) \\ -f(x_1, x_2) \end{bmatrix} \tag{9.12}$$

We may now write Eqs. (9.11) in the simple vector form

$$\dot{\mathbf{x}} = \mathbf{F}(\mathbf{x}) \tag{9.13}$$

An equilibrium point of this system, denoted $\mathbf{x}_e$, is simply any constant value of the state vector $\mathbf{x}$ for which $\mathbf{F}(\mathbf{x})$ is identically zero (called zero-phase velocity). Thus, the equilibrium point is any vector of constants that satisfies the relation.

$$\mathbf{F}(\mathbf{x}_e) = \mathbf{0} \tag{9.14}$$

In the linear case $\mathbf{F}(\mathbf{x}) = A\mathbf{x}$, where A is a nonsingular matrix of constants, there is only one solution of this equation, namely, $\mathbf{x} = \mathbf{0}$. However, the introduction of a nonlinear spring presents other solutions, as the examples that follow show.

The existence of multiple equilibria is similar to the existence of multiple roots for a nonlinear algebraic equation versus the single root for a linear algebraic equation. In the case of finding the roots (or solutions) of a nonlinear algebraic equation, the numerical solution depends on the initial guess we give the root-finding algorithm. In the case of the numerical integration of a nonlinear differential equation, the solution depends upon the initial conditions we give the numerical algorithm. If the initial conditions are near one equilibrium point, the solution is likely to oscillate around that point. Likewise, if a different initial condition is chosen, the solution may oscillate around it. We are not guaranteed that solutions will remain near the initial condition, and this dependence of the nature of the resulting oscillation on the value of the initial condition forms an important distinction between linear and nonlinear systems and leads to the issue of dynamic stability (a discussion beyond the scope of this introduction).

Sample Problem 9.4

Calculate the equilibrium positions for the soft nonlinear spring–mass system defined by Eq. (9.10), written in the state–variable form of Eq. (9.11) as

$$\dot{x}_1(t) = x_2(t)$$
$$\dot{x}_2(t) = x_1(t)(\beta^2 x_1^2(t) - \omega_n^2)$$

Solution   Following Eq. (9.14), the equilibrium solutions must satisfy the two coupled algebraic equations,

$$x_2 = 0$$

$$x_1(\beta^2 x_1^2 - \omega_n^2) = 0$$

There are three solutions of this system of algebraic equations, corresponding to the three equilibrium positions of a soft spring:

$$\mathbf{x}_c = \begin{bmatrix} x_1 \\ x_2 \end{bmatrix} = \begin{bmatrix} 0 \\ 0 \end{bmatrix}, \begin{bmatrix} \omega_n/\beta \\ 0 \end{bmatrix}, \begin{bmatrix} -\omega_n/\beta \\ 0 \end{bmatrix}$$

The motion of the soft spring could oscillate around any of these three equilibrium positions. The first equilibrium position is the zero point, corresponding to the linearized version of this system (i.e., when β is zero).

Sample Problem 9.5

Consider the pendulum of Sample Problem 9.3, and compute the equilibrium positions of the full nonlinear pendulum equation.

Solution   From the free-body diagram (recall Sample Problem 2.13 for the equation of motion), the state–space equations for the nonlinear pendulum (with $x_1 = \theta$ and $x_2 = \dot{\theta}$) are

$$\dot{x}_1 = x_2$$

$$\dot{x}_2 = -\frac{g}{l} \sin(x_1)$$

so that the vector equation $\mathbf{F}(\mathbf{x}) = \mathbf{0}$ yields the equilibrium solutions

$$x_2 = 0 \text{ and } x_1 = 0, \pi, 2\pi, 3\pi, 4\pi, 5\pi, \ldots$$

since $\sin(x_1)$ is zero for any multiple of π. Note that there are an infinite number of equilibrium positions, or vectors $\mathbf{x}_e$. These are all either the up position, corresponding to the odd values of π, or the down position, corresponding to even multiples of π, as indicated in the accompanying figure. These two alternatives form two distinct types of behavior: The response for initial conditions near the even values of π is a stable oscillation around the down position [part (b) of the figure], just as in the linearized case, while the response for initial conditions near even values of π [part (c)] moves away from the equilibrium position (and is therefore called unstable).

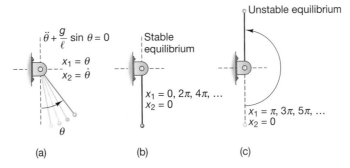

(a)   (b)   (c)

Sample Problem 9.6

Compute and plot the response of the pendulum of the previous example if $g/L = 10$ and the initial conditions are $\theta = \pi$ rad and $\dot{\theta} = 1$ rad/s. Repeat the calculation for the initial conditions $\theta = 0$ and $\dot{\theta} = 1$ rad/s. Compare the two solutions.

Solution

Using the Mathcad software, we plot the response from the state equations, as given in the accompanying figure. The first plot is for the initial condition at the top of the pendulum. A time step of 0.001 s is used, and 5000 points are utilized in the Euler integration. The value of $g/L = 10$ $(\text{rad/s})^2$.

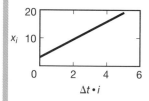

$$\Delta t = 0.001$$

$$i = 0\dots5000$$

$$\omega_2 = 10$$

$$\begin{bmatrix} x_0 \\ v_0 \end{bmatrix} = \begin{bmatrix} \pi \\ 1 \end{bmatrix}$$

$$\begin{bmatrix} x_i + 1 \\ v_i + 1 \end{bmatrix} = \begin{bmatrix} x_i + v_i \cdot \Delta t \\ v_i - \Delta t \cdot (\omega_2 \cdot \sin(x_i)) \end{bmatrix}$$

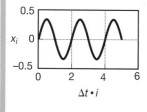

Note that in this case the pendulum is swinging around and around, so that the angle, denoted x_i in the plot, increases monotonically with time. Next, consider running the solution again with initial conditions near the equilibrium position at the bottom of the pendulum:

$$\begin{bmatrix} x_0 \\ v_0 \end{bmatrix} = \begin{bmatrix} 0 \\ 1 \end{bmatrix}$$

$$\begin{bmatrix} x_i + 1 \\ v_i + 1 \end{bmatrix} = \begin{bmatrix} x_i + v_i \cdot \Delta t \\ v_i - \Delta t \cdot (\omega_2 \cdot \sin(x_i)) \end{bmatrix}$$

In this case, the system oscillates back and forth around the $x_i = \theta = 0$ position, much as the linear pendulum does. Comparing the two plots, one can see how the response of a nonlinear system varies drastically, depending on the initial conditions used to start the motion.

Problems

9.1 A pendulum has a length of 0.25 mm. What is the natural frequency of the system in hertz?

9.2 The pendulum in Sample Problem 9.3 is required to oscillate once every second. What length should it be?

9.3 The approximation of $\sin\theta = \theta$ is reasonable for θ less than 10°. If a pendulum of length 0.5 m has an initial position of $\theta(0) = 0$, what is the maximum value of the initial angular velocity that can be given to the pendulum without violating the approximation of Sample Problem 9.3? (Make sure that you work in radians.)

9.4 Plot the solution of a linear spring–mass system with frequency $\omega_n = 2$ rad/s, $x_0 = 1$ mm, and $v_0 = 2.34$ mm/s for at least two periods.

9.5 Compute the natural frequency and plot the solution of a linear spring–mass system with a mass of 1 kg, a stiffness of 4 N/m, and initial conditions of $x_0 = 1$ mm and $v_0 = 0$ mm/s for at least two periods.

9.6 The amplitude of vibration of a linear undamped system is measured to be 1 mm. The phase is measured to be 2 rad, and the frequency is found to be 5 rad/s. Calculate the initial conditions that caused the vibration to occur.

9.7 A linear undamped system oscillates with a frequency of 10 Hz and an amplitude of 1 mm. Calculate the maximum amplitude of the system's velocity and acceleration.

9.8 Show that Eq. (9.5), the general solution of a linear spring–mass system, also can be written as $x(t) = B\sin(\omega_n t) + C\cos(\omega_n t)$, where B and C are now the constants of integration. Also, calculate B and C in terms of A and ϕ.

9.9 If the solution of Eq. (9.3) is given in the form $x(t) = B\sin(\omega_n t) + C\cos(\omega_n t)$, calculate the value of the constants B and C in terms of the initial conditions x_0 and v_0.

9.10 (a) A 0.5-kg mass is attached to a linear spring of stiffness 0.1 N/m. Determine the natural frequency of the system in hertz. (b) Repeat the calculation for a mass of 50 kg and a stiffness of 10 N/m, and compare your result with that of part (a).

9.11 Designing a linear spring–mass system is often a matter of choosing a spring constant such that the resulting natural frequency has a specified value. Suppose that the mass of a system is 4 kg and the stiffness is 100 N/m. How much must the spring stiffness be changed in order to increase the natural frequency by 10%?

9.12 Calculate the natural frequency of the system in Figure P9.12.

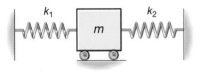

Figure P9.12

9.13 Calculate the natural frequency of the system in Figure P9.13.

Figure P9.13

9.14 Calculate the natural frequency of the system in Figure P9.14.

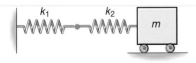

Figure P9.14

9.15 Calculate the natural frequency of the system in Figure P9.15.

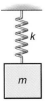

Figure P9.15

9.16 Calculate the natural frequency of the system in Figure P9.16.

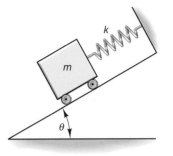

Figure P9.16

9.17 Calculate the natural frequency of the system in Figure P9.17.

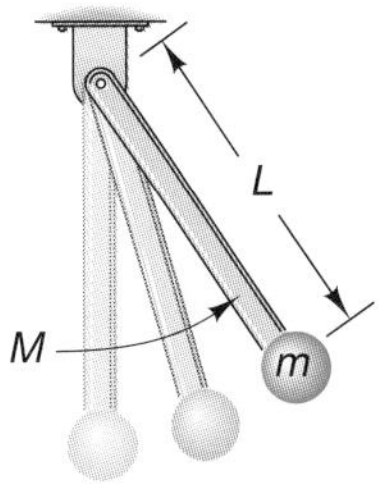

Figure P9.17

9.18 Calculate the natural frequency of the system in Figure P9.18.

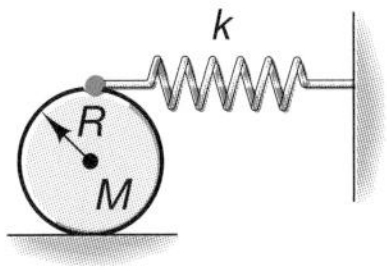

Figure P9.18

9.19 Calculate the natural frequency of the system in Figure P9.19. The disc has a mass m and a radius R.

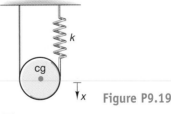

Figure P9.19

9.20 Compute the equilibrium positions of $\ddot{x} + \omega_n^2 x + \beta x^2 = 0$. How many are there?

9.21 Compute the equilibrium positions of $\ddot{x} + \omega_n^2 x - \beta^2 x^3 + \gamma x^5 = 0$. How many are there?

9.22 Consider the pendulum of Sample Problem 9.5 with of length 1 m and initial conditions of $\theta_0 = \pi/10$ rad and $\dot{\theta}_0 = 0$. Compare the difference between the response of the linear version of the pendulum equation (i.e., with $\sin(\theta) = \theta$) and the response of the nonlinear version of the equation by plotting the response of both for four periods.

🖥 **9.23** Repeat Problem 9.22 if the initial displacement is $\theta_0 = \pi/2$ rad.

9.24 If the pendulum of Sample Problem 9.5 is given initial conditions near the equilibrium position of $\theta_0 = 2\pi$ rad and $\dot\theta_0 = 0$, does it oscillate around this equilibrium point?

🖥 **9.25** Calculate the response of the system of Problem 9.21 for the initial conditions of $x_0 = 0.01$ m and $v_0 = 0$, for a natural frequency of 3 rad/s, and when $\beta = 100$ and $\gamma = 0$.

🖥 **9.26** Repeat Problem 9.25, and plot the response of the linear version of the system ($\beta = 0$) on the same

plot to compare the difference between the linear and nonlinear versions of this equation of motion.

🖥 **9.27** Consider the soft spring–mass system of Sample Problem 9.4 with an initial natural frequency of 3 rad/s and $\beta = 1$. Suppose the initial conditions are $x_0 = 1.01$ m and $v_0 = 0$. Which equilibrium position does the system oscillate around?

🖥 **9.28** Repeat Problem 9.27 for an initial condition of $x_0 = 0.01$ m and $v_0 = 0$, which is near the equilibrium at $(0, 0)$, and compare the response with that of the initial conditions $x_0 = 1.01$ m and $v_0 = 0$. Which equilibrium position does the response oscillate around?

9.3 DAMPED SINGLE-DEGREE-OF-FREEDOM SYSTEMS

The response of the spring–mass model of Eq. (9.1) and Figure 9.1 predicts that the system will oscillate indefinitely. However, everyday observation indicates that most freely oscillating systems eventually die out and reduce to zero motion. This experience suggests that the model needs to be modified to account for the decaying motion. The simplest choice of a representative model for the observed decay in an oscillating system is based partially on physical observation and partially on mathematical convenience in the linear case. The theory of differential equations suggests that adding a term of the form $c\dot{x}$, where c is a constant, to Eq. (9.3) will result in a solution $x(t)$ that dies out. This solution was presented in Section 1.5 and will be discussed in detail here. Physical observation agrees fairly well with such a model, and it is used very successfully to model the damping, or decay, in a variety of mechanical systems. This type of damping is called *viscous damping* and is a linear model of damping.

A variety of other models of damping also are used in vibrational analysis. Most of these are time dependent or nonlinear in nature, making closed-form solutions impossible. However, the solutions still can be determined numerically using simple Euler integration or Runge–Kutta methods employed throughout the text—in particular in the last section. Some examples of nonlinear damping terms are the damping induced by sliding friction (called Coulomb damping) and air resistance proportional to the square of the velocity. Both of these models were introduced in Section 1.5, repeated in numerous later problems, and are discussed in detail in the last part of this section.

9.3.1 LINEAR DAMPING

While the spring forms a physical model for storing kinetic energy and hence causing vibration, the *dashpot* (or *damper*) forms a physical model for dissipating energy and damping the response of a mechanical system. A dashpot is any mechanism that dissipates energy in the mathematical form $c\dot{x}(t)$, which is a linear model that captures energy dissipation. The constant c is called the *damping coefficient* and has units of N · s/m, or kg/s. The model is crude, but has the advantage of rendering a closed-form solution when it is used with a linear spring model. The constant c is determined by fluid forces in some cases, such as a shock absorber or air, and by equivalent effects occurring in the material forming

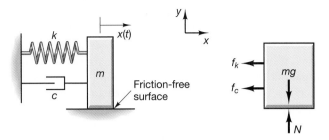

Figure 9.5

the device. Good examples are a block of rubber (which also provides stiffness f_k), such as an automobile motor mount, and the effects of air flowing around an oscillating mass.

In all cases where the damping force f_c is proportional to velocity, the schematic of a dashpot is used to indicate the presence of this force. Figure 9.5 is illustrative. In the linear case, the damping force, has the form

$$f_c = c\dot{x}(t) \tag{9.15}$$

Combined with the linear–spring relationship of the previous section, the equation of motion for the spring–mass–damper system is linear and can be solved easily by the methods of Section 1.5.

Balancing forces on the mass of Figure 9.5 in the x-direction, we find that the equation of motion for $x(t)$ becomes

$$m\ddot{x} = -f_c - f_k \tag{9.16}$$

or

$$m\ddot{x}(t) + c\dot{x}(t) + kx(t) = 0 \tag{9.17}$$

subject to the initial conditions $x(0) = x_0$ and $\dot{x}(0) = v_0$. Eq. (9.16) and Figure 9.5 are referred to as a *damped single-degree-of-freedom system*.

It is interesting to note that the damping coefficient of a system cannot be measured as simply as the mass or stiffness of a system can be. The stiffness and mass both can be determined by using static tests, such as the one shown in Figure 9.2. However, the damping force must be determined by some sort of dynamic experiment. This makes the field of damping an active one for researchers and engineers.

To solve the damped system of Eq. (9.17), the same method used for solving Eq. (9.2), as presented in Section 1.5, is employed. In fact, this provides the mathematical reason for choosing f_c to be of the form $c\dot{x}$. Suppose $x(t)$ has the form $x(t) = ae^{\lambda t}$. Substitution of this form into Eq. (9.17) yields

$$(m\lambda^2 + c\lambda + k)\,ae^{\lambda t} = 0 \tag{9.18}$$

Since $ae^{\lambda t} \neq 0$, this reduces to an algebraic equation in λ of the form

$$m\lambda^2 + c\lambda + k = 0 \tag{9.19}$$

called the *characteristic equation*. Eq. (9.19) is solved using the quadratic formula to yield the two solutions

$$\lambda_{1,2} = -\frac{c}{2m} \pm \frac{1}{2m}\sqrt{c^2 - 4km} \tag{9.20}$$

Examination of this expression indicates that the roots λ will be real or complex, depending on the value of the discriminant, $c^2 - 4km$. As long as m, c, and k are positive real numbers, λ_1 and λ_2 will be distinct negative real numbers if $c^2 - 4km > 0$. On the other hand, if this discriminant is negative, the roots will form a complex conjugate pair with negative real part. If the discriminant is zero, the two roots λ_1 and λ_2 are equal negative real numbers. Note that Eq. (9.19) reduces to the characteristic equation for the special undamped case (i.e., $c = 0$).

In examining these three cases, it is convenient to define the *critical damping* coefficient

$$c_{cr} = 2m\omega_n = 2\sqrt{km} \tag{9.21}$$

where ω_n is the undamped natural frequency. Furthermore, the nondimensional number ζ, called the *damping ratio* and defined by

$$\zeta = \frac{c}{c_{cr}} = \frac{c}{2m\omega_n} \tag{9.22}$$

can be used to characterize the three types of solutions of the characteristic equation. Rewriting the roots given by Eq. (9.20) in terms of ζ and ω_n yields

$$\lambda_{1,2} = -\zeta\omega_n \pm \omega_n\sqrt{\zeta^2 - 1} \tag{9.23}$$

where it is now clear that the damping ratio ζ determines whether the roots are complex or real. This in turn determines the nature of the response of the damped single-degree-of-freedom system. For positive mass, damping, and stiffness coefficients, there are three cases, which we delineate next.

Underdamped Motion In this case, the damping ratio ζ is less than 1 ($0 < \zeta < 1$) and the discriminant of Eq. (9.20) is negative, resulting in a complex conjugate pair of roots

$$\lambda_1 = -\zeta\omega_n - j\omega_n\sqrt{1 - \zeta^2} \tag{9.24}$$

and

$$\lambda_2 = -\zeta\omega_n + j\omega_n\sqrt{1 - \zeta^2} \tag{9.25}$$

where $j = \sqrt{-1}$ and

$$j\sqrt{1 - \zeta^2} = \sqrt{(1 - \zeta^2)(-1)} = \sqrt{\zeta^2 - 1} \tag{9.26}$$

Following the same argument made for the undamped response of Section 1.5, we find that the solution of Eq. (9.17) is then of the form

$$x(t) = e^{-\zeta\omega_n t}\left(a_1 e^{-\omega_n\sqrt{\zeta^2 - 1}\,t} + a_2 e^{+\omega_n\sqrt{\zeta^2 - 1}\,t}\right) \tag{9.27}$$

where a_1 and a_2 are arbitrary complex-valued constants of integration to be determined by the initial conditions. Using the Euler relations for the sine function, we can write Eq. (9.27) as

$$x(t) = Ae^{-\zeta\omega_n t}\sin(\omega_d t + \phi) \tag{9.28}$$

where A and ϕ are constants of integration and

$$\omega_d = \omega_n\sqrt{1 - \zeta^2} \tag{9.29}$$

is the *damped natural frequency*. The constants A and ϕ are evaluated using the initial conditions in exactly the same fashion as they were for the undamped system, as indicated in Eqs. (9.6) and (9.7). The result is

$$A = \sqrt{\frac{(v_0 + \zeta\omega_n x_0)^2 + (x_0\omega_d)^2}{\omega_d^2}}, \quad \phi = \tan^{-1}\left[\frac{x_0\omega_d}{v_0 + \zeta\omega_n x_0}\right] \quad (9.30)$$

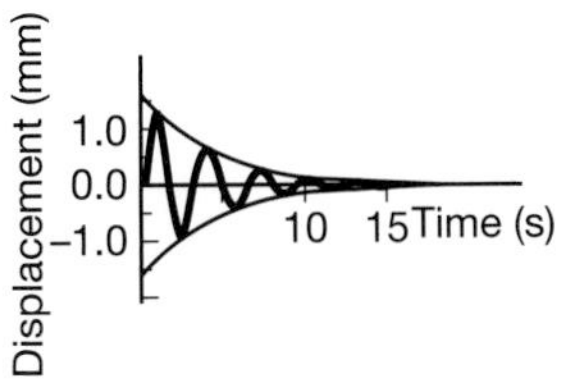

where x_0 and v_0 are the initial displacement and velocity, respectively. A plot of $x(t)$ versus t for this underdamped case $(0 < \zeta < 1)$ is given in Figure 9.6. Note that the motion is oscillatory with decaying amplitude. The damping ratio ζ determines the rate of decay. The response illustrated in the figure is exhibited in many mechanical systems and constitutes the most common case.

Figure 9.6

Overdamped Motion In this case, the damping ratio is greater than 1 $(\zeta > 1)$. The discriminant of Eq. (9.20) is positive, resulting in a pair of distinct real roots

$$\lambda_1 = -\zeta\omega_n - \omega_n\sqrt{\zeta^2 - 1} \quad (9.31)$$

and

$$\lambda_2 = -\zeta\omega_n + \omega_n\sqrt{\zeta^2 - 1} \quad (9.32)$$

The solution of Eq. (9.17) then becomes

$$x(t) = e^{-\zeta\omega_n t}\left(a_1 e^{-\omega_n\sqrt{\zeta^2-1}\,t} + a_2 e^{+\omega_n\sqrt{\zeta^2-1}\,t}\right) \quad (9.33)$$

which represents a nonoscillatory response. Again, the constants of integration a_1 and a_2 are determined by the initial conditions indicated in Eqs. (9.6) and (9.7). In this aperiodic case, the constants of integration are real valued and are given by

$$a_1 = \frac{-v_0 + (-\zeta + \sqrt{\zeta^2 - 1})\omega_n x_0}{2\omega_n\sqrt{\zeta^2 - 1}} \quad (9.34)$$

and

$$a_2 = \frac{-v_0 + (\zeta + \sqrt{\zeta^2 - 1})\omega_n x_0}{2\omega_n\sqrt{\zeta^2 - 1}} \quad (9.35)$$

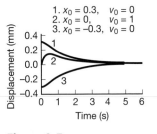

1. $x_0 = 0.3$, $v_0 = 0$
2. $x_0 = 0$, $v_0 = 1$
3. $x_0 = -0.3$, $v_0 = 0$

Figure 9.7

The response is plotted in Figure 9.7, in which it is clear that the motion does not involve oscillation. An overdamped system does not oscillate but, rather, returns to its rest position exponentially. The response is shown for two values of the initial displacement and zero initial velocity and one case with $x_0 = 0$ and $v_0 = 1$.

Critically Damped Motion In this last case, the damping ratio is exactly 1 $(\zeta = 1)$, and the discriminant of Eq. (9.20) is identically zero. This corresponds to the value of ζ that separates oscillatory motion from nonoscillatory motion. Since the roots are repeated, they have the value

$$\lambda_1 = \lambda_2 = -\omega_n \quad (9.36)$$

The solution takes the form

$$x(t) = (a_1 + a_2 t)e^{-\omega_n t} \quad (9.37)$$

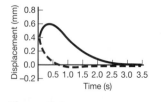

Figure 9.8

where, again, the constants a_1 and a_2 are determined by the initial conditions. Substituting the initial displacement into Eq. (9.37) and the initial velocity into the derivative of Eq. (9.37) yields

$$a_1 = x_0, \quad a_2 = v_0 + \omega_n x_0 \tag{9.38}$$

Critically damped motion is plotted in Figure 9.8 for two different values of initial conditions. Critically damped systems can be thought of in at least two ways: first, as systems with the smallest value of damping rate that yields aperiodic motion; and second, as the value of damping that separates nonoscillation from oscillation.

Sample Problem 9.7

Recall the small spring of Sample Problem 9.1 (for which $\omega_n = 132$ rad/s). Each material exhibits some internal damping. If the damping rate of the spring is measured to be 0.11 kg/s, calculate the damping ratio and determine whether the free motion of the spring–bolt system is overdamped, underdamped, or critically damped.

Solution   From Sample Problem 9.1, $m = 49.2 \times 10^{-3}$ kg and $k = 857.8$ N/m. Using the definition of the critical damping coefficient of Eq. (9.21) and the given values for m and k yields

$$c_{cr} = 2\sqrt{km} = 2\sqrt{(857.8 \text{ N/m})(49.2 \times 10^{-3} \text{ kg})}$$
$$= 12.993 \text{ kg/s}$$

If c is 0.11 kg/s, the critical damping ratio becomes

$$\zeta = \frac{c}{c_{cr}} = \frac{0.11 \text{ kg/s}}{12.993 \text{ kg/s}} = 0.0085$$

or 0.85% damping. Since this is less than unity, the system is underdamped. The motion that results when one gives the spring–bolt system a small displacement will be oscillatory.

The single-degree-of-freedom damped system of Eq. (9.17) is often written in a standard form. This form is obtained by dividing Eq. (9.17) by the mass m, yielding

$$\ddot{x} + \frac{c}{m}\dot{x} + \frac{k}{m}x = 0 \tag{9.39}$$

The coefficient of $x(t)$ is obviously ω_n^2, the square of the undamped natural frequency. A little manipulation illustrates that the coefficient of the velocity $\dot{x}(t)$ is $2\zeta\omega_n$. Thus, Eq. (9.39) can be written as

$$\ddot{x}(t) + 2\zeta\omega_n\dot{x}(t) + \omega_n^2 x(t) = 0 \tag{9.40}$$

In this standard form, the values of the natural frequency and the damping ratio are more obvious.

Sample Problem 9.8

Compute the natural frequency, damping ratio, and damped natural frequency for the system illustrated in the accompanying figure. Is this system overdamped, underdamped, or critically damped? If the system is underdamped, calculate the period of oscillation.

Solution   From the figure, the equation of motion is Eq. (9.17) with $m = 100$ kg, $k = 2000$ N/m, and $c = 25$ kg/s. From the definition of the natural frequency,

$$\omega_n = \sqrt{\frac{k}{m}} = \sqrt{\frac{2000 \text{ N/m}}{100 \text{ kg}}} = 4.472 \text{ rad/s} = 0.7117 \text{ Hz}$$

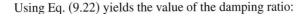

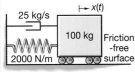

Using Eq. (9.22) yields the value of the damping ratio:

$$\zeta = \frac{c}{2\sqrt{km}} = \frac{25 \text{ kg/s}}{2\sqrt{(2000 \text{ N/m})(100 \text{ kg})}} = 0.028$$

This is less than unity, so the system is underdamped. The damped natural frequency is given by Eq. (9.29) and is

$$\omega_d = \omega_n\sqrt{1 - \zeta^2} = (4.472 \text{ rad/s})\sqrt{1 - (0.028)^2} = 4.470 \text{ rad/s} = 0.7110 \text{ Hz}$$

Note that the damped natural frequency and the undamped natural frequency are nearly the same for systems with a very small damping ratio. The period of oscillation of the system is the reciprocal of the frequency in hertz, or

$$T = \frac{2\pi}{\omega_d} = \frac{2\pi \text{ rad}}{4.470 \text{ rad/s}} = 14.1 \text{ s}$$

9.3.2 NONLINEAR DAMPING

The linear, viscous damping force given in Eq. (9.15) is a reasonable first approximation of damping in many situations. However, simple physical examples, such as sliding friction and air damping, often are not well approximated by that equation. Accordingly, we introduce more complete models of the damping force, which, though nonlinear, are more appropriate than the simple linear model in many applications. As in the case of linear versus nonlinear stiffness, the resulting equation of motion becomes nonlinear, so that nice, closed-form analytical solutions are not possible, and we must use numerical integration to calculate the solution. In addition, the nonlinear systems resulting from damping will have multiple equilibrium positions that must be accounted for in our analysis.

First, consider the effect of adding a frictional force due to sliding friction at the surface of the system of Figure 9.1. This situation, along with the free-body diagram, is illustrated in Figure 9.9. Recall from the study of friction in statics and from the review of sliding friction given in Section 2.2, that the direction of the friction force is opposite that of the direction of motion of the particle. The direction of motion is captured by the sign of the velocity term; hence, the form of the friction force is

$$f = \mu N \frac{v}{|v|} \tag{9.41}$$

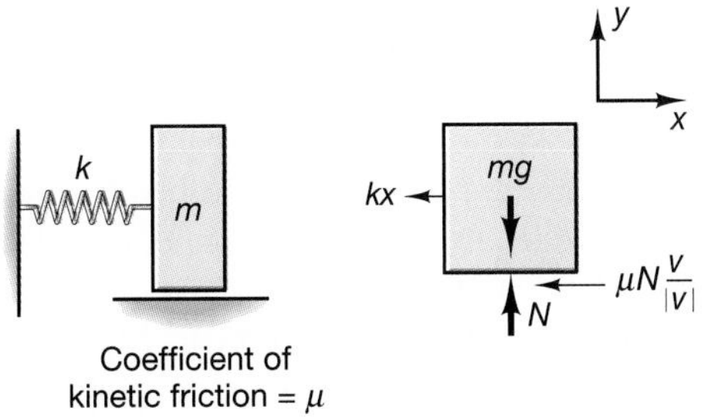

Figure 9.9

where v is the velocity and the ratio $v/|v|$ is just the sign of the velocity. From the free-body diagram, the static equilibrium in the y-direction yields $N = mg$. Using Eq. (9.41) and summing forces in the x-direction results in the equation of motion

$$m\ddot{x}(t) + \mu mg \frac{\dot{x}(t)}{|\dot{x}(t)|} + kx(t) = 0 \tag{9.42}$$

The middle term, representing sliding friction, is dependent on the velocity and, hence, denotes a damping force that dissipates energy. This force is called *Coulomb damping*, after the scientist who first characterized it. Because the expression is nonlinear, it must be solved numerically or in steps. Most introductory texts on vibration provide a stepwise analytical solution to Eq. (9.42). Here, we provide only a numerical solution, following the Euler or Runge–Kutta schemes as before. Because the equation is nonlinear, we expect that multiple equilibrium positions exist, and indeed, they do.

The equilibrium position of the system can be found by enforcing static equilibrium in the x-direction. The solution is a region defined by

$$x_2 = 0$$

$$-\frac{\mu mg}{k} < x_1 < \frac{\mu mg}{k}$$

Depending on the initial conditions, the response will end up at a value of $\mathbf{x}_e$ somewhere in this region. Thus, in this case, there are an infinite number of equilibrium positions.

The solution for a given set of initial conditions can be found by numerical integration using the techniques described in Section 1.5. The following sample problem illustrates both the numerical solution of a system with Coulomb damping and the effect of the initial conditions on the final rest position of the mass.

Sample Problem 9.9	Compute and plot the response of the system of Figure 9.9 with a coefficient of friction of $\mu = 0.3$, mass of $m = 100$ kg, and stiffness of $k = 500$ N/m for the two different initial conditions (a) $v_0 = 0$ and $x_0 = 4.5$ m and (b) $v_0 = 0$ and $x_0 = 5.0$ m.

Solution   Depending on the integration program used, the equations of motion need to be placed into first-order form for integrating. Defining $x_1 = x$ and $x_2 = v$ yields

$$\dot{x}_1 = x_2$$

$$\dot{x}_2 = -0.3(100)(9.81)\frac{x_2}{|x_2|} - 500x_1$$

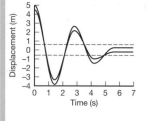

The plot that follows results from using MATLAB's Runge–Kutta routine. The top curve is part (b), and the bottom curve is part (a). Note that the response with the smallest initial position stops closer to zero (the linear–spring equilibrium position) than does the response of part (b), in agreement with intuition. The dashed line in the figure is the region of static equilibrium. That is, if the mass is given an initial displacement inside this region (with zero initial velocity), the mass will not move.

Note that the two different solutions come to rest at different times, as well as at different values of final displacement. This is again an indication of nonlinear behavior. For a linear system, the rest position is the same, regardless of the initial condition.

Sample Problem 9.10

Compute the solution of the following system, which models damping due to air viscosity acting against a spring–mass system, and plot the result:

$$m\ddot{x} + c\dot{x}|\dot{x}| + kx = 0$$

Here, $m = 50$ kg, $k = 200$ N/m, $c = 25$ kg/s, $x_0 = 0$, and $v_0 = 1$ m/s. Compare this result with that obtained from a system with linear viscous damping with the same damping coefficient.

Solution

Depending on the integration program used, the equations of motion need to be placed into first-order form for integrating. Defining $x_1 = x$ and $x_2 = v$ yields

$$\dot{x}_1 = x_2$$
$$\dot{x}_2 = -0.5x_2|x_2| - 4x_1$$

The plot that follows results from using an Euler method in Mathcad. The solution of the corresponding linear–viscous case given by

$$\dot{x}_1 = x_2$$
$$\dot{x}_2 = -0.5x_2 - 4x_1$$

also is plotted. First, we set up the time stepping by defining 10,000 steps of 0.001 s each:

$$\Delta t = 0.001$$
$$i = 0\ldots10{,}000$$

Next, we define the initial value of the position and velocity:

$$\begin{bmatrix} x_0 \\ v_0 \end{bmatrix} = \begin{bmatrix} 0 \\ 1 \end{bmatrix}$$

The Euler formula becomes

$$\begin{bmatrix} x_{i+1} \\ v_{i+1} \end{bmatrix} = \begin{bmatrix} x_i + v_i \cdot \Delta t \\ v_i - (\Delta t) \cdot (.5 \cdot v_i \cdot |v_i| + 4 \cdot x_i) \end{bmatrix}$$

The plot of the resulting solution is shown here (x_i stands for the position as a function of the running time $\Delta t \cdot i$).

The equivalent linear system is plotted next, using exactly the same values and initial conditions:

$$\begin{bmatrix} x_0 \\ v_0 \end{bmatrix} = \begin{bmatrix} 0 \\ 1 \end{bmatrix}$$

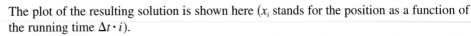

$$\begin{bmatrix} x_{i+1} \\ v_{i+1} \end{bmatrix} = \begin{bmatrix} x_i + v_i \cdot \Delta t \\ v_i - (\Delta t) \cdot (.5 \cdot v_i + 4 \cdot x_i) \end{bmatrix}$$

Note that the linear-system response definitely has a different look than the nonlinear response, which takes longer to die out than the linear response does.

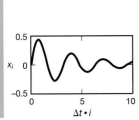

Problems

9.29 A spring–mass–damper system has a mass of 100 kg, stiffness of 3000 N/m, and damping coefficient of 300 kg/s. Calculate the undamped natural frequency, the damping ratio, and the damped natural frequency. Does the solution oscillate?

9.30 A spring–mass–damper system has mass of 150 kg, stiffness of 1500 N/m, and damping coefficient of 200 kg/s. Calculate the undamped natural frequency, the damping ratio, and the damped natural frequency. Is the system overdamped, underdamped, or critically damped? Does the solution oscillate?

9.31 The system of Problem 9.29 is given an initial velocity of 10 mm/s and an initial displacement of −5 mm. Calculate the form of the response and plot it for two cycles.

9.32 The system of Problem 9.29 is given a zero initial velocity and an initial displacement of 0.1 m. Calculate the form of the response and plot it for as long as it takes to die out.

9.33 The system of Problem 9.30 is given an initial velocity of 10 mm/s and an initial displacement of −5 mm. Calculate the form of the response and plot it for as long as it takes to die out. How long does it take to die out?

9.34 Choose the damping coefficient of a spring–mass–damper system with a mass of 150 kg and stiffness of 2000 N/m such that the response of the system will die out after about 2 s, given a zero initial position and an initial velocity of 10 mm/s.

9.35 Calculate the constants A and ϕ in terms of the initial conditions, and thus verify Eq. (9.30) for the underdamped case.

9.36 Calculate the constants a_1 and a_2 in terms of the initial conditions, and thus verify Eqs. (9.34) and (9.35) for the overdamped case.

9.37 Calculate the constants a_1 and a_2 in terms of the initial conditions, and thus verify Eq. (9.38) for the critically damped case.

9.38 Using the definitions of the damping ratio and the undamped natural frequency, derive Eq. (9.40) from Eq. (9.39).

9.39 Compute and plot the response of a system with Coulomb damping, as in Eq. (9.42), for the case where $x_0 = 0.5$ m, $v_0 = 0$, $\mu = 0.1$, $m = 100$ kg, and $k = 1500$ N/m. How long does it take for the vibration to die out?

9.40 One kind of nonlinear damping, called *material damping*, is modeled by a damping force of the form

$$f_c = \alpha \frac{\dot{x}}{|\dot{x}|} x^2$$

Compute and plot the response of the system of Figure 9.5 with a linear spring and this material-damping force for the case where $x_0 = 0.05$ m, $v_0 = 1$ m/s, $\alpha = 10{,}000$ N/m^2, $m = 100$ kg, and $k = 2500$ N/m. Plot the response for 3 s.

9.41 One kind of nonlinear damping, called *air damping* (usually associated with fast fluids), is modeled by a damping force of the form

$$f_c = \alpha \frac{\dot{x}}{|\dot{x}|} \dot{x}^2$$

Compute and plot the response of the system of Figure 9.5 with a linear spring and this air-damping force for the case where $x_0 = 0.05$ m, $v_0 = 1$ m/s, $\alpha = 100$ kg/m, $m = 100$ kg, and $k = 2500$ N/m. Plot the response for 5 s.

9.42 One kind of nonlinear damping, called *solid damping*, is modeled by a damping force of the form

$$f_c = \alpha \frac{\dot{x}}{|\dot{x}|} x$$

Compute and plot the response of the system of Figure 9.5 with a linear spring and this solid-damping force for the case where $x_0 = 0.5$ m, $v_0 = 0$, $\alpha = 5000$ N/m, $m = 100$ kg, and $k = 2500$ N/m. How long does it take for the vibration to die out?

9.43 A mass is subject to a nonlinear air-damping force of the form

$$f_c = c\dot{x} + \alpha \frac{\dot{x}}{|\dot{x}|} x$$

as the result of fluid forces acting on it. Compute and plot the response of the system of Figure 9.5 with a linear spring and this air-damping force for the case where $x_0 = 0.05$ m, $v_0 = 0$, $m = 100$ kg, $c = 30$ kg/s, $\alpha = 2000$ N/m, and $k = 2500$ N/m. How long does it take for the vibration to die out?

9.44 Compute and plot the response of the system of Figure P9.44 for the case where $x_0 = 0.1$ m, $v_0 = 0.1$ m/s, $\mu = 0.05$, $m = 250$ kg, $\theta = 20°$, and $k = 3000$ N/m. How long does it take for the vibration to die out?

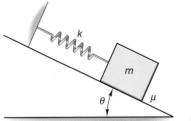

Figure P9.44

9.45 A mass moves in a fluid against sliding friction, as illustrated in Figure P9.45. Model the damping force as a fast fluid (see Problem 9.41) plus Coulomb friction (because of the sliding) with the following parameters: $m = 250$ kg, $\mu = 0.02$, $\alpha = 250$ N/m², and $k = 3000$ N/m. (a) Compute and plot the response to the initial conditions $x_0 = 0.1$ m and $v_0 = 0.1$ m/s. (b) Compute and plot the response to the initial conditions $x_0 = 0.1$ m and $v_0 = 1$ m/s. How long does it take for the vibration to die out in each case?

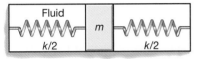

Figure P9.45

9.46 A mass moves in a fluid against sliding friction, as illustrated in Figure P9.45. Model the damping force as a slow fluid (i.e., linear, viscous damping) plus Coulomb friction (because of the sliding) with the following parameters: $m = 250$ kg, $\mu = 0.01$, $c = 25$ kg/s, and $k = 3000$ N/m. (a) Compute and plot the response to the initial conditions $x_0 = 0.1$ m and $v_0 = 0.1$ m/s. (b) Compute and plot the response to the initial conditions $x_0 = 0.1$ m and $v_0 = 1$ m/s. How long does it take for the vibration to die out in each case?

9.47 Consider the system of part (a) of Problem 9.46, and compute a new damping coefficient c that will cause the vibration to die out after one oscillation.

9.48 Consider the system of Problem 9.43, and compute a new damping coefficient c that will cause the vibration to die out after 2 s.

9.49 Repeat Problem 9.48 for the initial conditions $x_0 = 0.5$ m and $v_0 = 1$ m/s. Time for oscillations to die out: $t = 12$ s.

9.50 Consider Problem 9.43 with $\alpha = 500$ N/m, and find a value of the viscous damping coefficient c such that the vibration dies out in 2 s.

9.4 FORCED RESPONSE AND RESONANCE

One of the most important concepts in vibration, and perhaps in all of mechanical design, is resonance. Resonance may occur when a harmonic force is applied to a spring–mass or spring–mass–damper system oscillating at a frequency close to that of the vibrating system's natural frequency. In this situation, a very large amplitude and violent vibration may result, which causes the system to break or fatigue over time. Even away from resonance, the response of a system to harmonic excitation is important to analyze. For example, helicopters produce strong harmonic excitations as the blades rotate. These motions are transmitted to the cabin of the helicopter, causing annoying and potentially damaging vibration. Harmonic excitation refers to a sinusoidal external force of a single frequency applied to the system. Harmonic excitations are a very common source of external force applied to machines and structures. Rotating machines, such as fans, electric motors, reciprocating engines, etc., transmit a sinusoidally varying force to adjacent components.

A harmonic input also is chosen for study because the response of a single-degree-of-freedom system to a harmonic input forms the foundation for analyzing the response to other types of applied forces, for designing devices intended to protect machines from unwanted oscillation, and for designing transducers used in measuring vibration. Harmonic excitations are simple to produce in laboratories and, hence, are also very useful in studying damping and stiffness properties experimentally.

Almost any kind of force can be applied to a system, so the response of a system to an arbitrary input force also will be considered. In this section, the response to an input force to undamped, damped, linear, and nonlinear systems is examined with the use of numerical integration via simple computer codes, as has been the case throughout the text.

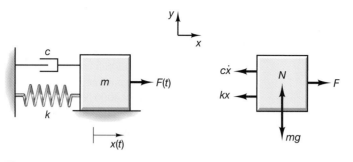

Figure 9.10

9.4.1 HARMONIC EXCITATION OF LINEAR SYSTEMS

Consider the system of Figure 9.10, a single-degree-of-freedom system acted on by an external force $F(t)$ and sliding on a friction-free surface. On the right is a free-body diagram of the friction-free spring–mass–damper system. For the case of harmonic excitation considered here, the applied force $F(t)$ has the form of a sine or cosine function at a single frequency. In particular, the driving force is chosen to be of the form

$$F(t) = F_0 \cos \omega_{\mathrm{dr}} t \tag{9.43}$$

where F_0 represents the magnitude, or maximum amplitude, of the applied force and ω_{dr} denotes the frequency of the applied force. The frequency ω_{dr} also is called the *input frequency*, *driving frequency*, or *forcing frequency*.

Undamped Forced Response   The sum of the forces in the y-direction yields $N = mg$, with the result that there is no motion in that direction. For the case of negligible damping ($c = 0$), summing forces on the mass in the x-direction reveals that the displacement $x(t)$ must satisfy

$$m\ddot{x}(t) + kx(t) = F_0 \cos \omega_{\mathrm{dr}} t \tag{9.44}$$

Note that this is a linear equation, and an analytical solution is possible. As in the homogeneous (unforced) case, it is convenient to divide this expression by the mass m, yielding

$$\ddot{x} + \omega_n^2 x(t) = f_0 \cos \omega_{\mathrm{dr}} t \tag{9.45}$$

where $f_0 = F_0/m$. A variety of techniques that commonly are studied in a first course in differential equations can be used to solve Eq. (9.45).

Recall from Section 1.5 and the properties of differential equations that Eq. (9.45) is a linear nonhomogeneous equation and that its solution is therefore the sum of the homogeneous solution (i.e., the solution for the case $f_0 = 0$) and a particular solution. The particular solution often can be found by assuming that it has the same form as the forcing function. This is also consistent with observation. That is, the oscillation of a single-degree-of-freedom system excited by $f_0 \cos \omega_{\mathrm{dr}} t$ is observed to be of the form

$$x_p = A_0 \cos \omega_{\mathrm{dr}} t \tag{9.46}$$

where x_p denotes the particular solution and A_0 is the unknown amplitude of the forced response.

Substitution of the assumed form of the solution from Eq. (9.46) into the equation of motion, (9.45), yields

$$-\omega_{\mathrm{dr}}^2 A_0 \cos \omega_{\mathrm{dr}} t + \omega_n^2 A_0 \cos \omega_{\mathrm{dr}} t = f_0 \cos \omega_{\mathrm{dr}} t \tag{9.47}$$

Dividing this expression by $\cos \omega_{dr} t (\neq 0)$ and solving for A_0 yields

$$A_0 = \frac{f}{(\omega_n^2 - \omega_{dr}^2)} \tag{9.48}$$

provided that $\omega_n \neq \omega_{dr}$. Thus, as long as the driving frequency and natural frequency are different (i.e., as long as $\omega_n \neq \omega_{dr}$), the particular solution will be of the form

$$x_p(t) = \frac{f_0}{(\omega_n^2 - \omega_{dr}^2)} \cos \omega_{dr} t \tag{9.49}$$

This approach of assuming that $x_p = A_0 \cos \omega_{dr} t$ to determine the particular solution is called the method of *undetermined coefficients*.

Since the system is linear, the total solution $x(t)$ is the sum of the particular solution of Eq. (9.49) plus the homogeneous solution given by Eq. (9.5). Recalling that $\sin(\omega_n t + \phi)$ can be represented as $A_1 \sin \omega_n t + A_2 \cos \omega_n t$, the total solution can be expressed in the form

$$x(t) = A_1 \sin \omega_n t + A_2 \cos \omega_n t + \frac{f_0}{(\omega_n^2 - \omega_{dr}^2)} \cos \omega_{dr} t \tag{9.50}$$

where it remains to determine the coefficients A_1 and A_2. These are found by enforcing the initial conditions. Let the initial position and velocity be given by the constants x_0 and v_0, as before. Then Eq. (9.50) yields

$$x(0) = A_2 + \frac{f_0}{(\omega_n^2 - \omega_{dr}^2)} = x_0 \tag{9.51}$$

and

$$\dot{x}(0) = \omega_n A_1 = v_0 \tag{9.52}$$

Solving Eqs. (9.51) and (9.52) for A_1 and A_2 and substituting these values into Eq. (9.50) yields the total response:

$$x(t) = \frac{v_0}{\omega_n} \sin \omega_n t + \left(x_0 - \frac{f_0}{(\omega_n^2 - \omega_{dr}^2)} \right) \cos \omega_n t + \frac{f_0}{(\omega_n^2 - \omega_{dr}^2)} \cos \omega_{dr} t \tag{9.53}$$

Figure 9.11 presents a plot of the total response of an undamped system to a harmonic excitation and specified initial conditions. The response is of the undamped system with $\omega_n = 1$ rad/s to harmonic excitation at $\omega_{dr} = 2$ rad/s, nonzero initial conditions of $x_0 = 0.01$ m and $v_0 = 0.01$ m/sec, and a force with magnitude $f_0 = 0.1$ N/kg. Note that the

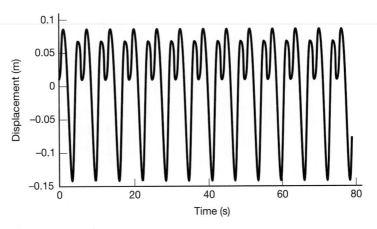

Figure 9.11

motion is the sum of two sine curves of different frequencies. Note also that the free-response coefficient A_2 is influenced by the applied force.

Sample Problem 9.11

Consider the forced vibration of a mass m connected to a spring of stiffness 2000 N/m, driven by a 20-N harmonic force at 10 Hz. The maximum amplitude of vibration is measured to be 0.1 m, and the motion is assumed to have started from rest ($x_0 = v_0 = 0$). Calculate the mass of the system.

Solution   From Eq. (9.53), the response with $x_0 = v_0 = 0$ becomes

$$x(t) = \frac{f_0}{(\omega_n^2 - \omega_{dr}^2)} [\cos \omega_{dr} t - \cos \omega_n t] \tag{9.54}$$

Using simple trigonometric identities, we find that this equation becomes

$$x(t) = \frac{2f_0}{\omega_n^2 - \omega_{dr}^2} \sin\left(\frac{\omega_n - \omega_{dr}}{2} t\right) \sin\left(\frac{\omega_n + \omega_{dr}}{2} t\right) \tag{9.55}$$

The maximum value of the total response is evident from this last expression, so that

$$\frac{2f_0}{(\omega_n^2 - \omega_{dr}^2)} = 0.1 \text{ m}$$

Solving for m from $\omega_n^2 = k/m$ and $f_0 = F_0/m$ yields

$$m = \frac{(0.1 \text{ m})(2000 \text{ N/m}) - 2(20 \text{ N})}{(0.1 \text{ m})(10)^2(2\pi \text{ rad/s})^2} = \frac{4}{\pi^2} = 0.405 \text{ kg}$$

Two very important phenomena occur when the driving frequency gets close to the system's natural frequency: beating and resonance. Consider the case where $(\omega_n^2 - \omega_{dr}^2)$ becomes very small. For zero initial conditions, the response is given by Eq. (9.55), which is plotted in Figure 9.12. In the figure, $f_0 = 10$ N, $\omega_n = 10$ rad/s, and $\omega_{dr} = 1.1\omega_n$ rad/s. The dashed line is a plot of $2f_0/(\omega_n^2 - \omega_{dr}^2) \sin[(\omega_n - \omega_{dr})t/2]$. Since $(\omega_n - \omega_{dr})$ is small, $(\omega_n + \omega_{dr})$ is large, and the term $\sin[(\omega_n - \omega_{dr})/2]t$ oscillates with a much longer period

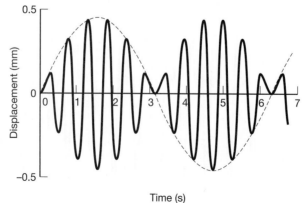

Time (s)

Figure 9.12

than does $\sin [(\omega_n + \omega_{dr})/2]t$. Recall that the period of oscillation, T, is defined as $2\pi/\omega_n$, or in this case, $2\pi/(\omega_n + \omega_{dr})/2 = 4\pi/(\omega_n + \omega_{dr})$. The resulting motion is a rapid oscillation with slowly varying amplitude and is called a *beat*. Figure 9.12 is a plot of the response of an undamped system for small $(\omega_n - \omega_{dr})$, illustrating the phenomenon of beats. In the figure, $f_0 = 0.1$ N/kg, $\omega_{dr} = 1$ rad/s, and $\omega_n = 1.1$ rad/s.

Because division by zero is not possible, as ω_{dr} becomes exactly equal to the system's natural frequency, the solution given in Eq. (9.55) is no longer valid. In this case, the choice of the function $A_0 \cos \omega_{dr}t$ for the particular solution fails because it is also a solution of the homogeneous equation. Therefore, a particular solution of the form

$$x_p(t) = tA_0 \sin \omega_{dr}t \tag{9.56}$$

is chosen, as explained, for instance, in Boyce and DiPrima (1986), *Elementary Differential Equations and Boundary Value Problems*, Wiley, New York. Substituting Eq. (9.56) into Eq. (9.45) and solving for A_0 yields

$$x_p(t) = \frac{f}{2\omega_n} t \sin \omega_{dr}t \tag{9.57}$$

Thus, the total solution is now of the form

$$x(t) = A_1 \sin \omega_n t + A_2 \cos \omega_n t + \frac{f_0}{2\omega_n} t \sin \omega_{dr}t \tag{9.58}$$

Evaluating the initial displacement x_0 and velocity v_0 as before yields

$$x(t) = \frac{v_0}{\omega_n} \sin \omega_n t + x_0 \cos \omega_n t + \frac{f_0}{2\omega_n} t \sin \omega_n t \tag{9.59}$$

where the subscripts on ω are all equal because $\omega_n = \omega_{dr}$. A plot of $x(t)$ is given in Figure 9.13, where it can be seen that $x(t)$ grows without bound. This defines the important phenomenon of *resonance*: If a spring–mass system is driven at a frequency equal to its natural frequency, the response grows without bound, which, of course, will cause the system to fail.

Figure 9.13 depicts the forced response of a spring–mass system driven harmonically at its natural frequency $(\omega_n = \omega_{dr})$, or *resonance.* For resonance, the amplitude of vibration becomes unbounded because $\omega_{dr} = \omega_n = \sqrt{k/m}$. This would cause the spring to fail and break. Note that Eq. (9.59) also can be obtained from Eq. (9.53) by taking the limit as ω_{dr} approaches ω_n, using limit theorems from calculus.

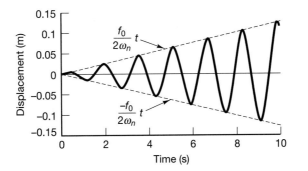

Figure 9.13

A machine part has a mass of 300 kg and is supported by a spring of stiffness 1200 N/m. A reciprocating engine is nearby that produces a 2-rad/s driving frequency applied to the part. A quick calculation illustrates that this system is experiencing resonance. If the force has a magnitude of 10 N, compute a new spring constant that will limit the amplitude of the forced response to 0.05 m for the case with zero initial conditions.

Solution   First note that the system with the original spring is in fact at resonance, since

$$\omega_n = \sqrt{\frac{k}{m}} = \sqrt{\frac{1200 \text{ N/m}}{300 \text{ kg}}} = 2 \text{ rad/s} = \omega_{dr}$$

From Eq. (9.55), the amplitude of the response for zero initial conditions is

$$x_{max} = 0.05 \text{ m} = \frac{2 f_0}{\omega_n^2 - \omega_{dr}^2} = \frac{2(10 \text{ N})/(300 \text{ kg})}{\left(\dfrac{k}{300 \text{ kg}}\right) - 2^2}$$

since the maximum value of the sine function is unity. Solving this expression for the value of k yields $k = 1600$ N/m. Designing the system with this spring constant will limit the amplitude of vibrations in the steady state at 2 rad/s to the desired value of 0.05 m.

Damped Forced Response   As mentioned before, most systems have some sort of energy dissipation associated with them. Using the model of viscous damping indicated in Figure 9.10, we see that the free-body diagram reveals the following equation of motion for the forced response of a spring–mass–damper system:

$$m\ddot{x} + c\dot{x} + kx = F_0 \cos (\omega_{dr} t) \tag{9.60}$$

Dividing by the mass m yields

$$\ddot{x} + 2\zeta\omega_n\dot{x} + \omega_n^2 x = f_0 \cos \omega_{dr} t \tag{9.61}$$

where $\omega_n = \sqrt{k/m}$ and $\zeta = c/(2m\omega_n)$, as before, and $f_0 = F_0/m$. The calculation of the particular solution for the damped case is similar to that for the undamped case and follows the method of undetermined coefficients.

From the study of differential equations, it is known that the forced response of a damped system is of the form of a harmonic function of the same frequency as the driving force, but with a different amplitude and phase. The phase shift is expected because of the effect of the damping force. Following the method of undetermined coefficients, we assume that the particular solution is of the form

$$x_p(t) = A_0 \cos (\omega_{dr} t - \theta) \tag{9.62}$$

To make the computations easy to follow, this equation is written in the equivalent form,

$$x_p(t) = A_s \cos \omega_{dr} t + B_s \sin \omega_{dr} t \tag{9.63}$$

where

$$A_0 = \sqrt{A_s^2 + B_s^2} \quad \text{and} \quad \theta = \tan^{-1}\left(\frac{B_s}{A_s}\right) \tag{9.64}$$

are the undetermined constants.

Taking derivatives of the assumed form of the solution given by Eq. (9.63) yields

$$\dot{x}_p(t) = -\omega_{dr} A_s \sin \omega_{dr} t + \omega_{dr} B_s \cos \omega_{dr} t \tag{9.65}$$

and

$$\ddot{x}_p(t) = -\omega_{dr}^2[A_s \cos \omega_{dr}t + B_s \sin \omega_{dr}t] \tag{9.66}$$

Substituting x_p, $\dot{x}_p$, and $\ddot{x}_p$ into the equation of motion (9.61) and grouping terms as coefficients of $\sin \omega_{dr}t$ and $\cos \omega_{dr}t$ gives

$$(-\omega_{dr}^2 A_s + 2\zeta\omega_n \omega_{dr}B_s + \omega_n^2 A_s - f_0) \cos \omega_{dr}t$$
$$+ (-\omega_{dr}^2 B_s - 2\zeta\omega_n\omega_{dr} A_s + \omega_n^2 B_s) \sin \omega_{dr}t = 0 \tag{9.67}$$

This equation must hold for all time and hence for $t = 2\pi/\omega_{dr}$, so that the coefficient of $\cos\omega_{dr}t$ must vanish. Likewise, the coefficient of $\sin\omega_{dr}t$ must vanish. The result is the two equations,

$$(\omega_n^2 - \omega_{dr}^2)A_s + (2\zeta\omega_n\omega_{dr})B_s = f_0$$
$$(-2\zeta\omega_n\omega_{dr})A_s + (\omega_n^2 - \omega_{dr}^2)B_s = 0 \tag{9.68}$$

in the two undetermined coefficients A_s and B_s. Solving the equations for these coefficients yields

$$A_s = \frac{(\omega_n^2 - \omega_{dr}^2)f_0}{(\omega_n^2 - \omega_{dr}^2)^2 + (2\zeta\omega_n\omega_{dr})^2}$$
$$B_s = \frac{2\zeta\omega_n\omega_{dr}f_0}{(\omega_n^2 - \omega_{dr}^2)^2 + (2\zeta\omega_n\omega_{dr})^2} \tag{9.69}$$

Substitution of these values into Eqs. (9.62) and (9.64) indicates that the particular solution is

$$x_p(t) = \frac{f_0}{\sqrt{(\omega_n^2 - \omega_{dr}^2)^2 + (2\zeta\omega_n\omega_{dr})^2}} \cos\left[\omega_{dr}t - \tan^{-1}\left(\frac{2\zeta\omega_n\omega_{dr}}{\omega_n^2 - \omega_{dr}^2}\right)\right] \tag{9.70}$$

Since the differential equation is linear, the total solution is again the sum of the particular solution and the homogeneous solution. For the underdamped case ($0 < \zeta < 1$), this becomes

$$x(t) = Ae^{-\zeta\omega_n t} \sin(\omega_{dr}t + \phi) + A_0 \cos(\omega_{dr}t - \theta) \tag{9.71}$$

where A_0 and θ are the coefficients of the particular solution as defined by Eq. (9.70), and A and ϕ are determined by initial conditions (different from those of the free response). Note that for large values of t, the first term (or the homogeneous solution) approaches zero, and the total solution approaches the particular solution. Thus, $x_p(t)$ is called the *steady–state response*, and the first term in Eq. (9.71) is called the *transient response*.

Observe that A and ϕ, the constants describing the transient response in Eq. (9.71), will be different from those calculated for the free-response case given in Eq. (9.30). This is because part of the transient term in Eq. (9.71) is due to the excitation force as well as the initial conditions.

It is common practice to ignore the transient part of the total solution given by Eq. (9.71) and focus only on the steady–state response; $A_0 \cos(\omega_{dr}t - \theta)$. The rationale for considering only the steady–state response is based on the value of the damping ratio ζ. If the system has relatively large damping, then the term $e^{-\zeta\omega_n t}$ causes the transient response to die out very quickly—perhaps in a fraction of a second. If, on the other hand, the system is lightly damped (with ζ very small), the transient part of the solution may last long enough to be significant and should not be ignored. The decision to ignore the transient part of the solution or take it into account should also be based on the application. In fact, in some applications (such as earthquake analysis and satellite analysis), the transient response may become even more important than the steady–state response.

It is of interest, then, to consider the magnitude A_0 of the steady–state response as a function of the driving frequency. Examining the form of Eq. (9.70) and comparing it with the assumed form $A_0 \cos(\omega_{dr} t - \phi)$ gives for the amplitude and the phase

$$A_0 = \frac{f_0}{\sqrt{(\omega_n^2 - \omega_{dr}^2)^2 + (2\zeta\omega_n\omega_{dr})^2}} , \quad \theta = \tan^{-1}\left[\frac{2\zeta\omega_n\omega_{dr}}{\omega_n^2 - \omega_{dr}^2}\right]$$

After factoring out ω_n^2 and dividing the magnitude by F_0/m, we can write these expressions as

$$\frac{A_0 k}{F_0} = \frac{A_0 \omega_n^2}{f_0} = \frac{1}{\sqrt{(1 - r^2)^2 + (2\zeta r)^2}} , \quad \theta = \tan^{-1}\left[\frac{2\zeta r}{1 - r^2}\right] \qquad (9.72)$$

Here, r is the frequency ratio ω_{dr}/ω_n, a dimensionless quantity. Figure 9.14 is a plot of the magnitude of the response as given by Eq. (9.72) versus the frequency ratio r, for several values of the damping ratio ζ.

Note from the figure that, as the driving frequency approaches the undamped natural frequency ($r \to 1$), the magnitude approaches a maximum value for those curves corresponding to light damping ($\zeta \leq 0.1$). Note also from the expression for θ that, as the driving frequency approaches the undamped natural frequency ($r = 1$), the phase shift crosses through $90°$. This defines *resonance* for the damped case. These two observations have important uses in both vibrational design and measurement. As ω_{dr} approaches zero, the amplitude approaches f_0/ω_n^2, and as ω_{dr} becomes very large, the amplitude approaches zero asymptotically.

It is important from the design point of view to note how the amplitude of steady–state vibration is affected by changing the damping ratio. This also is illustrated in the figure. As the damping ratio is increased, the peak in the magnitude curve decreases and eventually disappears. As the damping ratio decreases, however, the peak value increases and becomes sharper. In the limit as ζ goes to zero, the peak climbs to an infinite value, in agreement with the undamped response at resonance.

Note that resonance is defined to occur when $\omega_{dr} = \omega_n$, i.e., when the driving frequency becomes equal to the undamped natural frequency. This also corresponds with a phase shift of $90°$ ($\pi/2$). Resonance does not, however, exactly correspond with the value of ω_{dr} at which the peak value of the steady–state response occurs.

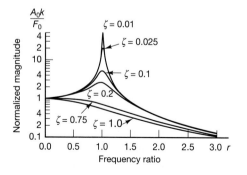

Figure 9.14

Sample Problem 9.13	Consider again the simple spring–mass system of Sample Problems 9.1 and 9.7, consisting of a spring and a bolt. Calculate the value of the steady–state response if $\omega_{dr} = 132$ rad/s for $f_0 = 10$ N/kg. Calculate the change in amplitude if $\omega_{dr} = 125$ rad/s.

Solution From Sample Problem 9.7, the natural frequency and damping ratio are $\omega_n = 132$ rad/s and $\zeta = 0.0085$, respectively. From the earlier expression for A_0, the magnitude of $x_p(t)$ is

$$|x_p(t)| = A_0 = \frac{f_0}{\sqrt{(\omega_n^2 - \omega_{dr}^2)^2 + (2\zeta\omega_n\omega_{dr})^2}}$$

$$= \frac{10}{[((132)^2 - (132)^2)^2 + [2(0.0085)(132)(132)]^2]^{1/2}} = 0.034 \text{ m}$$

If the driving frequency is changed to 125 rad/s, then the amplitude becomes

$$\frac{10}{[((132)^2 - (125)^2)^2 + [2(0.0085)(132)(125)]^2]^{1/2}} = 0.005 \text{ m}$$

So a slight change in the driving frequency from near resonance at 132 rad/s to 125 rad/s (about 5%) causes an order-of-magnitude change in the amplitude of the steady–state response. The closer the driving frequency gets to the actual value of resonance, the larger is the increase in amplitude. The farther away the driving frequency is from resonance, either larger or smaller, the lower the amplitude becomes.

Sample Problem 9.14	A machine is driven at resonance ($\omega_n = \omega_{dr} = 2$ rad/s). Design a damper (i.e., choose a value of c) such that the maximum deflection at steady state is 0.05 m. The machine has a stiffness of 2000 kg/m, and the harmonic excitation force has a magnitude of 100 N.

Solution Given the stiffness and the natural frequency, the mass can be computed from

$$\omega_n^2 = \frac{k}{m} = \frac{2000 \text{ kg}}{m} = 4 \text{ rad}^2/\text{s}^2$$

so that $m = 500$ kg. With $F_0 = 100$ N and $m = 500$ kg, the scaled force magnitude becomes $f_0 = 0.2$ N/kg. The expression for the maximum deflection at $\omega_n = \omega_{dr}$ then becomes

$$A_0 = 0.05 \text{ m} = \frac{f_0}{2\zeta\omega_n\omega_{dr}} = \frac{f_0}{2\left(\dfrac{c}{2\sqrt{mk}}\right)\omega_{dr}}$$

Solving this last expression for c yields $c = 1000$ kg/s.

9.4.2 GENERAL FORCED RESPONSE

The response to other inputs besides harmonic excitations of a single frequency may be computed by using numerical integration. In addition, the forced response of nonlinear systems may be computed numerically. Several important phenomena can be examined and investigated by analytical techniques applied to the various types of systems—linear undamped systems, linear damped systems, linear systems with periodic inputs, linear systems with aperiodic inputs, and nonlinear systems. Each of these is discussed in detail in a typical junior- or senior-year course in vibrations. Here we point out some of the basic concepts but focus mainly on the solution of problems involving these phenomena using numerical integration.

If the driving force of a linear system is continuous and repetitive (no matter how irregular), then the Fourier theorem can be used to express this driving force as the sum of sines and cosines. In the preceding paragraphs, the response of a linear system to a sinusoidal driving force was developed analytically. For the purpose of analysis, then, the linear system can be solved separately using Eq. (9.59) for each of the terms in the Fourier expansion. Because linear systems have the property of superposition, each of these solutions can be added to provide the total solution for any periodic input. Superposition, which holds only for linear systems, means that a linear combination of inputs (or driving forces) to a system produces the same linear combination of outputs (responses). Thus, for a periodic input to a linear system, the response will look like a summation of terms of the form of Eq. (9.59) in the undamped case and Eq. (9.71) in the underdamped case. Resonance can occur if any of the terms in the Fourier expansion contain a natural frequency of the system.

If the driving force applied to a system is not periodic, then the Fourier theorem cannot be applied, and other analytical methods must be used. The most common approach is to first consider the response of a linear spring–mass–damper system to an impact or impulse. An impulse is a short-duration force of very high magnitude. An example is hitting a drum with a drumstick or plucking a guitar string. Mathematically, this is handled by modeling the input force as a Dirac delta function, introduced in Section 2.3, Eq. (2.19), and in Figure 2.5. Once the response to an impulse is obtained, the concepts of superposition and convolution can be used to write an expression for the response of any linear system to any type of forcing function.

A physical argument may be used to obtain an expression for the vibrational response to an impulsive load. Suppose the mass of a spring–mass system is initially at rest, so that its initial velocity and displacement are zero. From the impulse–momentum relation (see Eq. (3.46)), the change in momentum during the impact will be $F\Delta t = mv_0 - 0$, where v_0 is the velocity just after impact, F is the magnitude of the impact, and Δt is the time during which the impulsive force is applied and during which the displacement does not change. Solving this equation for v_0 indicates that the impulsive load corresponds to the mass being given an initial velocity of

$$v_0 = \frac{F\Delta t}{m} = \frac{\hat{F}}{m}$$

where $\hat{F}$, in units of N $\cdot$ s, is the area under the curve of the plot of impulsive load versus time. Thus, the response to an impulse of magnitude $\hat{F}$ is simply given by Eq. (9.9) or Eq. (9.28) with the initial conditions $x_0 = 0$ and $v_0 = \hat{F}/m$, or

$$x(t) = \frac{\hat{F}e^{-\zeta\omega_n t}}{m\omega_d} \sin(\omega_d t) \tag{9.73}$$

for an underdamped system.

This approach, combined with the other discontinuity functions defined in Section 2.3, can be used to compute an analytical solution to various discontinuous forcing functions. In addition, Eq. (9.73) may be used along with the convolution integral to write the vibration response to any integrable force $F(t)$. The response is derived by writing Eq. (1.47) of Section 1.5 and passing the limit as Δt goes to zero, resulting in

$$x(t) = \frac{1}{m\omega_d} \int_0^t F(t - \tau)e^{-\zeta\omega_n\tau} \sin(\omega_d\tau)d\tau \tag{9.74}$$

Eq. (9.74) permits the analytical solution of any forced response for which the integral can be evaluated in closed form. The equation also can be used to evaluate the solution numerically, without recurring to any Euler or Runge–Kutta solutions of the differential equation.

Next, consider the numerical solution of a general vibration problem with nonlinear or linear stiffness and/or damping and any kind of driving force. The general equation of motion can be written as Eq. (1.49) and integrated numerically by putting the equations in first-order form as indicated in Eq. (1.51). Most mathematics programs require that the equations of motion be put into such form (Mathematica is the exception) for solution by Runge–Kutta or Euler methods. The general form is repeated here. Let v denote the velocity and x the displacement. Then the general vibration problem is to compute the solution of

$$m\ddot{x} + f(x, v, t) = F(t) \tag{9.75}$$

subject to the initial conditions x_0 and v_0, where $F(t)$ is the applied force. By defining the trival equation $\dot{x}(t) = v(t)$ and dividing Eq. (9.75) by the mass, the following two coupled first-order equations result:

$$\dot{x}(t) = v(t)$$
$$\dot{v}(t) = -\frac{f(x, v, t)}{m} + \frac{F(t)}{m} \tag{9.76}$$

These can be integrated using an Euler method with the form

$$\begin{bmatrix} x_{i+1} \\ v_{i+1} \end{bmatrix} = \begin{bmatrix} x_i + v_i\Delta t \\ v_i - \dfrac{\Delta t}{m}f(x_i, v_i, i\Delta t) + \dfrac{\Delta t}{m}F(i\Delta t) \end{bmatrix} \tag{9.77}$$

where $i\Delta t$ is the running time, in which Δt is the time increment, and i is an integer denoting the number of steps used in the integration.

Sample Problem 9.15

Let the system

$$150\ddot{x}(t) + 350\dot{x}(t) + 2000x(t) = \hat{F}\delta(t)$$

where $\delta(t)$ is the unit step function, have an applied impluse of 40 Ns. Compute the response of the system.

Solution From the equation of motion, $\omega_n = 3.65$ rad/s, $\zeta = 0.319$, and $\omega_d = 3.459$. According to Eq. (9.73), the response is

$$x(t) = \frac{\hat{F}e^{-\zeta\omega_n t}}{m\omega_d}\sin(\omega_d t) = 0.771e^{-1.164t}\sin(3.459t)$$

Sample Problem 9.16

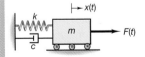

The damped linear system in the accompanying figure is subject to a pulse of the form

$$F(t) = \begin{cases} \sin(\pi t) & 0 \leq t \leq 1 \\ 0 & t > 1 \end{cases}$$

Compute and plot the response of the system to this input and the initial conditions $x_0 = 0.1$ m and $v_0 = 1$ m/s if the mass $m = 250$ kg, $k = 500$ N/m, and $c = 100$ kg/s.

Solution We will solve the system equation numerically using the Euler formulation given in Eq. (9.77). The equation of motion is

$$m\ddot{x} + c\dot{x} + kx = F(t)$$

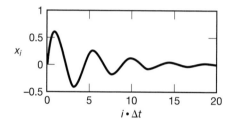

To place this equation into first-order form and into an Euler formula, $F(t)$ may be written using the Heaviside function $\Phi(t)$ introduced in Eq. (2.21). $F(t)$ then becomes

$$F(t) = \sin(\pi t)[\Phi(t) - \Phi(t-1)]$$

A plot of this applied force, (N) versus time (s), is shown here.

For this equation of motion, Eq. (9.77) becomes

$$\begin{bmatrix} x_{i+1} \\ v_{i+1} \end{bmatrix} = \begin{bmatrix} x_i + v_i \Delta t \\ v_i - \dfrac{\Delta t}{m}(cv_i + kx_i) + \dfrac{\Delta t}{m}\sin(\pi i \Delta t)[\Phi(i\Delta t) - \Phi(i\Delta t - 1)] \end{bmatrix}$$

Writing this relationship in Mathcad and plotting the response yields the following plot of displacement (m) versus time (s).

9.4.3 FORCED RESPONSE OF NONLINEAR SYSTEMS

The presence of nonlinear spring and damping forces makes analytical solutions of the forced response impossible and analysis and design difficult. We are unable to produce nice analytical formulas for the amplitude of vibration, such as that used in Sample Problem 9.14, to design a system so that it has a specific maximum amplitude of vibration. In addition, we are unable to use the superposition principle to add solutions of the homogenous and nonhomogenous parts of the response. Furthermore, new phenomena result from the nonlinear terms in the equations of motion, some of which cause undesirable motions. For instance, the period of oscillation of the response of a nonlinear system depends on the amplitude of the motion, which is not true in the linear case. In general, the response characteristics of a linear system do not depend on the initial conditions or the magnitude of the forcing function. However, the nature of the response of a nonlinear system depends greatly on the initial conditions and the level of the forcing amplitude. Some very interesting and exotic behavior can result in the forced response of a nonlinear system. Much of this behavior is well beyond the scope of the text, so we focus only on the numerical solution of these equations and attempt to use the simulations presented to point out some of the interesting phenomena.

An example of a uniquely nonlinear behavior is subharmonic resonance. For certain types of nonlinearity, it is possible for the response of a system to a harmonic excitation at a single frequency to have a component with large amplitude at a frequency of vibration lower (such as one-third) than the driving frequency. Since the linear design of vibrating systems typically consists of picking a value of the stiffness to avoid resonance, the prospect of having multiple "resonance" is alarming. In working with nonlinear systems, such possibilities need to be explored and understood.

Sample Problem 9.17

Consider the nonlinear vibration of an undamped, nonlinear spring–mass system modeled by

$$\ddot{x} + \omega_n^2 x - \beta x^3 = \Gamma \cos(\omega_{dr} t)$$

where the equation has been normalized by the mass, as before. Here, Γ is used to denote the mass-scaled magnitude of the forcing function, as is common in texts on nonlinear vibrations. Determine the response of the system with (a) $\beta = 0$ (the linear case); (b) $\beta = 1/6$, with an initial condition $x_0 = \sqrt{6}$ m and $v_0 = 0$ (near an unstable equilibrium); and (c) $\beta = 1/6$, with an initial condition $x_0 = 1$ m and $v_0 = 0$ (near a stable equilibrium). Compare the results for the parameters $\omega_n = 1$, $\omega_{dr} = 2.85$, and $\Gamma = 1.5$ N/kg.

Solution

The point of this problem is to note the different behavior caused by nonlinear terms in the equation of motion. In case (a), the linear version of the equation of motion, the response should be harmonic with two frequencies evident in the response, as predicted by Eq. (9.53). Using a Runge–Kutta method results in the plot of displacement (indicated by $Z_{i,1}$) versus time (indicated by $Z_{i,0}$):

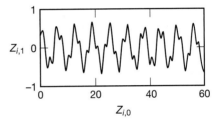

In case (b), no response can be determined numerically, because the response grows very quickly. The initial condition is at an unstable equilibrium, and the solution grows without bound.

In case (c), the identical system is given with an initial condition near a stable equilibrium, and the response is very similar to the linear case, as illustrated in the figure.

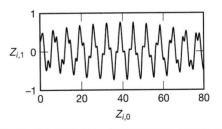

Numerically solving these three cases illustrates clearly that the nature of the solution of the nonlinear system depends critically on the initial conditions, in contrast to linear systems. In particular, the nonlinear system can have an unbounded solution even if the system is driven off resonance for certain initial conditions.

Sample Problem 9.18

Compute and plot the response of the system of Sample Problem 9.17 for the case where the system is driven at resonance ($\omega_n = 1 = \omega_{dr}$) with $\beta = -1/6$ and $\Gamma = 1.5$ N and for the initial conditions $x_0 = \sqrt{6}$ m and $v_0 = 0$.

Solution

Using a Runge–Kutta method, we plot the solution in a graph of displacement (indicated by $Z_{i,1}$) versus time (indicated by $Z_{i,0}$).

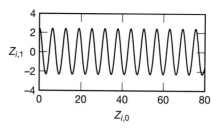

Note that this is a steady, bounded response at resonance! In the linear case, where $\beta = 0$, the response would grow without bound. Instead, in the nonlinear case, the nonlinear term causes the natural frequency to vary with the amplitude, so that as the amplitude changes, the resonance condition is never achieved. Note that this result depends on the nature of the parameters, and the dependence forms the focus of texts on nonlinear vibration.

Problems

9.51 Compute and plot the response of the harmonically driven, linear, undamped system of Eq. (9.44) with $m = 250$ kg, $k = 500$ N/m, $F_0 = 30$ N, $\omega_{dr} = 5$ rad/s, and initial conditions $x_0 = 0.03$ m and $v_0 = 0.05$ m/s. Plot the response for at least two periods.

9.52 Compute and plot the response of the harmonically driven, linear, undamped system of Eq. (9.44) with $m = 250$ kg, $k = 44$ N/m, $F_0 = 30$ N, $\omega_{dr} = 0.4$ rad/s, and initial conditions $x_0 = 0.03$ m and $v_0 = 0.05$ m/s.

9.53 Compute and plot the response of the harmonically driven, linear, undamped system of Eq. (9.44) with $m = 250$ kg, $k = 500$ N/m, $F_0 = 300$ N, $\omega_{dr} = 0.4$ rad/s, and initial conditions $x_0 = v_0 = 0$.

9.54 Compute and plot the response of the harmonically driven, linear, undamped system of Eq. (9.44) with $m = 250$ kg, $k = 500$ N/m, $\omega_{dr} = 0.4$ rad/s and initial conditions $x_0 = v_0 = 0$ for (a) $F_0 = 10$ N and (b) $F_0 = 500$ N. Plot the two responses on the same graph and compare them.

9.55 Consider the response of the harmonically driven, linear, undamped system of Eq. (9.44) with $m = 100$ kg, $F_0 = 100$ N, $\omega_{dr} = 2$ rad/s, and initial conditions $x_0 = v_0 = 0$. Calculate the value of spring stiffness such that the maximum value of the response will be less than 0.01 m.

9.56 Consider the response of the harmonically driven, linear, undamped system of Eq. (9.44) with $k = 3000$ N/m, $F_0 = 10$ N, $\omega_{dr} = 2$ rad/s, and initial conditions $x_0 = v_0 = 0$. Calculate the value of the mass such that the maximum value of the response will be less than 0.01 m.

9.57 Consider Sample Problem 9.11 and the general problem of calculating the mass to provide a specified maximum deflection x_{max} for a given stiffness and applied force. From the solution of the sample problem, note that a minus sign appears in the value for m. This means that for some choices of F_0, k, and x_{max} there will be no solution for m. Calculate the relationship between F_0, k, and x_{max} that must be satisfied for a solution to exist.

9.58 Compute and plot the response of the harmonically driven, linear, undamped system of Eq. (9.44) with $m = 525$ kg, $F_0 = 100$ N, $\omega_n = \omega_{dr} = 2$ rad/s, and initial conditions $x_0 = 0.1$ m and $v_0 = 0.5$ m/s.

9.59 Compute and plot the response of the harmonically driven, linear, undamped system of Eq. (9.44) with $k = 1000$ N/m, $F_0 = 100$ N, $\omega_n = \omega_{dr} = 4$ rad/s, and initial conditions $x_0 = 0.01$ m and $v_0 = 0.05$ m/s.

9.60 Resolve Problem 9.51 by first plotting the analytical expression Eq. (9.53), then using (Euler or Runge–Kutta) numerical integration, and plotting the result on the same graph. Take various step sizes Δt, and compare the plots, noting any differences in the responses.

9.61 Compute and plot the response of the harmonically driven, linear, damped system of Eq. (9.60) with $m = 250$ kg, $k = 500$ N/m, $F_0 = 100$ N, $c = 10$ kg/s, $\omega_{dr} = 0.4$ rad/s, and initial conditions $x_0 = v_0 = 0$. Plot the response for at least two periods.

9.62 Compute and plot the response of the harmonically driven, linear, damped system of Eq. (9.60) with $m = 250$ kg, $k = 500$ N/m, $F_0 = 100$ N, $c = 100$ kg/s, $\omega_{dr} = 0.4$ rad/s, and initial conditions $x_0 = v_0 = 0$. Plot the response for at least two periods.

9.63 Compute and plot the response of the harmonically driven, linear, damped system of Eq. (9.60) with $m = 250$ kg, $k = 500$ N/m, $F_0 = 100$ N, $c = 1000$ kg/s, $\omega_{dr} = 0.4$ rad/s, and initial conditions $x_0 = v_0 = 0$. Is this system underdamped, overdamped, or critically damped? Plot the response for at least two periods.

9.64 Compute and plot the response of the harmonically driven, linear, damped system of Eq. (9.60) with $m = 250$ kg, $k = 500$ N/m, $F_0 = 100$ N, $c = 100$ kg/s, $\omega_{dr} = 2$ rad/s, and initial conditions $x_0 = 0.1$ m and $v_0 = 1$ m/s. Is this system underdamped, overdamped, or critically damped? Plot the response for at least two periods.

9.65 Compute and plot the response of the harmonically driven, linear, damped system of Eq. (9.60) with $m = 250$ kg, $k = 500$ N/m, $F_0 = 100$ N, $c = 1000$ kg/s, $\omega_{dr} = 2$ rad/s, and initial conditions $x_0 = 0.1$ m and $v_0 = 1$ m/s. Is this system underdamped, overdamped, or critically damped? Plot the response for at least two periods, and compare the amplitude of the response with that in the case where $c = 100$ kg/s.

9.66 Reproduce Figure 9.14 from the text by plotting Eq. (9.72) for the values of ζ indicated in the figure.

9.67 Derive Eq. (9.64) from Eqs. (9.63) and (9.62).

9.68 Derive Eq. (9.69) from Eq. (9.68).

9.69 Derive Eq. (9.72) from the expressions for A_0 and θ given in the text.

9.70 Consider Sample Problem 9.14, and compute a value of the damping coefficient that will cause the system to have a maximum deflection of 0.01 m. Is the sytem underdamped, critically damped, or overdamped?

9.71 Consider the damped system with $m = 160$ kg of Eq. (9.60), driven at resonance with $\omega_{dr} = 5$, for the case where $F_0 = 10$ N and $k = 4000$ N/m, and compute a value of the damping coefficient that will cause the system's steady–state response to have a maximum deflection of 0.01 m. Is the system underdamped, critically damped, or overdamped?

9.72 Compute and plot the response of the system of Sample Problem 9.16 for the case where the damping rate is $c = 1000$ kg/s. Does the system oscillate?

9.73 The suspension system of an automobile driving over a curb may be modeled by the system of Sample Problem 9.16 with a forcing funtion given by

$$F(t) = 10[\Phi(t - 1) - \Phi(t - 2)] \text{ N}$$

where $\Phi(t)$ is the Heaviside step function. Compute and plot the response of the car with parameters $m = 1250$ kg, $k = 4 \times 10^5$ N/m, $c = 20{,}000$ kg/s, and with initial conditions $x_0 = 0$ and $v_0 = 0$.

9.74 The suspension system of an automobile driving over a curb may be modeled by the sytem of Sample Problem 9.16 with a forcing function given by

$$F(t) = 30[\Phi(t - 1) - \Phi(t - 2)] \text{ N}$$

Compute and plot the response of the car with parameters $m = 1000$ kg, $k = 4 \times 10^5$ N/m, $c = 5000$ kg/s, and with initial conditions $x_0 = 0$ and $v_0 = 0$.

9.75 Compute the smallest value of damping that can be used in Problem 9.74 such that the response decays in 6 s.

9.76 Repeat Problem 9.74 for the forcing function of Figure P9.76 (N versus s) and with parameters $m = 1000$ kg, $k = 4 \times 10^3$ N/m, $c = 1000$ kg/s, and with initial conditions $x_0 = 0$ and $v_0 = 0$.

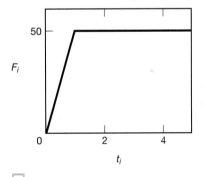

Figure P9.76

9.77 Repeat Problem 9.74 for the forcing function of Figure P9.77 (N versus s) and with parameters $m = 1000$ kg, $k = 4 \times 10^3$ N/m, $c = 1000$ kg/s, and with initial conditions $x_0 = 0$ and $v_0 = 0$. How long does it take for the response to die out?

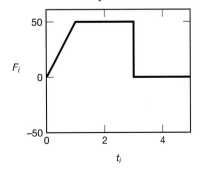

Figure P9.77

9.78 Determine a new value of the damping coefficient for Problem 9.77 so that the system dies out in about 12 s when excited with the force of Figure P9.77.

9.79 Repeat Problem 9.74 for the forcing function of Figure P9.77 (N versus s) and with parameters $m = 1000$ kg, $k = 4 \times 10^3$ N/m, $c = 1000$ kg/s, and with initial conditions $x_0 = 0$ and $v_0 = -1$ m/s. How long does it take for the response to die out?

9.80 Repeat Sample Problem 9.17 by examining case (b) and reducing the initial position by increments of 0.1 m until a stable response results. How far away from the unstable equilibrium position must you get before you can compute a solution?

9.81 Compute and plot the response for Sample Problem 9.17, parts (a) and (c).

9.82 Compute and plot the response for Sample Problem 9.18.

9.83 Compute and plot the response of the damped, nonlinear system

$$\ddot{x} + \omega_n^2 x + 2\zeta\omega_n \dot{x}|\dot{x}| - \beta x^3 = \Gamma\cos(\omega_{dr}t)$$

with parameter values $\beta = 1/6$, $\omega_n = 1$, $\omega_{dr} = 2.85$, $\zeta = 0.2$, $\Gamma = 1.5$ N, and with initial conditions $x_0 = 1$ m and $v_0 = 0$.

9.84 Find a value of ζ such that the response of the damped, nonlinear system

$$\ddot{x} + \omega_n^2 x + 2\zeta\omega_n \dot{x} - \beta x^3 = \Gamma\cos(\omega_{dr}t)$$

remains less then 0.25 m within 100 s for the parameter values $\beta = 1/6$, $\omega_n = 1$, $\omega_{dr} = 2.85$, $\Gamma = 1.5$ N, and with initial conditions $x_0 = 1$ m and $v_0 = 0$.

9.85 Consider again the system of Problem 9.83, and investigate the solution near the unstable equilibrium point at $x_0 = \sqrt{6}$ m and $v_0 = 0$. Does the addition of damping make this solution die out?

MASS MOMENTS OF INERTIA

In Chapter 6, the angular momentum about the center of mass of a rigid body was defined as

$$
\begin{aligned}
\mathbf{H}_{cm} = {}& [I_{xx}\omega_x + I_{xy}\omega_y + I_{xz}\omega_z]\hat{\mathbf{i}} \\
& + [I_{yx}\omega_x + I_{yy}\omega_y + I_{yz}\omega_z]\hat{\mathbf{j}} \\
& + [I_{zx}\omega_x + I_{zy}\omega_y + I_{zz}\omega_z]\hat{\mathbf{k}}
\end{aligned}
\tag{A.1}
$$

where ω is the angular velocity of the rigid body and the components of the mass-moment-of-inertia tensor are

$$
I_{xx} = \int_M (y^2 + z^2)\, dm \quad I_{yy} = \int_M (z^2 + x^2)\, dm \quad I_{zz} = \int_M (x^2 + y^2)\, dm \tag{A.2}
$$

$$
I_{xy} = I_{yx} = -\int_M xy\, dm \quad I_{yz} = I_{zy} = -\int_M yz\, dm \quad I_{zx} = I_{xz} = -\int_M zx\, dm
$$

The origin of the coordinates to determine the components of the mass-moment-of-inertia tensor are at the center of mass (cm) of the body, as shown in Figure A.1.

Equation (A.1) can be written in matrix notation as

$$
[H_{cm}] = [I_{cm}][\omega]
$$

$$
\begin{bmatrix} H_{cm_x} \\ H_{cm_y} \\ H_{cm_z} \end{bmatrix} = \begin{bmatrix} I_{xx} & I_{xy} & I_{xz} \\ I_{yx} & I_{yy} & I_{yz} \\ I_{zx} & I_{zy} & I_{zz} \end{bmatrix} \begin{bmatrix} \omega_x \\ \omega_y \\ \omega_z \end{bmatrix}
\tag{A.3}
$$

We saw in Chapter 2 that the linear momentum of a particle is, as introduced and defined by Newton,

$$
\mathbf{L} = m\mathbf{v} \tag{A.4}
$$

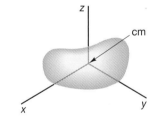

Figure A.1

where m, the inertial mass of the particle, is that property of the object which resists a change in its *linear* velocity. In a similar manner, $[I_{cm}]$ is that property of the object which resists a change in its *angular* velocity. The resistance of a rigid body to rotation depends not only on the mass of the body, but also on the distribution of that mass. The farther the mass is from the axis of rotation, the greater is the resistance of the body to rotation about the axis. The mass moment of inertia is a more complicated mathematical quantity then the simpler scalar quantity of mass. The mass moment of inertia is a second-order tensor. The diagonal elements of $[I_{cm}]$ are called the **mass moments of inertia about the x-, y-, and z-axes,** and the off-diagonal elements are called the **product moments of inertia.** We can increase our conceptual understanding of the mass moment of inertia by examining the rotation of a small mass dm about a fixed axis in the z-direction, as shown in Figure A.2.

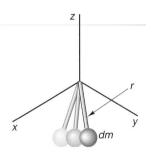

Figure A.2

We consider the small mass to be rotated by a bar of negligible mass and length r at a constant acceleration α. The moment required to accelerate the mass is

$$M_z = [r^2\,dm]\alpha \tag{A.5}$$

Therefore, the quantity $r^2 dm$ is a measure of the resistance (or inertia) of the mass to rotation about the z-axis. This property has led to the term being called the mass moment of inertia. For an object containing many masses, the moment of inertia depends upon the distribution of the masses about the z-axis. The resistance increases in proportion to the square of the distance from the axis. Consider the component of the mass moment of inertia about the z-axis, namely,

$$I_{zz} = \int_m (x^2 + y^2)\,dm = \int_m \rho^2\,dm \tag{A.6}$$

where ρ is the radius perpendicular to the z-axis around which the element of mass dm rotates. This component represents the resistance to rotation about the z-axis. In a similar manner, I_{xx} and I_{yy} represent the resistance to rotation about the x- and y-axes, respectively. The product moment-of-inertia terms represent the coupling of the rotations about two different axes. The units of the components of the mass moment of inertia are $\text{kg}\cdot\text{m}^2$ in SI units or $\text{lb}\cdot\text{ft}/\text{s}^2$ in U.S. customary units. Note that the mass moments of inertia are always positive, while the product moments of inertia may be positive, negative, or zero.

There are an infinite number of orthogonal coordinate axes with origin at the center of mass of a body. Let us consider the relationship of the components of a vector in one coordinate system with origin at the center of mass of the system to the components in another coordinate systems that is rotated relative to the first. Two such coordinate systems with origins at the center of mass of a body are shown in Figure A.3. In Chapter 8, using direction cosines, we developed the three-dimensional transformation matrix

$$[\mathbf{x'}] = [\mathbf{R}][\mathbf{x}] \tag{A.7}$$

where the *orthogonal transformation matrix* is

$$[R] = \begin{bmatrix} \lambda_{xx} & \lambda_{xy} & \lambda_{xz} \\ \lambda_{yx} & \lambda_{yy} & \lambda_{yz} \\ \lambda_{zx} & \lambda_{zy} & \lambda_{zz} \end{bmatrix} \tag{A.8}$$

in which $\lambda_{xy} = \cos(x', y)$, the cosine of the angle between the x'- and y'-axes. We also showed that the inverse of the orthogonal transformation matrix is equal to the transpose of the matrix. The components of the mass moment of inertia in the primed coordinate system can be expressed in terms of the unprimed coordinate system as

$$\begin{aligned}
[I'] &= [R][I][R]^T \\
&= \begin{bmatrix} \lambda_{xx} & \lambda_{xy} & \lambda_{xz} \\ \lambda_{yx} & \lambda_{yy} & \lambda_{yz} \\ \lambda_{zx} & \lambda_{zy} & \lambda_{zz} \end{bmatrix} \cdot \begin{bmatrix} I_{xx} & I_{xy} & I_{xz} \\ I_{xy} & I_{yy} & I_{yz} \\ I_{xz} & I_{yz} & I_{zz} \end{bmatrix} \cdot \begin{bmatrix} \lambda_{xx} & \lambda_{yx} & \lambda_{zx} \\ \lambda_{xy} & \lambda_{yy} & \lambda_{zy} \\ \lambda_{xz} & \lambda_{yz} & \lambda_{zz} \end{bmatrix}^T
\end{aligned} \tag{A.9}$$

Let us examine the mass moment of inertia about the center of mass to see whether there is a coordinate system such that the angular momentum vector is parallel to the angular velocity axis. The mass moment of inertia is

$$\mathbf{H}_{cm} = [I_{cm}]\omega \tag{A.10}$$

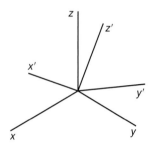

Figure A.3

PRINCIPAL MOMENTS OF INERTIA

If the angular momentum vector is parallel to the angular velocity vector, Eq. (A.10) can be written as

$$\begin{bmatrix} I_{xx} - \beta & I_{xy} & I_{xz} \\ I_{xy} & I_{yy} - \beta & I_{yz} \\ I_{xz} & I_{yz} & I_{zz} - \beta \end{bmatrix} \hat{\mathbf{n}} = 0 \tag{A.11}$$

where β is the ratio of the angular momentum vector to the angular velocity vector and $\hat{\mathbf{n}}$ is a unit vector in the direction of both of those vectors. The problem posed by Eq. (A.11) is called the *eigenvalue problem*, β is the *eigenvalue* or characteristic value, and $\hat{\mathbf{n}}$ is the *eigenvector* or characteristic vector. The eigenvalue problem has a solution if and only if the determinant of the coefficient matrix is zero, that is,

$$\begin{vmatrix} I_{xx} - \beta & I_{xy} & I_{xz} \\ I_{xy} & I_{yy} - \beta & I_{yz} \\ I_{xz} & I_{yz} & I_{zz} - \beta \end{vmatrix} = 0 \tag{A.12}$$

If the determinant is expanded, we will have a cubic equation for the eigenvalues,

$$\beta^3 - I_I\beta^2 + II_I\beta - III_I = 0 \tag{A.13}$$

where

$$I_I = (I_{xx} + I_{yy} + I_{zz})$$

$$II_I = \frac{1}{2}((I_{xx} + I_{yy} + I_{zz})^2 - (I_{xx}^2 + I_{yy}^2 + I_{zz}^2 + 2I_{xy}^2 + 2I_{xz}^2 + 2I_{yz}^2))$$

$$III_I = \det[I_{cm}]$$

These coefficients of the cubic equation are called the invariants of the mass-moment-of-inertia tensor, because it can be shown that they have the same value in all coordinate systems with origin at the center of mass.

It also can be shown that the three roots of the cubic equation will be real positive values. The three eigenvectors will form an orthogonal set of base vectors for a coordinate system. The axes along these eigenvectors are called the *principal axes,* and the associated eigenvalues are called the *principal mass moments of inertia.* The product moments of inertia about the principal axes are zero, and therefore, the angular movements about these three axes are independent of each other. The principal moment of inertia is a maximum about one of the principal axes, a minimum about another of the principal axes, and an intermediate value about the remaining axis. If the rigid body has an axis of symmetry, that axis will be a principal axis.

If X, Y, Z are principal axes of a rigid body, the angular momentum about these principal axes is

$$\begin{bmatrix} H_{cm_x} \\ H_{cm_y} \\ H_{cm_z} \end{bmatrix} = \begin{bmatrix} I_{XX} & 0 & 0 \\ 0 & I_{YY} & 0 \\ 0 & 0 & I_{ZZ} \end{bmatrix} \begin{bmatrix} \omega_X \\ \omega_Y \\ \omega_Z \end{bmatrix} \tag{A.14}$$

The Euler equations of motion may be written as three independent scalar equations; this is the basis of plane dynamic analysis.

Let $\hat{\mathbf{n}}$ be a unit vector in a particular direction. The mass moment of inertia about an axis in this direction may be represented by

$$I = \hat{\mathbf{n}}[I]\hat{\mathbf{n}} \tag{A.15}$$

Let the unit vector be expressed in terms of its components, which are direction cosines, as

$$\hat{\mathbf{n}} = \lambda_x \hat{\mathbf{i}} + \lambda_y \hat{\mathbf{j}} + \lambda_z \hat{\mathbf{k}} \tag{A.16}$$

Substituting Eq. (A.16) into Eq. (A.15) and expanding yields

$$I = I_{xx}\lambda_x^2 + I_{yy}\lambda_y^2 + I_{zz}\lambda_z^2 + 2I_{xy}\lambda_x\lambda_y + 2I_{yz}\lambda_y\lambda_z + 2I_{xz}\lambda_x\lambda_z \tag{A.17}$$

We next intoduce a vector

$$\rho = \frac{\hat{\mathbf{n}}}{\sqrt{I}} \tag{A.18}$$

Substituting Eq. (A.18) into Eq. (A.17) yields

$$I = I_{xx}\rho_x^2 + I_{yy}\rho_y^2 + I_{zz}\rho_z^2 + 2I_{xy}\rho_x\rho_y + 2I_{yz}\rho_y\rho_z + 2I_{xz}\rho_x\rho_z \tag{A.19}$$

Equation (A.19) is the equation of an ellipsoid in ρ-space, as shown in Figure A.4. This ellipsoid, which defines the mass moment of inertia with respect to any axis, is called the ***ellipsoid of inertia***. The coefficients in the quadratic defined in Eq. (A.19) change when the reference axes change, but the ellipsoid remains unaffected. It is well known that one can always transform the coordinate system to a coordinate system in which the inertial ellipsoid takes on its normal form; i.e.,

$$I = I_{11}\rho_1^2 + I_{22}\rho_2^2 + I_{33}\rho_3^2 \tag{A.20}$$

where

$$\rho_1 = x' \qquad \rho_2 = y' \qquad \rho_3 = z'$$

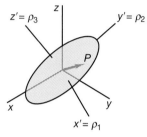

Figure A.4

The product moments of inertia with respect to the x'-, y'-, and z'-axes are zero, and therefore, these axes are the principal axes of inertia, and the mass moments of inertia about these three axes are the principal mass moments of inertia. The inertial tensor is diagonal when referenced to the principal axes. The principal moments of inertia determine the lengths of the axes of the inertial ellipsoid. If two of the eigenvalues of the inertial tensor are equal, the inertial ellipsoid has two equal axes and is an ellipsoid of revolution. If all three principal moments are equal, the inertial ellipsoid is a sphere. The inertial ellipsoid serves the same conceptual role as Mohr's circle does for two-dimensional examination of the symmetric second-order stress, strain, and second-moment-of-area tensors. The determination of the principal axes and principal moments of inertia are best obtained by eigenvalue analysis.

Sample Problem A.1

Consider the mass moment of inertia referenced to a given x, y, z coordinate system. Determine the principal moments of inertia and the principal axes if

$$I = \begin{bmatrix} 10 & -5 & 3 \\ -5 & 8 & 4 \\ 3 & 4 & 7 \end{bmatrix} \text{kg} \cdot \text{m}^2$$

Solution   Using Eq. (A.13) gives the cubic equation for the three principal moments of inertia:

$$\beta^3 - 25\beta^2 + 156\beta - 33 = 0$$

The easiest solution of the cubic equation is to first graph the equation to approximate the roots. We have

$$f(\beta) = \beta^3 - 25 \cdot \beta^2 + 156 \cdot \beta - 33$$
$$\beta = 0, 1 \ldots 15$$

yielding the accompanying plot.

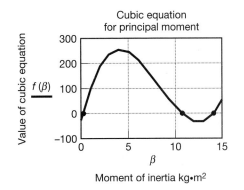

Cubic equation
for principal moment

The roots of the cubic equation may be determined by trial and error or with the use of a calculator or computational software and are

$$I_1 = 14.109 \text{ kg} \cdot \text{m}^2$$
$$I_2 = 10.672 \text{ kg} \cdot \text{m}^2$$
$$I_3 = 0.219 \text{ kg} \cdot \text{m}^2$$

The principal moments of inertia always must be positive, and in this case, the resistance to rotation about the third principal axis is very small compared with those about the other two principal axes. The eigenvector for each principal eigenvalue may be determined by substituting the principal values into three scalar equations from Eq. (A.11). For the first eigenvalue, we obtain

$$(10 - 14.109)x - 5y + 3z = 0$$
$$-5x + (8 - 14.109)y + 4z = 0$$
$$3x + 4y - (7 - 14.109)z = 0$$

We know that this system of linear equations is singular, since the determinant of the co-efficient matrix was set to zero to solve for the eigenvalues. We will solve for z from the first equation and substitute it into the second equation, yielding

$$z = 1.370x + 1.667y$$
$$0.480x + 0.559y = 0$$

Knowing that the third equation is linearly dependent on the first two, we will not involve that equation in the determination of the eigenvector. If we set $x = 1$, we obtain

$$y = -0.859$$
$$z = -0.062$$

A vector in the direction of the eigenvector is

$$\mathbf{A} = \begin{bmatrix} 1 \\ -0.859 \\ -0.062 \end{bmatrix}$$

Eigenvectors are always taken as unit vectors. Therefore, the eigenvector corresponding to the first eigenvalue is

$$\mathbf{n}_1 = \begin{bmatrix} 0.758 \\ -0.651 \\ -0.046 \end{bmatrix}$$

The other two eigenvectors are found in a similar manner and are

$$\mathbf{n}_2 = \begin{bmatrix} 0.415 \\ 0.426 \\ 0.804 \end{bmatrix} \qquad \mathbf{n}_3 = \begin{bmatrix} 0.503 \\ 0.628 \\ -0.593 \end{bmatrix}$$

These three eigenvectors serve as the base vectors for a coordinate system aligned with the principal axes. All computational software has operators that find the eigenvalues and eigenvectors of a symmetric tensor. (See the Computational Supplement for details.)

Problems

A.1 Determine the principal moments of inertia and the principal axes for a moment-of-inertia tensor given by

$$[I] = \begin{bmatrix} 12 & 6 & 4 \\ 6 & 12 & 1 \\ 4 & 1 & 12 \end{bmatrix} \text{kg} \cdot \text{m}^2$$

A.2 For the moment-of-inertia tensor given in Problem A.1, determine the tensor for a coordinate system rotated $30°$ about the z-axis.

A.3 Consider three masses connected by massless rods. Mass 1 is 5 kg and has coordinates of $(2, 1, 3)$ in the reference system, mass 2 is 3 kg and has coordinates of $(-2, 2, -2)$, and mass 3 is 4 kg and has coordinates of $(1, -3, -5)$. Determine the location of the center of mass of the system and the mass moment of inertia about the center of mass.

A.4 For the system in Problem A.3, determine the principal moments of inertia and the principal axes.

A.5 Determine the principal moments of inertia and the direction of the principal axes for the inertia tensor

$$[I] = \begin{bmatrix} 10 & -2 & -3 \\ -2 & 4 & 3 \\ -3 & 3 & 10 \end{bmatrix} \text{kg} \cdot \text{m}^2$$

A.6 Determine the orthogonal transformation matrix required to go from the reference system to the principal axes. Show that the tensor is in a diagonal form about the principal axes.

A.7 Consider a system of five equal masses attached together by massless links. If the coordinate positions of the five masses are given by

$$\mathbf{r}_1 = \begin{bmatrix} 3 \\ 5 \\ 8 \end{bmatrix} \text{m} \quad \mathbf{r}_2 = \begin{bmatrix} -1 \\ -2 \\ 3 \end{bmatrix} \text{m} \quad \mathbf{r}_3 = \begin{bmatrix} -4 \\ 5 \\ -6 \end{bmatrix} \text{m}$$

$$\mathbf{r}_4 = \begin{bmatrix} 2 \\ 0 \\ -3 \end{bmatrix} \text{m} \quad \mathbf{r}_5 = \begin{bmatrix} 0 \\ -3 \\ 4 \end{bmatrix} \text{m}$$

determine the center of mass and the mass-moment-of-inertia tensor about the reference coordinate system.

A.8 Determine the principal moments of inertia and the principal axes for the system in Problem A.7.

A.9 Determine the principal moment of inertia of a sphere of radius R and mass M.

A.10 Determine the principal moments of inertia of a homogeneous cylinder of radius R, length L, and mass M.

A.11 Determine the principal moments of inertia for a homogeneous cube with sides of length L and mass M.

A.12 Determine the principal moments of inertia for a homogeneous plate of length L, width W, mass M, and thickness t.

A.13 Determine the principal axes for a thin, homogeneous, circular plate of radius R, mass M, and thickness t.

RADIUS OF GYRATION

The definition of the mass moment of inertia indicates that this quantity's dimensions are mass multiplied by length squared. As a result, the mass moment of inertia is sometimes given by an equivalent distance at which the mass could have been concentrated, yielding the same mass moment of inertia. This distance is called the ***radius of gyration***, which is a convenient manner of expressing the moment of inertia of a body with a complex geometry. The radius of gyration, defined as

$$k = \sqrt{\frac{I}{m}} \tag{A.21}$$

does not have a physical interpretation and is only an alternative way of expressing the moment of inertia.

PARALLEL-AXES THEOREM FOR MOMENTS OF INERTIA

Figure A.5

Consider the body shown in Figure A.5, where the x-, y-, and z-axes have their origin at the center of mass. The x'-, y'-, and z'-axes are parallel to the x-, y-, and z-axes. The two coordinate systems are related by the equations

$$\begin{aligned}
\mathbf{r'} &= \boldsymbol{\rho} + \mathbf{r'_{cm}} \\
x' &= \rho_x + x'_{cm} \\
y' &= \rho_y + y'_{cm} \\
z' &= \rho_z + z'_{cm}
\end{aligned} \tag{A.22}$$

The moment of inertia about the z'-axis is defined as

$$I_{z'z'} = \int_m (x'^2 + y'^2)\,dm = \int_m (\rho_x^2 + \rho_y^2)\,dm + 2x'_{cm}\int_m \rho_x\,dm$$

$$+ \, 2y'_{cm}\int_m \rho_y\,dm + (x'^2_{cm} + y'^2_{cm})M \tag{A.23}$$

The first integral on the right side of Eq. (A.23) is the mass moment of inertia about the z-axis through the center of mass. The second and third integrals are zero from the definition of the center of mass. Therefore, Eq. (A.23) can be written as

$$I_{z'z'} = I_{cm_{zz}} + Md_z^2 \tag{A.24}$$

where

$$d_z^2 = x'^2_{cm} + y'^2_{cm}$$

In a similar manner, the product moment of inertia may be written

$$I_{x'y'} = -\int_M x'y'\,dm = I_{cm_{xy}} - Mx'_{cm}y'_{cm} \tag{A.25}$$

Equations (A.24) and (A.25) form the basis for the parallel-axis theorem for moments of inertia. Note that they allow computations of the mass moments and product moments of inertia about axes that are parallel to the center of mass axes of a rigid body. The parallel-axis theorem permits the calculation of the mass moment of inertia of composite bodies.

Table A.1
Moments of Inertia of Common Shapes

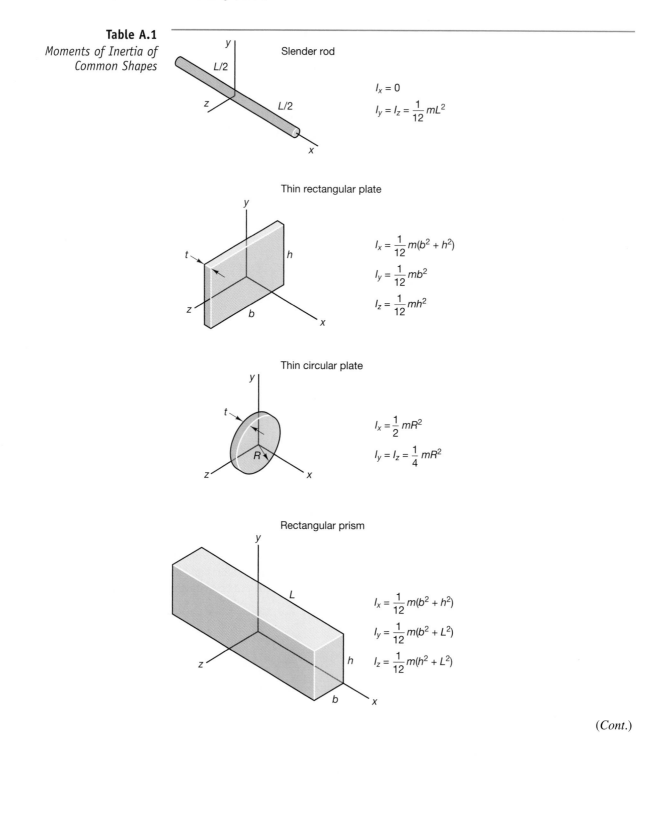

Slender rod

$$I_x = 0$$
$$I_y = I_z = \frac{1}{12} mL^2$$

Thin rectangular plate

$$I_x = \frac{1}{12} m(b^2 + h^2)$$
$$I_y = \frac{1}{12} mb^2$$
$$I_z = \frac{1}{12} mh^2$$

Thin circular plate

$$I_x = \frac{1}{2} mR^2$$
$$I_y = I_z = \frac{1}{4} mR^2$$

Rectangular prism

$$I_x = \frac{1}{12} m(b^2 + h^2)$$
$$I_y = \frac{1}{12} m(b^2 + L^2)$$
$$I_z = \frac{1}{12} m(h^2 + L^2)$$

(Cont.)

Table A.1
Moments of Inertia of Common Shapes

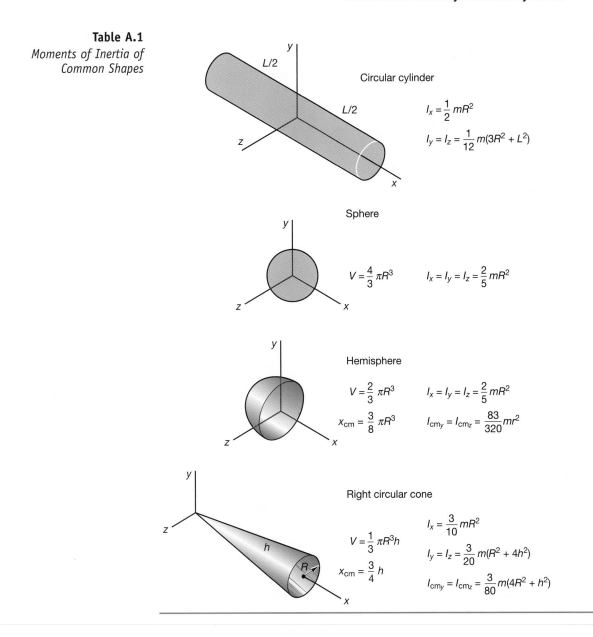

Circular cylinder

$$I_x = \frac{1}{2} mR^2$$

$$I_y = I_z = \frac{1}{12} m(3R^2 + L^2)$$

Sphere

$$V = \frac{4}{3} \pi R^3 \qquad I_x = I_y = I_z = \frac{2}{5} mR^2$$

Hemisphere

$$V = \frac{2}{3} \pi R^3 \qquad I_x = I_y = I_z = \frac{2}{5} mR^2$$

$$x_{cm} = \frac{3}{8} \pi R^3 \qquad I_{cmy} = I_{cmz} = \frac{83}{320} mr^2$$

Right circular cone

$$I_x = \frac{3}{10} mR^2$$

$$V = \frac{1}{3} \pi R^3 h$$

$$I_y = I_z = \frac{3}{20} m(R^2 + 4h^2)$$

$$x_{cm} = \frac{3}{4} h$$

$$I_{cmy} = I_{cmz} = \frac{3}{80} m(4R^2 + h^2)$$

Sample Problem A.2

Determine the mass moment of inertia about the z-axis through the center of mass of the barbell shown in the accompanying diagram. The bar is of length l and mass m_s and has two disks of radius r, thickness t, and mass m_d.

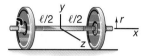

Solution The center of mass of the barbell is at its geometric center due to symmetry. The moment of inertia of the bar about its center of mass is

$$I_{bz} = \frac{1}{12} m_b l^2$$

The moment of inertia of the disks about a parallel z-axis through their center of mass is

$$I_{dz} = \frac{1}{4} m_d r^2$$

From the parallel axis theorem, the mass moment of inertia of the barbell is

$$I_z = \frac{1}{12} m_b l^2 + 2\left[\frac{1}{4}m_d r^2 + m_d \frac{l^2}{4}\right] = \frac{1}{12}(m_b + 6m_d)l^2 + \frac{1}{2}m_d r^2$$

Problems

A.14 Determine the mass moment of inertia of the rectangular plate shown in Figure PA.14. The plate has a uniform mass density $\rho = 0.08$ kg/mm³ about point A.

A.15 A pendulum is composed of a slender bar of mass m and a thin circular plate of mass M. (See Figure PA. 15.) Determine the mass moment of inertia with respect to point O.

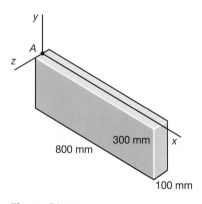

Figure PA.14

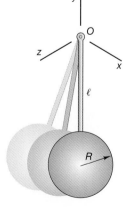

Figure PA.15

A.16 Determine the mass moment of inertia of an 8-mm-thick, 500-mm-square plate with a 100-mm-diameter hole at the center. (See Figure PA.16.) The mass density is 250 kg/mm³.

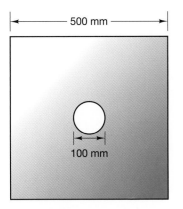

Figure PA.16

A.17 Determine the mass moment of inertia of the plate in Figure PA.16 about the upper-left corner.

A.18 Determine the mass moment of inertia of the plate in Figure PA.16 about a midpoint along an edge.

A.19 A long slender bar of mass m and length l is bent at its midpoint by an angle θ, as shown in Figure PA.19. Develop a general expression for the mass moment of inertia about the center of mass of the bar. Show that for an angle $\theta = 0$, the expression is the same as for a straight bar.

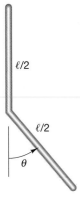

Figure PA.19

A.20 A shovel can be modeled as a bar and a plate. (See Figure PA.20.) Determine the mass moment of inertia about the handle point O. The mass of the bar is 2 kg, and the mass of the plate is 3 kg.

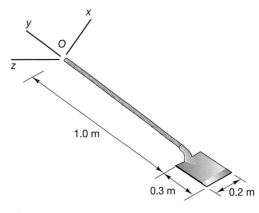

Figure PA.20

A.21 Determine the mass moment of inertia about point O of a pendulum composed of a sphere of mass 5 kg and radius 0.2 m, together with a 1-kg bar of length 1.5 m, as shown in Figure PA.21.

Figure PA.21

A.22 A thin 5-mm-thick steel plate is cut and bent to form a required part, as shown in Figure PA.22. If the mass density of steel is 8000 kg/m³, determine the mass moment of inertia of the part about the coordinate axes.

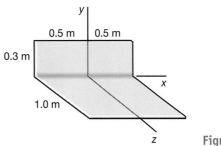

Figure PA.22

A.23 For the plate shown in Figure PA.22, determine the location of the center of mass and the principal moments of inertia about the center of mass.

A.24 Determine the principal moments of inertia about the center of mass of the shovel shown in Figure PA.20.

A.25 Determine the principal moments of inertia about the center of mass of the pendulum shown in Figure PA.21.

Appendix B

VECTOR CALCULUS AND ORDINARY DIFFERENTIAL EQUATIONS

VECTOR CALCULUS

Consider a position vector expressed in terms of rectilinear base vectors $\hat{\mathbf{i}}$, $\hat{\mathbf{j}}$, and $\hat{\mathbf{k}}$ as follows.

$$\mathbf{r}(t) = x(t)\hat{\mathbf{i}} + y(t)\hat{\mathbf{j}} + z(t)\hat{\mathbf{k}} \tag{B.1}$$

The rules of differentiation of a vector are

$$\frac{d}{dt}(\mathbf{A} + \mathbf{B}) = \frac{d\mathbf{A}}{dt} + \frac{d\mathbf{B}}{dt}$$

$$\frac{d}{dt}(\alpha\mathbf{A}) = \frac{d\alpha}{dt}\mathbf{A} + \alpha\frac{d\mathbf{A}}{dt}$$

$$\frac{d}{dt}(\mathbf{A} \cdot \mathbf{B}) = \frac{d\mathbf{A}}{dt} \cdot \mathbf{B} + \mathbf{A} \cdot \frac{d\mathbf{B}}{dt} \tag{B.2}$$

$$\frac{d}{dt}(\mathbf{A} \times \mathbf{B}) = \frac{d\mathbf{A}}{dt} \times \mathbf{B} + \mathbf{A} \times \frac{d\mathbf{B}}{dt}$$

The components of the position vector are products of a scalar function of time and the unit-base vector and therefore, the differentiation of Eq. (B.1) is

$$\mathbf{v}(t) = \frac{dx(t)}{dt}\hat{\mathbf{i}} + x(t)\frac{d\hat{\mathbf{i}}}{dt} + \frac{dy(t)}{dt}\hat{\mathbf{j}} + y(t)\frac{d\hat{\mathbf{j}}}{dt} + \frac{dz(t)}{dt}\hat{\mathbf{k}} + z(t)\frac{d\hat{\mathbf{k}}}{dt} \tag{B.3}$$

If the unit-base vectors are not base vectors for a rotating coordinate system, they are constants and their differentials are zero. However if the x, y, z coordinate system is rotating with an angular velocity of Ω, the time derivative of the unit vectors can be expressed as

$$\frac{d\hat{\mathbf{i}}}{dt} = \Omega \times \hat{\mathbf{i}} \qquad \frac{d\hat{\mathbf{j}}}{dt} = \Omega \times \hat{\mathbf{j}} \qquad \frac{d\hat{\mathbf{k}}}{dt} = \Omega \times \hat{\mathbf{k}} \tag{B.4}$$

We also encounter base vectors that are functions of time when a normal–tangential coordinate system is used or when curvilinear coordinates, such as cylindrical or spherical coordinates, are used. A problem is formulated in terms of normal and tangential coordinates because the velocity is always tangent to the path of motion and the velocity vector can be written as

$$\mathbf{v} = \frac{ds}{dt}\hat{\mathbf{e}}_t = \rho\frac{d\theta}{dt}\hat{\mathbf{e}}_t \tag{B.5}$$

where s is a parameter of measurement along the length of the curve, ρ is the radius of curvature of the curve, and θ is the angular change of the curve. The acceleration may be obtained by differentiation of Eq. (B.5) being aware that the tangential unit vector is a function of position and therefore of time.

$$\frac{d\hat{\mathbf{e}}_t}{dt} = \frac{d\theta}{dt} \hat{\mathbf{e}}_n \tag{B.6}$$

The acceleration in this coordinate system is

$$\mathbf{a} = \rho \frac{d^2\theta}{dt^2} \hat{\mathbf{e}}_t + \rho \left(\frac{d\theta}{dt}\right)^2 \hat{\mathbf{e}}_n \tag{B.7}$$

In cylindrical coordinates, the unit vectors $\hat{\mathbf{e}}_r$ and $\hat{\mathbf{e}}_\theta$ are function of θ.

$$\frac{d\hat{\mathbf{e}}_r}{d\theta} = \hat{\mathbf{e}}_\theta$$

$$\frac{d\hat{\mathbf{e}}_\theta}{d\theta} = -\hat{\mathbf{e}}_r \tag{B.8}$$

$$\frac{d\hat{\mathbf{e}}_z}{d\theta} = 0$$

The position vector in cylindrical coordinates is

$$\mathbf{r} = r\hat{\mathbf{e}}_r + z\hat{\mathbf{e}}_z \tag{B.9}$$

The velocity is

$$\mathbf{v} = \frac{d}{dt}\mathbf{r} = \frac{d}{dt}(r\hat{\mathbf{e}}_r + z\hat{\mathbf{e}}_z)$$

$$\mathbf{v} = \frac{dr}{dt}\hat{\mathbf{e}}_r + r\frac{d\hat{\mathbf{e}}_r}{dt} + \frac{dz}{dt}\hat{\mathbf{e}}_z = \dot{r}\hat{\mathbf{e}}_r + r\dot{\theta}\hat{\mathbf{e}}_\theta + \dot{z}\hat{\mathbf{e}}_z \tag{B.10}$$

Differentiating again yields the acceleration:

$$\mathbf{a} = \frac{d(\dot{r}\hat{\mathbf{e}}_r)}{dt} + \frac{d(r\dot{\theta}\hat{\mathbf{e}}_\theta)}{dt} + \frac{d(\dot{z}\hat{\mathbf{e}}_z)}{dt}$$

$$\mathbf{a} = \left[\ddot{r}\hat{\mathbf{e}}_r + \dot{r}\dot{\theta}\hat{\mathbf{e}}_\theta\right] + \left[\dot{r}\dot{\theta}\hat{\mathbf{e}}_\theta + r\ddot{\theta}\hat{\mathbf{e}}_\theta - r\dot{\theta}^2\hat{\mathbf{e}}_r\right] + \ddot{z}\hat{\mathbf{e}}_z \tag{B.11}$$

$$\mathbf{a} = (\ddot{r} - r\dot{\theta}^2)\hat{\mathbf{e}}_r + (2\dot{r}\dot{\theta} + r\ddot{\theta})\hat{\mathbf{e}}_\theta + \ddot{z}\hat{\mathbf{e}}_r$$

In spherical coordinates, the unit-base vectors are functions of the two angular coordinates.

$$\frac{\partial\hat{\mathbf{e}}_R}{\partial R} = 0 \quad \frac{\partial\hat{\mathbf{e}}_R}{\partial\phi} = \hat{\mathbf{e}}_\phi \quad \frac{\partial\hat{\mathbf{e}}_R}{\partial\theta} = \sin\phi\,\hat{\mathbf{e}}_\theta$$

$$\frac{\partial\hat{\mathbf{e}}_\phi}{\partial R} = 0 \quad \frac{\partial\hat{\mathbf{e}}_\phi}{\partial\phi} = -\hat{\mathbf{e}}_R \quad \frac{\partial\hat{\mathbf{e}}_\phi}{\partial\theta} = \cos\phi\,\hat{\mathbf{e}}_\theta \tag{B.12}$$

$$\frac{\partial\hat{\mathbf{e}}_\theta}{\partial R} = 0 \quad \frac{\partial\hat{\mathbf{e}}_\theta}{\partial\phi} = 0 \quad \frac{\partial\hat{\mathbf{e}}_\theta}{\partial R} = -\sin\phi\,\hat{\mathbf{e}}_R - \cos\phi\,\hat{\mathbf{e}}_\phi$$

The position vector in spherical coordinates is

$$\mathbf{r} = R\hat{\mathbf{e}}_R \tag{B.13}$$

The velocity is

$$\mathbf{v} = \dot{R}\hat{\mathbf{e}}_R + R\dot{\phi}\hat{\mathbf{e}}_\phi + R\sin\phi\dot{\theta}\hat{\mathbf{e}}_\theta \tag{B.14}$$

The acceleration is

$$\mathbf{a} = \left[\ddot{R} - R\dot{\phi}^2 - R\sin^2\phi\,\dot{\theta}^2\right]\hat{\mathbf{e}}_R$$
$$+ \left[R\ddot{\phi} + 2\dot{R}\dot{\phi} - R\dot{\theta}^2\sin\phi\cos\phi\right]\hat{\mathbf{e}}_\phi \tag{B.15}$$
$$+ \left[R\ddot{\theta}\sin\phi + 2\dot{R}\dot{\theta}\sin\phi + 2R\dot{\phi}\dot{\theta}\cos\phi\right]\hat{\mathbf{e}}_\theta$$

ORDINARY DIFFERENTIAL EQUATIONS

Two types of differential equations frequently are encountered in engineering: ordinary differential equations and partial differential equations. If a differential equation only has one independent variable, it is called an ordinary differential equation. The differential equation of motion for a particle in rectilinear motion is in the form:

$$\frac{d^2x}{dt^2} = F\left[t, x(t), \frac{dx(t)}{dt}\right] \tag{B.16}$$

When the acceleration is given as a function of only one other dynamic variable—time, position, or velocity—the differential equation is a separable differential equation. If the acceleration is a function of time only, the differential equation may be separated and integrated directly.

$$\int_{v_0}^{v} dv = \int_{0}^{t} a(t)\,dt \tag{B.17}$$

This first integral of the motion is the basis of an impulse–momentum analysis of the motion. The displacement then is determined from the velocity.

$$v = \frac{dx}{dt}$$
$$\int_{x_0}^{x} dx = \int_{0}^{t} v(t)\,dt \tag{B.18}$$

If the acceleration is given as a function of displacement only, the acceleration may be written as

$$a = \frac{dv}{dt} = \frac{dv}{dx}\frac{dx}{dt} = v\frac{dv}{dx}$$

or $\tag{B.19}$

$$a\,dx = v\,dv$$

The velocity can be obtained as a function of position by integration:

$$a = f(x)$$
$$\int_{v_0}^{v} v\,dv = \int_{x_0}^{x} f(x)\,dx \tag{B.20}$$

This yields a first integral of the motion and is the basis of a work–energy analysis of the motion. The time can be found as a function of the displacement using the velocity determined in Eq. (B.20).

$$v = \frac{dx}{dt}$$

Rewriting this definition gives

$$dt = \frac{dx}{v(x)} \tag{B.21}$$

and integrating both sides yields

$$\int_0^t dt = t(x) = \int_{x_0}^x \frac{dx}{v(x)}$$

Finally, if the acceleration is given as a function of the velocity $a = f(v)$, the equation may again be separated.

$$\int_{x_0}^x dx = \int_{v_0}^v \frac{v\,dv}{f(v)}$$

$$x(v) = x_0 + \int_{v_0}^v \frac{v\,dv}{f(v)} \tag{B.22}$$

Note the independent variable in this case is the velocity. Time also will be determined as a function of velocity, and both velocity and time are not necessarily single-valued functions.

$$a = \frac{dv}{dt}$$

Rewriting:

$$dt = \frac{dv}{f(v)} \tag{B.23}$$

Integration of both sides gives time as a function of velocity

$$\int_0^t dt = \int_{v_0}^v \frac{dv}{f(v)} \quad \text{or} \quad t(v) = \int_{v_0}^v \frac{dv}{f(v)}$$

It usually is required to invert these relations to obtain the displacement–time function.

Another special case that yields analytical solutions to first-order differential equations is when the acceleration is an explicit function of the velocity and time only. If the differential equation of motion is linear, it can be written as

$$\frac{dv}{dt} + p(t)v = f(t) \tag{B.24}$$

This form of linear first-order differential equation frequently can be solved by the use of an integrating factor, defined as

$$\lambda(t) = e^{\int p(t)\,dt} \tag{B.25}$$

The general solution of Eq. (B.24) is

$$v(t) = \frac{\int \lambda(t)f(t)\,dt + C}{\lambda(t)} \tag{B.26}$$

It is not always possible to analytically perform the foregoing integrations, and in those cases, a numerical solution to the differential equation is required.

NUMERICAL SOLUTION OF ORDINARY DIFFERENTIAL EQUATIONS

Most differential equations encountered in dynamics are nonlinear and analytical solutions are not possible. In these cases, numerical solutions are required. All computational software programs have some form of differential equation solver and usually these are based upon a Runge–Kutta method. This method was originated by the German mathematician and physicist Carl David Runge (1856–1927) and extended to systems of equations by the German mathematician and aerodynamicist M. Wilhelm Kutta (1867–1944) and has a truncation error of Δt^5 where the older method developed by Euler has a truncation error of Δt. After you have become familiar with a numerical solution of differential equations using the Euler or tangent line method, you should examine solutions using the more accurate Runge–Kutta method. We have used the Euler method throughout the text and in the computational supplements to develop a better feel for the solution. The Euler method of solution of the equation of motion is based upon the fact that the acceleration is the time derivative of the velocity and the velocity is the time derivative of the displacement. Since the dynamics problem is a initial value problem, the values of the velocity and position at time zero are known. We therefore start with those values and "walk" up the velocity–time and displacement–time curves.

To illustrate the Euler method, consider the first-order differential equation

$$\frac{dy}{dt} = f(y, t) \tag{B.27}$$

with the initial conditions $y(0) = y_0$.

The slope of the curve $y(t)$ is the first derivative or $f(y_0, 0)$ and the value of the function at a time increment Δt is

$$y(\Delta t) = y_0 + f(y_0, 0)\Delta t \tag{B.28}$$

This simple method of numerical integration can be generalized to determine the function at the $i + 1$ in terms of the i value as

$$y_{i+1} = y_i + f(y_i, t_i)\Delta t \tag{B.29}$$

where $t_i = i\Delta t$.

The smaller the time increment, the more accurate the numerical integration. Note that the limit as the time increment approaches zero is defined as analytical integration. However, the smaller the time increment, the more numerical steps are necessary to determine the solution over a period of time. For example, if we wish the solution during a period of 2 seconds, and choose a time increment $\Delta t = 0.01$ s, 200 numerical steps are needed. If we choose the time increment to be one millisecond, 2000 steps would be needed. However, the personal computers available today can compute this number of steps in less than a second. This is, of course, what has made numerical solution to dynamic problems possible.

Now, let us consider the second-order differential equation of motion, when the acceleration may be a function of position, velocity, and time. We will couple the two Euler integrations to determine both the velocity and position as a function of time. Consider the acceleration given as

$$a = f(v, x, t) \tag{B.30}$$

The acceleration may be a linear or nonlinear function of the velocity and position. If the function is nonlinear, the time increment must be taken smaller to obtain the required accuracy. You should vary Δt to investigate the sensitivity of the solution to time increment. The format for solving the equation of motion is as follows:

Define the time increment:

$$\Delta t = 0.001 \text{ s}$$

Determine the number of steps required for 2 s:

$$i = 0 \dots 2000$$

Define time at any step:

$$t_i = i\Delta t$$

Define the acceleration as a function of velocity, position, and time:

$$a = f(v, x, t)$$

Define the initial values of the velocity and position:

$$\begin{bmatrix} v_0 \\ x_0 \end{bmatrix} = \begin{bmatrix} \text{initial velocity} \\ \text{initial position} \end{bmatrix}$$

Numerical solution:

$$\begin{bmatrix} v_{i+1} \\ x_{i+1} \end{bmatrix} = \begin{bmatrix} v_i + f(v_i, x_i, t_i)\Delta t \\ x_i + v_i\Delta t \end{bmatrix}$$

The position now can be plotted as a function of time yielding the desired motion. After you have become familiar with the Euler method, consult a differential equation text for the mathematical basis of the Runge–Kutta, and then study the manual for the computational software of your choice to obtain more accurate solutions to the equations of motion.

Dynamics Index Dictionary

acceleration: The time rate of change of the velocity. (Pages 15–24)

> **average acceleration:** the change in the velocity vector over a finite change in time. $\mathbf{a}_{ave} = \dfrac{\Delta \mathbf{v}}{\Delta t}$. (Page 15)

> **Coriolis acceleration:** The acceleration term arising from the change in the relative velocity in a rotating coordinate system caused by rotation of the coordinate system; named after the French military engineer, G. G. Coriolis who first examined this term $2\Omega \times (\mathbf{v}_{B/A})_{xyz}$. This acceleration accounts for the counterclockwise movement of the wind around a low-pressure center in the northern hemisphere and the clockwise movement of the wind around a low-pressure center in the southern hemisphere. (Page 243)

> **gravitational acceleration:** The acceleration due to gravitational attraction of an object in free fall designated by g. $g = 9.806$ m/s^2 or 32.2 ft/s^2. (Page 40)

> **instantaneous acceleration:** The time derivative of the velocity vector $\mathbf{a} = \dfrac{d\mathbf{v}}{dt}$.

> **normal acceleration:** The acceleration due to the change in direction of the velocity vector proportional to the magnitude of the velocity squared and inversely proportional to the radius of curvature of the path of motion. This acceleration is directed toward the center of curvature of the path of motion at any point. $a_n = \dfrac{v^2}{\rho}$. (Page 47)

> **relative acceleration:** The acceleration of one point or particle relative to another point or particle. The relative acceleration is equal to the time derivative of the relative velocity. $\mathbf{a}_{B/A} = \mathbf{a}_B - \mathbf{a}_A$. (Pages 284–293)

> **relative acceleration between two points on a rigid body:** The relative acceleration between two points on a rigid body is $\mathbf{a}_{B/A} = \boldsymbol{\alpha} \times \mathbf{r}_{B/A} + \boldsymbol{\omega} \times (\boldsymbol{\omega} \times \mathbf{r}_{B/A})$. (Pages 284–293)

> **tangential acceleration:** The acceleration due to the change in the magnitude of the velocity vector directed parallel to the velocity vector and tangent to the path of motion. $a_t = \dfrac{d|\mathbf{v}|}{dt} = \dfrac{d^2 s}{dt^2}$ where s is a parameter of distance along the path of motion. (Page 246)

center of mass: The center of mass of a system of particles is defined by $\mathbf{r}_{cm} = \dfrac{\Sigma m_i \mathbf{r}_i}{\Sigma m_i}$. (Page 219)

central force motion: If the only force acting upon a particle is directed toward a fixed point is space, the particle undergoes what is called central force motion. The angular momentum of the particle about this point is a conserved quantity. Planetary motion and Kepler's laws are examples of central force motion. (Pages 150–159)

coefficient of friction:

> **coefficient of static friction:** The ratio of the maximum friction force to the normal force before slipping occurs. The static friction force would almost never be equal to this maximum value, as the system would be unstable at this point. (Pages 102–108)

> **coefficient of kinetic friction:** The ratio of the actual friction force to the normal force when there is slip between the surfaces. The coefficient of kinetic friction is 20–25% lower than the coefficient of static friction. This coefficient is taken as a constant but is actually dependent on the relative velocity between the surfaces. (Pages 102–108)

conservative force: A force satisfying the condition $\nabla \times \mathbf{F} = 0$. The work done by a conservative force is independent of the path of movement and depends only on the initial and final positions. (Page 189)

constraint to the motion: A restriction of one of the degrees of freedom of movement. (Pages 253–263)

control volume: A fixed volume in space into and out of which mass flows. (Page 232)

curvilinear translation of a particle: The general motion of a particle in space such that the position is described by a position vector $\mathbf{r}(t)$. (Page 240)

d'Alembert's principle: A formulation of dynamics problem developed by Jean le Rond d'Alembert in which the acceleration terms in Newton's equations of motion are grouped with the force terms and are considered as inertial forces. This method offers little to the traditional approach to the solution of plane motion dynamics but is important in the development of the Lagrange equations of motion. (Pages 382–386)

damper or dashpot: A physical model for dissipating energy and damping the response of a mechanical system. (Page 462)

degrees of freedom: The number of types of motion that an object can have. The maximum is six: three translations and three rotations.

del operator: The directed derivative vector defined by $\nabla = \hat{\mathbf{i}}\dfrac{\partial}{\partial x} + \hat{\mathbf{j}}\dfrac{\partial}{\partial y} + \hat{\mathbf{k}}\dfrac{\partial}{\partial z}$. (Page 190)

differential equation: An equation with an unknown function (the displacement) and containing one or more of its derivatives (velocity and acceleration) in terms of an independent variable (time). (Pages 13–24)

> **ordinary differential equation:** A differential equation that has only one independent variable, for example, $\dfrac{d^2x}{dt^2} = F\left[t, x(t), \dfrac{dx(t)}{dt}\right]$. (Page 14)

> **partial differential equation:** A differential equation having more than one independent variable, for example, $\dfrac{\partial^2 u(x, y)}{\partial x^2} + \dfrac{\partial^2 u(x, y)}{\partial y^2} = 0$. (Page 14)

> **order of a differential equation:** The order of a differential equation is the order of the highest derivative that appears in the equation. (Page 14)

> **separable differential equation:** A linear or nonlinear first-order differential equation that can be written in the form: $M(x) + N(y)\dfrac{dy}{dx} = 0$. (Pages 15–24)

> **numerical solution:** *See* Euler's method.

> **vector differential equation:** An equation with the unknown function and its derivatives are vectors. This equation can be separated into a system of linear or nonlinear, coupled or uncoupled, scalar differential equations. (Pages 37–45)

direct dynamics problem (forward problem): Use of Newton's law to determine the motion of a particle or rigid body when the forces and moments acting on the object are known. The problem is formulated in terms of the differential equation of motion. (Page 6)

Euler equations of motion: A system of equations used to determine the motion of a rigid body moving in three dimensions based upon a rotating coordinate system with origin at the center of mass of a rigid body and aligned with the principal axes of the rigid body. (Page 440)

Euler method or tangent line method: The oldest and simplest method of solution of an initial-value differential equation based on the fact that the derivative of a function is

$$\frac{d^2y}{dt^2} = f\left(\frac{dy}{dt}, y, t\right)$$

(Pages 320, 440)

the slope of the function curve at any point.

$$\begin{bmatrix} \dfrac{dy}{dt_0} \\ y_0 \end{bmatrix} = \text{Given}$$

$$\begin{bmatrix} \dfrac{dy}{dt_{i+1}} \\ y_{i+1} \end{bmatrix} = \begin{bmatrix} \dfrac{dy}{dt_i} + f\left(\dfrac{dy}{dt_i}, y_i, t_i\right)\Delta t \\ y_i + \dfrac{dy}{dt_i}\Delta t \end{bmatrix}$$

Eulerian angles: The angles of rotation for three successive orthogonal rotations used to describe the orientation of a rigid body in space. These angles are sequence dependent. (Pages 427–432)

general motion of a rigid body: The general motion of a rigid body in space has six degrees of freedom, three translations, and three rotations. (Pages 242–243)

general plane motion: Motion of a rigid body that is restricted to a single plane. The body has three degrees of freedom, two translations in the plane, and rotation about an axis perpendicular to the plane of motion. (Page 241)

Heaviside step function or unit step function: A singularity function that is equal to zero for all value of the variable less than a specific value and equal to one for all values greater than the specified value.

$$\langle t - t_1 \rangle^0 = \begin{cases} 0 & x < t_1 \\ 1 & x > t_1 \end{cases}$$

(Page 124)

impact: Impact is a collision between two or more particles such that the internal forces produced during that collision are large compared to the external forces. Therefore, only the impulse of the internal force is considered in consideration of an impulse–momentum analysis. (Page 204)

impulse: The integral of the force over a specified time period. A first integral of Newton's equation yields the principle of impulse and momentum. Impulse = change in the linear momentum. (Page 124)

inertial reference system: A coordinate system that is considered to be fixed in space, that is, one whose absolute velocity is zero.

instantaneous center of rotation in plane motion: For any rigid body in plane motion, a point on the body or the body extended can be found that has zero velocity at an instant and about which the body is rotating at that instant. (Page 243)

instantaneous helical axis (screw axis): At any instant of time, a rigid body can be considered to be rotating about an axis in space and translating along that axis. (Page 317)

inverse dynamics problem: A dynamics problem in which the position vector is given as a function of time and the forces, and moments that produce the motion are determined by differentiation of the position vector. (Pages 7–12)

joint coordinate system: A nonorthogonal coordinate system based upon Euler angles used to describe the angular movement of joint in the human body. (Pages 433–434)

kinematics: The study of motion without consideration of the causes of that motion. The study of kinematics usually is divided into two parts: kinematics of particles and kinematics of rigid bodies. (Pages 97–173)

kinetic energy: Energy due to motion defined as $T = \dfrac{1}{2}m\mathbf{v} \cdot \mathbf{v}$ for linear motion and $T = \dfrac{1}{2}I\omega \cdot \omega$ for angular motion. The S.I. units of work are Joules (J). (Page 176)

mass moment of inertia: A second-order tensor $[I] = \begin{bmatrix} I_{xx} & I_{xy} & I_{xx} \\ I_{xy} & I_{yy} & I_{yz} \\ I_{xz} & I_{yz} & I_{zz} \end{bmatrix}$ that relates the angular velocity of a rigid body to the angular momentum of the body $[H] = [I][\omega]$. The elements of the mass-moment-of-inertia tensor are $I_{xx} = \displaystyle\int_M (y^2 + z^2)dm \quad I_{xy} = -\displaystyle\int_M xydm$. (Page 345)

mechanical efficiency: The ratio of the work done by a machine to the work done on the machine.

$$e = \frac{U_{\text{output}}}{U_{\text{input}}}$$

(Pages 179–189)

momentum:

    angular momentum: The angular momentum of a particle about a point 0 is defined as: $\mathbf{H}_0 = \mathbf{r} \times m\mathbf{v}$ (Pages 149–150)

    linear momentum: The linear momentum is the basic dynamic measure as defined by Newton and is: $\mathbf{L} = m\mathbf{v}$. Newton's second law relates the net force acting on a particle to the linear momentum: $\mathbf{F} = \dfrac{d\mathbf{L}}{dt}$ (Page 176)

natural frequency of vibration: The frequency or cycles per second of a linear system in free oscillation.

Newton's laws:

1. Every body or particle continues in a state of rest or in uniform motion (constant velocity) in a straight line, unless it is compelled to change that state by forces impressed upon it. (Page 98)

2. The change of motion of a body or particle is proportional to the net external force acting on the body or particle in the direction of the net external force. (Page 98)

3. If one body or particle exerts a force on a second body or particle, then the second exerts a force on the first that is equal in magnitude to, opposite in direction to, and collinear with the given force. (Page 98)

4. Any two particles are attracted toward each other with a force whose magnitude is proportional to the product of the gravitation masses and inversely proportional to the square of the distance between them. (Page 98)

normal and tangential coordinates: A coordinate system that is attached to a particle or rigid body such that the tangential coordinate is tangent to the path of motion and therefore in the direction of the velocity. The normal coordinate is directed toward the center of curvature of the path of motion. The velocity and acceleration can be written as

$$\mathbf{v}(t) = \frac{ds}{dt}\hat{\mathbf{e}}_t$$

$$\mathbf{a}(t) = \frac{d^2s}{dt^2}\hat{\mathbf{e}}_t + \frac{1}{\rho}\left(\frac{ds}{dt}\right)^2\hat{\mathbf{e}}_n$$

(Pages 45–55)

orthogonal tranformation matrix: A (3×3) matrix comprised of the direction cosines between two orthogonal coordinate systems used to rotate one coordinate system to the other coordinate system. The determinant of the matrix is equal to one and the inverse of the matrix is the transpose of the matrix.

osculating plane: The plane formed by the normal and tangential unit vectors in general three-dimensional motion of a particle. (Page 52)

binormal vector: A unit vector that is perpendicular to the osculating plane defined by the cross product of the tangential unit vector with the normal unit vector. (Page 53)

polar coordinates (radial and transverse coordinates): A two-dimensional coordinate system where the position given by a radial coordinate and the angle the radial coordinate makes with the x-axis. (Pages 55–62)

position vector: A vector from the origin of a fixed reference system to a particle or a point on a rigid body. (Page 35)

relative position vector: The position of one point or particle relative to another point or particle defined by $r_{B/A} = r_B - r_A$. In general, these position vectors are a function of time. (Page 240)

potential energy: The potential function of a conservative force, V and related to the force by $F \cdot dr = -dV$. The potential energy is a scalar measure of the force's capability to do work. Potential energy is usually the result of the position of a mass in the gravitational field or stored by elastic deformation of a body or spring. (Pages 189–194)

power: The rate of doing work equal to the time rate of change of the kinetic energy. Power is measured in the S.I. unit Watt (W). (Page 177)

primary inertial reference system: A reference coordinate system considered fixed in space, that is, not translating or rotating. (Pages 45–55)

principal axes: A set of orthogonal axes usually with origin at the center of mass of a rigid body about which the mass moment of inertia tensor is in diagonal form. (Page 346)

principal mass moments of inertia: The three mass moments of inertia with respect to the principal axes. These three moments of inertia will be a maximum and minimum of all moments of inertia about any axis passing through the center of mass. (Page 346)

projectile motion: The motion of a particle under only the influence of the gravitational attraction of the earth. If the motion is close to the surface of the earth, the acceleration of the particle will be constant during the movement and directed toward the center of the earth. (Pages 39–45)

rectilinear motion: Motion in a straight line, that is, motion that can be describe by a single rectangular coordinate. (Page 3)

Rodrigue's formula: For a body moving in general plane motion, the angular velocity of the body can be obtained by a direct vector calculation as

$$\omega = \frac{r_{B/A} \times v_{B/A}}{r_{B/A} \cdot r_{B/A}}$$

(Pages 263–275)

rolling without sliding: A condition in which a wheel or cylinder rolls along a surface where there is sufficient friction that there is no sliding at the contact point. (Page 356)

simple harmonic oscillator: An undamped single-degree-of-freedom system undergoing simple harmonic motion or oscillatory motion. (Page 455)

steady mass flow: A steady mass flow is a condition in which the rate of change of the mass entering the control volume equals the rate of change of the mass leaving the control volume. (Pages 232–234)

trajectory of the motion: The curve or path that a particle or a point on a rigid body traces in space over a period of time.

velocity: The directed time rate of change of position. (Pages 4–6)

average velocity: The change in position during a finite length of time.

$$v_{ave} = \frac{\Delta r}{\Delta t}$$

(Page 4)

instantaneous velocity: The time derivative of the position vector.

$$v = \frac{dr}{dt}$$

(Page 4)

relative velocity: The velocity of one point or particle relative to another point or particle. In general, both points will be moving. The velocity of B relative to A is $v_{B/A} = v_B - v_A$. (Pages 76–82)

relative velocity of two points on a rigid body: The relative velocity between two points on a rigid body must be equivalent to the rotation of one point about the other point, as the distance between them is constant. $v_{B/A} = \omega \times r_{B/A}$. (Pages 76–82)

vibration: The study of repetitive motion of objects relative to some stationary frame of reference or nominal position (usually the equilibrium position). (Pages 451–486)

work: To produce results or to perform a function. Mathematically defined as the scalar product between a force and the displacement through which the force moves.

$$U_{1 \to 2} = \int_{r_1}^{r_2} F \cdot dr$$

(Pages 175–214)

ANSWERS TO ODD-NUMBERED PROBLEMS

Chapter 1

1.1 $\dfrac{\Delta v}{\Delta t} = 9.565 \text{ ft/s}^2$

1.3 $x(0) = 5 \text{ m}$ $x(3) = 26 \text{ m}$
$v(0) = -2 \text{ m/s}$ $v(3) = 25 \text{ m/s}$
$a(0) = 0$ $a(3) = 18 \text{ m/s}^2$
Total distance traveled 23.18 m

1.5 Acceleration is zero at $t = 0, \dfrac{n\pi}{2} \text{ s}$

1.7 $v(t) = 3e^{-t}[-\sin(10t) + 10\cos(10t)]$
$a(t) = 3e^{-t}[-99\sin(10t) - 20\cos(10t)]$

1.9 a) $v(t) = 0.6\cos(2t)$
     $a(t) = -1.2\sin(2t)$
b) $d = 0.6 \text{ m}$
c) $x\left(\dfrac{\pi}{2}\right) = 0$ The distance traveled is the integral
of the absolute velocity.
d)

1.11 $v_{i+1}: = \dfrac{x_{i+1} - x_i}{\Delta t}$

$a_{i+1}: = \dfrac{v_{i+1} - v_i}{\Delta t}$

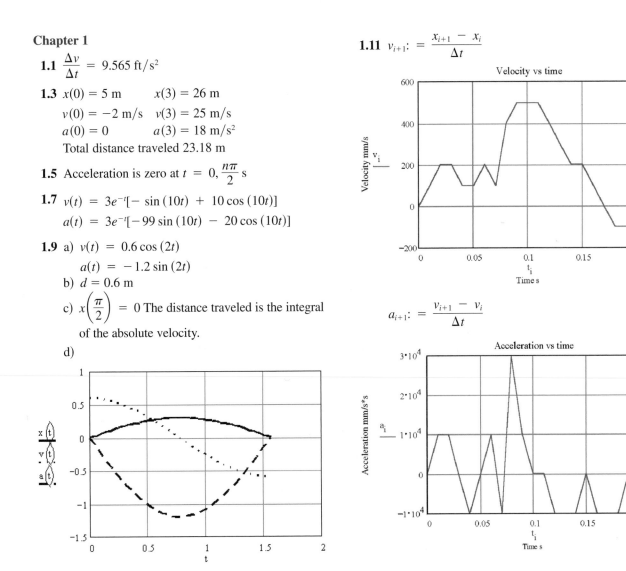

1.13 $y(5) = -22.625$ m

$v(5) = -29.05$ m/s

$a(5) = -9.81$ m/s^2

Distance traveled $= 63.4$ m

1.15 $x(t) = 3t^3 - 2t^2 + 5$

$v(t) = 9t^2 - 4t$

$a(t) = 18t - 4$

$x(0) = 5 \qquad v(0) = 0 \qquad a(0) = -4$

1.17 $y(t) = 3t^2 - 20$

$v(t) = 6t$

$a(t) = 6$

$y(0) = -20 \qquad v(0) = 0 \qquad a(0) = 6$

1.19 $y(t) = 5t - e^{-t}(3t)$

$v(t) = 5 + 3e^{-t}(t - 1)$

$a(t) = -(3t - 6)e^{-t}$

$y(0) = 0 \qquad v(0) = 2 \qquad a(0) = 6$

1.21 Ave accel $= 17.6$ ft/s^2 $= 5.364$ m/s^2 $= 0.547$ g

1.23 $v_0 := 0$

$v_1 := 0.2$

$v_2 := 0.27$

$v_3 := 0.3$

$v_4 := 0.3$

$v_5 := 0.3$

$v_6 := 0.3$

$v_7 := 0.3$

$v_8 := 0.35$

$v_9 := 0.4$

$v_{10} := 0.5$

$$a_n := \frac{v_{n+1} - v_n}{0.1}$$

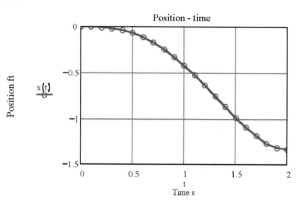

$$x_n := (v_{n+1} + v_n) \cdot \frac{0.1}{2}$$

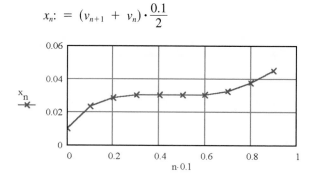

1.25 2nd order linear

1.27 1st order linear

1.29 2nd order nonlinear; nonlinear terms $\dot{\theta}^2$, $\sin \theta$

1.31 2nd order linear

1.33

Position - time

1.35 Sprinter accel $= 3.33$ m/s^2

Car accel $= 4.47$ m/s^2

1.37 $a = 6.25 \times 10^{12}$ m/s^2 $\qquad \Delta t = 7.98 \times 10^{-7}$ s

1.39 67.87 s

1.41 $v(x) = \sqrt{v_0^2 - \dfrac{2c}{3} x^3}$

1.43 $y(\text{max}) = 2575$ km

1.45 $x(t) = \dfrac{v_0}{c}(1 - e^{-ct})$

1.47 $v(4) = 0.305$ ft/s

1.49 $t = 1.365$ s

$v(1.365) = 43.95$ ft/s

1.51 a) 171.78 ft

b) 96.63 ft

c) 42.95 ft

1.53 Time $= 15.18$ s   Distance $= 2227$ ft

1.55 $x(120) = 297,500$ m $v(120) = 7750$ m/s

1.57 $v(t) = (t - 1) + e^{-t}$

$x(t) = \dfrac{t^2}{2} - t - e^{-t} + 2$

1.59 $x(t) = -\dfrac{11}{48}e^{-4t} + \dfrac{5}{3}e^{-t} + \dfrac{3}{4}t - \dfrac{15}{16}$

$v(t) = \dfrac{11}{10}e^{-4t} - \dfrac{5}{3}e^{-t} + \dfrac{3}{4}$

1.61 $c = 15,000$

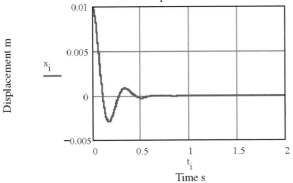

1.63

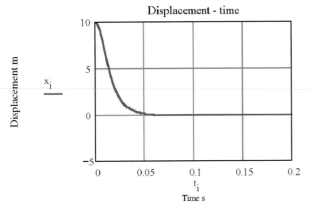

1.65 One form of the solution is $c = 6$ and $k = 40$

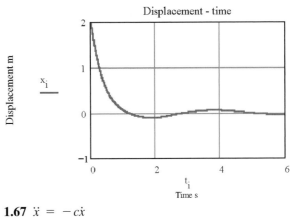

1.67 $\ddot{x} = -c\dot{x}$

$\ddot{y} = -c\dot{y} - g$ linear and uncoupled

$\ddot{z} = -c\dot{z}$

1.69 $\ddot{x} = 3t^2$

$\ddot{y} = -\sin(\pi t)$

$\ddot{z} = xz$

1.71 $d = 1.564$ m

1.73 $\theta = 45°$

1.75 $y(\text{max}) = 39.6$ ft

1.77 There are two solutions:

High arc $59.2° \le \theta \le 66.2°$

Low arc $40.9° \le \theta \le 55.6°$

1.79

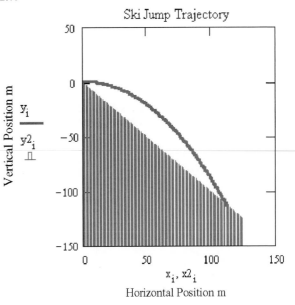

Landing coordinates $(110, -110)$ and distance
$d = 155.6$ m

1.81 $v0 = 306.2$ ft/s or 209 mph

1.83 251 yds

1.85 64.8°

1.87 28.89 ft/s $\leq v0 \leq$ 41.4 ft/s

1.89 $|a| = \sqrt{[6 \sin (\pi t)]^2 + \left[-\dfrac{18}{\pi^2} (1 - \cos \pi t)^2 \right]^2}$

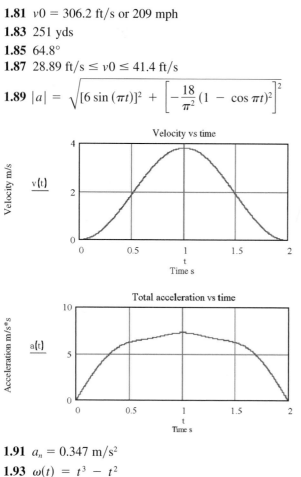

1.91 $a_n = 0.347$ m/s²

1.93 $\omega(t) = t^3 - t^2$

$\theta(t) = \dfrac{t^4}{4} - \dfrac{t^3}{3}$

1.95 $\theta(10) = 2.06$ rad

1.97

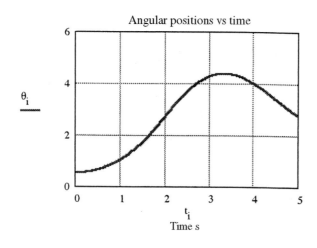

1.99 $\hat{e}_t(3) = \begin{bmatrix} -0.085 \\ -0.594 \\ 0.8 \end{bmatrix}$ $\hat{e}_n(3) = \begin{bmatrix} 0.99 \\ -0.1 \\ 0 \end{bmatrix}$

$\hat{e}_b(3) = \begin{bmatrix} 0.113 \\ 0.792 \\ 0.6 \end{bmatrix}$ $\rho = 8.333$

1.101 $\rho = 8.333$ m constant

1.103 $\begin{bmatrix} v_r \\ v_\theta \end{bmatrix} = \begin{bmatrix} t \\ 0.5(1 + t^2)2\pi t \end{bmatrix}$

$\begin{bmatrix} a_r \\ a_\theta \end{bmatrix} = \begin{bmatrix} 1 - 0.5(1 + t^2)4\pi^2 t^2 \\ 0.5(1 + t^2)2\pi + 2t(2\pi t) \end{bmatrix}$

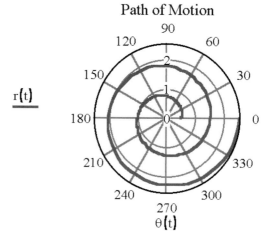

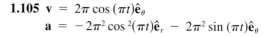

1.105 $\mathbf{v} = 2\pi \cos (\pi t)\hat{\mathbf{e}}_\theta$

$\mathbf{a} = -2\pi^2 \cos^2(\pi t)\hat{\mathbf{e}}_r - 2\pi^2 \sin (\pi t)\hat{\mathbf{e}}_\theta$

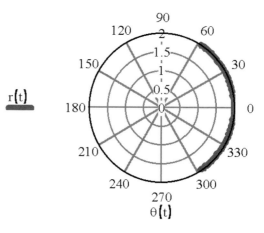

The path of motion reverses at 1/2 s and follows it-self to 1 s, continuing in the negative direction until 1.5 s and then positive back to original position.

Distance traveled = 8 m

1.107 $\begin{bmatrix} v_r \\ v_\theta \end{bmatrix} = \begin{bmatrix} -2e^{-t} + 2 \\ (2e^{-t} + 2t - 1)\pi t \end{bmatrix}$ m/s

$\begin{bmatrix} a_r \\ a_\theta \end{bmatrix} = \begin{bmatrix} 2e^{-t} - (2e^{-t} + 2t - 1)\pi^2 t^2 \\ (2e^{-t} + 2t - 1)\pi + 2(-2e^{-t} + 2)\pi t \end{bmatrix}$ m/s²

1.109 $\mathbf{a}(t) = -4\pi^2[r_\theta + 2r_a \sin(2\pi t)]\hat{\mathbf{e}}_r$
$+ 8\pi^2 r_a \cos(2\pi t)\hat{\mathbf{e}}_\theta$ m/s²

1.111 $a(v,r,\omega,\theta,t) = 2t^2 + r\omega^2$

$\alpha(v,r,\omega,\theta,t) = \dfrac{1}{r}(\cos \pi t - 2v\omega)$

1.113 The differential equations are decoupled and may be solved independently.

1.115 $\mathbf{v} = -15\pi \sin(\theta)\hat{\mathbf{e}}_r + (25 + 15 \cos \theta)\pi\hat{\mathbf{e}}_\theta$

$\mathbf{a} = -(100 + 120 \cos \theta)\dfrac{\pi^2}{16}\hat{\mathbf{e}}_r$

$- 120 \sin \theta \dfrac{\pi^2}{16}\hat{\mathbf{e}}_\theta$

1.117 $a_r(t) := ddr(t) - r(t) \cdot \pi^2$
$a_\theta(t) := 2 \cdot dr(t) \cdot \pi$

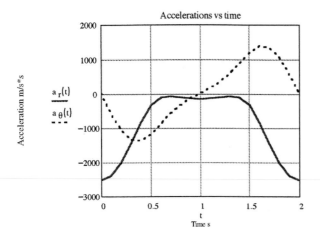

1.119 $\mathbf{v}(t) = 1.5\pi\hat{\mathbf{e}}_\theta - \pi \sin(2\pi t)\hat{\mathbf{e}}_z$
$\mathbf{a}(t) = -1.5\pi^2\hat{\mathbf{e}}_r - 2\pi^2 \cos(2\pi t)\hat{\mathbf{e}}_z$

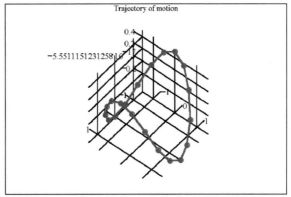

1.121 $\mathbf{v} = 0.1R\hat{\mathbf{e}}_r + 0.1Rt(2\pi)\hat{\mathbf{e}}_\theta - 0.1h\hat{\mathbf{e}}_z$
$\mathbf{a} = -0.1Rt(4\pi^2)\hat{\mathbf{e}}_r + 2(0.1R)(2\pi)\hat{\mathbf{e}}_\theta$
Time to reach bottom = 10 s. Particle will circle 10 times.

1.123

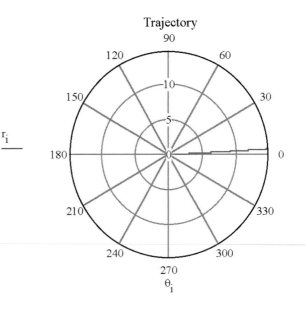

1.125

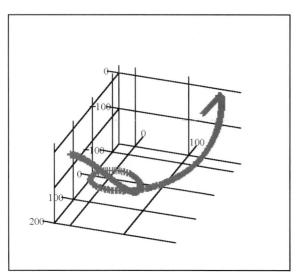

x, y, z

1.127 $\beta(t) - \beta_0 \cos \sqrt{\frac{g}{R}} t + \sqrt{\frac{g}{R}} \dot{\beta}_0 \sin \frac{g}{R} t$ radians

1.129 $x_B = 3t^2 + 4t + 9$ m
$v_B = 6t + 4$ m/s
$a_B = 6$ m/s^2
$a_{B/A} = 6$ m/s^2

1.131 $t = 40$ s $\quad d = 700$ m

1.133 Yes, they collide.
$v_{B/A} = -108.5$ ft/s $= 74$ mph

1.135 $a_B = 9$ m/s^2
$x_B = 4.5t^2 + 11$ m

1.137 $a = 15.9$ m/s^2

1.139

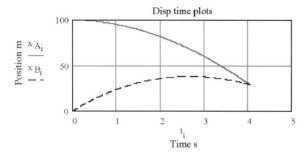

Disp time plots

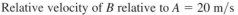

Relative velocity of *B* relative to *A* = 20 m/s

1.141 $t = 125$ s

1.143 Orientation S 10.2° E $\qquad$ Velocity = 232 mph

1.145 $\theta = 56.4°$ $\qquad t = 60.3$ s

1.147 From diagram and references, determine distances and wind direction and plot course.

1.149 Use data from 1.147

1.151 $\theta = 7.66°$

1.153
$\mathbf{v}_{G/B} = (4 \cos 30 - 2\pi \cos \pi t)\hat{\mathbf{i}} + (4 \sin 30 - 5)\hat{\mathbf{j}}$
$\mathbf{a}_{G/B} = 2\pi^2 \sin (\pi t)\hat{\mathbf{i}}$

1.155
$v_{B/G} = (2\pi \cos \pi t - 4 \cos 30)\hat{\mathbf{i}} +$
$\left[5 - 4 \sin 30 \sin \frac{\pi t}{2} - 4t \sin 30 \frac{\pi}{2} \cos \frac{\pi t}{2} \right]\hat{\mathbf{j}}$
$a_{B/G} = (-2\pi^2 \sin \pi t)\hat{\mathbf{i}}$
$+ \left[-4\pi \sin 30 \sin \frac{\pi t}{2} + t \sin 30 \pi^2 \sin \frac{\pi t}{2} \right]\hat{\mathbf{j}}$

1.157

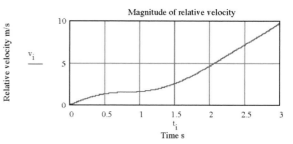

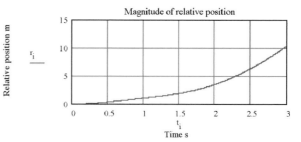

1.159

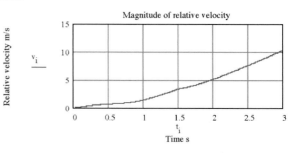

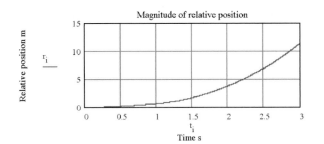

1.161 $a_A = -a_B$

1.163 $v_A = 30$ ft/s down

1.165 $v = 0.4$ m/s

1.167 $v_A + 2v_B + \frac{1}{2}v_C = 0$

1.169 $0 = \dot{y}_A + \dfrac{x_B \dot{x}_A}{\sqrt{d^2 + x_B^2}}$

1.171

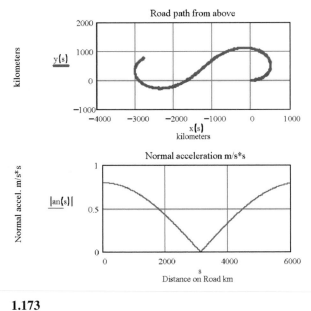

1.173

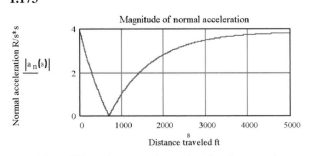

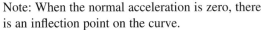

Note: When the normal acceleration is zero, there is an inflection point on the curve.

1.175

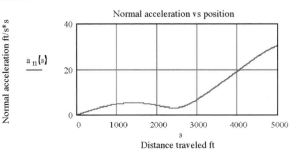

Note: The maximum normal acceleration is approximately g and the direction of the acceleration should be examined to see if the car will remain on the road.

1.177

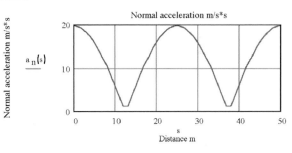

Chapter 2

2.1 $t = 7.73$ s

2.3 $a_{C/T} = -5.17$ ft/s^2

2.5 a) 1181 N   b) 731 N

2.7 $a_A = 2.65$ m/s^2

$a_B = 1.77$ m/s^2

$T = 23.5$ N

2.9 $v = 2.89$ m/s

2.11 $T_{BC} = \dfrac{2}{3}P$ $T_{AB} = \dfrac{1}{3}P$

2.13 a) 0.721 s

b) $x = 386$ m $v = 248$ m/s

c) $d = 8223$ m

2.15 System moves as a whole; $a = 1.99$ m/s^2

2.17 $a(x) = g - \dfrac{kx}{m}\left[1 - \dfrac{l_0}{\sqrt{l_0^2 + x^2}}\right]$

2.19

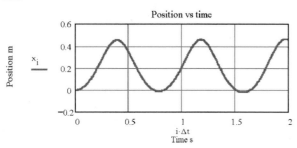

2.21 a) Yes, she will slip.
    b) $v = 31.45$ ft/s (21.4 mph)
       Yes, she would be injured.

2.23 a) $t = 0.549$ s
    b) $T = 56.8$ N

2.25 $a = \dfrac{\mu_s g}{\cos\theta - \mu_s \sin\theta}$

2.27 $t = 0.76$ s
    $T_A = 46.2$ N
    $T_B = 23.1$ N

2.29 System moves as one. $a = 1.862$ m/s²

2.31 $t = 0.184$ s

2.33 $L = 0.2$ m $k = 40$ N/m $\mu = 0.2$

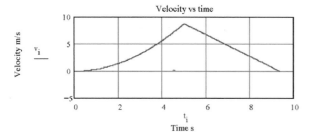

2.35 $A = 16.4$ m²

2.37

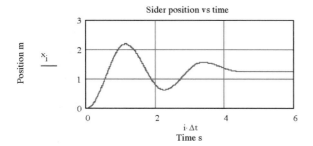

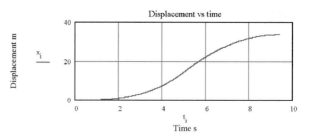

2.39

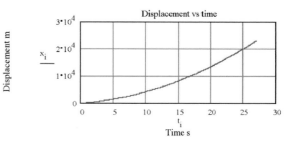

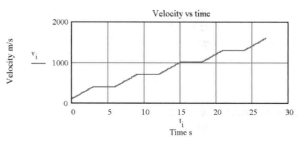

2.41

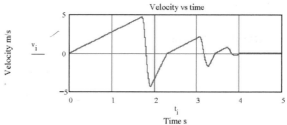

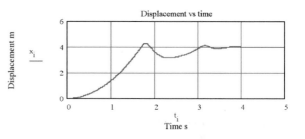

2.43 Maximum velocity of bag $= 70$ m/s

2.45 443 lb

2.47 $v = \sqrt{\rho g \dfrac{\sin\beta + \mu\cos\beta}{\cos\beta - \mu\sin\beta}}$

2.49 6551 ft

2.51 $v = \sqrt{2gR(\sin\theta - \sin\theta_0)}$

2.53 The solution is independent of the mass.
$\mu = 0.15 \quad R = 0.3$ m

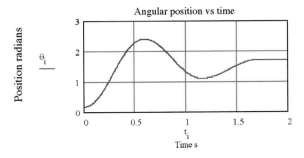

2.55 $\mu_s = \dfrac{r\alpha}{g}\sqrt{1 + \alpha^2 t^4}$

2.57 $\theta = \cos^{-1}\left(\dfrac{g}{R\omega^2}\right)$

2.59 N = 6.5 mg   This is high enough to cause injury.

2.61 $(m_A + m_B\sin^2\theta)\ddot\theta + m_B\dot\theta^2\sin\theta\cos\theta$
$\quad + (m_A + m_B)\dfrac{g}{l}\sin\theta = 0$

2.63
$\ddot\theta = -\dfrac{(m_A + m_B)g\sin(\beta+\theta)\cos\beta + m_B l\dot\theta^2\sin(\beta+\theta)\cos(\beta+\theta)}{l[m_A + m_B\sin^2(\beta+\theta)]}$

2.65

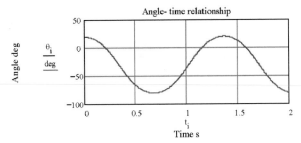

2.67 Velocity with friction = 4.227 m/s
Velocity without friction = 6.668 m/s
Time with friction = 0.917 s

2.69 Coefficient of friction = 0.493
Time on slide = 6.7 s   L = 40 m $\alpha = 44°$

2.71

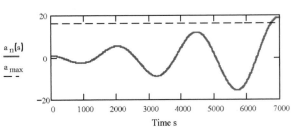

Yes, the passenger will slip near the end of the ride.

2.73 $\mathbf{v}(t) = -0.100\pi\sin(\pi t)\hat{\mathbf{e}}_r$
$\quad - [0.3 + 0.1\cos(\pi t)]\dfrac{\pi^2}{6}\cos(\pi t)\hat{\mathbf{e}}_\theta$

$a(t) = \left[-0.1\pi^2\cos\pi t\right.$
$\quad\left. - (0.3 + 0.1\cos\pi t)\dfrac{\pi^4}{36}\cos^2\pi t\right]\hat{\mathbf{e}}_r$
$\quad + \left[(0.3 + 0.1\cos\pi t)\left(-\dfrac{\pi^3}{6}\sin\pi t\right)\right.$
$\quad\left. - 0.2\pi\sin(\pi t)\dfrac{\pi^2}{6}\cos\pi t\right]\hat{\mathbf{e}}_\theta$

2.75

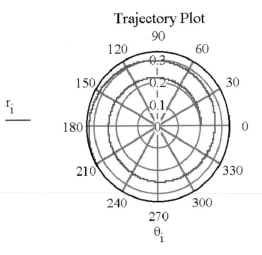

2.77 $N = 191.4 - 126.4\cos\theta$

2.79 m = 3.6 kg or if m > 10.6

2.81 $v_0 = \sqrt{5gR}$

2.83 $\dfrac{d^2\theta}{dt^2} = \dfrac{k}{2m}\left[1 - \dfrac{1}{\sqrt{5 + 4\cos\theta}}\right]\sin\theta$

2.85 For numerical solution:

$$F_s = \frac{kR}{2}(\sqrt{5 + 4\cos\theta} - 1)$$

$$L_s = \frac{R}{2}\sqrt{5 + 4\cos\theta}$$

$$\sin\beta = \frac{R}{2L_s}\sin\theta$$

$$mR\alpha = F_s\sin\beta - \mu_k|N|\frac{\omega}{|\omega|}$$

2.87 $v = 7614$ m/s   **2.89** Altitude = 35,790 km

2.91 $v_1 \rightarrow v_2$ $7.670 \times 10^3 \rightarrow 10.070 \times 10^3$ m/s

$v_3 \rightarrow v_4$ $1.619 \times 10^3 \rightarrow 3.075 \times 10^3$ m/s

2.93 17,015 km

2.95 $v_1 \rightarrow v_2$ $7.846 \times 10^3 \rightarrow 7.932 \times 10^3$ m/s

$v_3 \rightarrow v_4$ $7.598 \times 10^3 \rightarrow 7.687 \times 10^3$ m/s

2.97

$$\ddot\theta = \frac{g}{R}(\sin\beta\cos\beta - \mu\cos^2\beta) - \mu\dot\theta^2\cos\beta$$

$$\ddot z = g(1 + \cos^2\beta + \mu\sin\beta\cos\beta) + \mu R\dot\theta^2\sin\beta$$

2.99 $\ddot\theta = -\dfrac{\mu R\dot\theta^3}{\sqrt{R^2\dot\theta^2 + \dot z^2}}$

$\ddot z = -g - \dfrac{\mu R\dot\theta^2\dot z}{\sqrt{R^2\dot\theta^2 + \dot z^2}}$

2.101 $v_0 = \dfrac{\sqrt{5gR}}{\cos\beta}$

2.103

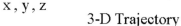

Angular position vs time

Angle degrees $\dfrac{\theta_i}{\deg}$

Time s t_i

Position down sluiceway

Z position m $\dfrac{z_i}{}$

Time s t_i

2.105

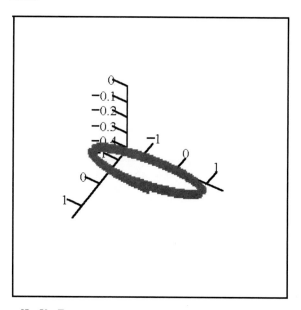

x , y , z

3-D Trajectory

2.107

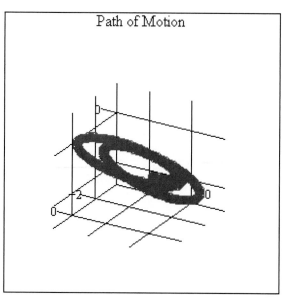

Path of Motion

x , y , z

2.109 The coordinates of the particle when it leaves the hemisphere are $\phi = 41.5°$ and $\theta = 75.7°$.

2.111

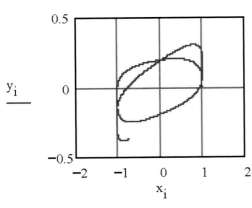

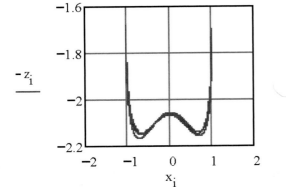

2.113

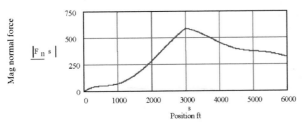

2.115 $v = 11$ m/s

Chapter 3

3.1 14 m/s

3.3 0.309 m

3.5 $v = \sqrt{2gR(\sin\theta - \sin\theta_0)}$

3.7 0.693 m

3.9 2.276 radians   130.4°

3.11 180°

3.13 201.4 ft

3.15 Bottom of slide 6.26 m/s
At the floor 8.86 m/s

3.17 $v_A = 8.744$ m/s $v_B = 9.667$ m/s
$v_C = 10.51$ m/s $v_D = 11.29$ m/s

3.19 49.4°

3.21 0.271 m

3.23 5.535 m/s

3.25 0.72 m/s

3.27 1.219×10^3

3.29 1.937 m

3.31 $k = \dfrac{mv^2}{\Delta^2}$

3.33 Since the total work done on the ball is zero, the final velocity is v.

3.35 6.552 ft/s

3.37 14.15 m/s

3.39 5.82 m/s

3.41 Deformation of net is 4.471 ft, so net should be 5 to 6 feet above floor.

3.43 11.55 m/s

3.45 2521 kW

3.47 $v = \sqrt{2gh}$ for each case

3.49 0.83 m

3.51 Deformation of upper spring = 0.112 m
Max dist = 0.912 m

3.53 $V = \dfrac{1}{n+1}kx^{n+1}$

3.57 $V = -\dfrac{1}{2}\ln(x^2 + y^2) + C$

3.59 $v = \sqrt{2gl\sin\theta}$

3.61 0.27 m

3.63 0.721 m/s

3.65 Numerical example: $k = 200$ N/m $m = 4$ kg
$h = 2$ m $\Delta = 0.732$ m

3.67 $v = \sqrt{2gh}$

3.69 Force is conservative.
$$V = -\frac{r^3}{3}\sin\theta - rz^2 + C$$

3.71 7.533 s

3.73 $v_f = \dfrac{v_0}{3}$

3.75 1.768 m/s

3.77 7.746 s

3.79 -0.296 ft/s

3.81 62.8 mph   19.23°

3.83 1.09 m

3.85 32.36 m

3.87 472,500 N

3.89 $v = \sqrt{\dfrac{2m_A^2 gl}{m_B^2 + m_A m_B}}$

3.91 20.9 N

3.93 a) 1.387 s

b) 6.128 s

c) 5.857 m/s

3.95 $v_A' = 5.92$ m/s $v_B' = 26.64$ m/s

a) 102×10^3 J b) 101×10^3 J

3.97 $v_C = \dfrac{1}{4}(1 + e)^2 V$

3.99 a) $v_A' = v_B' = \dfrac{1}{2}V$

b) $v_A' = 0$ $v_B' = V$

c) $v_A' = \dfrac{V}{4}$ $v_B' = \dfrac{3}{4}V$

3.101 a) 50%

b) 37.5%

c) 0%

3.103 $\beta = \tan^{-1}(e \tan \theta)$

3.105 $v_1' = \begin{bmatrix} 0.1 \\ -0.083 \\ 0 \end{bmatrix}$ m/s $v_2' = \begin{bmatrix} 0.333 \\ -0.200 \\ 0 \end{bmatrix}$ m/s

Vector components are parallel and perpendicular to the padded wall.

3.107 $v' = \begin{bmatrix} -0.336 \\ 0.135 \\ 0.329 \end{bmatrix}$ m/s

3.109 a) $v' = \begin{bmatrix} -0.317 \\ 0.161 \\ 0.365 \end{bmatrix}$ m/s

b) $v' = \begin{bmatrix} -0.409 \\ 0.03 \\ 0.183 \end{bmatrix}$ m/s

3.111 $v_s' = \begin{bmatrix} 2.155 \\ -0.432 \\ 0 \end{bmatrix}$ m/s

3.113 $d = 2.018$ m

3.115

$$v_W = -\dfrac{m_S V}{m_W + m_S \sin^2 \beta}[\cos \theta + e \sin \beta \sin(\theta + \beta) + \cos \beta \cos(\theta + \beta)]$$

Chapter 4

4.1 Momentum is conserved, therefore 5th ball moves out with the same velocity as 1st ball on impact.

4.3 $L_2 = L_1 = nmv$

4.5 $v_A''' = 4.16$ ft/s

$v_B''' = 6.61$ ft/s

$v_C''' = 32.4$ ft/s

4.7 $v_f = 13.76$ ft/s

4.9 $r_{cm} = \begin{bmatrix} 5.857 \\ 2.286 \\ 0.429 \end{bmatrix}$ $H_{cm} = \begin{bmatrix} 54 \\ 49.714 \\ 6.857 \end{bmatrix}$

$v_{cm} = \begin{bmatrix} -0.143 \\ -0.857 \\ -0.286 \end{bmatrix}$

4.11 $r = \begin{bmatrix} -1.705 \\ -1.364 \\ 0.795 \end{bmatrix}$ m/s

4.13 $v = -0.27$ m/s (minus to left)

4.15 $H_{0x} = 0$ Therefore there is no r that will produce this condition.

4.17 5 ft

4.19 $r_{cm} = \begin{bmatrix} 1.667 \\ 1.5 \\ 0 \end{bmatrix}$ m $a_{cm} = \begin{bmatrix} 33.33 \\ -26.67 \\ 0 \end{bmatrix}$ m/s^2

$dH_{cm} = \begin{bmatrix} 0 \\ 0 \\ -73.33 \end{bmatrix}$ kg·m^2/s

4.21 $(20.39, -1.50, 0)$

4.23 19,490 ft lb

4.25 85.5 J

4.27 $v_B = \dfrac{V \sec \theta \tan \theta}{2 \tan^2 \theta + 3}$

$v_A = -\dfrac{V \tan^2 \theta}{2 \tan^2 \theta + 3}$

4.29 $v_C = 1.877$ m/s

$v_8 = 0.436$ m/s

$v_{13} = 2.299$ m/s

4.31 $v_A = 2$ m/s $\quad v_B = 1.732$ m/s

$\quad v_C = 3$ m/s $\quad$ (solution)

$\quad v_A = 0$ m/s $\quad v_B = 3.464$ m/s

$\quad v_C = 2$ m/s $\quad$ (not possible because it violates conservation of momentum during the first collision)

4.33 $v_A = 3.5$ m/s $\quad v_B = 3.031$ m/s $\quad v_C = 5.25$ m/s

4.35

$$\alpha = \frac{1}{m(l^2 - 2al + 2a^2)} \{mg(l - 2a)\cos\theta - c\omega\}$$

4.37 6.251 Nm

4.39 $\omega = \sqrt{\dfrac{2g}{l}}\sin\theta$

4.41 $ml^2\alpha = m\,\mathrm{lg}\cos\theta - \mu_k(ml\omega^2 - mg\sin\theta)\dfrac{\omega}{|\omega|}$

4.43 $F = \rho Vv(1 - \cos\theta)$

4.45 9120 N

4.47 2940 lb

4.49 $\mathbf{f} = 1.5\hat{\mathbf{i}} + 3.54\hat{\mathbf{j}}$ N

4.51 5.46 m/s^2

4.53 $\mathbf{F} = -20784\hat{\mathbf{i}} - 1200\hat{\mathbf{j}}$ N

4.55 $v = \sqrt{gL}$

4.57 $\mathbf{F} = -\rho Vv(\cos\theta\hat{\mathbf{i}} + \sin\theta\hat{\mathbf{j}})$

Chapter 5

5.1 a) 7.09 rad/s

$\quad$ b) 3.545 s

5.3 $\alpha = 94.25$ rad/s^2

$\quad \omega = 3600$ rpm

5.5 -23.6 rad/s^2 $\qquad$ 120 rev

5.7 Accelerations in m/s^2

$$a_C = \begin{bmatrix} -3.9 \cdot 10^4 \\ -3.6 \cdot 10^4 \\ 0 \end{bmatrix} \qquad a_D = \begin{bmatrix} -6.15 \cdot 10^4 \\ 2.75 \cdot 10^4 \\ 0 \end{bmatrix}$$

5.9 $\mathbf{v}_A = 2000(-\sin\theta\,\hat{\mathbf{i}} + \cos\theta\hat{\mathbf{j}})$

$\quad \mathbf{a}_A = 1000(-\sin\theta\,\hat{\mathbf{i}} + \cos\theta\hat{\mathbf{j}})$

$\qquad - 20{,}000(\cos\theta\,\hat{\mathbf{i}} + \sin\theta\hat{\mathbf{j}})$

5.11 $\omega_E = 72.9 \times 10^{-6}$ rad/s $\qquad v = 260$ m/s

5.13 $v = \begin{bmatrix} 75.6 \\ 17.1 \\ -25.2 \end{bmatrix} \qquad a = \begin{bmatrix} -771.525 \\ 2.158 \times 10^3 \\ -849.96 \end{bmatrix}$

5.15 0.376 s

5.17 Solution graphs should appear as

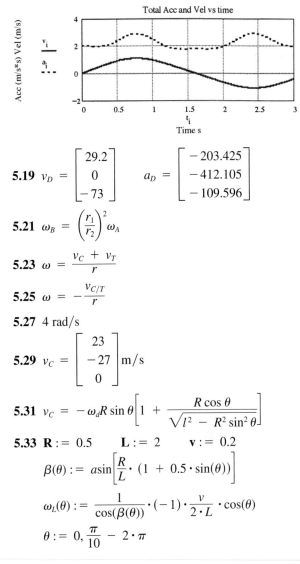

5.19 $v_D = \begin{bmatrix} 29.2 \\ 0 \\ -73 \end{bmatrix} \qquad a_D = \begin{bmatrix} -203.425 \\ -412.105 \\ -109.596 \end{bmatrix}$

5.21 $\omega_B = \left(\dfrac{r_1}{r_2}\right)^2 \omega_A$

5.23 $\omega = \dfrac{v_C + v_T}{r}$

5.25 $\omega = -\dfrac{v_{C/T}}{r}$

5.27 4 rad/s

5.29 $v_C = \begin{bmatrix} 23 \\ -27 \\ 0 \end{bmatrix}$ m/s

5.31 $v_C = -\omega_d R\sin\theta\left[1 + \dfrac{R\cos\theta}{\sqrt{l^2 - R^2\sin^2\theta}}\right]$

5.33 $\mathbf{R} := 0.5 \qquad \mathbf{L} := 2 \qquad \mathbf{v} := 0.2$

$\quad \beta(\theta) := a\sin\left[\dfrac{R}{L}\cdot(1 + 0.5\cdot\sin(\theta))\right]$

$\quad \omega_L(\theta) := \dfrac{1}{\cos(\beta(\theta))}\cdot(-1)\cdot\dfrac{v}{2\cdot L}\cdot\cos(\theta)$

$\quad \theta := 0, \dfrac{\pi}{10} - 2\cdot\pi$

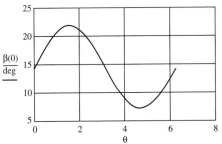

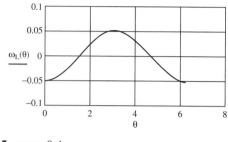

5.35 $\omega := 0.4$

$$\omega_R(\theta) := \frac{\omega}{L} \frac{\cos(\theta)}{\cos(\beta(\theta))}$$

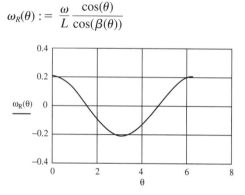

Since the angular velocity of the disk is four times faster then in Problem 5.33 and in the opposite direction, the two curves compare exactly.

5.37 ω

5.39 -4 rad/s (c.w.)

5.41 0.025 rad/s

5.43 $\omega = -5.714$ rad/s $\qquad v_C = 285.7$ mm/s

5.45 1000 mm/s to the left

5.47 1000 mm/s to the right

5.49 $\omega = -1.25$ rad/s $\qquad v_C = 250$ mm/s

5.51 $\omega = 29.33\hat{\mathbf{k}}$

5.53 $v_C \begin{bmatrix} 8 \\ -1 \\ 0 \end{bmatrix}$ m/s

5.55 The contact point of the yo-yo with the surface.

5.57 $\mathbf{r}_{IC/A} = (R\cos\theta + l\cos\beta)\hat{\mathbf{i}}$
$\qquad\qquad + (R\sin\theta + l\cos\beta\tan\theta)\hat{\mathbf{j}}$

5.59 $\mathbf{r}_{J/B} = \begin{bmatrix} 0.495 \\ 0.228 \\ 0 \end{bmatrix}$ m

5.61 $a_C = -536.8$ mm/s^2 $\qquad \alpha_{BC} = -0.011$ rad/s^2

5.63 $\alpha = 0.05$ rad/s^2 $\qquad a_C = -10$ mm/s^2

5.65 $v_C = \begin{bmatrix} 0.86 \\ -1.4 \\ 0 \end{bmatrix}$ m/s $\qquad a_C = \begin{bmatrix} 1.085 \\ -0.036 \\ 0 \end{bmatrix}$ m/s^2

5.67 $\mathbf{v}_D = 1000\hat{\mathbf{i}}$ mm/s $\quad \mathbf{a}_D = 400\hat{\mathbf{i}} - 1250\hat{\mathbf{j}}$ mm/s^2

5.69 -1.5 rad/s^2

5.71 -1.5 rad/s^2

5.73 $\alpha = 0.72$ rad/s^2

5.75 $\mathbf{v} = 35.75\hat{\mathbf{i}} + 50\hat{\mathbf{j}}$ mm/s
$\qquad \mathbf{a} = -185.7\hat{\mathbf{i}} + 10.225\hat{\mathbf{j}}$ mm/s^2

5.77 $\alpha_{BC} = -0.814$ rad/s^2 $\qquad \alpha_{CD} = -0.982$ rad/s^2

5.79 $\omega_{AB} = -\dfrac{v\cos\theta}{3r\cos\beta}$ where $3\sin\beta = (1 + \sin\theta)$

5.81 $v_C = -\dfrac{\omega r\sin(\theta - \gamma)}{\cos(\beta - \gamma)}$

where $\gamma = \beta - \sin^{-1}\left[\dfrac{1}{3}\sin(\theta - \beta) + 0.5\right]$

5.83 $L2 \cdot \cos(\beta_i) - L1 \cdot \cos(\theta_i) + L3 \cdot \sin(\gamma_i) = d$
$\qquad L2 \cdot \sin(\beta_i) - L1 \cdot \sin(\theta_i) - L3 \cdot \cos(\gamma_i) = 0$

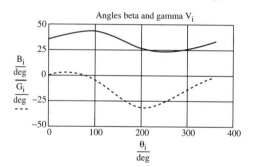

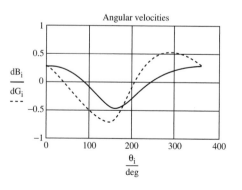

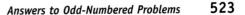

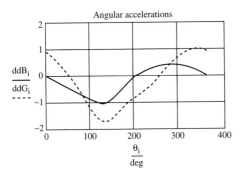

5.85

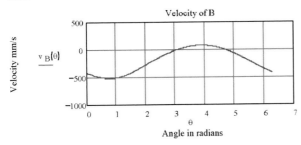

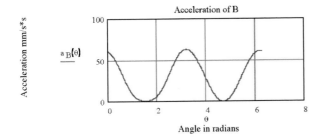

5.87

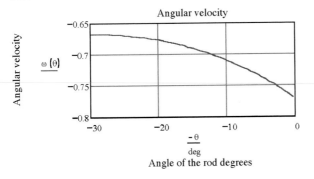

5.89

$$\dot{\beta}(t) = -\frac{\sin (0.1t^2)0.2t}{3 \cos \left[\sin^{-1} \left(\frac{1}{3} \cos (0.1t^2) \right) \right]}$$

$$\ddot{\beta}(t) = \frac{1}{\cos \beta} \left[\frac{1}{3} (\omega^2 \cos \theta - \alpha \sin \theta) + \dot{\beta}^2 \sin \beta \right]$$

where $\beta = \sin^{-1} \left[\frac{1}{3} \cos \theta \right]$

5.91 Let γ be the angle BC makes with the horizontal and β be the angle CD makes with the vertical.

$$\omega_{BC} = \frac{4}{7} \frac{\sin (\theta - \beta)}{\cos (\beta + \gamma)}$$

The angular acceleration may be obtained by differentiation.

5.93 Let θ be the ccw angle of the disk, β be the angle of AB with the vertical, and γ be the angle BCD makes with the horizontal.

$$BC \cdot \cos(\gamma_i) + AB \cdot \sin(\beta_i) - R \cdot \sin(\theta_i) = BC$$
$$BC \cdot \sin(\gamma_i) + AB \cdot \cos(\beta_i) + R \cdot \cos(\theta_i) = AB + R$$

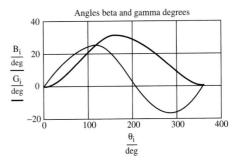

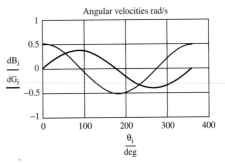

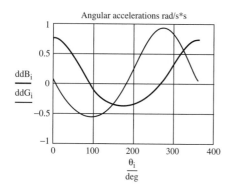

Angular accelerations rad/s*s

$\dfrac{ddB_i}{ddG_i}$

$\dfrac{\theta_i}{deg}$

5.95 Yes, the direct vector method can be used.

$$\omega = \begin{bmatrix} 0.2 \\ -0.2 \\ 0.5 \end{bmatrix} \text{ rad/s}$$

5.97 $\alpha = \begin{bmatrix} 0 \\ -0.4 \\ 0.3 \end{bmatrix} \text{ rad/s}^2$

5.99 $a_x = 9.0 \text{ m/s}^2$ $d = 0.824 \text{ m}$

5.101

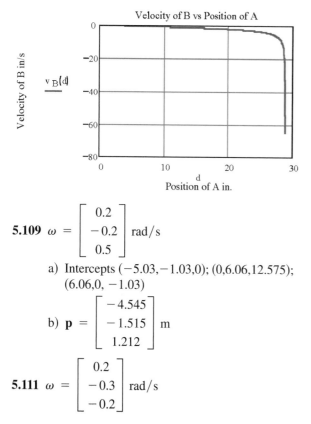

Velocity of B vs Position of A

$v_B(d)$

Velocity of B in/s

d
Position of A in.

5.109 $\omega = \begin{bmatrix} 0.2 \\ -0.2 \\ 0.5 \end{bmatrix} \text{ rad/s}$

a) Intercepts $(-5.03, -1.03, 0)$; $(0, 6.06, 12.575)$; $(6.06, 0, -1.03)$

b) $\mathbf{p} = \begin{bmatrix} -4.545 \\ -1.515 \\ 1.212 \end{bmatrix} \text{ m}$

5.111 $\omega = \begin{bmatrix} 0.2 \\ -0.3 \\ -0.2 \end{bmatrix} \text{ rad/s}$

5.113 $\omega = \begin{bmatrix} 7.972 \times 10^{-3} \\ 2.708 \times 10^{-3} \\ 0.306 \end{bmatrix} \text{ rad/s}$

$$\mathbf{p}_{H/D} = \begin{bmatrix} 1.485 \times 10^{-3} \\ 0.022 \\ -2.323 \times 10^{-4} \end{bmatrix} \text{ m}$$

5.115

$$\mathbf{a}_D = \begin{bmatrix} -1r[\Omega^2 + \omega^2] \sin\theta \\ r\Omega^2 \cos\beta(\cos\beta\cos\theta - 3\sin\beta) + r\omega^2\cos\theta \\ r\Omega^2 \sin\beta(\cos\beta\cos\theta - 3\sin\beta) - 3r\omega^2 \end{bmatrix}$$

5.117

$$\mathbf{v}_D = \begin{bmatrix} -(2\Omega + \omega\sin\theta) \\ \omega\cos\theta \\ -\Omega(3 + \cos\theta) \end{bmatrix} \text{ m/s}$$

$$\mathbf{a}_D = \begin{bmatrix} -[\omega^2\sin\theta + \Omega^2(3 + \cos\theta)] \\ -\omega^2\sin\theta \\ 2[\Omega^2 + \Omega\omega\sin\theta] \end{bmatrix}$$

5.121

$\mathbf{v}_B = -\dot{\Omega}t(vt - R)\cos\alpha\,\hat{\mathbf{k}} + v(\cos\alpha\,\hat{\mathbf{i}} - \sin\alpha\,\hat{\mathbf{j}})$

$\mathbf{v}_B = -\dot{\Omega}(3vt - R)\cos\alpha\,\hat{\mathbf{k}} - (\dot{\Omega}t)^2(vt - R)\cos\alpha\,\hat{\mathbf{k}}$

5.123

$\mathbf{v}_{S/E} = (R + h)\omega(-\sin\theta\,\hat{\mathbf{I}} + \cos\theta\,\hat{\mathbf{J}}) + h\Omega\cos\theta\,\hat{\mathbf{K}}$

$\mathbf{a}_{S/E} = -(R + h)\omega^2(-\cos\theta\,\hat{\mathbf{I}} - \sin\theta\,\hat{\mathbf{J}})$
$\qquad - h\Omega^2\cos\theta\,\hat{\mathbf{I}} - 2(R + h)\Omega\omega\sin\theta\,\hat{\mathbf{K}}$

Chapter 6

6.1 57.22 ft/s^2

6.3 7.5 m/s^2

6.5 $a_{cm} = g\sin\theta$

6.7 13.33 ft/s^2

6.9 $\omega = \sqrt{\dfrac{10}{7}\dfrac{g}{(R - r)}(\sin\theta - \sin\theta_0)}$

6.11 $\dfrac{1}{3}ml^2\ddot{\theta} = -mg\dfrac{1}{2}\sin\theta$
$\qquad + kl^2\left(\sqrt{2(1 - \sin\theta)} - 1\right)$
$\qquad \dfrac{\cos\theta}{\sqrt{2(1 - \sin\theta)}}$

6.13 $F_t = 0.64\,mg\sin\theta$
$\qquad F_n = mg[1.72\cos\theta - 0.72\cos\theta_0]$

6.15 $\alpha = \dfrac{3g}{2l} \sin\theta - \dfrac{3\mu_k r}{l^2} \left| \dfrac{l\omega^2}{2} - g\cos\theta \right| \dfrac{\omega}{|\omega|}$

6.17 $\dfrac{T}{A} = \rho r^2 \omega^2$

6.19

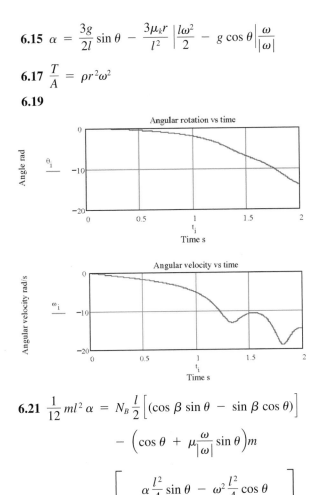

6.21 $\dfrac{1}{12} ml^2 \alpha = N_B \dfrac{l}{2} \Big[(\cos\beta \sin\theta - \sin\beta \cos\theta) \Big]$

$\qquad - \left(\cos\theta + \mu \dfrac{\omega}{|\omega|} \sin\theta \right) m$

$$\begin{bmatrix} \alpha \dfrac{l^2}{4} \sin\theta - \omega^2 \dfrac{l^2}{4} \cos\theta \\[2mm] - \dfrac{l^2}{2} \tan^2\beta (\alpha\cos\theta - \omega^2\sin\theta) \end{bmatrix}$$

where

$N_B = \dfrac{1}{\cos\beta + \mu \dfrac{\omega}{|\omega|} \sin\beta} \left\{ mg - m \left[\left(\dfrac{\alpha l}{2} - \dfrac{\omega^2 l}{2} \right) \right. \right.$

$\qquad \left(\cos\theta + \mu \dfrac{\omega}{|\omega|} \sin\theta \right)$

$\qquad \left. \left. - \mu \dfrac{\omega}{|\omega|} \tan\beta (\alpha l \cos\theta - \omega^2 l \sin\theta) \right] \right\}$

6.25

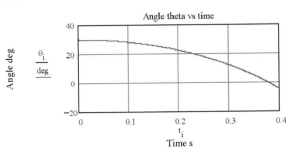

Note that although we have developed the solution using the general solution from 6.20 taking the co-efficient of kinetic friction as zero. In this case, the rod rotates in the opposite direction, that is, to an upright position.

6.27 a) 0.35 s

b) $A_t = 8.57$ N $\qquad A_n = 18$ N

6.29 a) 161.3 rad/s²

b) 53.7 rad/s²

c) 23.0 rad/s²

6.31 0.577

6.33 $\alpha = -16.35$ rad/s² $\qquad T = 8.175$ N

6.35 $\alpha_B = \dfrac{M_A}{\left(2\dfrac{r_A}{r_B} l_B + \dfrac{r_B}{r_A} l_A \right)}$

6.37 $\alpha = -\dfrac{4(M+m)}{4M + m(4 - 3\cos^2\theta)}$

$$\begin{bmatrix} \dfrac{3g}{2l} \sin\theta + \dfrac{3m}{4(M+m)} \omega^2 \sin\theta \cos\theta \end{bmatrix}$$

$a = \dfrac{4(M+m)}{4M + m(4 - 3\cos^2\theta)}$

$$\begin{bmatrix} \dfrac{3mg}{4} \sin\theta \cos\theta + \dfrac{ml}{2} \omega^2 \sin\theta \end{bmatrix}$$

6.39 $M(\theta) = m_{CD} r^2 \sin\theta \cos\theta \, \omega^2$

6.41 a) -2.252 N

b) -2.252 N

6.43 $\alpha_A = 35.3 \text{ rad/s}^2 \text{ ccw}$

$\alpha_B = -9.42 \text{ rad/s}^2 \text{ cw}$

6.45

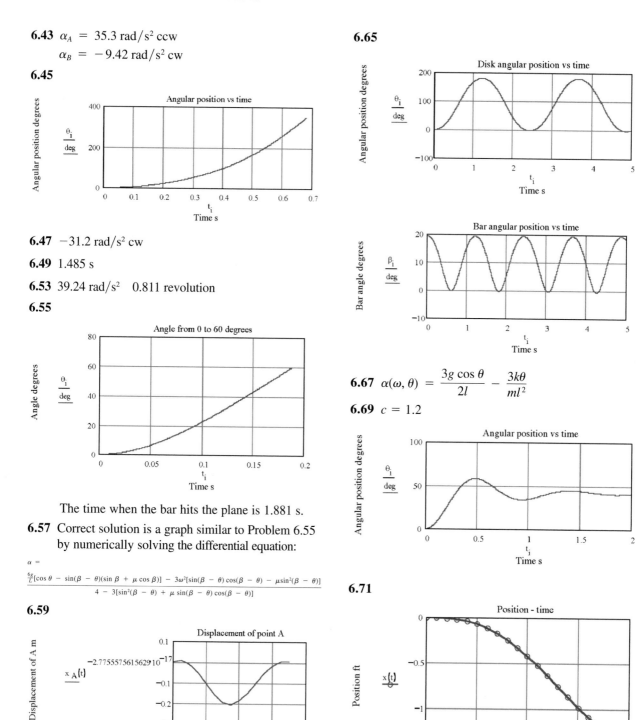

Angular position vs time

6.47 $-31.2 \text{ rad/s}^2 \text{ cw}$

6.49 1.485 s

6.53 39.24 rad/s^2 0.811 revolution

6.55

Angle from 0 to 60 degrees

The time when the bar hits the plane is 1.881 s.

6.57 Correct solution is a graph similar to Problem 6.55 by numerically solving the differential equation:

$$\alpha = \frac{\frac{6g}{L}[\cos\theta - \sin(\beta - \theta)(\sin\beta + \mu\cos\beta)] - 3\omega^2[\sin(\beta - \theta)\cos(\beta - \theta) - \mu\sin^2(\beta - \theta)]}{4 - 3[\sin^2(\beta - \theta) + \mu\sin(\beta - \theta)\cos(\beta - \theta)]}$$

6.59

Displacement of point A

-2.775557561562910^{-17}

6.63 0.526 s

6.65

Disk angular position vs time

Bar angular position vs time

6.67 $\alpha(\omega, \theta) = \dfrac{3g\cos\theta}{2l} - \dfrac{3k\theta}{ml^2}$

6.69 $c = 1.2$

Angular position vs time

6.71

Position - time

6.73 5.996 rad/s

6.75

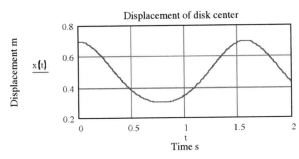

Displacement of disk center

6.77 a) Car cannot climb hill

b) Car cannot climb hill

c) $a = 0.122\ g$

6.79 $5.64\ \text{ft/s}^2$

6.81 $a = g\cot 30°$

6.83 $-19.62\ \text{rad/s}^2$

6.85 $\alpha - \dfrac{2\mu_k g \sin \beta}{r(\mu_k + \sin 2\beta)}$

6.87 $88.9\ \text{rad/s}^2 \qquad 0.19\ \text{m/s}^2$

6.89 $\mu_s = \dfrac{2}{7} \tan \beta$

Chapter 7

7.1 $3.721\ \text{m/s}$

7.3 $5.421\ \text{rad/s}$

7.5 $1.944\ \text{m/s}$

7.7 $0.95\ \text{m}$

7.9 $44.29\ \text{rad/s}$

7.11 $\omega = \sqrt{\dfrac{6k \sin \theta}{m} + \dfrac{3g}{l}(1 - \cos \theta)}$

7.13 $\omega = \sqrt{\dfrac{3g}{l}(1 - \cos \theta)}$

7.15 $\omega = \sqrt{\dfrac{6}{ml^2}(mgl - fr\pi)}$

7.17 $\dfrac{34}{11} mg$

7.19 Transendental equation for θ:
$$-M_f\theta + mRg(1 - \cos \theta) = 0$$

7.21 $\omega = \sqrt{\dfrac{36\pi M}{m(a^2 - ab + b^2)}}$

7.23 $12.8\ \text{rad/s}$

7.25 $0.463\ \text{rad}\ (0.074\ \text{rev})$

7.27 $k = \dfrac{2mg}{l}$

7.29 $\omega = \sqrt{\dfrac{3 \cos \theta_0}{4l}}$

7.31 $\sqrt{\dfrac{3mg}{\left(\frac{7}{3}m + \frac{3}{2}M\right)l^2}}$

7.33 $-13.133\ \text{rad/s}$

7.35 $38.9\ \text{N}$

7.37 $1.82\ \text{m/s}$

7.39 $20.96\ \text{rad/s}$

7.41 $\omega = \sqrt{\dfrac{3g}{2l} - \dfrac{3k}{m}}$

For the bar to reach the verticle position $k \le \dfrac{mg}{2l}$.

7.43 $\omega = \sqrt{\dfrac{3 \sin \theta}{ml}(mg - kl \sin \theta)}$

7.45 $\omega_B = \sqrt{\dfrac{2M\theta_A}{I_B + I_C + \left(\frac{r_B}{r_A}\right)^2 I_A}}$

7.47 $v_{cm} = \dfrac{2Pt}{3m}$

7.49 a) $30.65\ \text{rad/s}$

b) $20.44\ \text{rad/s}$

7.51 a) $\omega_p = 8.38\ \text{rad/s}$

b) $\omega_t = 8.45\ \text{rad/s}$

7.53 $v_1 = -8\omega_1 r$

7.55 $v = \dfrac{2}{3}gt$

7.57 $v = \dfrac{5\mu_k \cos \alpha v_0}{7\ \mu_k \cos \alpha - 2 \sin \alpha}$

$\omega = \dfrac{5\mu_k \cos \alpha\ v_0}{r\ (7\mu_k \cos \alpha - 2\sin \alpha)}$

7.59 $T_B = 190\ \text{N}$

$T_{AB} = 160\ \text{N}$

7.61 $-8.4\ \text{rad/s}$

7.63 $-112.1\ \text{rad/s}$

7.65 $c = \dfrac{2}{3}l$

7.67 $R_x\Delta t = -0.717\ \text{Ns}$

7.69 $\dfrac{(1.75 + 0.75e)m_A v_A}{r\ (0.75m_A + 1.5m_B)}$

7.71 $v_B = -0.076$ m/s
$v_{Acm} = 0.559$ m/s
$\omega = 0.725$ rad/s

7.73 $v' = -e\sqrt{\dfrac{3gl}{8}}$

7.75 $c = \dfrac{2}{3}l$

7.77

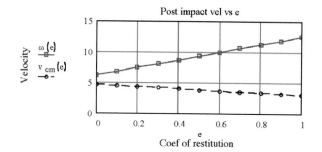

7.79

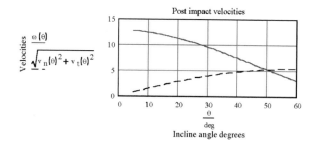

7.81 $F\Delta t = v(1 + e)\cos \beta$

7.83 $\omega = \dfrac{12v}{7l}$

7.85 $F\Delta t = m(1 + e)v$

7.87 $\omega'_{bar} = \dfrac{m_{bar} - 3em_{disk}}{m_{bar} + 3m_{disk}}\sqrt{\dfrac{3g}{l}}$

Chapter 8

8.3 $A \cdot B = \begin{bmatrix} 0.5 & 0.866 & 0 \\ -0.75 & 0.433 & 0.5 \\ 0.433 & -0.25 & 0.866 \end{bmatrix}$

8.5 $\begin{bmatrix} \cos(\beta) & \sin(\theta)\cdot\sin(\beta) & -\cos(\theta)\cdot\sin(\beta) \\ 0 & \cos(\theta) & \sin(\theta) \\ \sin(\beta) & -\sin(\theta)\cdot\cos(\beta) & \cos(\theta)\cdot\cos(\beta) \end{bmatrix}$

8.7 For infinitesimal transformation:

$$\begin{bmatrix} 1 & 0 & -d\beta \\ 0 & 1 & d\theta \\ d\beta & -d\theta & 1 \end{bmatrix}$$

8.11 There are two solutions:
$\quad 0 \qquad 19.95° \qquad\quad -19.59°$
$\phi = 45.18° \quad\text{or}\quad 225.2°$
$\psi \quad 30.1° \qquad\qquad 010.1°$

8.13

$$\begin{bmatrix} \cos(\gamma)\cdot\cos(\beta) - \sin(\gamma)\cdot\sin(\alpha)\cdot\sin(\beta) & \sin(\gamma)\cdot\cos(\beta) + \cos(\gamma)\cdot\sin(\alpha)\cdot\sin(\beta) & -\cos(\alpha)\cdot\sin(\beta) \\ -\sin(\gamma)\cdot\cos(\alpha) & \cos(\gamma)\cdot\cos(\alpha) & \sin(\alpha) \\ \cos(\gamma)\cdot\sin(\beta) + \sin(\gamma)\cdot\sin(\alpha)\cdot\cos(\beta) & \sin(\gamma)\cdot\sin(\beta) - \sin(\gamma)\cdot\sin(\alpha)\cdot\cos(\beta) & \cos(\alpha)\cdot\cos(\beta) \end{bmatrix}$$

8.15 $\alpha_x = \ddot{\phi}\sin\theta\sin\psi + \ddot{\theta}\cos\psi + \dot{\phi}\dot{\theta}\cos\theta\sin\psi$
$\qquad + \dot{\phi}\dot{\psi}(\sin\theta\cos\psi - \sin\psi)$

$\alpha_y = \ddot{\phi}\sin\theta\cos\psi - \ddot{\theta}\sin\theta\psi + \dot{\phi}\dot{\theta}\cos\theta\cos\psi$
$\qquad - \dot{\phi}\dot{\psi}(\sin\theta\sin\psi - \cos\psi)$

$\alpha_z = \ddot{\phi}\cos\theta + \ddot{\psi} - \dot{\phi}\dot{\theta}\sin\theta$

8.21 $A_z = B_z = \dfrac{1}{4}mR\omega\Omega$

8.23

$\omega(\theta): =$

$$\sqrt{\dfrac{g\cdot\sin(\theta)\cdot L}{\dfrac{L^2}{2}\cdot\cos(\theta)\cdot(\sin(\theta) + 1) + \dfrac{1}{6}\cdot(L^2 + 3\cdot r^2)\cdot\sin(\theta)\cdot\cos(\theta)}}$$

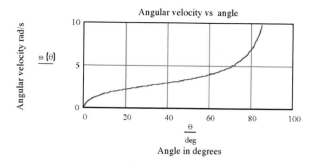

8.25 $M_x = \dfrac{1}{4}mR^2\dot{\Omega}$

8.27 $\beta = \cos^{-1}\left(\dfrac{4g}{5R\omega^2}\right)$

8.29

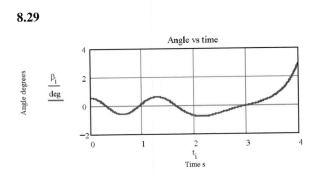

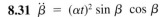

8.31 $\ddot{\beta} = (\alpha t)^2 \sin \beta \, \cos \beta$

Chapter 9

9.1 $\omega_n = 32.5$ kHz

9.3 $\omega_0 = 0.77$ rad/s

9.5 $\omega_n = 2$ rad/s

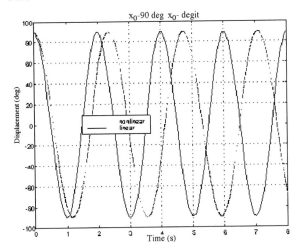

9.7 $v_{\max} = 62.83$ mm/s and $a_{\max} = 3.947$ m/s^2

9.9 $C = x_0 \qquad B = \dfrac{v_0}{\omega_n}$

9.11 Increase spring stiffness by 21%.

9.13 $\omega_n = \sqrt{\dfrac{k_1 + k_2 + k_3}{m}}$

9.15 $\omega_n = \sqrt{\dfrac{k}{m}}$

9.17 $\omega_n = \sqrt{\dfrac{g}{l}\left(\dfrac{3M + 6m}{2M + 6m}\right)}$

9.19 $\omega_n = \sqrt{\dfrac{2k}{3m}}$

9.21 Five equilibrium corresponding to $x_{e1} = 0$:

$$x_{e2,3,4,5} = \pm\frac{1}{\sqrt{2}}\sqrt{\frac{\beta^2 \pm \sqrt{\beta^4 - 4\gamma\omega_n^2}}{\gamma}}$$

9.23

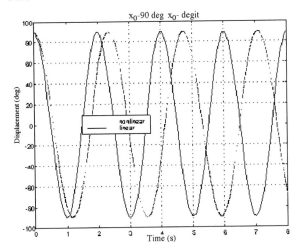

9.25

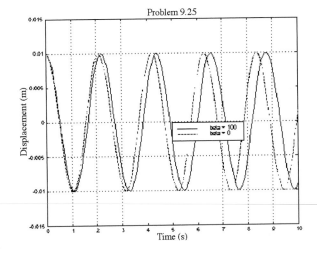

9.27 $x_{e1} = 0$, $x_{e2} = -3$, $x_{e3} = 3$

The system oscillates about x_{e1}.

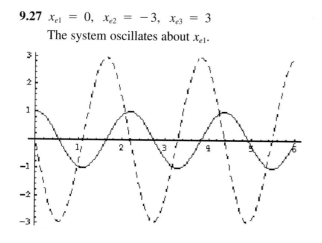

9.29 $\omega_n = 5.48$ rad/s

$\omega_d = 5.27$ rad/s $\zeta = 0.27$

The system oscillates.

9.31 $A = 0.005$ m and $\phi = -1.47$ rad

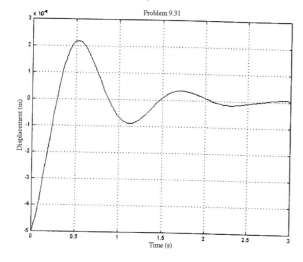

9.33 $\omega_n = 4.47$ rad/s

$\omega_d = 4.36$ rad/s $\zeta = 0.22$

$A = 6$ mm $\phi = 0.97$ rad

$x(t) = Ae^{-\zeta\omega_n t} \sin(\omega_d t + \phi)$

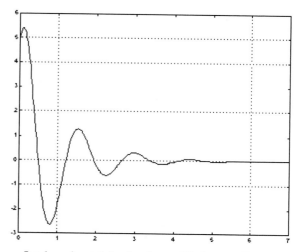

It takes about 6.5 s for the oscillations to die out.

9.35 $A = \sqrt{\dfrac{(v_0 + \zeta\omega_n x_0)^2 + (x_0\omega_n)^2}{\omega_d^2}}$

$\phi = \tan^{-1}\left(\dfrac{x_0 w_n}{v_0 + \zeta\omega_n x_0}\right)$

9.37 $a_1 = x_0$, $a_2 = v_0 + \omega_n x_0$

9.39 The oscillation dies out after about 3.2 seconds.

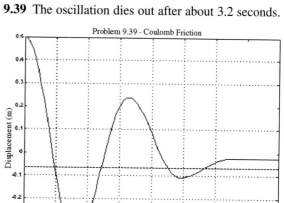

9.41

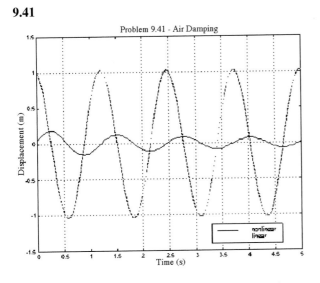

9.43 The oscillations die out after about 37 seconds.

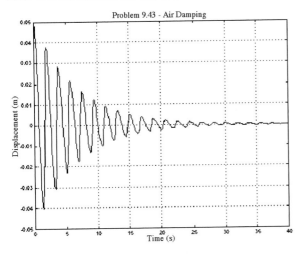

9.45

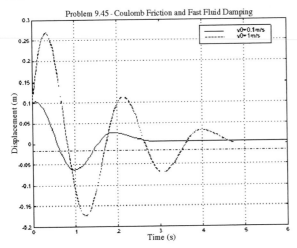

9.47 Damping Coefficient: $c \equiv 200$ kg/s

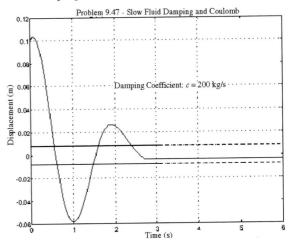

9.49 For $c = 100$ kg/s the oscillations die out after 12 seconds.

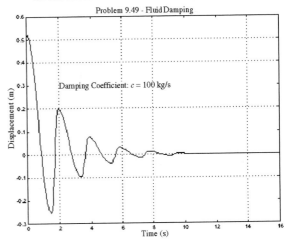

9.51

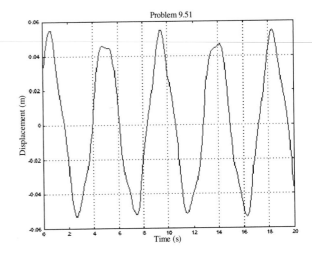

9.53

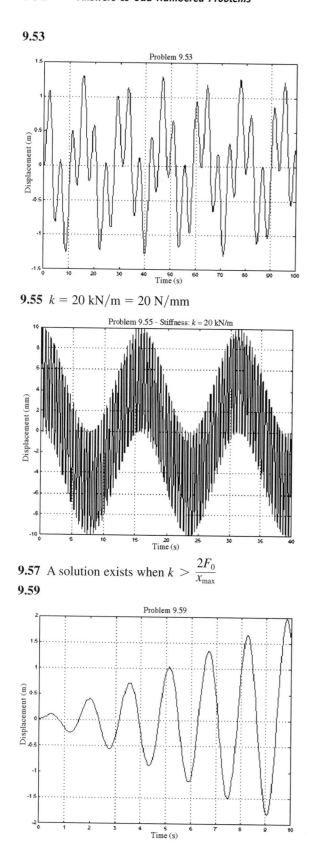

Problem 9.53

9.55 $k = 20$ kN/m = 20 N/mm

Problem 9.55 - Stiffness: $k = 20$ kN/m

9.57 A solution exists when $k > \dfrac{2F_0}{x_{max}}$

9.59

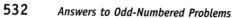

Problem 9.59

9.61

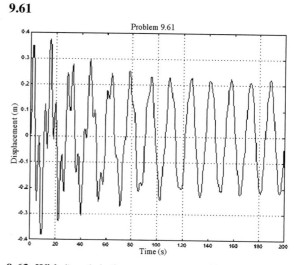

Problem 9.61

9.63 With $\zeta = 1.4$, the response resembles an over-damped system.

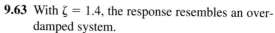

Problem 9.63

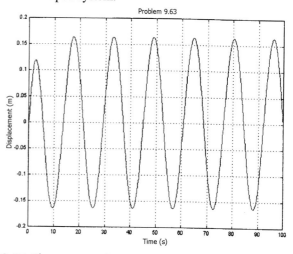

9.65 $\zeta|_{c=100 \text{ kg/s}} = 0.14 \Rightarrow$ underdamped

$\zeta|_{c=1000 \text{ kg/s}} = 1.4 \Rightarrow$ overdamped

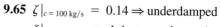

Problem 9.65

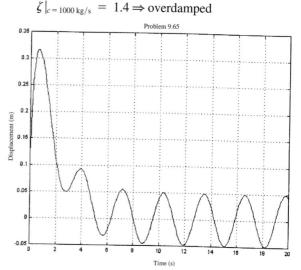

9.67 $A_0 = \sqrt{A_s^2 + B_s^2}, \quad \theta = \tan^{-1}\left(\dfrac{B_s}{A_s}\right)$

9.69

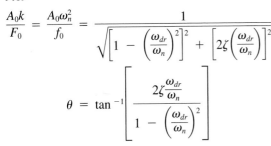

$$\frac{A_0 k}{F_0} = \frac{A_0 \omega_n^2}{f_0} = \frac{1}{\sqrt{\left[1 - \left(\dfrac{\omega_{dr}}{\omega_n}\right)^2\right]^2 + \left[2\zeta\left(\dfrac{\omega_{dr}}{\omega_n}\right)\right]^2}}$$

$$\theta = \tan^{-1}\left[\frac{2\zeta\dfrac{\omega_{dr}}{\omega_n}}{1 - \left(\dfrac{\omega_{dr}}{\omega_n}\right)^2}\right]$$

9.71 $c = 200$ and $\zeta = 0.125$. This system is under-damped.

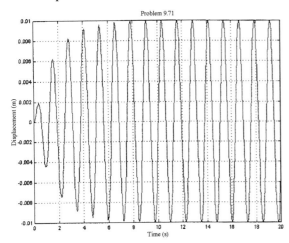

9.73

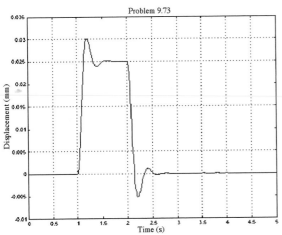

9.75 With $c = 2500$ kg/s the oscillations die out after 6 seconds.

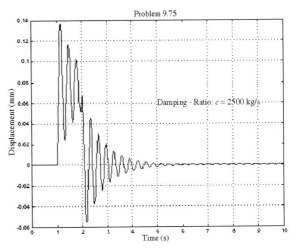

9.77 It takes about 14 seconds for the oscillations to die out.

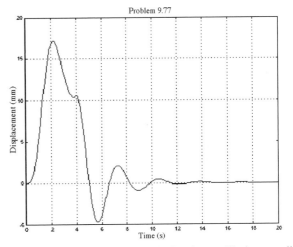

9.79 It takes about 14 seconds for the oscillations to die out.

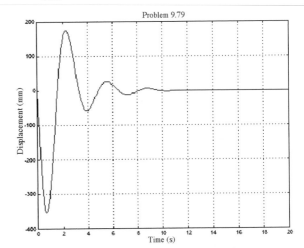

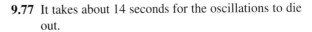

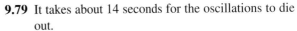

9.81

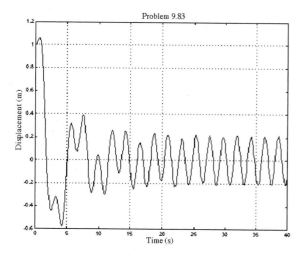

9.83

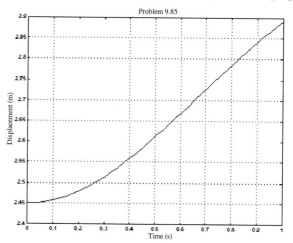

9.85 The system remains unstable, despite the damping!

INDEX